A £54-17

AF745243

WITHDRAWN

306129

Subjective Probability

Subjective Probability

Edited by
George Wright
University of Strathclyde in Glasgow, UK
and
Peter Ayton
City University, London, UK

BOLTON INSTITUTE OF HIGHER EDUCATION
THE LIBRARY – DEANE CAMPUS
Accession No. 306129
Classification No. B 519.2 SUB

JOHN WILEY & SONS
Chichester · New York · Brisbane · Toronto · Singapore

Copyright © 1994 by John Wiley & Sons Ltd,
Baffins Lane, Chichester,
West Sussex PO19 1UD, England

Telephone (+44) 243 779777

All rights reserved.

No part of this book may be reproduced by any means, or transmitted, or translated into a machine language without the written permission of the publisher.

Other Wiley Editorial Offices

John Wiley & Sons, Inc., 605 Third Avenue,
New York, NY 10158-0012, USA

Jacaranda Wiley Ltd, 33 Park Road, Milton,
Queensland 4064, Australia

John Wiley & Sons (Canada) Ltd, 22 Worcester Road,
Rexdale, Ontario M9W 1L1, Canada

John Wiley & Sons (SEA) Pte Ltd, 37 Jalan Pemimpin #05–04,
Block B, Union Industrial Building, Singapore 2057

Library of Congress Cataloging-in-Publication Data

Subjective probability / edited by George Wright and Peter Ayton.
p. cm.
Includes bibliographical references and index.
ISBN 0-471-94443-2
1. Social sciences—Statistical methods. 2. Probabilities.
3. Subjectivity. I. Wright, George, *1952—* . II. Ayton, Peter.
HA29.S8264 1994
519.2—dc20 93–49631
CIP

British Library Cataloguing in Publication Data

A catalogue record for this book is available from the British Library

ISBN 0-471-94443-2

Typeset in 10/12 pt Times by
Mathematical Composition Setters Ltd., Salisbury, Wiltshire
Printed and bound in Great Britain by Bookcraft (Bath) Ltd

Contents

Contributors

PETER AYTON
Department of Psychology, City University, Northampton Square, London EC1V 0HB, UK

JONATHAN BARON
Department of Psychology, University of Pennsylvania, Philadelphia, PA 19104-6196, USA

LEE ROY BEACH
Department of Management and Policy, University of Arizona, Tucson, AZ 85721, USA

P. GEORGE BENSON
Department of Information and Decision Sciences, Rutgers University, PO Box 2101, New Brunswick, NJ 08903, USA

FERGUS BOLGER
Department of Psychology, University College London, Gower St, London WC1E 6BT, UK

CHARLES J. BRAINERD
Program in Education, College of Education, University of Arizona, Tucson, AZ 85721, USA

GARY P. BRAUN
Department of Management and Policy, University of Arizona, Tucson, AZ 85721, USA

WIBECKE BRUN
Cognitive Unit, Department of Psychology, University of Bergen, Sydneshaugen 2 5007, Bergen, Norway

DOMINIC A. CLARK
Imperial Cancer Research Fund Laboratories, PO Box No. 123, Lincoln's Inn Fields, London WC2A 3PX, UK

SHAWN P. CURLEY
Department of Information & Decision Sciences, University of Minnesota, 271 19th Avenue South, Minneapolis, MN SJ455, USA

RUMA FALK
Department of Psychology, The Hebrew University, 91905 Jerusalem, Israel

WILLIAM R. FERRELL
Systems and Industrial Engineering, The University of Arizona, Tucson, AZ 85721, USA

JOHN FOX
Imperial Cancer Research Fund Laboratories, PO Box No 123, Lincoln's Inn Fields, London WC2A 3PX, UK

DEBORAH FRISCH
Department of Psychology, University of Oregon, Eugene, OR 97405, USA

GERD GIGERENZER
Department of Psychology, University of Chicago, 5848 S. University Avenue, Chicago IL 60637, USA

NIGEL HARVEY
Department of Psychology, University College London, Gower Street, London WC1E 6BT, UK

KEES VAN DER HEIJDEN
Strathclyde Graduate Business School, 199 Cathedral Street, Glasgow G4 0QU, UK

COLIN HOWSON
Department of Philosophy, Logic and Scientific Method, London School of Economics and Political Science, Houghton Street, London WC2A 2AE, UK

GIDEON KEREN
Department of Philosophy and Social Sciences, Eindhoven University of Technology, Den Dolech 2, Postbus 513, 5600 MB Eindhoven, The Netherlands

CLIFFORD KONOLD
Scientific Reasoning Research Institute, Hasbrouck Laboratory, University of Massachusetts, Amherst, MA 01003, USA

PAUL J. KRAUSE
Imperial Cancer Research Fund Laboratories, PO Box No 123, Lincoln's Inn Fields, London WC2A 3PX

DENNIS LINDLEY
2 Periton Lane, Minehead, Somerset TA24 8AQ, UK

ABIGAIL LIPSON
Bureau of Study Counsel, Harvard University, 5 Linden Street, Cambridge, MA 02128, USA

ALASTAIR MCCLELLAND
Department of Psychology, University College London, Gower Street, London WC1E 6BT, UK

VALEIR F. REYNA
Program in Education, College of Education, University of Arizona, Tucson, AZ 85721, USA

GLENN SHAFER
132 Patton Avenue, Princeton, New Jersey 08540, USA

PATRICK SUPPES
Institute for Mathematical Studies in the Social Sciences, Ventura Hall, Stanford University, CA 94305, USA

KARL HALVOR TEIGEN
University of Tromso, Department of Psychology, Asgardveien, 900 Tromso, Norway

PETER URBACH
Department of Philosophy, Logic and Scientific Method, The London School of Economics and Political Science, Houghton Street, London WC2A 2AE, UK

WILLEM A. WAGENARR
Department of Psychology, Faculty of Social Sciences, Rijks Universitet Leiden, PO Box 9555, 2300 RB Leiden, Netherlands

George Wright
Strathclyde Graduate Business School, 199 Cathedral Street, Glasgow G4 0QU, UK

J. Frank Yates
The University of Michigan, Department of Psychology, 330 Packard Road, Ann Arbor, MI 48104-2994, USA.

Preface

This book presents a broad ranging view of subjective probability. Chapters range from the discussions of the philosophy of axiom systems through to studies in the psychological laboratory and then to the real world of business decision-making.

Topics covered include subjective probability in statistical inference and expert systems, the treatment of causality in laboratory studies and in scenario planning, whether Man is a Bayesian thinker or a frequentist thinker, descriptive and normative theories of subjective probability, confidence and performance, and subjective probability in gambling and court-room decisions—a wide range of topics. Nevertheless, underpinning all the topics and approaches is a fundamental desire, on the part of the authors, to analyse and document the human ability to deal with uncertainty—no easy task. The multidisciplinary nature of this volume—which includes authorities who are psychologists, philosophers, statisticians, management scientists, educationalists, and corporate planners—illustrates the essentially human challenge of this enormous project.

Fundamentally, we believe that a reconceptualization of the base issues will be prompted by, and benefit from, the exchange of knowledge across disciplinary boundaries. The challenge of facilitating this flow of knowledge was the driving force behind this book. We have commissioned chapters from those individuals who possess both the subject expertise and the ability to write in an accessible way. Our hope is that readers of this collection will be stimulated to apply fresh insights to their own disciplinary endeavours and perhaps also be inspired to contribute to the development of new material to add to the evolving corpus of knowledge and debate.

The book is organized into four major parts. The first, *Background*, provides the philosophical and statistical foundations. The second, *Studies in the Psychological Laboratory*, overviews theory and research in cognitive and developmental psychology. The third, *Accuracy of Probability Judgements*, focuses on theories and models that allow assessment of the quality of assessed probabilities. The fourth, *Real World Studies*, reviews subjective probability judgement in situations that have material consequences for the decision-makers. A detailed chapter-by-chapter overview follows next.

In the first chapter of the *Background* Part, Dennis Lindley introduces the concept of probability and the probability laws and demonstrates that, logically, probability is inevitable.

Subjective probability depends upon two things, the event whose uncertainty is contemplated, and the knowledge that you have at the time. The calculus of probability leads to results that may not agree with common sense because, Lindley argues, common sense is often not capable of the calculations. There is no suggestion that the probability rules describe how people behave in the face of uncertainty. The probability calculus is a norm which intuitive ideas on uncertainty *should* follow. Lindley shows that subjective probability leads directly to utility and thus to the procedure of maximization of expected utility as the optimal decision criterion.

Patrick Suppes provides a systematic discussion of the major aspects of the subjective theory of probability. As he notes, a central question for a set of axioms of *qualitative* probability is what formal comparative relation the expression "more probable than" must have in order to be represented by a *numerical* probability measure over events. Suppes argues that the subjective theory of probability provides necessary but not sufficient conditions for success in probabilistically predicting future events, such as tomorrow's weather. Next, Suppes considers a thought experiment and argues that individuals may ignore hypotheses that are ultimately true. It is unrealistic, Suppes argues, to have a positive prior opinion for *all* plausible hypotheses. Similarly, he debates the issue of whether there are situations in which it is not sensible to make *exact* probability estimates about possible events about which little is known. He argues that extension of the theory of subjective probability to such situations is desirable.

Colin Howson and Peter Urbach review the background to the development of a theory of inductive inference. As Howson and Urbach note, the apparent impossibility of determining objective prior probabilities in any non-arbitrary manner has been a powerful factor in convincing many people that a probabilistic theory of inductive inference was impossible. However, as these authors show, debates about methods for *determining* priors fall outside Bayesian theory. Next, Howson and Urbach consider classical statistical inference and argue that, relative to Bayesianism, it has no proper foundation and that apparently objective inferential statements are, in fact, illusory. The principles of significance testing and estimation are, they argue, simply wrong and beyond repair.

Developing this theme, Glenn Shafer argues that subjective probability is integral to all applications of supposedly objectivistic applications. Focusing on statistical tests, Shafer argues that subjectivity enters into probability both in the way in which belief and frequency are unified and in the way that this unification is applied to a practical problem. Shafer demonstrates that statistical testing often uses instances of what he terms "the informal story" (the unification) simply as standards against which to rate performance. The divide between the "frequentists" and the "subjectivists", he argues, has a foundational rigidity that may have been useful in the past, when subjectivists had few successful, practical Bayesian applications, but is unnecessary today.

By contrast, John Fox argues that, in addition to subjective probability, there is a family of distinct theories of uncertainty that can be shown to have sound mathematical foundations and that these theories capture *different* intuitions about uncertainty and belief. He questions the general adequacy of the "probability paradigm" and, by introducing issues from artificial intelligence (AI), argues that subjective probability does *not* have universal applicability. He contends that a number of alternative uncertainty formalisms, deriving from research in AI which has attempted to formalize intuitive concepts of "common sense", provide an alternative framework. AI has become the latest stage on which the probability debate is being conducted.

In the first chapter of the Part on *Studies in the Psychological Laboratory*, Lee Roy Beach and Gary Braun identify 1967 as the year in which experimentation began in earnest. They describe early work which focused on identifying whether probability theory was a descriptive behavioural model of individuals' judgements of subjective probability. Beach and Braun evaluate whether probability theory is, in fact, the appropriate standard for evaluating the quality of subjective probabilities. They argue that knowledge-based reasoning as well as probability-law-based reasoning may give rise to subjective probability. Whilst experimenters have generally assumed that the domain covered by the problems they posed was properly addressed by probability theory, the subjects of the experiments, they argue, have frequently thought otherwise.

Gerd Gigerenzer extends this discussion and argues that, from a strong frequency view of probability, observed "biases" in probabilistic reasoning are not errors, since probability theory simply does not apply to *single* events. In his chapter, Gigerenzer focuses on the usefulness of a distinction between single-event probabilities and frequencies and draws on evidence from both the history of probability and from experimental work in the psychological laboratory. He argues that empirical demonstrations of errors are not stable and cognitive "illusions" disappear when single-event probabilities are changed to frequencies. He concludes that the untutored mind has more of a frequentist design than a Bayesian one.

Peter Ayton and George Wright consider this latest view of probability, as explicated by Gigerenzer in the preceding chapter. Ayton and Wright suggest that the well-known gambler's fallacy can also be made to "disappear". But there is some evidence that judgements of frequencies can also show the characteristic bias found for confidence judgements.

A number of questions are raised by Gigerenzer's critique. How should we view human judgement under uncertainty? For biases to be capable of being made to disappear they have to be there in the first place: why do the conjunction fallacy and base-rate neglect occur when they do? We note that the evidence for, and generality of, the availability and anchor-and-adjust heuristics proposed by Kahneman and Tversky appear untouched by the critique. Aside from the theoretical debate as to the nature of human probabilistic judgement there are also questions as to its general competence. Should we now believe that human probabilistic judgement is satisfactory and that the evidence for incompetence was illusory?

Shawn Curley and George Benson focus on the construction of subjective probability estimates. They argue that the process is dominated by the construction of reasoned arguments and develop a cognitive theory of probability construction. Next, they utilize this theory to analyse the results of earlier experimentation. They argue that influence diagrams and knowledge maps, often used in the initial phases of decision analysis, are an indication of a new focus on the *process* of belief assessment rather than the *output* of the judgement. The latter topic, as Beach and Braun note in Chapter 6, has been the primary concern of psychologists to date. However, Curley and Benson argue that investigation of the reasoning underlying judgemental assessments is critical and, as a first step, they develop a model of "belief processing" that supports the construction of subjective probabilities. Next they utilize this model to elucidate judgemental heuristics in probability assessment.

Karl Teigen argues that subjective judgements of probability are arrived at by a number of different processes which may, or may not, cause them to differ from the experimenters' rules. Teigen does not debate whether or not there should be a correspondence between subjects' assessments and experimenters' evaluation of what constitutes a good (often relative to normative standards) performance. This issue was already debated by Beach and Braun in Chapter 6. Instead, Teigen evaluates the conditions under which judgements and norms show agreement and when they do not. Teigen suggests some *simple* probability rules which, he argues, people generally seem to accept as valid. Understanding and application of other rules, for example the product rule for arriving at the probability value of a conjunction, is less commonplace. Teigen argues that, when people start judging probabilities on an intuitive basis, judgement is dependent upon a richer source of subjectively available concepts and strategies. Teigen concludes that we are very sophisticated probabilists in most respects *except* the quantitative one.

Valerie Reyna and Charles Brainerd document the developmental studies of probability judgement that have been conducted and argue that these provide an important perspective on adult conceptions of probability. Knowledge of developmental stabilities and changes supply an independent body of evidence that can be used to select among competing theories of (non-)probabilistic thinking in adulthood. Their developmental analysis deals successively with Piagetian information processing and intuitive reasoning approaches. Reyna and Brainerd debate the evidence for precocity on the one hand and late emergence on the other. They conclude that young children can perform advanced processing operations under certain conditions. Of necessity, the question of adult rationality is central to developmental research—if cognitive development is seen as progress toward rationality. Reyna and Brainerd argue that reasoners understand probability at an early age but that they increasingly rely on intuitive processes as they grow older.

Jonathan Baron and Deborah Frisch analyse what they term the "ambiguity effect", where individuals often prefer to bet on gambles with a stated and known chance of winning, as opposed to gambles where the chance of winning is unstated and unknown but is *formally* identical from the perspective of expected utility theory. This phenomenon, which has demonstrated subjects' aversion to ambiguity, has led to empirical and theoretical research on the causes and effects of ambiguity. In their paper, Baron and Frisch discuss the implications of ambiguity avoidance for expected utility theory. They develop their own theory of ambiguity as missing information and address issues concerning the practicality of dealing with situations in which information is missing.

Wibecke Brun focuses on the perception of risk. Some conceptualizations define risk as a product of the probability of a loss and its magnitude. Other definitions are concerned with lay perceptions and aim to describe how lay persons intuitively understand the term. Nevertheless, probability is one of the main components in most definitions of risk. Does a discrepancy between "actual" risk measures, like statistical fatality estimates, and subjectively perceived risk constitute a problem? Brun identifies two major research issues in risk perception studies. The first has to do with gaining knowledge of what public concerns are. The second has to do with identifying and explaining attitudes and reactions toward hazards. Brun differentiates experimental studies in the psychological laboratory and psychometric or questionnaire studies of risk attitudes in the lay population. Brun concludes that the uncertainty component of a risk is multidimensional, involving intuitive probability concepts such as those described by Teigen in Chapter 9.

Nigel Harvey discusses the relationship between how well people perform skilled tasks and the confidence that they have in their performance. Most of the work that Harvey evaluates has to do with motor skills and cognitive skills. The major issues are whether confidence accurately reflects performance and whether changes in performance produce changes in confidence. Here confidence is treated as an effect of performance. These issues, and especially the former one, are similar to those addressed by cognitively orientated studies of probability judgement accuracy reviewed in a later section of this volume. In addition, however, theoretical concerns have led to an examination of how confidence influences performance. Harvey argues that confidence and performance are coupled together as a dynamical system where neither should be seen as just a cause nor as just an effect. Here, motivational factors are of major concern. Harvey concludes that there is a need for the theoretical integration of cognitive and motivational accounts.

Ruma Falk, Abigail Lipson and Clifford Konold investigate how we cope with the search for a target object in a finite field of locations, if one's initial uncertainty is met with a series of negative results. Falk, Lipson and Konold examine the nature of probabilistic reasoning in such situations along with the Bayesian solution. But, descriptively, how is conflict resolved between the diminishing long-term hope and the increasing immediate hope implied by the diminishing finite number of locations? Do we overestimate our chance of success, thus wasting time in futile search, or underestimate our chances, giving up too early in frustration and despair?

In the first chapter of the Part on *Accuracy of Probability Judgements*, J. Frank Yates introduces the issues and analyses the variety of accuracy measures that have been proposed. Such measures, often implemented as scoring rules, can be used in the form of feedback to the judge as well as direct assessment of aspects of the judge's (in)accuracy. Yates gives several detailed examples of accuracy measurement in such contexts as pneumonia diagnosis, intensive-care prognoses, and studies of cross-national variations in propensity to engage in probabilistic thinking.

William R. Ferrell discusses practical issues in subjective probability from the standpoint of decision analysis, and he extends discussion of the elicitation of subjective probabilities from individuals and subsequent accuracy measurement. As Ferrell notes, the quality of a decision analysis is critically dependent on the quality of assessed probabilities. He develops Yates' discussion of probability accuracy, discusses the consequences of poor accuracy for decision analysis and argues that a common form of inaccuracy is overconfidence. Next, Ferrell describes his model of probability accuracy based on signal-detection theory and he demonstrates that the model can explain a variety of experimental data drawn from subjective probability judgement tasks.

Alastair McClelland and Fergus Bolger focus on the accuracy or calibration of subjective probability judgements. They review theories and models of calibration that have appeared since the time of the last review, in 1982. As McClelland and Bolger note, there are *two* distinct views of the locus of observed biases in calibration and other measures of probabilistic reasoning. The "pessimists" believe that biases are in people whilst the "optimists" believe that biases are in-built in the experimental tasks utilized by researchers. Theories and models of calibration can also be located within these two distinct views of the quality of probabilistic judgement. In their review of seven models of subjective probability calibration, McClelland and Bolger conclude that the "optimists" provide the most satisfactory explanation of calibration with *general-knowledge* items.

In the first chapter in the Part on *Real World Studies*, Gideon Keren describes gamblers' conceptions of uncertainty in light of a distinction between two different and irreducible fundamental modes of thought. One mode is based on the abstract rules of logic (e.g. the probability laws), the other is phenomenological in nature and is based on action and associated conscious experience. As he notes, the fact that people gamble at all in face of negative expected values is one of the main paradoxes of gambling behaviour. However, as Keren argues, this paradox exists only under the rules of the logic of the probability calculus. Gamblers' probabilistic assessments are contaminated by their desires and associated emotions. Clearly, the formal theory of probability is not descriptive of how gamblers deal with uncertainty.

Paul Krause and Dominic Clark discuss the ways in which uncertainty and subjective probability have been represented in AI systems. As the authors note, *ad hoc* uncertainty calculi have often been used in AI systems because they are computationally efficient. Bayesian updating of a system's knowledge base, on the other hand, is complex and posed problems of combinational explosion *both* for the elicitation of subjective probabilities from human experts and for the numerical computation of updated probabilities. However, new efficient algorithms have been recently developed for rapid belief updating which match the computational capability of earlier *ad hoc* approaches. However, most probabilistic expert systems are still dependent on the elicitation from experts of the majority of the required conditional probabilities. Nevertheless, such expert systems are in use, for example in the diagnosis of congenital heart disease. Krause and Clark describe such a system and argue that there are situations where aspects of imprecision and vagueness may be more effectively addressed with alternative calculi to that of subjective probability. This approach supplements and extends Chapter 5 by Fox.

Willem Wagenaar focuses on the courtroom criterion of "beyond reasonable doubt". As he notes, there is a paradox implicit in the criterion since if a judge is not absolutely certain then this must mean that there is a logical possibility that the accused is innocent. Why then is this possibility not a reason for doubt? Wagenaar analyses the way in which judges approach the task of probability assessment. One descriptive theory proposes that the acceptance of good causal stories, given by defence or prosecution, is taken as diagnostic of truth. Of course, presented evidence should underpin good stories. Wagenaar argues that insight into how judges and juries deal with probabilities may lead to better courtroom procedures and better laws.

Kees van der Heijden analyses the advantage of probabilistic planning compared to scenario planning. He notes that probabilistic planning is based on axiom-based theory, whilst scenario planning, a more intuitive approach to dealing with uncertainty, derives from the world of decision-making practice.

Scenario planning, van der Heijden argues, addresses more adequately the needs of managerial decision-makers, downplays the decision-maker's poor ability to think probabilistically and promotes managerial ability to create causal stories of plausible futures. In general, this approach complements that described by Wagenaar in the previous chapter. Scenario-planning techniques promote the generation of action options and aid the manager's desire to close the gap between expected and desired futures. The contrast with probabilistic thinking is stark.

Part One

Background

Chapter 1

Foundations

Dennis Lindley
Minehead, Somerset

1.1 UNCERTAINTY

The object of our study is uncertainty, the situation that arises when we do not know whether a given statement is true or false. Uncertainty is everywhere about us; all the future is uncertain, and so is much of the present and the past. We are all uncomfortable with uncertainty and try to avoid it as much as possible. But it will not go away, so let us face up to the fact that not everything is known and study the phenomenon of uncertainty.

It will be enough if consideration is confined in the first instance to statements which it would be reasonable to describe as either true or false, if only we knew which. Such statements will be termed events, though sometimes other descriptions, such as hypotheses, might be appropriate. It is a purely technical problem to extend our study from uncertain events to uncertain quantities; a problem that will not be discussed here. Thus we might take the event of "rain tomorrow", rather than consider the amount of rain. Events will be denoted by capital letters, thus A, B, etc.

Some events are more uncertain than others. For example, we are fairly sure that the sun will rise tomorrow, less sure that it will rain then, and very unsure whether a tossed coin will fall heads uppermost. Let us suppose that any degree of uncertainty can be described by a number. This is a major assumption and we will return to consider it later. For the moment, let us just see where this reduction of a complex notion to a single value leads us.

Subjective Probability. Edited by G. Wright and P. Ayton.
© 1994 John Wiley & Sons Ltd.

In describing the uncertainty of A by a number, it must be recognized that the number may depend on the person contemplating A and also on how much that person knows. We will refer to the person in the state of uncertainty concerning A as "you". The knowledge that you have will be denoted K. If K changes, so might your uncertainty. Indeed, one of our main tasks will be to quantify the change. Our task is to measure your uncertainty about A when your knowledge base is K.

1.2 PROBABILITY

Underlying all measurement is the concept of comparison with a standard. To say the length of an object is 2.4 m is to compare the length with a standard metre. So we need a standard for uncertainty. There are many possibilities but an easy one to handle is that of an urn containing a number of balls, identical except for the fact that some are white and the remainder black. Suppose that your knowledge base, H, contains the information that there are w white and b black balls in the urn, $w + b = n$. You now draw a ball from the urn in such a way that you think any one ball is as likely to be withdrawn as any other. (This concept can be made precise without circularity in the argument.) We then say that the uncertainty concerning the event W, that the ball is white, is w/n. It is called your probability of W, given H, and written $p(W \mid H)$. The notation does not incorporate reference to you, since we shall here always be dealing with one person.

We now have a standard for uncertainty, called subjective probability, though the adjective will be omitted. The subject is you. If you now contemplate another, general event, A, when your knowledge base is K, it has probability w/n if the uncertainty is the same as that for the ball being white. Thus, for given n, the value of w can be selected to make the two events equally uncertain. Remember, we are supposing that all uncertainties can be described numerically, so that this comparison is feasible. Notice that you might not be able to do the direct comparison of A, given K, with W, given H, any more than you would actually use the standard metre to measure length. All that is being said is that you would regard it as possible in principle, just as, in principle, the standard metre might be employed. Although this procedure only provides rational numbers, it is a purely technical matter to extend it to real numbers. As a special case, if K includes the information that A is true, then you would choose $w = n$ to make W true, and hence assign a probability of one.

We thus have the first rule of probability, usually called the

Convexity rule. For any uncertain event A and state of knowledge K, the probability of A, given K, $p(A \mid K)$, lies between 0 and 1 and assumes the value 1 if K includes the knowledge that A is true.

There are two important features of any measurement process: how the measurement is made, and how different measurements combine. The first part is solved in principle by comparison with a standard, though this is rarely practical. Let us take the second part, the rules of combination. For probability there are two of these. To describe them, some notation is required. If A and B are two events, then the event which is only true when both A and B are true, the conjunction, is written AB. Similarly, if your knowledge base is enlarged from K to include the additional information that A is true, we write AK for the new base. If it is impossible, on K, for both A and B to be true, we say they are exclusive on K and write $AB = \emptyset$, the empty event, which is known to be false. If $AB = \emptyset$, we write $A + B$ for the event which is true whenever one of A or B is true, the disjunction. The two rules can now be stated.

Addition rule. If $AB = \emptyset$ on K, then $p(A + B \mid K) = p(A \mid K) + p(B \mid K)$.
Multiplication rule. $p(AB \mid K) = p(A \mid K)p(B \mid AK)$.

It is an easy matter to establish these rules by comparing all the probabilities with the standard of balls in an urn. In addition to supposing some balls are white, some black, to compare with A, as above, it will be necessary to suppose some of the balls spotted, some plain, to compare with B. For the addition rule, there will be no balls that are both white and spotted, corresponding to $AB = \emptyset$. The rules are then simply a reflection of the fact that the proportions of balls in the urn obey these rules, which are thereby transferred to probabilities generally. The rules of probability are just those of proportions.

1.3 THE INEVITABILITY OF PROBABILITY

Probability is therefore a measure of uncertainty obeying the three rules of convexity, addition and multiplication. It is not the only way to measure uncertainty. For example, gamblers use odds. But odds are merely a transform of probability. If the probability of A, given K, is p; then the odds against A, given K, are $(1 - p)/p$ to 1. Elsewhere in this chapter, the odds on, $p/(1 - p)$, will be used and referred to simply as odds. In many statistical calculations it is convenient to use the logarithm of odds. We usually prefer probability because the rules are simpler in that form, though Bayes' rule, below, will demonstrate an advantage of odds.

However, although we can switch from probability to functions, like odds, any measure of uncertainty must be a function of probability. Workers in fuzzy logic argue for rules expressed, not in terms of addition and multiplication, but in terms of maxima and minima. This is not possible. Here is an outline of a demonstration of this fact, due to de Finetti (1974). The account also suggests a way of measuring uncertainty that is more practical than the comparison with a standard, just used.

Suppose, still accepting that uncertainty can be described by a number, that you are asked to provide a numerical value for the uncertainty of some event A, given K, and provide, using whatever method you like, the answer, p. Then let you be scored an amount $(1-p)^2$, if A is later found to be true, and p^2 if A is false. The score is to be regarded as a penalty and you wish to minimize your total score. It is easy to see that, under these conditions, you would provide a value of p between 0 and 1, and that you would give the value 1 if you knew A to be true. This is the convexity rule. Using some beautiful, geometric arguments of great simplicity, de Finetti showed that the numbers that you would give must satisfy the other two rules as well. If they did not, then you would necessarily incur a larger score: not simply expect to get a bigger value than by using probability, but actually get one. It is not an exaggeration to say that it would be foolish to provide numbers that did not satisfy the three rules. It may be objected that this depends on the particular method of scoring used by de Finetti. However, it can be shown that if other scores are used, one of two things can happen. The first possibility is that you will give a transform of probability, for example, odds. Which transform will be a function of the scoring system. The second is that you will always give one of only two numbers, say either 1 or 0, which is patently absurd. All sensible rules lead back, via a possible transformation, to probability. Probability is inevitable.

There are other approaches, all of which lead to probability. There is none that provides alternative rules like those of fuzzy logic. Jeffreys (1961) does an analysis that might appeal to a scientist. Another method, due to Ramsey (1926), is based on decision-making, a topic we will discuss below. From modest assumptions, Ramsey was able to describe the class of reasonable decision procedures. They are all based on probability. A more modern, and detailed, development along similar lines is due to Savage (1954).

Subjective probability for you depends on two things, the event whose uncertainty is contemplated, and the knowledge that you have at the time. We say it is a function of two arguments. It is common to refer to the probability of A. This is strictly wrong and can lead to misunderstandings by omission of the conditions, K. Also two people considering the same event, with the same knowledge base, can have different probabilities. De Finetti expressed this vividly by saying that "Probability does not exist". Does not exist, that is, outside of an individual. The correct form is your probability of A, given K. Practical experience suggests that there are bases where most people agree. For example, on being presented with a coin similar to others that you have seen over the years, most people would say that, if they were to toss it, the probability of heads for them is $\frac{1}{2}$. Other probabilities, like those concerning the winner of a political election, are much more subjective.

1.4 EXTENSION OF THE CONVERSATION

Let us now look at the rules in more detail. The whole of probability theory evolves from the three rules. Other rules are mathematical consequences of them. The rules are sometimes modified, but only in small ways. For example, the addition rule has been stated for two events and can easily be extended to any finite number. But it is usual to suppose that it also holds for an enumerable infinity of events. Similarly, in the convexity rule, it is sometimes supposed that a probability is only 1 if the knowledge base implies the truth of the event. Notice that, although the rules are simply those of proportions, probabilities combine in two different ways, by addition and by multiplication. Lengths, for example, only combine by addition; multiplication yields a new concept, area. This means that the calculus of probability is extremely rich. It also means that the results do not always agree with common sense, for common sense is often not capable of the calculation.

For any event A, the event which is true (false) whenever A is false (true) is called the complement of A and will be written A^c. Since $AA^c = \emptyset$ and the event $A + A^c$ is surely true, it follows that $p(A^c \mid K) = 1 - p(A \mid K)$. The addition rule shows that, since $A = AB + AB^c$, $p(A \mid K) = p(AB \mid K) + p(AB^c \mid K)$. Use of the multiplication rule enables this to be written

$$p(A \mid K) = p(A \mid BK)p(B \mid K) + p(A \mid B^cK)p(B^c \mid K).$$

This most useful formula is known as the extension of the conversation (from A to include B). It is also known as the generalized addition rule. Its merit lies in the fact that the probabilities of A on the right-hand side are often easier to evaluate than that on the left, because you knowledge base is larger there.

1.5 BAYES' RULE

If, in the multiplication rule, the roles of the two events, A and B, are interchanged, we easily get that

$$p(A \mid K)p(B \mid AK) = p(B \mid K)p(A \mid BK),$$

so that if $p(B \mid K) \neq 0$,

$$p(A \mid BK) = \frac{p(B \mid AK)p(A \mid K)}{p(B \mid K)}.$$

This is known as Bayes' rule. It is more easily appreciated in its odds form. Write $O(A \mid K) = p(A \mid K)/p(A^c \mid K)$, the odds on A, given K, as in Section

1.3. With Bayes' rule, both as written above and with A^c replacing A, we have

$$O(A \mid BK) = \frac{p(B \mid AK)}{p(B \mid A^c K)} O(A \mid K).$$

In this form it clearly displays the effect on the odds for A of learning the truth of B. The original odds are multiplied by

$$\frac{p(B \mid AK)}{p(B \mid A^c K)}$$

to obtain the new odds. The multiplier is known as the likelihood ratio. Here is an application that might clarify the rule.

Consider a trial in a criminal court and let G be the event that the defendant is truly guilty of the charge. Write G^c as I, innocence. At some point in the course of the trial, let K denote the knowledge that the court has. Finally, let E denote a new piece of evidence before the court, additional to K. Then with G for A and E for B, Bayes' rule says

$$O(G \mid EK) = \frac{p(E \mid GK)}{p(E \mid IK)} O(G \mid K).$$

(Since K occurs in every probability, the reader may temporarily like to omit it as an aid in appreciating the formula.) On the left-hand side, there is the odds for guilt given the new evidence (and K); on the other side is the odds without that evidence. The latter is multiplied by the likelihood ratio to obtain the former. This, according to probability theory, is the procedure that should be adopted in the court. As the trial proceeds, the odds are continually updated by multiplication by the relevant likelihood ratio.

Let us consider this ratio in more detail. The probabilities involved are not those of guilt, but of the new evidence. Furthermore, these have to be taken both on the assumption of innocence, and on that of guilt, though only their ratio matters. The court therefore has to ask itself how probable is the evidence were the defendant guilty, and how probable is it were he innocent. Generally, whenever there are two competing hypotheses, here guilt and innocence, one needs to assess the uncertainties of the evidence on the bases of both, and compare them.

1.6 INVERSION, SUPPOSITION AND DESCRIPTION

There is another important feature of Bayes' rule. On the left-hand side we have $p(G \mid EK)$, expressed in odds: on the right there is $p(E \mid GK)$. Here the roles of E and G, are reversed. People often experience difficulty in distinguishing between these two probabilities, yet they are essentially different.

Statisticians find it useful to distinguish the two ideas by using different words. Omitting K for the moment, $p(A \mid B)$ is the probability of A, given B; whereas, as a function of B, it is termed the likelihood of B, given A. Thus in the legal application of Bayes' rule, the odds refer to the probability of guilt; the ratio refers to the likelihood of guilt.

Bayes' rule also brings out another important feature of probability. The probability $p(E \mid GK)$ uses the knowledge base GK. But the court never knows G to be true. What is being studied here is the uncertainty of the evidence on the supposition that the defendant is guilty. Supposition replaces fact. Strictly, one should write $p(A \mid B:K)$, the probability of A, supposing B to be true and knowing K. This complication is not needed if the assumption is made that

$$p(A \mid BC:K) = p(A \mid B:CK).$$

This says that, replacing the supposition that C is true by the knowledge that it is true, makes no difference to the uncertainty of A. In acquiring the knowledge that an event is true, you often acquire other knowledge as well. If you do not, then the assumption seems reasonable. If made, there is no need to distinguish between supposition and fact, and the previous, simpler notation may be used. This is almost universal practice.

There is no suggestion, in the development given, that probability describes how people currently behave in the face of uncertainty. All that is being said is that you would wish to behave in accordance with probability, if you could. The calculus of probability is there to help you to do this. The method is said to be normative: it provides a norm by which your ideas might be expressed. A description of how people actually behave might look very different. There is no suggestion that courts of law nowadays use Bayes' rule in reaction to new evidence; only that they should.

1.7 NUMERICAL ASSIGNMENT AND COHERENCE

If you are going to use the probability calculus, you have got to input numbers for some probabilities. How are these to be obtained? First, there are some probabilities that are easy to calculate. The probability of $\frac{1}{2}$ for a coin falling heads, or 1/6 for a die showing an ace, are natural. Your probability that Jean celebrates her birthday on 15 March is 1/365, with refinements for leap years. Generally, from easily obtained probabilities, it is possible to evaluate others using the calculus. For example, if, to the one about Jean's birthday, you include similar statements about other people, and assume all the judgements independent, then it is possible to show that the probability, in a room of 23 people, that there are two who share the same celebration, is about $\frac{1}{2}$; a value that is often found to be surprising. Generally, from some values you can calculate others by use of the rules.

Another way to influence your probability evaluations is through scoring rules. Faced with a sequence of events and the knowledge that you are to be scored by some sensible system, you will usually make a better job of the task than you would without the check provided by the score. Certainly it encourages coherence as explained in the next paragraph. More experience is needed with this method to determine which scoring rule is best.

Perhaps the most important way of calculating probabilities, and certainly one that is always available, is through coherence. You are said to be coherent in your approach to uncertainty if all the values you give obey the rules of probability. For a single event, there is only one rule, convexity. Coherence only comes into play with several events, and the more there are, the more powerful it is. For example, suppose that you contemplate the event A that your party will win the next election, knowing K, $p(A \mid K)$. The state of the economy is surely relevant to the party's fortunes, so you might consider extending the conversation to include B, the event that the economy is favourable. This will involve $p(A \mid BK)$ and $p(A \mid B^c K)$, contemplating election, knowing the state of the economy. Other events, like those concerning foreign policy, can be added. The procedure can be inverted and you may contemplate $p(B \mid AK)$, the probability of a sound economy were your party elected. From evaluations already made, you can calculate others and see whether you like them. Computer programs exist which do the calculations and provide ranges between which unstated probabilities might lie, given what has been input.

1.8 FREQUENCY IDEAS

There is class of situations in which it is often easy to determine your belief numerically. This is where K includes frequency information relevant to the uncertain event being considered. For example, if you learn that, in a recent survey, 15% of people carried a certain gene, you might assert that the probability that Tom carries the gene is 0.15. It is important to notice that subjective probability, as developed here, has nothing to do with frequency. It is merely a numerical expression of your belief. In this example, the frequency belongs to K and is transferred to the uncertain event by a judgement of a connection between the frequency and the event. If you learned that Tom's mother carried the gene, then the frequency of 15% would have much less relevance. The connection between the knowledge base and the event has been destroyed by the additional information about the mother. Frequencies have a useful role to play in the evaluation of beliefs but it is wrong to interpret probability as frequency.

Frequency ideas surface in another context. Suppose that, over many days, you forecast tomorrow's weather by each day giving your probability of rain tomorrow. After a long period, you look at all the occasions on which you have given a specific value for the probability, say 0.25. Then one might expect that on 25% of these occasions, it will have been found to have rained. If this match between belief and actuality obtains for all probabilities, you are said to be calibrated and there is agreement between belief and frequency. Calibration is often said to be a desirable feature. It is easy to see that it is not necessary, by thinking of the meteorologist who always provides probabilities of 0 or 1 for rain tomorrow. Every time 0 is stated, it rains; every 1 is followed by a dry day. This is an excellent forecaster even though quite uncalibrated; it is only necessary to do the opposite of what his forecast suggests.

1.9 DECISION ANALYSIS

The discussion so far has been entirely in terms of beliefs. But why do we have beliefs? Why do we want to calculate with probabilities? The usual answer is, in order to take action in the face of uncertainty; to decide in a situation where not all is known. A belief does not have to be associated with a decision, but it must have the potentiality for action if needed. You need not bet on the fall of the coin, but the $\frac{1}{2}$ would be useful if you did. If the gene could have dangerous consequences for his children, Tom might find the 0.15 very relevant to a decision whether to have children. It is easy to extend subjective probability to encompass decision-making. This is done by the introduction of utility.

Decisions and the resulting actions lead to consequences, which are uncertain if the events are. Suppose that, amongst all the consequences that might arise there is one that is more desirable than the others, or at least is very good. Write this C_1 and give it a utility of 1. Similarly take a consequence that is very bad, C_0, and give it utility 0. Now take any consequence C whose merit lies between these two extremes. The utility of C can be constructed as follows. Consider a choice between an action that will lead to C for sure, and another action that will yield C_1 with probability u and C_0 with probability $1 - u$. Since C is intermediate between the two extremes, there will be a value of u that will make the uncertain action equivalent, in your mind, to the certain C. This is called the utility of C and will be written $u(C)$. The choice of 0 and 1 above is arbitrary and work with utility is unaffected by changes of origin or scale.

Take any decision and suppose that it can lead to one of a number of consequences c_i, $i = 1,2, \ldots n$, with utilities $u(c_i)$, their probabilities being $p(c_i)$, omitting reference to K. By the way in which utility was derived, c_i can be replaced by C_1, the highly desirable outcome, with probability $u(c_i)$, and otherwise C_0. Hence the decision can be thought of as always leading to one of the extreme consequences. Let us evaluate the probability of getting the better C_1 rather than the worse. By the extension of the conversation (see above) this is

$$p(C_1) = \Sigma_i p(C_1 \mid c_i) p(c_i) = \Sigma_i u(c_i) p(c_i).$$

Since you would wish to maximize your probability of getting the best of all possible worlds, C_1, you can achieve this by maximizing the right-hand side of the last result. The sum is called the expected utility of the decision, obtained in the usual way with an expectation by multiplying the respective utilities by their probabilities and adding. Hence subjective probability, expressing your beliefs about uncertain events, leads directly to utility and thus to the procedure of maximization of expected utility (MEU) as the proper criterion for action. We do not use the term, risk, referring to undesirable outcomes. Utility embraces the good and the bad equally and no distinction need be made beyond the numerical value. Risk is sometimes used when probabilities are unknown. Since our usage of probability refers to your knowledge, or lack of it, it is always known in principle, though sometimes hard to determine.

It is important to notice that utility is not an arbitrary measure of the worth of a consequence. It is a measure on the scale of probability. A glance at the way that it was derived above shows that the concept of a gamble, and hence of probability, is basic to the concept. Furthermore, since it was derived from probability, the extension of the conversation can be used to demonstrate that expected utility is the correct quantity to maximize in order to optimize your decision-making. Just as a single number describes uncertainty, so one value, expected utility, is enough to decide. Actually, all utilities are really expected, since the worth you attach to a consequence is what you expect to obtain from it.

It is necessary to insert a caveat here. The whole edifice concerns a single individual, called you. It does not describe how a group of people should act. Nor does it say how two people in conflict should behave, either in the play of a game or in a situation that may lead to war. But for one decision-maker, contemplating an uncertain world, MEU is the only sensible way to proceed.

The concept of utility is a subtle one and requires care in its use. It applies to any consequence and, in contemplating the consequence, you can take account of anything that you consider relevant. For example, suppose that you are in a gambling situation where the outcomes are monetary. Then you may wish to think solely in terms of money, when all you need do is to take your utility function for money. But you may perceive a consequence in terms of more than just cash. Many people feel that £100 received as by right, or almost certainly, is different from £100 had unexpectedly, or with small probability. In that case, utility of money is inadequate for your contemplation of gambles, and you will need to add an extra dimension to your consideration of the consequences. This type of analysis lies behind the resolution of several paradoxes, like those of Allais, that have appeared in the literature.

1.10 UNCERTAINTY AS NUMBER

Let us return to the strong assumption made at the beginning of the chapter to the effect that you would wish to measure your uncertainty by a single number. People have often felt unhappy with this bold assertion, feeling that such a complicated idea cannot be reduced to something as simple as number. Here is an example of where it may be inadequate. If you are asked for the probability that a coin will fall heads when tossed, then, under normal circumstances, you will confidently announce $\frac{1}{2}$. If asked whether the political party you support will win the next election, you may also provide a probability of $\frac{1}{2}$, but will feel less confident of its value. Here are two values of $\frac{1}{2}$, but you feel more assured about one than the other. You might think that another number would be needed to express this confidence. Some writers have suggested the use of upper and lower probabilities, reflecting the range of reasonable values. In our examples, these might be (0.49, 0.51) for the coin but more separated values, such as (0.35, 0.65) for the election. An argument is now presented to suggest that this complication is not necessary.

Consider an urn that you know contains two balls, identical except for colour. There are two scenarios:

(1) You know one ball is white, the other black,
(2) You know that there are three possibilities, two black, two white or one of each colour, and you think that all three are equally likely.

A ball is removed in such a way that you think it is as likely to be one as the other. In both scenarios, the probability that the withdrawn ball is white is $\frac{1}{2}$. Yet presented with the two scenarios, most people prefer the first over the second because it contains less uncertainty. This preference is not reflected in the probabilities, which are $\frac{1}{2}$ in both cases. A decision, based on a single expected utility, would be the same in the two cases. The single value of $\frac{1}{2}$ may be inadequate.

Now consider a second drawing from the urn, the first ball not having been replaced, and contemplate the uncertain event that the two withdrawn balls match. This has probability 0 in the first scenario, but $\frac{2}{3}$ in the second. In other words, belief based on a single number is capable of distinguishing the two scenarios when it is necessary. It was not probability that was inadequate when only one ball was taken, it was the fact that decision analysis did not require any distinction. When two were taken, the distinction was essential and was met by belief based on a single number. There have been several attempts to produce paradoxes based on the use of a single number to describe uncertainty, but all, in my opinion, can be answered in a similar fashion to that just used in the urn illustration. An excellent defence of the idea of using upper and lower probabilities is given by Walley (1991).

1.11 PROBABILITY AND LOGIC

Probability theory is an extension of logic. The latter deals with truth and falsehood. Probability deals with uncertainty where the two extremes are truth and falsehood, with probabilities 1 and 0 respectively. Two situations can look the same logically, yet be different when the uncertain element is introduced. The following example has arisen in the literature. There are a number of cards. Each card has a letter on one side and a number on the other: thus (D,3). On a table, this card will appear as either (D,·) or (·,3), depending on which face is showing. Suppose that it is a question of whether the rule "D implies 3" applies. Logic says that, presented with four cards

$$(D,\cdot)\ (F,\cdot)\ (\cdot,3)\ (\cdot,7),$$

only the first, with D showing, and the last, with 7 showing, need be investigated to test the rule.

If there is a set of cards, of which these are just four, then the probability of whether the rule obtains would be changed, through Bayes' rule, by turning up any of the cards, especially that with the 3 showing. Equally there are cases, where your knowledge base is different, where the card with a 7 showing would not be worth consideration. This case is often known as the paradox of the swans. Let D correspond to "swan" and 3 to "white", so that the rule under investigation is that all swans are white. But no one looking at a black object, 7, and finding it was a jug, F, would think that this supported the rule.

1.12 SUMMARY

If every uncertainty is to be measured by a number, then it must be in terms of numbers that obey the rules of probability. The beliefs so generated are in an appropriate form for decision analysis and, with the concept of utility, yield the principle of maximization of expected utility. For a single decision-maker, in a state of uncertainty, the theory seems adequate. There do not appear to be difficulties caused by the restriction to a single number. The calculus provides a generalization of logic. The problem of measurement is substantial and coherence is possibly the most important tool, though frequency consideration are often useful.

REFERENCES

De Finetti, B. (1974) *Theory of Probability, a Critical Introductory Treatment*, Vol. 1, Wiley, Chichester.
Jeffreys, H. (1961) *Theory of Probability*. Clarendon Press, Oxford.
Ramsey, F.P. (1926) Truth and probability. In *The Foundations of Mathematics and other Logical Essays*. Routledge & Kegan Paul, London.
Savage, L.J. (1954) *The Foundations of Statistics*. Wiley, New York.
Walley, P. (1991) *Statistical Reasoning with Imprecise Probabilities*. Chapman & Hall, London.

Chapter 2

Qualitative Theory of Subjective Probability

Patrick Suppes
Stanford University

Because we want to use probability concepts to talk about everything from the chance of drawing four aces in a hand of bridge to the probability of rain tomorrow or the probability distribution of position in a quantum-mechanical experiment, it is hardly surprising that no simple categorical theory of probability can be found. The subjective theory of probability accepts this diversity of applications, and, in fact, utilizes it to argue that the many ways in which information must be processed to obtain a probability distribution do not admit of categorical codification. Consequently, two reasonable men in approximately the same circumstances can hold differing beliefs about the probability of an event as yet unobserved. For example, according to the subjective theory of probability, two meterorologists can be presented with the same weather map and the same history of observations of basic meterological variables such as surface temperature, air pressure, humidity, upper air pressures, wind, etc., and yet still differ in the numerical probability they assign to the forecast of rain tomorrow morning. I hasten to add, however, that the term "subjective" can be misleading. It is not part of the subjective theory of probability to countenance every possible diversity in the assignment of subjective probabilities. It is a proper and important part of subjective theory to analyze, e.g., how classical relative-frequency data are to be incorporated into proper estimates of subjective probability. Such data are obviously

Subjective Probability. Edited by G. Wright and P. Ayton.
© 1994 John Wiley & Sons Ltd.

important in any systematic estimate of probabilities, as we can see from examination of the scientific literature in which probabilities play a serious part. It is also obvious that the principles of symmetry naturally applied in the classical definition of probability play an important role in the estimation of subjective probabilities whenever they are applicable.

Bayes' theorem provides an example of the sort of strong constraints to be placed on any subjective theory. The prior probability distributions selected by different investigators can differ widely without violating the subjective theory; but if these investigators agree on the method of obtaining further evidence, and if common observations are available to them, then these commonly accepted observations will often force their beliefs to converge.

2.1 DE FINETTI'S QUALITATIVE AXIOMS

Let us turn now to a more systematic discussion of the major aspects of the subjective theory. For a more detailed treatment of many questions the reader is referred to the historically important article of de Finetti (1937), which has been translated in Kyburg and Smokler (1964), and also to de Finetti's treatise (1974; 1975). The 1937 article of de Finetti's is one of the important pieces of work in the foundations of probability in this century. Probably the most influential work on these matters since 1950 is the book by Savage (1954). Savage extends de Finetti's ideas by paying greater attention to the behavioral aspects of decisions, but this extension cannot be examined in any detail in this chapter.

Perhaps the best way to begin a systematic analysis of the subjective theory is by a consideration of de Finetti's axioms for qualitative probability. The spirit of these axioms is to place restraints on qualitative judgments of probability which will be sufficient to prove a standard representation theorem, i.e. to guarantee the existence of a numerical probability measure in the standard sense. From this standpoint the axioms may be regarded as a contribution to the theory of measurement with particular reference to comparative judgments of probability. The central question for such a set of axioms is how complicated must be the condition on the qualitative relation *more probable than* in order to obtain a numerical probability measure over events.

The intuitive idea of using a comparative qualitative relation is that individuals can realistically be expected to make such judgments in a direct way, as they cannot when the comparison is required to be quantitative. On most occasions I can say unequivocally whether I think it is more likely to rain or not in the next four hours at Stanford, but I cannot in the same direct way make a judgment of how much more likely it is not to rain than rain. Generalizing this example, it is a natural move on the subjectivist's part to next ask what formal properties a qualitative comparative relation must have in order to be represented by a standard probability measure. (Later we shall review some of the experimental literature on whether people's qualitative judgments do have the requisite properties.)

We begin with the concept of a qualitative probability structure, the axioms for which are very similar formally to those for a finitely additive probability space. The set-theoretical realizations of the theory are triples $(\Omega, \mathfrak{F}, \succeq)$ where Ω is a nonempty set, $\mathfrak{F}$ is a family of subsets of Ω, and the relation $\succeq$ is a binary relation on $\mathfrak{F}$. We follow here the discussion given in Luce & Suppes (1965).

Definition 1 A structure $\Omega = (\Omega, \mathfrak{F}, \succeq)$ is a *qualitative probability structure* if the following axioms are satisfied for all A, B, and C in $\mathfrak{F}$:

S1. $\mathfrak{F}$ is an algebra of sets on Ω;
S2. If $A \succeq B$ and $B \succeq C$, then $A \succeq C$;
S3. $A \succeq B$ or $B \succeq A$;
S4. If $A \cap C = \emptyset$ and $B \cap C = \emptyset$, then $A \succeq B$ if and only if $A \cup C \succeq B \cup C$;
S5. $A \succeq \emptyset$;
S6. Not $\emptyset \succeq \Omega$.

The first axiom on $\mathfrak{F}$ is the same as the first axiom of finitely additive probability spaces. Axioms S2 and S3 just assert that $\succeq$ is a weak ordering of the events in $\mathfrak{F}$. Axiom S4 formulates in qualitative terms the important and essential principle of additivity of mutually exclusive events. Axiom S5 says that any event is (weakly) more probable than the impossible event, and Axiom S6 that the certain event is strictly more probable than the impossible event. Defining the strict relation $\succ$ in the customary fashion:

$$A \succ B \text{ if and only if not } B \succeq A,$$

we may state the last axiom as: $\Omega \succ \emptyset$.

To give a somewhat deeper sense of the structure imposed by the axioms, we state some of the intuitively desirable and expected consequences of the axioms. It is convenient in the statement of some of the theorems to use the (weakly) less probable relation, defined in the usual manner.

$$A \preceq B \text{ if and only if } B \succeq A.$$

The first theorem says that $\preceq$ is an extension of the subset relation.

Theorem 1 If $A \subseteq B$, then $A \preceq B$.

Proof. Suppose on the contrary, that not $A \preceq B$, i.e. that $A \succ B$. By hypothesis $A \subseteq B$, so there is a set C, disjoint from A such that $A \cup C = B$. Then, because $A \cup \emptyset \neq A$, we have at once

$$A \cup \emptyset = A \succ B = A \cup C,$$

whence by contraposition of Axiom S4, $\emptyset \succ C$, which contradicts Axiom S5.

Q.E.D.

Some other elementary properties follow.

Theorem 2

(i) If $\emptyset \prec A$ and $A \cap B = \emptyset$, then $B \prec A \cup B$;
(ii) if $A \succeq B$, then $-B \succeq -A$;
(iii) if $A \succeq B$ and $C \succeq D$ and $A \cap C = \emptyset$, then $A \cup C \succeq B \cup D$;
(iv) if $A \cup B \succeq C \cup D$ and $C \cap D = \emptyset$, then $A \succeq C$ or $B \succeq D$;
(v) if $B \succeq -B$ and $-C \succeq C$, then $B \succeq C$.

Because it is relatively easy to prove that a qualitative probability structure has many of the expected properties, as reflected in the preceding theorems, it is natural to ask the deeper question whether or not it has all of the properties necessary to guarantee the existence of a numerical probability measure P such that for any events A and B in $\mathfrak{F}$

$$P(A) \geqslant P(B) \text{ if and only if } A \succeq B. \tag{I}$$

If Ω is an infinite set, it is moderately easy to show that the axioms of Definition 1 are not strong enough to guarantee the existence of such a probability measure. General arguments from the logical theory of models in terms of infinite models of arbitrary cardinality suffice; a counterexample is given in Savage (1954, p. 41). De Finetti stressed the desirability of obtaining an answer in the finite case. Kraft, Pratt & Seidenberg (1959) showed that the answer is also negative when Ω is finite; in fact, they found a counterexample for a set Ω having five elements, and, thus, 32 subsets. The gist of their counterexample is the following construction. Let $\Omega = \{a,b,c,d,e\}$, and let ϕ be a measure (not a probability measure) such that

$$\begin{aligned}
\phi(a) &= 4 - \varepsilon \\
\phi(b) &= 1 - \varepsilon \\
\phi(c) &= 2 \\
\phi(d) &= 3 - \varepsilon \\
\phi(e) &= 6,
\end{aligned}$$

and

$$0 < \varepsilon < \tfrac{1}{3}.$$

Now order the 32 subsets of Ω according to this measure—the ordering being, of course, one that satisfies Definition 1. We then have the following strict inequalities in the ordering:

$$\begin{aligned}
\{a\} &\succ \{b,d\} \text{ because } \phi(a) = 4 - \varepsilon > 4 - 2\varepsilon = \phi(b) + \phi(d) \\
\{c,d\} &\succ \{a,b\} \text{ because } \phi(c) + \phi(d) = 5 - \varepsilon > 5 - 2\varepsilon = \phi(a) + \phi(b) \\
\{b,e\} &\succ \{a,d\} \text{ because } \phi(b) + \phi(e) = 7 - \varepsilon > 7 - 2\varepsilon = \phi(a) + \phi(d)
\end{aligned}$$

We see immediately also that any probability measure P that preserves these three inequalities implies that

$$\{c, e\} \succ \{a, b, d\},$$

as may be seen just by adding the three inequalities. In the case of ϕ

$$\phi(c) + \phi(e) = 8 > 8 - 3\varepsilon = \phi(a) + \phi(b) + \phi(d).$$

However, no set A different from $\{c, e\}$ and $\{a, b, d\}$ has the property that

$$\phi(\{c, e\}) \geqslant \phi(A) \geqslant \phi(\{a, b, d\}).$$

Thus we can modify the ordering induced by ϕ to the extent of setting

$$\{a, b, d\} \succ \{c, e\} \tag{II}$$

without changing any of the other inequalities. But no probability measure can preserve (II) as well as the three earlier inequalities, and so the modified ordering satisfies Definition 1 but cannot be represented by a probability measure.

Of course, it is apparent that by adding special structural assumptions to the axioms of Definition 1 it is possible to guarantee the existence of a probability measure satisfying (I). In the finite case, for example, we can demand that all the atomic events be equiprobable, although this is admittedly a very strong requirement to impose.

Fortunately, a simple general solution of the finite case has been found by Scott (1964). (Necessary and sufficient conditions for the existence of a proability measure in the finite case were formulated by Kraft, Pratt and Seidenberg, but their mutliplicative conditions are difficult to understand. Scott's treatment represents a real gain in clarity and simplicity.) The central idea of Scott's *formulation* is to impose an algebraic condition on the indicator (or characteristic) functions of the events. Recall that the indicator function of a set is just the function that assigns the value 1 to elements of the set and the value 0 to all elements outside the set. For simplicity of notation, if A is a set we shall denote by A^i its indicator function. Thus if A is an event

$$A^i(x) = \begin{cases} 1 & \text{if } x \in A, \\ 0 & \text{otherwise.} \end{cases}$$

Scott's conditions are embodied in the following theorem, whose proof we do not give.

Theorem 3 (Scott's representation theorem). Let Ω be a finite set and $\succeq$ a binary relation on the subsets of Ω. Necessary and sufficient conditions that there exists a probability measure P on Ω satisfying (I) are the following: for all subsets A and B of Ω,

(1) $A \succsim B$ or $B \succsim A$;
(2) $A \succsim 0$;
(3) $\Omega \succ 0$;
(4) for all subsets $A_0, \ldots, A_n, B_0, \ldots, B_n$ of Ω, if $A_i \succsim B_i$ for $0 \leqslant i < n$, and for all x in Ω

$$A_0^i(x) + \cdots + A_n^i(x) = B_0^i(x) + \cdots + B_n^i(x),$$

then $A_n \precsim B_n$.

To illustrate the force of Scott's condition (4), we may see how it implies transitivity. First, necessarily for any three indicator functions

$$A^i + B^i + C^i = B^i + C^i + A^i,$$

i.e. for all elements x

$$A^i(x) + B^i(x) + C^i(x) = B^i(x) + C^i(x) + A^i(x).$$

By hypothesis $A \succsim B$ and $B \succsim C$, whence by virtue of condition (4),

$$C \precsim A,$$

and thus by definition, $A \succsim C$, as desired. The algebraic equation of condition (4), just requires that any element of Ω, i.e. any atomic event, belong to exactly the same number of A_i and B_i, for $0 \leqslant i \leqslant n$. Obviously, this algebraic condition cannot be formulated in the simple set language of Definition 1 and thus represents quite a strong condition.

2.2 GENERAL QUALITATIVE AXIOMS

In the case that Ω is infinite, a number of strong structural conditions have been shown to be sufficient but not necessary. For example, de Finetti (1937) and independently Koopman (1940a, 1940b, 1941) use an axiom to the effect that there exist partitions of Ω into arbitrarily many events equivalent in probability. This axiom, together with those of Definition 1, is sufficient to prove the existence of a numerical probability measure. Related existential conditions are discussed in Savage (1954). A detailed review of these various conditions is to be found in Chapters 5 and 9 of Krantz et al. (1971). However, as is shown in Suppes & Zanotti (1976), by going slightly beyond the indicator functions, simple necessary and sufficient conditions can be given for both the finite and infinite case.

In the present case the move is from an algebra of events to the algebra $\mathfrak{F}^*$ of extended indicator functions relative to $\mathfrak{F}$. The algebra $\mathfrak{F}^*$ is just the smallest semigroup (under function addition) containing the indicator functions of all events in $\mathfrak{F}$. In other words, $\mathfrak{F}^*$ is the intersection of all sets with the property that if A is in $\mathfrak{F}$ then A^i is in $\mathfrak{F}^*$ and if A^* and B^* are in $\mathfrak{F}^*$, then $A^* + B^*$ is in $\mathfrak{F}^*$; It is easy to show that any function A^* in $\mathfrak{F}^*$ is an integer-valued function defined on Ω. It is the extension from indicator functions to integer-valued functions that justifies calling the elements of $\mathfrak{F}^*$ extended indicator functions.

The qualitative probability ordering must be extended from $\mathfrak{F}$ to $\mathfrak{F}^*$, and the intuitive justification of this extension must be considered. Let A^* and B^* be two extended indicator functions in $\mathfrak{F}^*$. Then, to have $A^* \succeq B^*$ is to have the expected value of A^* equal to or greater than the expected value of B^*. As should be clear, extended indicator functions are just random variables of a restricted sort. The qualitative comparison is now not one about the probable occurrences of events, but about the expected value of certain restricted random variables. The indicator functions themselves form, of course, a still more restricted class of random variables, but qualitative comparison of their expected values is conceptually identical to qualitative comparison of the probable occurrences of events.

There is more than one way to think about the qualitative comparisons of the expected value of extended indicator functions, and so it is useful to consider several examples.

(1) Suppose Smith is considering two locations to fly to for a weekend vacation. Let A_j be the event of sunny weather at location j and B_j be the event of warm weather at location j. The qualitative comparison Smith is interested in is the expected value of $A_1^i + B_1^i$ versus the expected value of $A_2^i + B_2^i$. It is natural to insist that the utility of the outcomes has been too simplified by the sums $A_j^i + B_j^i$. The proper response is that the expected values of the two functions are being compared as a matter of belief, not value or utility. Thus it would seem quite natural to bet that the expected value of $A_1^i + B_1^i$ will be greater than that of $A_2^i + B_2^i$, no matter how one feels about the relative desirability of sunny versus warm weather. Put another way, within the context of decision theory, extended indicator functions are being used to construct the subjective probability measure, not the measure of utility.

Note that if Smith prefers the country ($j = 1$) to the city ($j = 2$) when it is warm and sunny, then even if

$$A_1^i + B_1^i \approx A_2^i + B_2^i$$

in belief, his choice of country or city could vary depending on the degree of belief or expectation: with high expectation go to the country; with low expectation go to the city.

(2) Consider a particular population of n individuals, numbered $1, \ldots, n$. Let A_j be the event of individual j going to Hawaii for a vacation this year, and let B_j be the event of individual j going to Acapulco. Then define

$$A^* = \sum_{i=1}^{n} A_j^i \text{ and } B^* = \sum_{i=1}^{n} B_j^i.$$

Obviously A^* and B^* are extended indicator functions—we have left implicit the underlying set Ω. It is meaningful and quite natural to qualitatively compare the expected values of A^* and B^*. Presumably such comparisons are in fact of definite significance to travel agents, airlines, and the like.

We believe that such qualitative comparisons of expected value are natural in many other contexts as well. What the representation theorem below shows is that very simple necessary and sufficient conditions on the qualitative comparison of extended indicator functions guarantee existence of a strictly agreeing, finitely additive measure in the sense of (I), whether the set Ω of possible outcomes is finite or infinite.

The axioms are embodied in the definition of a qualitative algebra of extended indicator functions. Several points of notation need to be noted. First, Ω^i and $\emptyset^i$ are the indicator or characteristic functions of the set Ω of possible outcomes and the empty set $\emptyset$, respectively. Second, the notation nA^* for a function in $\mathfrak{F}^*$ is just the standard notation for the (functional) sum of A^* with itself n times. Third, the same notation is used for the ordering relation on $\mathfrak{F}$ and $\mathfrak{F}^*$, because the one on $\mathfrak{F}^*$ is an extension of the one on $\mathfrak{F}$: for A and B in $\mathfrak{F}$,

$$A \succeq B \text{ iff } A^i \succeq B^i.$$

Finally, the strict ordering relation $\succ$ is defined in the usual way: $A^* \succ B^*$ iff $A^* \succeq B^*$ and not $B^* \succeq A^*$.

Definition 2 Let Ω be a nonempty set, let $\mathfrak{F}$ be an algebra of sets on Ω, and let $\succeq$ be a binary relation on $\mathfrak{F}^*$, the algebra of extended indicator functions relative to $\mathfrak{F}$. Then the qualitative algebra $(\Omega, \mathfrak{F}, \succeq)$ is *qualitatively satisfactory* if and only if the following axioms are satisfied for every A^*, B^*, and C^* in $\mathfrak{F}^*$:

Axiom 1 The relation $\succeq$ is a weak ordering of $\mathfrak{F}^*$;
Axiom 2 $\Omega^i \succ \emptyset^i$;
Axiom 3 $A^* \succeq \emptyset^i$;
Axiom 4 $A^* \succeq B^*$ iff $A^* + C^* \succeq B^* + C^*$;

Axiom 5 If $A^* > B^*$ then for every C^* and D^* in $\mathfrak{F}^*$ there is a positive integer n such that

$$nA^* + C^* \succeq nB^* + D^*.$$

These axioms should seem familiar from the literature on qualitative probability. Note that Axiom 4 is the additivity axiom that closely resembles de Finetti's additivity axiom for events: If $A \cap C = B \cap C = \emptyset$, then $A \succeq B$ iff $A \cup C \succeq B \cup C$. As we move from events to extended indicator functions, functional addition replaces union of sets. What is formally of importance about this move is seen already in the exact formulation of Axiom 4. The additivity of the extended indicator functions is unconditional—there is no restriction corresponding to $A \cap C = B \cap C = \emptyset$. The absence of this restriction has far-reaching formal consequences in permitting us to apply without any real modification the general theory of extensive measurement. Axiom 5 has, in fact, the exact form of the Archimedean axiom used in Krantz et al. (1971, p. 73) in giving necessary and sufficient conditions for extensive measurement.

Theorem 4 Let Ω be a nonempty set, let $\mathfrak{F}$ be an algebra of sets on Ω and let $\succeq$ be a binary relation on $\mathfrak{F}$. Then a necessary and sufficient condition that there exists a strictly agreeing probability measure on $\mathfrak{F}$ is that there be an extension of $\succeq$ from $\mathfrak{F}$ to $\mathfrak{F}^*$ such that the qualitative algebra of extended indicator functions $(\Omega, \mathfrak{F}^*, \succeq)$ is qualitatively satisfactory. Moreover, if $(\Omega, \mathfrak{F}^*, \succeq)$ is qualitatively satisfactory, then there is a unique strictly agreeing expectation function on $\mathfrak{F}^*$ and this expectation function generates a unique strictly agreeing probability measure on $\mathfrak{F}$.

Proof. The main tool used in the proof is from the theory of extensive measurement: necessary and sufficient conditions for existence of a numerical representation, as given in Krantz *et al.* (1971, pp. 73–74). In particular, let A be a nonempty set, $\succeq$ a binary relation on A, and $\circ$ a binary operation closed on A. Then there exists a numerical function ϕ on A unique up to a positive similarity transformation (i.e. multiplication by a positive real number) such that for a and b in A

$$\text{(i) } a \succeq b \text{ iff } \phi(a) \geqslant \phi(b),$$
$$\text{(ii) } \phi(a \circ b) = \phi(a) + \phi(b)$$

if and only if the following four axioms are satisfied for all a, b, c, and d in A:

E1. The relation $\succeq$ is a weak ordering of A;
E2. $a \circ (b \circ c) \approx (a \circ b) \circ c$, where $\approx$ is the equivalence relation defined in terms of $\succeq$;
E3. $a \succeq b$ iff $a \circ c \succeq b \circ c$ iff $c \circ a \succeq c \circ b$;

E4. If $a \succ b$ then for any c and d in A there is a positive integer n such that $na \circ c \succeq nb \circ d$, where na is defined inductively.

It is easy to check that qualitatively satisfactory algebras of extended indicator functions as defined above satisfy these four axioms for extensive measurement structures. First, we note that functional addition is closed on $\mathfrak{F}^*$. Second, Axiom 1 is identical to E1. Extensive Axiom E2 follows immediately from the associative property of numerical functional addition: for any A^*, B^*, and C^* in $\mathfrak{F}^*$

$$A^* + (B^* + C^*) = (A^* + B^*) + C^*$$

and so we have not just equivalence but identity. Axiom E3 follows from Axiom 4 and the fact that numerical functional addition is commutative. Finally, E4 follows from the essentially identical Axiom 5.

Thus, for any qualitatively satisfactory algebra $(\Omega, \mathfrak{F}^*, \succeq)$ we can infer there is a numerical function ϕ on Ω such that for A^* and B^* in $\mathfrak{F}^*$.

$$\text{(i) } A^* \succeq B^* \text{ iff } \phi(A^*) \geqslant \phi(B^*),$$
$$\text{(ii) } \phi(A^* + B^*) = \phi(A^*) + \phi(B^*),$$

and ϕ is unique up to a positive similarity transformation.

Second, since for every A^* in $\mathfrak{F}^*$

$$A^* + \emptyset^i = A^*$$

we have at once that from (ii)

$$\phi(\emptyset^i) = 0.$$

Since $\Omega^i \succ \emptyset^i$ by Axiom 2, we can choose

$$\phi(\Omega^i) = 1.$$

And thus have a standard (unique) expectation function E for extended indicator functions:

$$\text{(i) } E(\emptyset^i) = 0$$
$$\text{(ii) } E(\Omega^i) = 1$$
$$\text{(iii) } E(A^* + B^*) = E(A^*) + E(B^*).$$

But such an expectation function for $\mathfrak{F}^*$ defines a unique probability measure P on $\mathfrak{F}$ when it is restricted to the indicator functions in $\mathfrak{F}^*$, i.e. for A in $\mathfrak{F}$, we define

$$P(A) = E(A^i).$$

Thus the axioms are sufficient, but it is also obvious that the only axioms, Axioms 2 and 3, that go beyond those for extensive structures are also necessary for a probabilistic representation. From the character of extended indicator functions, it is also clear that for each probability measure there is a unique extension of the qualitative ordering from $\mathfrak{F}$ to $\mathfrak{F}^*$.

Q.E.D.

The proof just given, even more than the statement of the theorem itself, shows what subset of random variables defined on a probability space suffices to determine the probability measure in a natural way. Our procedure has been to axiomatize in qualitative fashion the expectation of the extended indicator functions. There was no need to consider all random variables, and, on the other hand, the more restricted set of indicator functions raises the same axiomatic difficulties confronting the algebra of events.

2.3 QUALITATIVE CONDITIONAL PROBABILITY

One of the more troublesome aspects of the qualitative theory of conditional probability is that $A \mid B$ is not an object—in particular it is not a new event composed somehow from events A and B. Thus the qualitative theory rests on a quaternary relation $A \mid B \succeq C \mid D$, which is read: event A given event B is at least as probable as event C given event D. There have been a number of attempts to axiomatize this quaternary relation (Koopman, 1940a, 1940b; Aczél, 1961, 1966, p. 319; Luce, 1968; Domotor, 1969; Krantz et al., 1971; and Suppes, 1973). The only one of these axiomatizations to address the problem of giving necessary and sufficient conditions is the work of Domotor, which approaches the subject in the finite case in a style similar to that of Scott (1964).

By using indicator functions or, more generally, extended indicator functions, the difficulty of $A \mid B$ not being an object is eliminated, for $A^i \mid B$ is just the indicator function of the set A restricted to the set B, i.e. $A^i \mid B$ is a partial function whose domain is B. In similar fashion if X is an extended indicator function, $X \mid A$ is that function restricted to the set A. The use of such partial functions requires care in formulating the algebra of functions in which we are interested, for functional addition $X \mid A + Y \mid B$ will not be well defined when $A \neq B$ but $A \cap B \neq \emptyset$. Thus, to be completely explicit we begin with a nonempty set Ω, the probability space, and an algebra $\mathfrak{F}$ of events, i.e. subsets of Ω, with it understood that $\mathfrak{F}$ is closed under union and complementation. Next we extend this algebra to the algebra $\mathfrak{F}^*$ of extended indicator functions, i.e. the smallest semigroup (under function addition) containing the indicator functions of all events in $\mathfrak{F}$. This latter algebra is now extended to include as well all partial functions on Ω that are extended indicator functions restricted to an event in $\mathfrak{F}$. We call this algebra of partial extended indicator functions

$\mathcal{RF}^*$, or, if complete explicitness is needed, $\mathcal{RF}^*(\Omega)$. From this definition it is clear that if $X|A$ and $Y|B$ are in $\mathcal{RF}^*$, then

(i) $A = B$, $X|A + Y|B$ is in $\mathcal{RF}^*$.
(ii) If $A \cap B = \phi$, $X|A \cup Y|B$ is in $\mathcal{RF}^*$.

In the more general setting of decision theory or expected utility theory there has been considerable discussion of the intuitive ability of a person to directly compare his preferences or expectations of two decision functions with different domains of restriction. Without reviewing this literature, we do want to state that we find no intuitive general difficulty in making such comparisons. Individual cases may present problems, but not necessarily because of different domains of definition. In fact, we believe comparisons of expectations under different conditions is a familiar aspect of ordinary experience. In the present setting the qualitative comparison of restricted expectations may be thought of as dealing only with beliefs and not utilities. The fundamental ordering relation is a weak ordering $\succeq$ of $\mathcal{RF}^*$ with strict order $\succ$ and equivalence $\approx$ defined in the standard way.

Following Suppes & Zanotti (1982), we give axioms that are strong enough to prove that the probability measure constructed is unique when it is required to cover expectation of random variables. It is worth saying something more about this problem of uniqueness. The earlier papers mentioned have all concentrated on the existence of a probability distribution, but from the standpoint of a satisfactory theory it seems obvious for many different reasons that one wants a unique distribution. For example, if we go beyond properties of order and have uniqueness only up to a convex polyhedron of distributions, as is the case with Scott's axioms for finite probability spaces, we are not able to deal with a composite hypothesis in a natural way, because the addition of the probabilities is not meaningful.

Definition 3 Let Ω be a nonempty set, let $\mathcal{RF}^*(\Omega)$ be an algebra of partial extended indicator functions, and let $\succeq$ be a binary relation on $\mathcal{RF}^*$. Then the structure $(\Omega, \mathcal{RF}^*, \succeq)$ is a *partial qualitative expectation structure* if and only if the following axioms are satisfied for every X and Y in $\mathcal{F}^*$ and every A, B and C in $\mathcal{F}$ with $A, B \succ \emptyset$:

Axiom 1 The relation $\succeq$ is a weak ordering of $\mathcal{RF}^*$;
Axiom 2 $\Omega^i \succ \emptyset^i$;
Axiom 3 $\Omega^i|A \succeq C^i|B \succeq \emptyset^i|A$;
Axiom 4 If $X_1|A \succeq Y_1|B$ and $X_2|A \succeq Y_2|B$ then

$$X_1|A + X_2|A \succeq Y_1|B + Y_2|B;$$

Axiom 5 If $X_1|A \preceq Y_1|B$ and $X_1|A + X_2|A \succeq Y_1|B + Y_2|B$ then

$$X_2|A \succeq Y_2|B;$$

Axiom 6 If $A \subseteq B$ then

$$X|A \succeq Y|A \text{ iff } X\cdot A^i|B \succeq Y\cdot A^i|B;$$

Axiom 7 (Archimedean). If $X|A \succ Y|B$ then for every Z in $\mathfrak{F}^*$ there is a positive integer n such that

$$nX|A \succeq nY|B + Z|B.$$

The axioms are simple in character and their relation to the axioms of Definition 2 is apparent. The first three axioms are very similar. Axiom 4, the axiom of addition, must be relativized to the restricted set. Notice that we have a different restriction on the two sides of the inequality. The really new axiom is Axiom 6. In terms of events and numerical probability, this axiom corresponds to the following: If $A \subseteq B$, then

$$P(C|A) \geqslant P(D|A) \text{ iff } P(C \cap A|B) \geqslant P(D \cap A|B).$$

Note that in the axiom itself, function multiplication replaces intersection of events. (Closure of $\mathfrak{F}^*$ under function multiplication is easily proved.) This axiom does not seem to have previously been used in the literature. Axiom 7 is the familiar and necessary Archimedian axiom.

We now state the main theorem. In the theorem we refer to a strictly agreeing expectation function on $\mathcal{R}\mathfrak{F}^*(\Omega)$. From standard probability theory and conditional expected utility theory, it is evident that the properties of this expectation should be the following for $A, B \succ \emptyset$:

(1) $E(X|A) \geqslant E(Y|B)$ iff $X|A \succeq Y|B$,
(2) $E(X|A + Y|A) = E(X|A) + E(Y|A)$,
(3) $E(X\cdot A^i|B) = E(X|A)E(A^i|B)$ if $A \subseteq B$,
(4) $E(\emptyset^i|A) = 0$ and $E(\Omega^i|A) = 1$.

Using primarily (3), it is then easy to prove the following property, which occurs in the earlier axiomatic literature mentioned above:

$$E(X|A \cup Y|B) = E(X|A)E(A^i|A \cup B) + E(Y|B)E(B^i|A \cup B),$$

for $A \cap B = \emptyset$.

Theorem 5 Let Ω be a nonempty set, let $\mathfrak{F}$ be an algebra of sets on Ω, and let $\succeq$ be a binary relation on $\mathfrak{F} \times \mathfrak{F}$. Then a necessary and sufficient condition that there is a strictly agreeing conditional probability measure on $\mathfrak{F} \times \mathfrak{F}$ is that there is an extension $\succeq^*$ of $\succeq$ from $\mathfrak{F} \times \mathfrak{F}$ to $\mathcal{R}\mathfrak{F}^*(\Omega)$ such that the structure $(\Omega, \mathcal{R}\mathfrak{F}^*(\Omega), \succeq^*)$ is a partial qualitative expectation structure. Moreover, if $(\Omega, \mathcal{R}\mathfrak{F}^*(\Omega), \succeq^*)$, is a partial qualitative expectation structure, then there is a unique strictly agreeing expectation function on $\mathcal{R}\mathfrak{F}^*(\Omega)$ and this expectation generates a unique strictly agreeing conditional probability measure on $\mathfrak{F} \times \mathfrak{F}$.

The proof is given in Suppes & Zanotti (1982).

2.4 GENERAL ISSUES

I now want to turn to a number of general issues that arise in evaluating the correctness or usefulness of the subjective view of probability.

Use of symmetries

A natural first question is to ask how subjective theory utilizes the symmetries that are such a natural part of the classical, Laplacian definition of probability. If we think in terms of Bayes' theorem the answer seems apparent. The symmetries that we all accept naturally in discussing games of chance are incorporated immediately in the subjective theory as prior probabilities. Thus, for example, if I wish to determine the probability of getting an ace in the next round of cards dealt face up in a hand of stud poker, given the information that one ace is already showing on the board, I use as prior probabilities the natural principles of symmetry for games of chance, which are a part of the classical definition. Of course, if I wish to introduce refined corrections I could do so, particularly corrections arising from the fact that in ordinary shuffling, the permutation groups introduced are groups of relatively small finite order, and, therefore, there is information carry-over from one hand to another. These second-order refinements with respect to shuffling are simply an indication of the kind of natural corrections that arise in the subjective theory and that would be hard to deal with in principle within the framework of the classical definition of probability. On the other hand, I emphasize that the principles of symmetry used in the classical definition are a natural part of the prior probabilities of an experienced card player. The extent to which these symmetries are compelling is a point I shall return to later.

Use of relative frequencies

It should also be clear that the proper place for the use of relative-frequency data in the subjective theory is in the computation of posterior probabilities. It is clear what is required in order to get convergence of opinion between observers whose initial opinions differ. The observers must agree on the method of sampling, and, of course, they must also agree on the observations that result from this sampling. Under very weak restrictions, no matter how much their initial opinions differ, they can be brought arbitrarily close to convergence on the basis of a sufficient number of sampled observations. The obvious requirement is that the individual observations be approximately independent. If, for example, the observations are strongly dependent, then many observations will count for no more than a single observation.

Reflection upon the conditions under which convergence of beliefs will take place also throws light on the many situations in which no such convergence occurs. The radically differing opinions of men about religion, economics, and politics are excellent examples of areas in which there is a lack of convergence; no doubt a main source of this divergence is the lack of agreement on what is to count as evidence.

Problem of the forcing character of information

As already indicated, it is an important aspect of the subjective theory to emphasize that equally reasonable men may hold rather different views about the probability of the same event. But the ordinary use of the word "rational" seems to go beyond what is envisaged in the subjective theory of probability. Let us consider one or two examples of how this difference in usage may be expressed.

The first kind of example deals with the nonverbal processing of information by different individuals. One man is consistently more successful than another in predicting tomorrow's weather. At least before the advent of powerful mathematical methods of predicting weather, which are now just beginning to be a serious forecasting instrument, it was the common observation of experienced meteorologists that there was a great difference in the ability of meteorologists with similar training and background and with the same set of observations in front of them to predict successfully tomorrow's weather in a given part of the world. As far as I can see, in terms of the standard subjective theory as expressed, for example, by de Finetti, there is no very clear way of stating that on a single occasion the better predictor is in some sense more rational in his processing of information than the other man; yet in common usage we would be very inclined to say this. It is a stock episode in novels, and a common experience in real life for many people, to denigrate the intelligence or rationality of individuals who continue to hold naive beliefs about other people's behavior in the face of much contrary, even though perhaps subtle, evidence.

But successive predictions can be studied like any other empirical phenomena, and there is a large literature on evaluating the performance of forecasters, an important practical topic in many arenas of experience. Examination of quantitative methods of evaluation of subjective, as well as objective, forecasts lies outside the scope of this chapter. The *Journal of Forecasting* is entirely devoted to the subject. See also, for example, Makridakis *et al.* (1984) and Dawid (1986).

Contrary to the tenor of many of de Finetti's remarks, it seems fair to say that the subjective theory of probability provides necessary but not sufficient conditions of rationality.

Bayesian probabilities and the problem of concept formation

An important point revolving around the notion of *mistaken* belief is involved in the analysis of how information is processed. In common usage, a belief is often said to be mistaken or irrational when later information shows the belief to be false. According to the subjective theory of probability, and much sensible common usage in addition, this view of mistaken beliefs is itself a mistake. A belief is not shown to be mistaken on the basis of subsequent evidence not available at the time the belief was held. Proper changes in belief are reflected in the change from a prior to a posterior probability on the basis of new information. The important point for subjective theory is that the overall probability measure does not itself change, but rather we pass from a prior to a posterior conditional probability. Applications of Bayes' theorem illustrate this well enough. The following quotation from de Finetti (1937/1964, page 146) illustrates this point beautifully.

> Whatever be the influence of observation on predictions of the future, it never implies and never signifies that we *correct* the primitive evaluation of the probability $P(E_{n+1})$ after it has been *disproved* by experience and substitute for it another $P^*(E_{n+1})$ which conforms to that experience and is therefore probably *closer to the real probability*; on the contrary, it manifests itself solely in the sense that when experience teaches us the result A on the first n trials, our judgment will be expressed by the probability $P(E_{n+1})$ no longer, but by the probability $P(E_{n+1} \mid A)$, i.e., that which our initial opinion would already attribute to the event E_{n+1} considered as conditioned on the outcome A. Nothing of this initial opinion is repudiated or corrected; it is not the function P which has been modified (replaced by another P^*), but rather the argument E_{n+1} which has been replaced by $E_{n+1} \mid A$, and this is just to remain faithful to our original opinion (as manifested in the choice of the function P) and coherent in our judgment that our predictions vary when a change takes place in the known circumstances.

In spite of the appeal of what de Finetti says, there seems to be a wide class of cases in which the principles he affirms have dubious application. I have in mind all those cases in which a genuinely new concept is brought to bear on a subject. I do not mean necessarily the creation of a new scientific concept, but rather any situation in which an individual suddenly becomes aware of a concept that he was not previously using in his analysis of the data in front of him.

Suppose an organism has the sensory capability to recognize at least 100 features, but does not know how to combine the features to form the concepts being forced upon it by experience. Assuming the features have only two values (presence or absence), even with this drastic simplification it does not make sense from a computational standpoint to suppose the organism has a prior distribution that is positive for each of the 2^{100} possible patterns that might be nature's choice.

I am not entirely certain what subjectivists like de Finetti would say about this kind of example. I cannot recall reading anywhere a systematic discussion of concept formation, or even identification, by one of the main proponents of subjective probability. It is my own view that no adequate account of concept formation can be given within the framework of subjective probability, and that additional more complicated and detailed learning processes in organisms must be assumed in order to provide an adequate account. This is not to denigrate the theory of subjective probability, but to be realistic about its limitations.

Problem of unknown probabilities

Another feature of the subjective theory of probability that is in conflict with common usage of probability notions is the view that there are no unknown probabilities. If someone asks me what is the probability of rain in the Fiji Islands tomorrow, my natural inclination is to say, "I don't know," rather than to try to give a probability estimate. If another person asks me what I think the probability is that Stanford University will have an enrollment of at least 50 000 students 500 years from now, I am naturally inclined simply to say, "I haven't the faintest idea what the probability or likelihood of this event is." De Finetti insists on the point that a person always has an opinion, and, therefore, a probability estimate about such matters, but it seems to me that there is no inherent necessity of such a view. It is easy to see one source of it. The requirement that one always have a probability estimate of any event, no matter how poor one's information about the circumstances in which that event might occur may be, arises from a direct extension of two-valued logic. Any statement is either true or false, and, correspondingly, any statement or event must always have a definite probability for each person interrogated. From a formal standpoint it would seem awkward to have a logic consisting of any real number between 0 and 1, together with the quite disparate value, "I don't know."

A little later we shall examine the view that one can always elicit a subjective probability for events about which the individual has very little background information by asking what sort of bet he will make concerning the event. Without anticipating that discussion, I still would like to insist that it does not really seem to be a proper part of the subjective theory to require an assignment of a probability to every imaginable event. In the same spirit with which we reply to a question about the truth of a statement by saying that we simply don't know, we may also reply in the same fashion to a request for an estimate of a probability. This remark naturally leads to the next problem I want to consider.

Inexact probability estimates

There are many reasons, some of which were just mentioned, for being skeptical of one's own or other people's ability to make sensible exact probability estimates of possible events about which little is known. A retreat, but within the general subjective framework, is to give upper and lower probability estimates. So, in what we might think of as a state of nearly total ignorance about the possible occurrence of an event E, we assign upper probability $P^*(E) = 1$, and lower probability $P_*(E) = 0$. Note that the positions of the stars distinguishes in a natural way upper from lower probabilities.

Development of this idea is pursued in some detail in Chapter 4 by Glenn Shafer. There is, however, one important concept I want to mention here. The use of upper and lower probabilities developed by Dempster and Shafer, for example, assumes a supporting probability, i.e. the upper and lower probabilities assigned to an algebra of events are consistent with the existence of at least one probability measure P such that for every event E in the algebra

$$P_*(E) \leqslant P(E) \leqslant P^*(E).$$

But it is easy to envisage realistic incoherent inexact probabilities, incoherent in the sense that they are not compatible with the existence of a probability measure. As a simple hypothetical example, consider the person whose partial beliefs about the economy of his country ten years from now are expressed in part by the following correlations. Let $u =$ high unemployment, $p =$ at least moderate prosperity, and $d =$ at least fairly high deficit. Let these three events be represented by random variables **U**, **P** and **D** respectively, with value $+1$ for occurrence and -1 otherwise, let the subjective expectations of all three be 0, and let the subjective correlations satisfy the three inequalities $\rho(\mathbf{U}, \mathbf{P}) < -0.5$, $\rho(\mathbf{D}, \mathbf{P}) < -0.5$ and $\rho(\mathbf{U}, \mathbf{D}) < 0.0$. I think subjective correlations of this sort are not unlikely for many triples of events. But then there can be no upper and lower probability of the kind envisaged by Dempster and Shafer to express these beliefs, for there is no possible joint probability distribution of the three random variables satisfying the expectations and the correlation inequalities—here I am assuming, for simplicity, exact subjective probabilities for the marginal distribution of each of the three pairs of random variables.

On the other hand, there can be a nonmonotonic upper probability compatible with any pairwise distribution satisfying the correlation inequalities. Such an upper probability P^* satisfies for any two events A and B such that $A \cap B \neq \emptyset$

$$P^*(A \cup B) \leqslant P^*(A) + P^*(B),$$

with, of course,

$$P^*(\Omega) = 1 \qquad \text{and} \qquad P^*(\emptyset) = 0.$$

But the necessary nonmonotonicity means there are events A and B such that $A \subseteq B$, but

$$P^*(B) < P^*(A).$$

which is not possible for any probability measure.

It seems desirable to extend the theory of subjective probability to situations in which it is natural to start with an incoherent or inexact prior because of lack of knowledge—or the opposite problem of too much—with accompanying computational problems. (For a natural application of such nonmonotonic upper probabilities to physics, see Suppes and Zanotti, 1991.)

Decisions and the measurement of subjective probability

It is commonplace to remark that a man's actions or decisions, and not his words, are the true mark of his beliefs. As a reflection of this commonly accepted principle, there has been considerable discussion of how one may measure subjective probabilities on the basis of decisions actually made. This is a complex subject, and I shall not attempt to give it a fully detailed treatment.

The classical response in terms of subjective probability is that we may find out the subjective probabilities a man truly holds by asking him to place wagers. For example, if he thinks the probability of rain tomorrow is really $\frac{1}{2}$, then he will be willing to place an even-money bet on this occurrence. If he thinks that the probability of snow tomorrow has a subjective probability of 0.1, then he will bet against snow at odds of 1 : 9. It is also clear how this same procedure may be used to test precise statements. For example, if a man says the probability of rain tomorrow is at least $\frac{1}{2}$, then presumably he will accept any bet that provides odds at least this favorable to him.

Unfortunately, there is a central difficulty with this method of measuring subjective probability. Almost all people will change the odds at which they will accept a bet if the amount of money varies. For example, the man who will accept an even-money bet on the probability of rain tomorrow with the understanding that he wins the bet if in fact it does rain, will not accept an even-money bet if the amount of money involved moves from a dollar on each side to a hundred dollars on each side. Many people who will casually bet a dollar will not in the same circumstances and at the same odds be willing to bet a hundred dollars, and certainly not a thousand dollars. The man who will accept an even-money bet on its raining tomorrow will perhaps be willing to accept odds of two to one in his favor only if the wager is of the order of a hundred dollars, while he will accept still more favorable odds for a bet involving a larger sum of money. What then are we to say is his true estimate of the probability of rain tomorrow if we use this method of wagers to make the bet?

In spite of this criticism, there have been a number of interesting empirical studies of the measurement of subjective probability using the simple scheme we have just described. A review of the older literature is to be found in Luce & Suppes (1965). An excellent review of recent literature on risk-sensitive models, together with new results, are to be found in Luce & Fishburn (1991).

ACKNOWLEDGMENT

Dennis Lindley suggested several clarifying improvements in an earlier draft.

REFERENCES

Aczél, J. (1961) Über die Begründung der Additions- und Multiplikationsformeln von bedingten Warscheinlichkeiten. *Magyar Tud. Akad. Mat. Kutató Int Közl.* **6**, 110–22.

Aczél, J. (1966) *Lectures on Functional Equations and Their Applications.* Academic Press, New York.

Dawid, A.P. (1986) Probability forecasting. In S. Kotz, N.L. Johnson, and C.B. Read (eds.), *Encyclopedia of Statistical Sciences*, Vol. 7. Wiley, New York, pages 210–18.

deFinetti, B. (1937) La prévision: ses lois logiques, ses sources subjectives. *Ann. Inst. H. Poincaré*, **7**, 1–68. Translated into English in H.E. Kyburg, Jr., & H.E. Smokler (eds.), *Studies in Subjective Probability*, 1964, pages 93–158.

de Finetti, B. (1974) *Theory of Probability*, Vol. 1. Wiley, New York.

de Finetti, B. (1975) *Theory of Probability*, Vol. 2. Wiley, New York.

Domotor, Z. (1969) Probabilistic Relational Structures and Their Applications. Technical Report No. 144, Stanford University, Institute for Mathematical Studies in the Social Sciences.

Koopman, B.O. (1940a) The axioms and algebra of intuitive probability. *Annals of Math.* **42**, 269–92.

Koopman, B.O. (1940b) The bases of probability. *Bulletin of the American Mathematical Society.* **46**, 763–74. Reprinted in H.E. Kyburg, Jr., & H.E. Smokler (eds.), *Studies in Subjective Probability.* Wiley, New York, 1964, pages 159–72.

Koopman, B.O. (1941) Intuitive probability and sequences. *Ann. of Math.* **42**, 169–87.

Kraft, C.H., Pratt, J.W., & Seidenberg, A. (1959) Intuitive probability on finite sets. *Ann. Math. Statist.* **30**, 408–19.

Kranz, D.H., Luce, R.D., Suppes, P. & Tversky, A. (1971) *Foundations of Measurement*, Vol. I. Academic Press, New York.

Kyburg, H.E. Jr. & Smokler, H.E. (eds) (1964) *Studies in Subjective Probability*, Wiley, New York.

Luce, D. (1968) On the numerical representation of qualitative conditional probability. *Ann. Math. Statist.* **39**, 481–91.

Luce, D. & Fishburn, P. (1991) Rank- and sign-dependent linear utility models for finite first-order gambles. *Journal of Risk and Uncertainty.* **4**, 29–59.

Luce, R.D. & Suppes, P. (1965) Preference, utility and subjective probability. In R.D. Luce, R.R. Bush and E. Galanter (eds.), *Handbook of Mathematical Psychology*. Vol. 3, Wiley, New York, pages 249–410.

Makridakis, S., Andersen, A., Carbone, R., Fildes, R., Hibon, M., Lewandowski, R., Newton, J., Parzen, E. & Winkler, R. (1984). *The Forecasting Accuracy of Major Time Series Methods*. Wiley, Chichester.

Savage, L.J. (1954) *The Foundations of statistics*, Wiley, New York.

Scottt, D. (1964) Measurement models and linear inequalities. *Journal of Mathematical Psychology*, **1**, 233–47.

Suppes, P. (1973) New foundations of objective probability: axioms of propensities. In P. Suppes, L. Henkin, G.C. Moisil, & A. Joja (eds.), *Logic, Methodology and Philosophy of Science IV*, Proceedings of the 4th International Congress for Logic, Methodology and Philosophy of Science, Bucharest, 1971. Amsterdam: North Holland, pages 515–29.

Suppes, P. & Zanotti, M. (1976) Necessary and sufficient conditions for existence of a unique measure strictly agreeing with a qualitative probability ordering. *Journal of Philosophical Logic*. **5**, 431–8.

Suppes, P. & Zanotti, M. (1982) Necessary and sufficient qualitative axioms for conditional probability. *Zeitschrift für Wahrscheinlichkeitstheorie und verwandte Gebiete*. **60**, 163–9.

Suppes, P. & Zanotti, M. (1991) Existence of hidden variables having only upper probabilities. *Foundations of Physics*. **12**, 1479–99.

Chapter 3

Probability, Uncertainty and the Practice of Statistics

Colin Howson and Peter Urbach
London School of Economics

3.1 INTRODUCTION

If we have observed that all the swans in a sample of swans are white, it is generally accepted that from this we cannot infer with certainty that all swans are white, or even that the next swan to be observed will be white. Depending on the sample, we may be more or less certain, but never entirely certain. But if we could somehow know *a priori* that whiteness is an essential property of swans, then we would know that necessarily all swans were white, and so we would not have to worry about whether the sample of swans we have actually observed is representative of the class of all swans in terms of colour properties. We need not in principle observe anything, to know everything.

This is a beguiling idea, and it beguiled Plato, and later Descartes, Leibniz and Spinoza. The idea that factual knowledge can simply be excogitated in an armchair, or even a philosopher's cell, seems fantastic to us now, but we should remember that these writers had concrete evidence, or so they thought, to the contrary. For the mathematics they were acquainted with, that is to say Euclidean, and later Cartesian, geometry, did appear to generate exact and certain *factual* knowledge *a priori*, more exact and certain than any knowledge derived from observation. Indeed, deviations from its predictions in the world of space and time were plausibly put down to spatio-temporal lines, planes,

Subjective Probability. Edited by G. Wright and P. Ayton.
© 1994 John Wiley & Sons Ltd.

solids and angles being imperfect exemplars of the Ideal Euclidean lines, planes, solids and angles. Only quite recently was it realized that Euclid's geometry is not a body of synthetic *a priori* knowledge at all but an axiomatic theory, and that real lines approximate to ideal ones only insofar as they approximately satisfy its axioms. We now know, or think we know, that an ideal line is an imperfect representation of a real one, rather than conversely, and alternative geometries were developed precisely because the ideal lines, planes etc. in Euclid's geometry were *too* imperfect representations of real ones.

Long before this realization dawned, however, the Platonic doctrine, later known as rationalism, that all true knowledge is quasi-mathematical, *a priori*, had succumbed to the charge of sterility and emptiness, a verdict reinforced by the spectacular development in the seventeenth and eighteenth centuries of the new empirical science of physics, which rapidly assumed—and still does to many—the status of role model for scientific knowledge. This status was ratified by the new epistemological doctrine of empiricism, according to which all knowledge is derived, in a sense that was never made very precise, from observation.

Empiricism seemed to be a fruitful epistemology, but it was one with a price. That price, as we saw, is uncertainty. Certain deductive inferences are replaced by uncertain inductive ones. But the lack of information implicit in saying that something is uncertain is mitigated if we can say exactly how certain it is. And more importantly, can an inductive inference ever approach deductive certainty?

Hume is celebrated for answering these questions firmly in the negative. But even while he was proclaiming total inductive scepticism, an inductive logic was being developed which gave entirely different, positive answers, and claimed to do so, moreover, with finality and exactitude. It was because of this that Hume's sceptical arguments evoked little contemporary interest; they had, it was thought, and incorrectly as we shall see, already been answered.

The new inductive logic was the theory of probability, based on the then recently developed mathematics of combinatorial algebra and analysis, and first identified as a distinct discipline in the late seventeenth century. The equation of mathematical probability with degree of certainty was made almost immediately. It is already explicit in James Bernoulli's *Ars Conjectandi* (1715). Actual computations of inductive probability proceeded by means of three auxiliary principles. The first of these is an easily proved consequence of the mathematical theory, known as Bayes' theorem, which states that the conditional probability of a hypothesis H on data E (the so-called *posterior probability* of H) is proportional, as H varies through some partition $\{H_i\}$, to the unconditional probability of H (the *prior probability* of H) multiplied by the conditional probability of E given H:

$$P(H \mid E) \propto P(E \mid H)P(H)$$

where the constant of proportionality is $P(E) = \Sigma P(E \mid H_i)P(H_i)$.

The importance of this theorem lies in the fact that if H is deterministic and the data arise from a well-designed experiment, then in the ideal state of affairs (very ideal, but never mind) E will either be the outcome predicted by H or one inconsistent with H; in the former case $P(E \mid H) = 1$ and in the latter $P(E \mid H) = 0$. If H is a statistical hypothesis and E an event in the corresponding outcome space, then the second principle, which used to be called the *principle of direct probability*, authorizes equating $P(E \mid H)$ with the probability which H assigns E.

The third principle tells one how to compute the prior probabilities $P(H)$. Known as the *principle of insufficient reason*, and later the *principle of indifference*, it is a symmetry principle stating that if nothing is known about a quantity X save that it takes one of n possible values, then the *a priori* probability that it will take any one is constant, and hence by the finite additivity property of probability functions is equal to n^{-1}. Hence the probability that its value will be in any r-membered subset is r/n (this is of course Laplace's celebrated "favourable cases to possible cases" ratio (1820)).

There is a natural extension of this principle to a bounded real-valued random variable X: if nothing is known about X except that its range of values is an open or closed interval of length k, then the *a priori* probability density $f(x)$ is equal to k^{-1}. In the middle of the eighteenth century Thomas Bayes, in a celebrated Memoir to the Royal Society of London, used such a uniform density over the values of a variable Q representing the objective chance of a specified event, to derive a posterior probability distribution for Q, conditional on the data that the event in question occurred r times out of n.

Despite apparent successes like this, the new logic based on these three principles proved far from satisfactory and by the end of the nineteenth century was widely regarded as discredited. The problem lay with the third principle, the principle of indifference, which is difficult if not impossible to implement consistently. Suppose, for example, that you know nothing about a variable X except that it is confined within the closed unit interval, say. In that case all you know about $Y = X^2$ is that it has the same range of values. So the principle of indifference appears to demand that X and Y have exactly the same *a priori* probability density, namely $f(x) = f(y) = 1$, which is impossible.

It might seem that this is a problem only for continuously distributed variables, but that is not true. Consider the hypothesis H_X: $X = x$, for a particular value x of the variable X above. H_X itself can be regarded as a quantity taking two possible values, 1 (i.e. true) and 0 (false). This is all you know about H_X, moreover. So the principle of indifference would seem to require that $P(H_X = 1) = \frac{1}{2}$. But the principle of indifference also "said" that the probability $P(X = x)$ was equal to 0 (since it "said" that X has a constant probability density at each point equal to 1). Yet of course $X = x$ is equivalent to $H_X = 1$.

Though people wrestled with the problem of the correct interpretation of the principle of indifference, it gradually dawned on them that there was no uniquely "correct" one. There are many, usually infinitely many, partitions of logical space containing some specified cell in that space. There is no reason why, on the basis of a presumptively null knowledge state, any one of these should be any more privileged than any other. To put the matter another way, while it is possible to define a function representing neutrality between the cells in any given partition of logical space, it is not possible to define one that has this property for all partitions. A *uniform* neutrality is not possible, in other words: neutrality over one partition will mean bias over another.

The principle of indifference had to go. But to abandon the idea of a neutral prior distribution brought into question the enterprise of an objective inductive logic securely based on probability theory. Alternative rules for determining or partially determining prior probabilities have from time to time been appealed to, but none has been uncontroversial. Jeffreys (1961) and others have argued that simpler hypotheses have higher prior probabilities, while E.T. Jaynes, following earlier suggestions by Jeffreys, recommended using the requirement of invariance under a suitable group to determine prior probabilities (1973), or alternatively choosing the prior distribution which maximizes entropy, if there is one, subject to whatever constraints are deemed appropriate (1957, 1967). These suggestions all share with the principle of indifference problems of consistent application, as well as possessing their own peculiar difficulties. The choice of the appropriate group with respect to which which the prior is to be invariant is usually a fairly arbitrary matter, as is the choice of which of the many non-equivalent explications of simplicity to adopt, and entropy-maximizing distributions may not exist or may not be unique.

The apparent impossibility of determining prior probabilities in any non-arbitrary manner has been a powerful factor in convincing many people that a probabilistic theory of inductive inference was impossible. This was true in the case of R.A. Fisher, who repudiated Bayesian probabilism in favour of a theory of inference allegedly based on the logic of refutation. In this he was followed by Popper, who is today better known, at any rate outside statistics, as the principal proponent of falsificationism. Their attempt, though it continues to be influential, cannot nevertheless be regarded as successful, as we shall see in due course.

3.2 THE MODERN SUBJECTIVE THEORY

While Fisher and Popper were dismissing inductive probability, important discoveries were being made about the probability axioms which suggested an entirely different way in which a probabilistic inductive logic could be constructed, in which objectivity is achieved without recourse to debatably "objective" priors. We shall look briefly at just one of these, because it is both simple to explain and convincing in its claim to show that the probability axioms can be interpreted as a complete logic of consistent partial belief. Different ways of arriving at substantially the same conclusion are to be found in Ramsey (1931), Savage (1954), Jeffrey (1965), and Lindley (1982).

Suppose you agree to indicate your degree of confidence in the truth of a proposition A in the traditional manner of stating the odds you currently think would be the fair ones in any bet on A, were the truth of A to be decided after the bet by a competent authority. It is actually more convenient to measure that confidence not in terms of the odds on A directly but in terms of the derivative betting quotient, which we shall suggestively symbolize by $P(A)$. The betting quotient P is obtained from the odds by the bijective mapping (allowing odds to be infinite): $P = \text{odds}/(1 + \text{odds})$, with its familiar inverse: $\text{odds} = P/(1 - P)$. The scale of betting quotients has the advantage that it is bounded and that the point of indifference between A and $-A$ (not-A), corresponding to odds of 1, is its midpoint. Your fair odds we shall take to be odds which you believe give no advantage to either side of the bet. Ramsey and de Finetti independently showed that if a set of betting quotients does not satisfy the finitely additive probability axioms, then the odds determined by them could be combined into a certainly winning or certainly losing betting strategy, where a betting strategy is a set of n decisions of the form "bet on/against A_i with stake S_i", $i = 1, \ldots, n$. The finitely additive probability axioms are (1) $P(A) \geqslant 0$, (2) $P(T) = 1$ where T is the certain proposition, and (3) $P(A \vee B) = P(\text{A}) + P(B)$, where $A \vee B$ is the disjunction A or B and A and B are mutually exclusive. If we define a conditional bet on A given B to be one that proceeds in the normal way if B is true and is called off if not, then it is also true that if $P(A \mid B)$ is the betting quotient on A in a conditional bet, and $P(A \mid B)$ is not equal to $P(A\&B)/P(B)$, where $P(A\&B)$, $P(B)$ are betting quotients on $A\&B$ and B respectively and $P(B) > 0$, then a betting strategy can be devised which will deliver a sure loss or gain.

The proof that if (1)–(3) are not satisfied then a betting strategy exists which if implemented leads to inevitable gain or loss is not difficult (an elementary proof is given in Howson & Urbach, 1993). A set of betting quotients with this pathological property is colloquially said to be vulnerable to a Dutch book, and the theorem above has consequently become known as the Dutch book theorem. That theorem shows that a necessary condition for a set of degrees of belief, measured as the agent's fair betting quotients, to be internally consistent is that they satisfy the finitely additive probability axioms, and a straightforward extension of the argument for (3) shows that they must in addition be countably additive. Henceforward when we mention the probability axioms we shall include among them the principle of countable additivity, or continuity as it is sometimes called. The converse to the Dutch book theorem, that if a set of betting quotients does satisfy the axioms then there is no betting strategy that can be guaranteed in advance to generate a loss (or gain) independently of the truth values of the propositions bet on, is also easily demonstrated.

Since they are necessary and sufficient for avoidance of vulnerability to certain loss or gain, the probability axioms are in effect a complete set of consistency constraints, and in our opinion that fact precisely restricts the domain of objectively valid reasoning about uncertainty to the deductive closure of the axioms. This has important consequences. In particular, it means that even were it capable of a consistent formulation, the principle of indifference, or indeed any other method of determining priors, has no claim to legitimacy, for it is not a consequence of the axioms. But accepting that the probability axioms are a complete theory of valid probabilistic inference means accepting that the priors in any Bayes's theorem calculation of posterior probabilities are essentially indeterminate within the theory. This has seemed to some, and not a few of them Bayesians, an abdication of responsibility, and an admission of incompleteness insofar as allegedly more liberal criteria of rationality are concerned.

This view, common though it is, is mistaken. If we look at the paradigm of reasonable cognitive activity, namely science, we find a great diversity of opinion about new theories, sometimes, as with Einstein's doggedly negative attitude to quantum mechanics, persisting over long periods. In fact, of course, a diversity of opinion is an efficient way of managing uncertainty at the group level, for it allows the exploration of alternatives to the dominant view, alternatives which may well show that that view is actually only of temporary and conditional validity. It is widely appreciated that the suppression of deviant opinions in the long run does much more harm than good, if it ever does good, and would certainly have retarded the growth of scientific knowledge had it been more successfully practised.

Tolerance of alternative opinions is all very well, however, but the fact remains that as empiricists we must allow the accumulation of observational evidence to exert an increasingly strong pressure against diversity. But this is exactly what the probability calculus, in the form of various convergence theorems, "predicts" will occur. The convergence is generally with probability one, so even in the appropriate conditions the posterior distributions do not necessarily converge for all data sequences. But, granted that they agree on which hypotheses are to be assigned a positive probability, there are cases where the posterior probability functions actually converge to certainty on the true hypothesis (for example, Halmos, 1950, p. 213. Theorem B; Earman, 1992 contains a clear and up-to-date discussion of these Bayesian convergence theorems).

So the fact that the Bayesian theory does not fix "rational" priors neither condemns it as undesirably incomplete nor as explanatorily empty. Nevertheless it is still felt in many quarters that it is too liberal in what it fails to prohibit. But criticism of this sort simply begs the question as to what is and is not rational cognitive behaviour. Nobody has yet shown where the boundaries of the rational and the irrational lie, and all the evidence points against there being any uniquely rational cognitive policy for each possible state of background information.

We have tried to show that the Bayesian theory is not vulnerable to the commonly made charge of an excessive reliance on subjective opinion. It is now time to switch from defence to attack, and look more closely at those alternative theories which purport to rest on an unimpeachably objective and secure foundation. We shall see that this is very far from being the case.

3.3 THE OBJECTIVIST IDEAL

How convenient it would be if one could arrive at firm theoretical conclusions by simple logical deduction from observations. For if that were possible, since direct observational evidence is presumably true (setting aside a philosopher's extreme sceptical doubts), the theoretical conclusions drawn would, of logical necessity, also be true. Hume's demonstration that such inferences are not in general legitimate left philosophers of science pondering the problem (the "problem of induction") of what, then, may legitimately be said regarding the truth of theories, on the basis of facts of observation.

We regard the Bayesian answer to this question as extremely satisfactory, but the fact that inductive probabilities are subjective or personal has galvanized opposition to the Bayesian idea. Scientific judgement, critics say, should have nothing subjective about it, but should be perfectly objective. Lakatos (1978, Vol. 1, p. 1) put this objection with admirable clarity: "The cognitive value of a theory has nothing to do with its psychological influence on people's minds ... [but] depends only on what objective support it has in facts."

The objectivist ideal implicit in such objections to subjective Bayesianism is greatly appealing; it would be nice if disagreements in science could be resolved by impartially measuring the "objective cognitive values" of contending hypotheses. But what does the suggestive phrase "objective cognitive value" really mean? Lakatos never said. And since the goal which he postulated for the scientific enterprise is shrouded in such a haze of imprecision, it is scarcely surprising that Lakatos never succeeded in showing how to reach that goal. We find this weakness in all of the well-known attempts to formulate a non-Bayesian methodology of science, where rules for processing experimental evidence are offered, and conclusions purporting to contain objective information about the cognitive status of certain theories are drawn: on closer examination, those conclusions turn out to have no cognitive meaning at all, and to create the misleading impression that they do only by the pregnant language in which they are couched. Methodologies of this (as we shall argue) ineffectual kind have been developed both by philosophers of science and by statisticians.

3.4 POPPER'S METHODOLOGY FOR DETERMINISTIC THEORIES

Karl Popper is a prominent representative of the first. He started from the fact that general, non-statistical hypotheses (simple example: "all swans are white") can often be decisively falsified by observational evidence (e.g. "this is a black swan"). And his famous thesis (or definition?) is that only hypotheses that are falsifiable by "possible or conceivable observations" are "scientific". A hypothesis may also have empirical implications whose truth can be checked in appropriate experiments; and if any such implication is verified, Popper describes the hypothesis as thereby "corroborated". Now according to the *Oxford English Dictionary*, "to corroborate" means to strengthen or make strong; to support or confirm. So Popper appears to be saying that you can strengthen a hypothesis simply by verifying one of its implications, a process involving no subjective probabilities but logic and observation alone. But what could this strengthening possibly signify? It is easy to think of a bridge being strengthened, or the Tower of Pisa being supported; but how can that thought be stretched to include an abstract thing like a hypothesis? To this question, crucial to his thesis, Popper gives no adequate answer. He acknowledges that the hypothesis is not conclusively proved in the corroboration process; nor is it in any sense "partially" proved; nor is its objective probability augmented (there seems to be no such thing as a theory's objective probability). Popper sometimes says that it is "rational to prefer" a corroborated hypothesis over one that is not, on the grounds that the hypothesis is "better tested"; but this turns out, disappointingly, to be a circumlocution for "better corroborated". The dismal fact is, as Popper more or less concedes, that saying that a hypothesis is corroborated in the sense he defines, implies nothing as to the cognitive or epistemic standing of the hypothesis, and certainly does not clarify those hazy notions.

3.5 CLASSICAL METHODOLOGIES FOR STATISTICAL THEORIES

Statistical theories crop up frequently in science, in quantum mechanics, genetics, psychology, economics, and the rest, and they play a more humble but ubiquitous role with regard to experimental error. A statistical hypothesis attributes statistical probabilities, or chances, to events; these are not the subjective degree-of-belief probabilities already discussed, but objective properties of repeatable experiments. The fair-coin hypothesis mentioned earlier is a simple example of a statistical hypothesis; it states of a particular coin that it is "fair", i.e. that it has equal statistical probabilities of a half of landing heads and tails. And what this is standardly taken to mean is that if the coin were tossed repeatedly, the relative frequency of heads in the resulting sequence of outcomes would tend to $\frac{1}{2}$ as the number of throws increased to infinity.

Popper's approach does not even begin to deal with statistical theories, which can neither be falsified. nor corroborated (in Popper's sense of the term), since they make no categorical predictions. Thus, the fair-coin hypothesis would not be falsified, however many times the coin in question was tossed and landed heads. For the hypothesis does not say you *can't* get a million heads in a row, nor that you *must* get about 50 heads in a 100 throws; like all statistical theories, it merely assigns larger or smaller probabilities to such outcomes.

Classical statistical inference has two main branches: the testing of hypotheses using "significance tests" and the estimation of parameters; both have acquired highly technical refinements but their essential principles (and failings) can be appreciated, and are best appreciated, through the simplest examples.

The theory of significance tests purports to show how statistical hypotheses can be tested. Here is a simple example: the hypothesis to be tested (the "null" hypothesis) asserts that the coin before us is fair. An experiment is performed in which the coin is tossed a predetermined number of times, say 20, and the resulting number of heads recorded. The outcome space of the experiment comprises the 21 possibilities, ranging from no heads and 20 tails to 20 heads and no tails, and the significance test requires the statistical probability of each, relative to the null hypothesis, to be calculated. One must then choose a region of the outcome space—usually in one or both tails of the probability distribution, which is such that the probability that any actual experimental outcome would fall within in it, if the null hypothesis were true, is fairly small—0.05 has established itself as an acceptably small value. Finally, if the outcome obtained in the experiment does lie in the critical region, it is said to be "significant at the 5% significance level".

All this merely defines significance and significance level, but does not yet tell us anything about the truth, or "cognitive value" of the null hypothesis. This last, crucial step is effected, according to advocates of this approach, by saying that a result significant at the 5% level requires the null hypothesis to be "rejected at the 5% level".

What exactly does this mean? You can, of course, reject a hypothesis in the sense of denying its truth, but how can your rejection be pitched at a

THE LIBRARY
BOLTON INSTITUTE OF HIGHER
EDUCATION - DEANE CAMPUS

percentage level? Some statisticians carelessly suggest that such rejections amount to logical refutations, for they speak of statistical hypotheses being "disproved", or "contradicted" in significance tests. Fisher (1956, p. 39), on the other hand, held that the "force of a test of significance" resides in the following dichotomy: *either* an extremely improbable result has occurred, *or* the null hypothesis is false. But in fact the dichotomy has no force: all it says is that the null hypothesis is either true or false.

Another way to interpret "rejection" in the significance test sense is due to Neyman and Pearson, who invented the currently standard form of a significance test, which is slightly more complicated than that given above, in that rivals to the null hypothesis are brought into the picture. Neyman and Pearson (1933, p. 142) suggested that although we may not conclude that the null hypothesis is false when it has been "rejected at the such-and-such level", we should act in our practical life as *if* we believed just that. This oft-repeated advice is always justified by saying that if you performed significance tests repeatedly on the same or different hypotheses, and if you decided to act as if you believed the null hypothesis was false each time the result was "significant at the 5% level", only "around" 5% of your decisions would be wrong.

The argument has a specious plausibility. It is fallacious, though. It is based on the fact that a test carried out using a 5% significance level would lead to the "rejection" of a true null hypothesis with probability 0.05. But as we have stressed, from the probability of an event you cannot deduce the frequency, or even the approximate frequency, with which that event will appear in any actual run of trials, however long.

We must conclude that "significant at such-and-such level" is a phrase which says nothing about the truth or the cognitive status of any hypothesis. Like Popper's "corroboration" notion, it is precisely defined, suggestively named, yet cognitively empty.

The second great branch of classical statistical inference is known as estimation; we shall consider that part of the theory of estimation known as confidence interval estimation, where the aim is not to test hypotheses but is the purportedly different one of estimating parameter values. We again consider the simplest case: the task is to estimate the mean height, μ, of some population, whose standard deviation, σ, is known. Evidence is obtained by measuring the mean height M of a random sample drawn from the population. If the experiment was designed to select n elements, so that each possible sample is necessarily of that size, the probability distribution over the outcome space is normal with standard deviation $s = \sigma/\sqrt{n}$; it follows from this that with probability 0.95, $-1.96s \leqslant M - \mu \leqslant +1.96s$. Rearranging these inequalities gives the result that with probability 0.95: $M - 1.96s \leqslant \mu \leqslant M + 1.96s$.

Let M' be the value of m that is actually observed in the experimental sample. Then, since s may be computed, so may the terms $M' - 1.96s$ and

$M' + 1.96s$; the range covered by the resulting values is called a 95% *confidence interval for m*, 0.95 being the *confidence coefficient*.

All this merely defines confidence interval and confidence coefficient, but does not yet tell us anything about the value of the parameter. This last, crucial step is effected, according to advocates of this approach, by saying that you should be "95% confident" that the 95% confidence interval measured in any experiment includes the parameter in question.

Textbooks regularly remind readers not to interpret this degree of "confidence" as a probability, whether objective or subjective; but they never say what in fact it is, nor why they believe we are entitled to any given level of confidence on the basis of a confidence interval. The standard interpretation of confidence intervals is nevertheless not without some plausibility; but this plausibility seems to derive from a line of reasoning, implicit in many expositions, which while tempting, is, we argue, invalid.

The reasoning seems to rest on the rule of inference we referred to earlier, namely the *principle of direct probability*, often also called the *principal principle*, which is used extensively in Bayesian statistics. It states that if the objective, physical probability of a random event (in the sense of its limiting relative frequency) were known to be r, and if no other relevant information were available, then the appropriate subjective degree of belief that the event will occur on any particular trial would also be r. If, for example, the event in question is a, the principal principle says that $P^*(a_t \mid P(a) = r) = r$, where a_t describes the occurrence of the event on a particular trial; $P(a)$ is its objective probability; and P^* is a subjective probability function.

Thus since the physical probability of getting more than 5 heads in 20 throws of a fair coin is 0.86, the principal principle states that your confidence that any particular sequence of 20 throws with the coin will produce more than 5 heads is also 0.86: that is, $P^*([K > 5]_t \mid P(K > 5) = 0.86) = 0.86$.

Now suppose the coin is tossed 20 times and produces 2 heads. To apply this to the principal principle and conclude that we should now be 86 percent confident that 2 is greater than 5 would of course be absurd, and fallacious. For the principal principle does not assert a general rule for each number K from zero to 20; the K-term is not in fact a number, but a function which takes different values depending on the outcome of the underlying experiment. So it is impermissible to substitute numbers for K in the principal principle.

But this is precisely the substitution required in the standard interpretation of confidence intervals. It is true that the objective probability of m being enclosed by experimentally determined 95% confidence intervals is 0.95. By the principal principle $P^*([l < \mu < l']_t \mid P(l < \mu < l') = 0.95) = 0.95$; and this tells us to be 95% confident that any particular performance of the experiment will produce an interval that contains m. Suppose ℓ and ℓ' are the values of l and l' obtained from a particular experiment; the standard interpretation says that we should now be 95 percent confident that $\ell < \mu < \ell'$. But as the simple counterexample above shows, this is a fallacious inference.

3.6 CONCLUSION

The Bayesian theory, we have argued, has a solid logical foundation; it affords a unified approach to deterministic and statistical theories, and to testing and estimation; it is objective and rigorous where objectivity and rigour are appropriate; and where they are not, it accommodates the personal judgements of scientists in an explicit and controlled way.

Other methodologies which have been developed in conscious reaction to subjective Bayesianism have, by contrast, no proper foundation and are quite inadequate to the task of accounting for scientific reasoning. The leading examples of such methodologies are Popper's corroboration idea, and the theories of significance tests and confidence intervals; they all issue in apparently objective statements, couched in a deceptive terminology which creates the impression that some important, objective theoretical evaluation is being achieved. But these appearances are quite illusory. "Corroborating" a hypothesis does not strengthen it, a "significant" result has no significance for the truth of the hypothesis it is supposedly testing, and a "95% confidence interval" has no legitimate power to impart confidence, let alone 95%'s worth, to any estimate. The principles of significance testing and estimation are simply wrong, and clearly beyond repair. They are the phlogiston and alchemy of twentieth century statistics; and statisticians in the next century will look back at them in sheer wonderment.

REFERENCES

Bernoulli, J. (1715) *Ars Conjectandi*. Basel.

Earman, J. (1992) *Bayes or Bust? A Critical Examination of Bayesian Confirmation Theory*. MIT Press, Cambridge.

Fisher, R.A. (1956) *Statistical Methods and Statistical Inference*. Oliver and Boyd, Edinburgh.

Halmos, P. (1950) *Measure Theory*. Van Nostrand Reinhold, New York.

Howson, C. and Urbach, P.M. (1989) *Scientific Reasoning: The Bayesian Approach*. Open Court, LaSalle.

Jaynes, E.T. (1957) Information theory and statistical mechanics, I. *Phys. Rev.* **106**, 620–30; II. *Phys. Rev.* **108**, 171–90.

Jaynes, E.T. (1967) Foundations of probability theory and statistical mechanics. In M. Bunge (ed.), *Delaware Seminar in the Foundations of Physics*. Springer, New York.

Jaynes, E.T. (1973) The well-posed problem. *Foundations of Physics*, **3**, 477–92.

Jeffrey, R.C. (1965) *The Logic of Decision*. McGraw-Hill, New York.

Jeffreys, H. (1961) *Theory of Probability* (3rd edn.) Clarendon Press, Oxford.

Lakatos, I. (1978) *Philosophical Papers* (2 vols.) J. Worrall and G. Currie (eds.), Cambridge University Press, Cambridge.

Laplace, S. D. (1820) *Essai philosophique sur les probabilités*. Paris.

Lindley, D.V. (1982) Scoring rules and the inevitability of probability. *International Statistical Review*. **50**, 1–26.

Neyman, J. and Pearson. E. (1933) On the problem of the most efficient tests of statistical hypotheses. *Philosophical Transactions of the Royal Society*, **231A**, 289–337. (Page references are to the reprint in Neyman and Pearson's, *Joint Statistical Papers*, 1967, Cambridge University Press, Cambridge.)

Ramsey, F.P. (1931) *The Foundations of Mathematics and Other Logical Essays*: Routledge and Kegan Paul, London.

Savage, L.J. (1954) *The Foundations of Statistics*. Wiley, New York.

Chapter 4

The Subjective Aspect of Probability

Glenn Shafer

Is subjective probability a kind of probability, corresponding to a particular interpretation of the mathematical calculus of probability? Or is subjectivity always an integral aspect of probability, even in applications such as statistical testing, where the objective aspects of probability are usually emphasized? In this chapter, I argue that subjectivity is an aspect of all applications of probability. When we enunciate clearly the subjective aspects of supposedly objectivistic applications, the subjectivist critique of these applications loses its force. It is not necessary that these applications be rejected or be replaced with more complicated Bayesian procedures. It is only necessary that they be properly understood.

When we learn the mathematics of probability, we learn an informal story in which belief and frequency are unified. This story has many variations, but it usually involves a sequence of experiments in which known odds simultaneously define fair prices, warranted degrees of belief, and long-run frequencies. Different ways of using probability are understood most clearly when seen as different ways of using this informal story. Thus subjectivity enters into probability in two ways. First, subjectivity is part of the informal story itself. The probabilities in the story are, *inter alia*, the beliefs of some person, real or imaginary. Second, it is up to us to bring the informal story to bear on a practical problem. In doing so, we construct an argument, which must itself be criticized and subjectively evaluated.

Subjective Probability. Edited by G. Wright and P. Ayton.
© 1994 John Wiley & Sons Ltd.

In previous essays, I have described the unified informal story of probability and argued for its primacy over any particular axiomatization of probability. I have also made the general point that different applications of probability use the informal story in different ways. Here I review and refine these arguments with emphasis on a particular class of applications: statistical tests. In many cases, as we shall see, statistical tests use instances of the informal story simply as standards against which to rate the performance of a forecaster or method of prediction. This is very different from using the informal story as a representation (model or map) of a problem. By saying this clearly, we can dispel much of the confusion and controversy that now besets statistical testing.

The larger point of this chapter is that proponents of subjective probability can afford to recognize the diversity of ways in which the informal story of probability can be used. Most frequentists, deeply influenced by the empiricism of the late nineteenth and early twentieth centuries, consider an application of mathematical probability legitimate only if each probability number is mapped to an empirical frequency. Despite the anti-realism of de Finetti, subjectivists have tended to adopt an equally rigid understanding of the relation between theory and application: an application is legitimate only if each probability number is mapped to a belief or betting rate (actual or perhaps only proposed) about a practical question. This foundational rigidity may have been helpful when subjectivists had few practical Bayesian applications to their credit, but it is not necessary today. The self-confidence of today's subjectivists should allow them to lay claim to the subjective nature and legitimacy of all uses of probability.

This chapter is divided into two sections. The first section reviews the argument for the unified understanding of the informal story of probability. The second section relates this story to some simple examples of statistical testing.

4.1 THE UNIFIED INFORMAL STORY

Subjectivists and frequentists each have their own informal stories about probability, stories that they take to underly and justify the formal theory. The subjectivist story is about the betting rates of ideal rational agents, while the frequentist story is about the properties of exceptionally complex and unpredictable (i.e. random) sequences. The informal story I have in mind combines the subjectivist and frequentist stories. It involves both a sequence and a person who has a certain limited kind of knowledge about the sequence. This unified story is familiar in its basics; we learn it inadvertently when our teachers slide back and forth between subjectivist and frequentist ideas in order to persuade us to accept the various rules of probability. But it has not received much philosophical attention. Those who could give it such attention have usually chosen instead to defend one of the narrower stories.

In order to understand the unified informal story fully, we must first describe it in its own terms and then relate it to its various axiomatizations, each of which captures or emphasizes only certain of its aspects. I have made a beginning on these tasks in earlier essays.[1] There is not enough space here to discuss axiomatizations, but I will briefly recount the story and explain why I prefer it to the narrower stories.

4.1.1 A Brief Recounting of the Story

Since it must capture the frequency aspects of probability, an adequate recounting of the unified informal story must have some representation for a sequence of events. The simplest and perhaps oldest such representation is the event tree.[2] Figure 4.1 is an example. As we see in this figure, the events in an event tree result from a sequence of experiments, and the experiment performed in a given situation may depend on what has happened so far. The figure uses circles for situations in which an experiment is performed and octagons (stop signs) for situations in which experimentation has stopped.

The unified informal story also involves a spectator, who observes the outcome of each experiment as it is performed. This spectator begins with

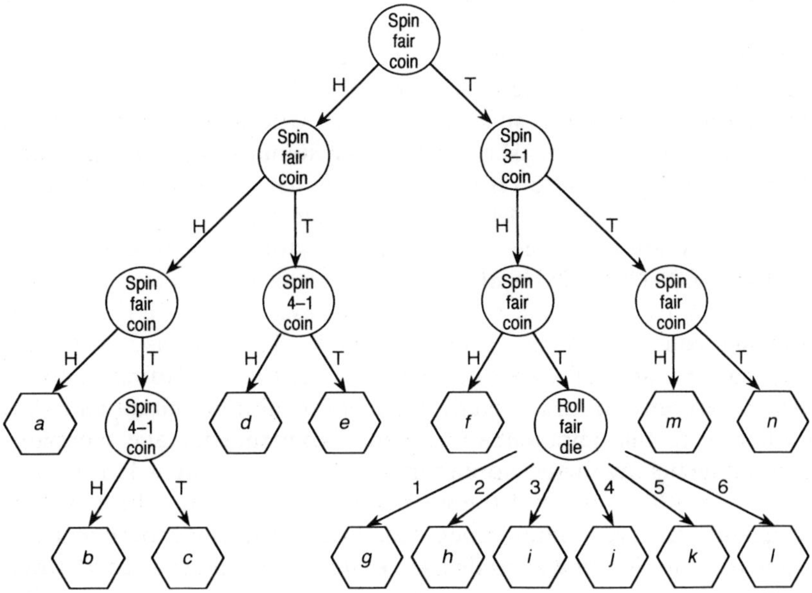

Figure 4.1 An event tree

some limited knowledge about how the experiments will turn out; she can make certain predictions about what will happen on average, but she cannot go beyond this to predict reliably the outcomes of individual experiments.

Each experiment has several possible outcomes, and a probability is specified for each outcome. These probabilities have several roles. They define fair odds, warranted degrees of belief, and long-run frequencies. The odds corresponding to the probabilities are fair because the spectator knows that if she makes many small bets at these odds—say a small bet on the outcome of each experiment as she moves down the tree—she will approximately break even. She also knows that she has no way of finding a strategy for betting at these odds that can give her any reasonable expectation of substantially multiplying her initial stake. Since she is willing to bet at these odds, the probabilities may be considered her degrees of belief, and since the odds are fair, her degrees of belief may be considered warranted. Finally, in a limited way, she interprets the probabilities as frequencies: she knows that if she bets on the outcome of each successive experiment, the frequency with which she wins will approximately equal, in the long run, the average of the probabilities for the outcomes on which she bets. (Notice that this "frequency interpretation" does not involve repeatedly going down the tree. It refers to the spectator's single trip down the tree. It is only an interpretation of certain average probabilities, however; it is not an interpretation of each and every probability in the tree.)

In order for our assertion about the spectator breaking even to be reasonably accurate, every path down the tree must go through many (a few hundred at least) situations before coming to a stop sign, and the spectator must specify a complete strategy for laying bets. For each situation, she must specify how she will, if she arrives in that situation, bet on the experiment performed there, subject to the constraint that she will have the money to pay off the bet. (How much she has in the situation is determined by her initial stake together with her strategy, for the strategy determines what she will win and lose on the way down to the situation.) When we say the spectator will approximately break even, we mean that she will approximately break even no matter what path she takes down the tree and what strategy she chooses. After she has gone down the tree, she will see ways she could have laid her bets so as to win heavily, but she has no way of choosing such a strategy in advance, and she is practically certain that any strategy she does choose will be of no avail.

In addition to the outcomes of individual experiments, the spectator can also bet on events involving more than one trial. In Figure 4.1, for example, she can bet on the event that the path down the tree will end up in the set $\{a, d, e\}$, and this event may depend on three different experiments, those performed in the situations labelled U, V and W in Figure 4.2. (If the spin[3] in U yields tails, the event fails. If it yields heads, then we move on to the spin in V. If the spin in V yields tails, the event happens; if it yields heads, we move

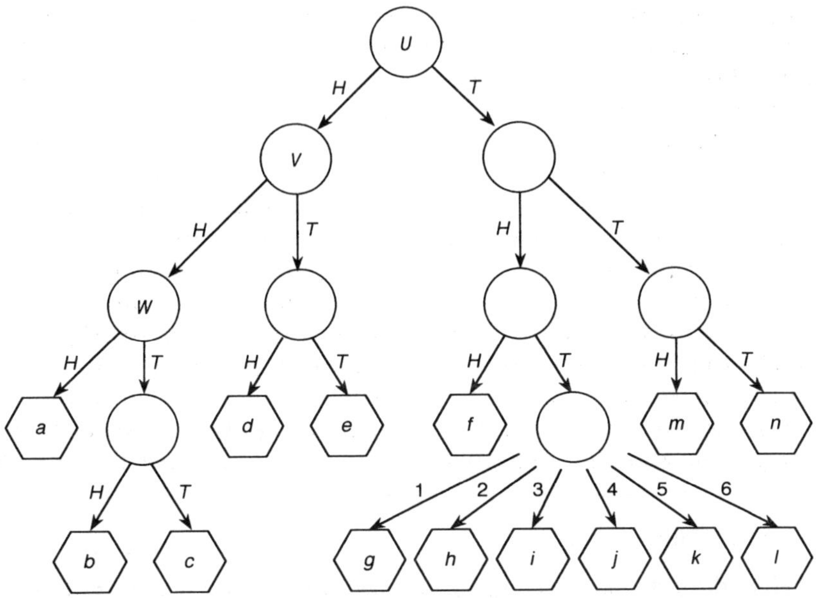

Figure 4.2 The event $\{a, d, e\}$ depends on experiments in U, V and W

on to the spin in W. If the spin in W yields heads, the event happens; if it yields tails, the event fails.) In general, any set of stop signs is an event, and a bet on any such event can be compounded from bets on individual experiments, so that its fair price is determined by the fair prices for bets on the individual experiments. In other words, the probabilities for the individual experiments determine probabilities for all events in the tree—probabilities for all sets of stop signs.

The spectator's probabilities change as events move down the tree. Her knowledge unfolds with events; she sees the outcome of each experiment as it is performed. So as she moves on to the next situation, she changes her probabilities for the experiment she just saw performed, giving probability one to the outcome she actually observed. Since more complicated events are compounded from events involving the individual experiments, she also changes her probabilities for them as she moves down the tree. So when we speak about the spectator's probabilities, we must, in general, specify the situation to which we are referring—the situation in which she has those probabilities. When we talk about the probabilities for the outcomes of an experiment performed in a given situation, we usually mean the probabilities in that situation. But in general, we can talk about the probability for any event in any situation.

Among the events in the tree are events that correspond to the assertion that a given strategy will approximately break even. Thus this assertion itself has a probability. In the initial situation, before any experiments are performed, this probability is close to one, expressing what we have described as the spectator's knowledge or practical certainty that she will approximately break even. We can similarly express her practical certainty that she cannot substantially multiply her initial stake: her probability in the initial situation that a given strategy will multiply her stake by k or more is never more than $1/k$.

It is part of the story that these practical certainties match realities in the spectator's situation. The story is about more than the spectator's inner life. According to the story, she really does move down a tree of experiments, and her ability to predict the outcomes really is limited. She really is unable to pick out a winning strategy. Any strategy that she does choose for placing small bets on successive experiments really will approximately break even. The story is a story about knowledge—a story about the relation between fact and belief.

4.1.2 Why This Story?

Why should we be interested in this unified story? Why not instead base our understanding of probability and its applications on the separate but narrower stories of the subjectivists and the objectivists?

The shortcomings of the objectivistic story have been exhaustively discussed during the past several decades. Here let me simply point out that these shortcomings lie not in the coherence of the story itself, but in the difficulty of applying it to a broad range of practical problems. Indeed, proponents of the objectivistic story are usually outspoken about the need to restrict application. Some argue that probability should only be used in cases where data is generated by random mechanisms (Freedman *et al.*, 1991). Others find objectivity in the mathematical theory of infinite sequences and leave us to puzzle over how application to finite problems can ever be justified.

Criticisms of the subjectivistic story also center on the difficulty of using it. It is argued that we often have inadequate information on which to base the betting rates that would make us like the ideal rational agents in the story. My own interest the theory of belief functions, which uses non-additive numerical degrees of belief (Shafer 1990b) has encouraged me to push the criticism one step further: it is only in the unified story that we have grounds for calling our betting rates fair and hence using them both for buying and selling.

The standard expositions of the subjectivistic story do not place event trees in the foundation of the theory. Sequences of events are seen merely as one thing about which we can have beliefs. But it turns out that sequences of events are needed in order to justify the idea of belief change by conditional probability; without the "protocol" for new information represented by an event tree, we are led into paradox (Shafer, 1985). Thus even the internal logic of the subjectivistic story pushes it in the direction of the unified story for which I am arguing (Dawid, 1982).

My purpose in this chapter is to show that there are practical reasons for favouring the unified story that go beyond these general arguments. There are some applications of probability that can be understood in terms of the unified story but not in terms of the narrower stories.

4.2 EVALUATING CATEGORICAL PREDICTIONS

Statistical testing, presented with little theory, is often very persuasive. There is no reason to hire a forecaster who does no better than chance. When one treatment does better than another no more often than might be expected by chance, its performance provides no evidence that it is better. When an additional variable improves the performance of a prediction equation no more than might be expected by chance, the improvement is a poor argument for adding it to the equation.

When we turn to theoretical accounts of testing, on the other hand, we find confusion and controversy. Every teacher of elementary statistics knows how confused students are by the objectivistic accounts we teach, and every theoretical statistician is familiar with the mockery these accounts evoke from subjectivists and other skeptics.[4] Bayesian elaborations of the objectivistic accounts are also controversial; they add to the complexity of the objectivistic accounts and correct only some of their shortcomings.

Why is it so difficult to make theoretical sense of statistical testing? The difficulty, I believe, lies in an unspoken but powerful assumption about how probability theory should be related to practice. We assume, without reflection, that any probability model we formulate to study a phenomenon must be a model for—a representation of—that phenomenon. So when we undertake to explain a statistical test (or rather, to improve the apparently shallow explanation we first found persuasive), we begin by trying to construe the probability model involved in the test as a representation of the phenomenon being tested. We try to make the informal story corresponding to the model a story about that phenomenon—a story about the behavior of the forecaster or what is forecasted, a story about the effect of the treatment, or a story about the effect of the additional variable. We forget that the model and the story originally stood apart from the forecaster, treatment, or variable, as an independent standard to which to compare their performance.

The unified informal story of probability can help us keep our hands on the knowledge that testing involves comparison rather than representation. This unified story can serve as a clear standard for comparison in a way that its objectivistic and subjectivistic cousins cannot, for within the unified story there is a spectator with clearly delimited powers of prediction, and it is to this spectator that we compare our forecaster or our prediction equation.

Any particular statistical test involves, of course, a particular story; we compare our forecaster not with the unified informal story in general but with a particular instance of it, an instance with a particular event tree and particular probabilities. For brevity, let us call an instance of the unified informal story a "stochastic story". For clarity, let us reserve the name "forecaster" for the real forecaster we wish to evaluate (as opposed to the "spectator" in the stochastic story), whether it be a person, a prediction equation, or an expert system. (We may speak of forecasting or prediction even when we are dealing with assertions about the past or present.[5] We require only that after the forecaster makes a prediction we are able to classify it as right or wrong.)

We deliberately construct the stochastic story that serves as our standard for comparison. We often construct several. We may begin by comparing the forecaster's performance to what can be achieved by a nearly clueless spectator in a very austere stochastic story. If the forecaster can do better than this spectator, then we may move on to a stochastic story whose spectator is more (or perhaps merely differently) advantaged. Continuing in this way if necessary, we may (or may not) find a stochastic story in which the performance of the spectator roughly matches the performance of our forecaster. But none of this requires us to go beyond the idea of rating the forecaster's performance. At no point are we required to think of the forecaster herself or of the phenomenon being forecasted as part of a stochastic story.

There are some general principles that can guide our search for an appropriate stochastic story. We must make the story and the situation of the forecaster comparable without contriving to force any particular conclusion. No general principle can guarantee, however, that the comparison with the stochastic story will be persuasive. In the end, this comparison is only an argument, and like any other nondemonstrative argument, it is open to criticism and counterargument. A particular stochastic story will not be persuasive unless equally natural stochastic stories give similar or consistent results.

I will discuss two simple examples of categorical prediction. In both examples, as we will see, the success of the prediction can be evaluated by comparison with a stochastic story. The two evaluations can be extended to deal with problems that are usually treated as statistical testing problems. The first corresponds to testing whether a binomial parameter is equal to $\frac{1}{2}$, and the second corresponds to testing independence in a 2×2 table.

The analysis of these simple examples falls short of a general theory of statistical testing. But there are some obvious ways to extend the analysis. In order to extend it to the kinds of problems that are usually treated by goodness-of-fit tests and and tests of independence in larger tables, we will need to use the ideas on the evaluation of probability forecasting developed by Dawid (1984, 1985, 1986, 1990) and Vovk (1993). In order to deal with the conventional normal-theory tests, will need to adapt the ideas of Freedman & Lane (1983a,b) and Beaton (1981).

4.2.1 Evaluating Melinda's Performance

A crude way of scoring the performance of a person who makes categorical predictions is simply to count how often she is right. But even this crude score will be meaningful only in relation to some baseline. The following example shows how a stochastic story can provide that baseline.

> Melinda claims some insight into the behavior of the local train. She claims that at 7:30 she can predict whether the 8:05 train will be on time. As a demonstration, she makes predictions on 100 successive days, and we find that 55 of her predictions are correct. What does this meager success tell us? Does it provide any evidence that Melinda knows what she is talking about?

It appears that Melinda does not know what she is talking about, because she is right barely half the time. Why is being right only half the time so unimpressive? Because we could do as well spinning a coin. Suppose Mary, who knows nothing about the train's behavior, predicts whether it will be on time by spinning a fair coin. In spite of her ignorance, Mary can expect to be right about half the time, too. In fact, Mary has a probability of about one-sixth of being right 55 or more times out of 100.

The simplicity of this example allows us to see clearly that the stochastic story is serving only as a standard for comparison. We compare Melinda's performance to the story, but the story is not about Melinda. I could tell a stochastic story about Melinda if I wanted. I might tell one of these stories:

(1) I might claim that Melinda's own knowledge about the train is such that she can be described as a spectator in a stochastic story. For example, she might know that the train is on time about half the time, without being able at all to predict which half.
(2) I might claim that my knowledge about Melinda's behavior is such that I can be described as a spectator in a stochastic story. For example, I might know that Melinda will predict correctly about half the time, without being able at all to predict which half.

But there is no basis for these stories in what I have told you about Melinda and the train. There is no basis for the first story, because I told you nothing about how often the train is on time. (I only said that Melinda predicted correctly 55 times out of 100. This is consistent with the train always, never, or sometimes being on time.) There is no basis for the second story, because I told you nothing about my knowledge. (Perhaps by 7:29 I always know what Melinda is going to predict and whether she is going to be right.)

The stochastic story is only a thought experiment. The force of the comparison depends, however, on the fact that we could implement the thought experiment if we wished. We are unimpressed by Melinda because we really could predict equally well by spinning a coin (or by using computer-generated random numbers).

A statistical test always involves some score or "test statistic". Melinda's score is the number of times she predicts correctly. This is an obvious way to score her performance, quite independently of our invention of a stochastic story. The purpose of the stochastic story is to calibrate the score. How large does Melinda's score, say t, have to be in order to provide evidence that Melinda has some insight? We answer this question by calculating, for various values of t, the probability that Mary's score, say T, will be at least as large as t. Table 4.1 gives $P(T \geqslant t)$ for a few values of t. As this table indicates, Mary has a reasonable chance of scoring as well as 55, but she is quite unlikely to score as well as 65. Had Melinda predicted correctly 65 times or more, we would have said that she did better than Mary could reasonably expect to do, and that her performance therefore provides some evidence that she knows more than Mary. (This would say nothing, of course, about the nature of Melinda's knowledge. She may have a way of identifying days on which the train will have difficulties, or she may know that the train is late about 65% of the time and take advantage of this knowledge by always predicting that it will be late.)

Of course, our imaginary Mary is only one example of a person who knows nothing about the train. Perhaps someone else who knows nothing about the train could find a more effective way of predicting its performance than flipping a coin. So even if Melinda does better than we could hope for Mary to do, the comparison with Mary is only an argument for Melinda having some insight or knowledge. The argument is a strong one, however. We have had much experience with stochastic stories as standards for comparison, and we do not expect to find a person who is totally ignorant about the train and yet knows how to predict better than Mary.

The probability $P(T \geqslant t)$, where t is the value of the score actually recorded, is called the "P-value" in the usual accounts of statistical testing. When the P-value is small (less than the conventional values 5% or 1%, say),

Table 4.1 P-values from spinning a fair coin 100 times[6]

t	55	60	65	70
$P(T \geqslant t)$	0.16	0.03	0.002	0.00005

the observed score t is called significant—i.e. significantly better than could be expected by chance. I have used P-values in this example in a familiar-looking way, but I have not explained them in the usual way. The usual explanation, which is due to R.A. Fisher, talks about "rejecting a null hypothesis". The null hypothesis asserts that the data (and hence the score t) was produced by chance, in accordance with a particular objectivistic probability model. We are supposed to reject the null hypothesis when t is large and $P(T \geqslant t)$ is therefore small, on the grounds that it is easier to disbelieve the hypothesis than to believe that the event $T \geqslant t$, which actually happened, is so unlikely.

Subjectivists often criticize Fisher's logic on the grounds that it does not justify attention to the event $T \geqslant t$.[7] What actually happened in Fisher's story was $T = t$. If we want to claim that the null hypothesis makes what actually happened too surprising, the critics say, we should look at the probability of $T = t$, without amalgamating it with $T > t$, which did not happen. This is not a criticism of my logic. In my story, $T = 55$ does not happen (T is Mary's score; 55 is Melinda's score), and attention to the event $T \geqslant 55$ is justified even before the stochastic story is invented. I observe Melinda's score of 55. I ask myself whether someone who knows nothing about the train can hope to do as well—i.e., can hope for a score T such that $T \geqslant 55$. I invent the stochastic story precisely in order to study the chances of $T \geqslant 55$ for one person (Mary) who knows nothing about the train.

In order to put Melinda into Fisher's objectivistic framework, we would have to tell an objectivistic stochastic story about her predictions: they are independent and each is correct with constant probability p. We then test the null hypothesis $p = \frac{1}{2}$, which seems to correspond to Melinda having no real ability to predict. (This null hypothesis is an objectivistic version of the second of the two stochastic stories about Melinda that I listed earlier.) If Melinda gets 65 predictions out of 100 right, we can reject this story; if she gets only 55, we cannot. The difficulty with this talk, of course, is that the objectivistic story is so ungrounded. Who told us that Melinda's ability to predict is constant from day to day? Why should we accept inferences that seem to depend on such an assumption?

Though the simplicity of this example is atypical of the practice of statistical testing, the lack of grounding for the objectivistic story is quite typical. Statisticians often excuse this lack of grounding by drawing an analogy to the shortcomings of scientific theories, which can be useful even if they simplify reality and remain unconfirmed in many respects. Perhaps our stochastic story about Melinda is a simplification of a more adequate stochastic story, and perhaps analysis of this more adequate story would give the same results. But subjectivists, who tend to doubt the meaningfulness of even the simplest of these objectivistic stories, are not comforted by the thought of making them more complex.

I believe that statisticians take simple statistical tests seriously not because they take the corresponding objectivistic stories seriously as representations of reality, but because they see these stories as standards for comparison. The account of testing I am giving here makes this explicit. This account has not been articulated clearly in the past primarily because the unified informal story of probability, which it uses in an essential way (the spectator must be in the story so that we can compare Melinda with her), has lacked respectability.

Notice that I am arguing for a subjectivist interpretation of the standard test, not for a Bayesian replacement. Bayesian analyses of testing, while vaunting their emphasis on subjectivity, usually take the Fisherian objectivistic story as their starting point. Like the Fisherian analysis, they assume that an objectivistic model generates the data by chance, without reference to any observer. We enter as observers only after this objectivistic model has done its job, and we remain outside the model; our job is to decide whether to believe it.

Before leaving the example, we should note the comparison of Melinda with Mary does not touch on the question of whether the future will be like the past. If Melinda's performance gives evidence that she knew something that helped her predict during the past 100 days, then we may wish to infer that she will continue to know something and continue to make effective predictions during the next 100 days. But this inference goes beyond what we have learned by comparing Melinda with a stochastic story. Neither the story about Melinda nor the stochastic story made any assumption about the 100 days we observed being like other days in the future or the past. In particular, we did not assume that these 100 days were drawn at random from a larger population of days.

4.2.2 Evaluating a Treatment

It is a short step from Melinda to examples that appear in statistics textbooks.

> Amanda, who wants to add a new razor blade to her line of toiletries, is trying to decide which of two types of razor blade will be most popular among users, type A or type B. She gives 100 users one blade of each type, and asks them to report back which they prefer. When they do so, 65 report that they prefer type A. Is this strong evidence in favour of type A?

We can deal with this example just as we dealt with Melinda. Melinda made 100 binary predictions. Here, too, we have 100 binary predictions. We can think of the labels on each pair of blades as Amanda's prediction that the blade labelled "type A" will be preferred. Then we can ask whether the success (albeit limited) of these predictions indicates some genuine insight about the superiority of A. In order to rate Amanda's performance, we compare her with Anna, who cannot tell the two types of blades apart and predicts which blade in each pair will be preferred by spinning a fair coin. Anna, we know, can scarcely hope to do as well as Amanda has done. According to Table 4.1, the probability she will do as well is 0.002. So Amanda's knowledge must be helping her predict. In other words, something about type *A* blades makes them more widely preferred.

We may be giving Amanda an unfair advantage in this comparison. We are talking as if Amanda thought type *A* was better and organized the study to prove the point. If this is so, then the comparison with Anna is fair. But another possibility is that Amanda was uncertain which, if either, of the blades was better, and that she was simply trying to find out. In this case, Amanda has an unfair advantage over Anna. For a fair comparison, we should compare Amanda with Amy, who spins a fair coin in order to label the blades in each pair "*A*" and "*B*" and then waits to see how the 100 people's preferences turn out before deciding whether her prediction was that *A*s would be preferred or that *B*s would be preferred. Amy's chance of doing as well as Amanda is twice Anna's, or 0.004. (The comparison with Anna is a "one-sided test", while the comparison with Amy is a "two-sided test".)

The objectivistic treatment of this example follows the same path as the objectivistic treatment of Melinda. We posit that each of the 100 people has the same probability p of preferring *A* over *B*, and that the preference of each person is independent of the preference of the others. Then we test the null hypothesis that $p = \frac{1}{2}$. Is this probability model any better grounded, any more plausible, or any more meaningful here than in the story about Melinda? I think not.

The comparison of the razor blades with Anna is more complicated than the comparison of Melinda with Mary, because it involves an additional step. First we relate the merit of the razor blades to Amanda's ability to predict, and then we compare Amanda's ability with Anna's or Amy's. But otherwise the issues are the same. The comparison of Amanda with Anna again makes explicit the real role of the stochastic story; it is really only serving as a standard for comparison.

Here, as in the case of Melinda, we have not touched on whether the future will be like the past. We want, of course, to take the next step and conclude that the majority of future customers will prefer blade *A*. But our argument based on the comparison with Anna or Amy has no bearing on this next step. Had the 100 people testing the blades been chosen at random from the population of potential future customers, probability arguments might help us make the step into the future, but that is another story.

4.2.3 Evaluating Lucinda's Ability to Discriminate

Our rating of Melinda, though instructive, was rather crude. We compared Melinda to Mary, who knew absolutely nothing about the train. Mary is easy to beat; if Melinda knows the train is usually late, then she can beat Mary simply by always predicting it will be late. Let us turn, therefore, to a more subtle question about Melinda's performance. For clarity, we will discuss this question for a different forecaster, named Lucinda.

> Lucinda claims that by 7:30 she can tell (though she sometimes makes mistakes) whether the 8:05 will be late or not. As a demonstration, she makes predictions on 100 successive days. As it turns out, she predicts 60 times that the train will be late, and she predicts 40 times that it will be on time. We find that 70 of her 100 predictions are right. She was right 55 of the 60 times she said the train would be late, and she was right 15 of the 40 times she said it would be on time. Does this performance provide evidence that Lucinda can tell days the train will be late from days it will be on time?

Table 4.2 displays the joint performance of Lucinda and the train. The train was late 80 times. Lucinda was right only 70 times, so she could have scored better overall by always predicting the train would be late. But her performance does seem to provide evidence that she can tell a difference between days. The train was late 91.7% of the times she said it would be late (55 out of 60) and only 62.5% of the times (25 out of 40) she said it would be on time.

How might we score Lucinda's performance in distinguishing between days? I just suggested one reasonable score: how much more often the train is late when Lucinda says it will be. This is

$$\frac{55}{60} - \frac{25}{40} \simeq 0.29 \tag{4.1}$$

Alternatively, we might measure how much more often Lucinda says the train will be late when it is; this is

$$\frac{55}{80} - \frac{5}{20} \simeq 0.44 \tag{4.2}$$

There are many other possibilities as well; any "measure of association" for the 2×2 table would do. But to interpret any such score we need some kind of baseline or calibration. We need a stochastic story.

Table 4.2 Lucinda and the train

	Lucinda says train will be late	Lucinda says train will be on time	Total
Train is late	55	25	80
Train is on time	5	15	20
Total	60	40	100

Here is a stochastic story that will do. Suppose Lois, who knows nothing about the train, is told that it was late 80 of the last 100 days (this information is not going to help her, but it helps set the stage). And suppose she is asked to guess 60 of these days. Lacking any other information, Lois uses random numbers produced by her personal computer to choose 60 out of the 100 days, with all the possible choices being equally likely. What are the chances, under these circumstances, for Lois to do as well or better than Lucinda did? In other words, what are the chances that the 60 days she chooses will include 55 or more days on which the train is late? The answer, as it turns out, is about 0.0002. Lucinda has done much better in identifying days on which the train will be late than we could expect from someone who has no knowledge that would help her discriminate.

Notice that the stochastic story has simplified the scoring. Instead of using (4.1) or (4.2), we score Lois simply by the number, say T, of her 60 guesses that turn out right. Thus our P-value is $P(T \geqslant 55)$. We would get the same P-value using (4.1) or (4.2), since $T \geqslant 55$ is equivalent to

$$\frac{T}{60} - \frac{80-T}{40} \geqslant \frac{55}{60} - \frac{25}{40} \quad \text{or} \quad \frac{T}{80} - \frac{60-T}{20} \geqslant \frac{55}{80} - \frac{5}{20}$$

Almost any other measure of association in the 2×2 table will also give the same P-value; since the row and column totals of Lois's table are the same as Lucinda's, Lois can do better only by making T greater than 55.

This example illustrates how comparison with a stochastic story can be effective even though we make arbitrary choices in setting the story up. In order to make Lois's performance comparable to Lucinda's, we asked Lois to guess exactly 60 days. This did not weaken the force of the comparison, because it did nothing to put Lois at a disadvantage relative to Lucinda.

How is the P-value of 0.0002 computed? Readers familiar with combinatorial probability will see that the probabilities for T are hypergeometric:

$$P(T=x) = \frac{\binom{80}{x}\binom{20}{60-x}}{\binom{100}{60}} \tag{4.3}$$

We can find $P(T \geqslant 55)$ by adding these probabilities as x goes from to 55 to 60. An approximation using the chi-squared distribution is also available.[8]

Let us now consider the textbook approach to testing Lucinda's performance. There are a number of ways we might proceed, all involving different objectivistic stochastic stories. We might model the behavior of the train, so that we can test whether it behaves differently on days Lucinda thinks are different. We might model the behavior of Lucinda, so that we can test whether she predicts differently on days that are different for the train. Or we might model both together.

Here is a way to model the train. Let X_1 be the number of times the train is late out of the 60 times Lucinda says it will be late, and let X_2 be the number of times it is late out of the 40 times she says it will be on time. Assume that the train is late with probability p_1 on the days Lucinda says it will be late, that it is late with probability p_2 on the days she says it will not be, and that whether it is late on a given day is independent of its performance on preceding days. Under these assumptions, X_1 and X_2 are independent binomial random variables; X_1 has parameters 60 and p_1, and X_2 has parameters 40 and p_2. The question whether the days are different has become the question whether $p_1 \neq p_2$. We test the null hypothesis of no difference: $p_1 = p_2$. Under this hypothesis, X_1 and X_2 are independent binomials with a common parameter $p = p_1 = p_2$. As our test statistic, we take the difference

$$\frac{60}{X_1} - \frac{X_2}{40} \tag{4.4}$$

This is the score (4.1) we considered earlier. It turns out that the probability that it will equal or exceed the value we observed for Lucinda, 0.29, is approximately 0.0002, the same as the *P*-value we obtained by comparing Lucinda with Lois.[9]

Instead of computing the probability of (4.4) exceeding its observed value unconditionally, it may be better, according to Fisher,[10] to compute its probability of doing so conditionally, given the observed marginal totals in Table 4.2. The resulting test is called Fisher's exact test. It is a better test, according to Fisher, because it brings the population of potential repetitions with which we are comparing the actual result closer to that result, and also because it simplifies the analysis. In the unconditional model, the choice of the statistic (4.4) is somewhat arbitrary, but, as we noted earlier, once the margins of the table are fixed, all measures of association are essentially equivalent. Moreover, the computation of the *P*-value is simplified. In fact, the conditional probabilities are precisely the hypergeometric probabilities given in (4.3); Fisher's exact test comes out exactly the same as our comparison of Lucinda with Lois.

It would delay us too long to explore here the other objectivistic models that I have mentioned; suffice it to say that they give similar results and also reduce to Fisher's exact test conditionally. We should also note that there is yet another justification for Fisher's exact test for the 2×2 table in situations where an experimenter is able distribute units over one of the classifications (over the rows or over the columns) of the table randomly. Fisher preferred this justification, but it is obviously inapplicable to Lucinda.

What should we say about the objectivistic approach? Does it make sense? Here, as in the case of Melinda, the objections are obvious. Who told us that the behavior of the train is stochastic? That the probability is the same on every day that Lucinda says the train will be on time? That its behavior on one day is independent of its behavior on another? There are no grounds for these assumptions. Surely a stochastic story can only be justified here as a standard for comparison.

Taking the stochastic story as a standard for comparison allows us make sense of Fisher's intuitions about conditionality, intuitions that many of his objectivistic successors have found puzzling. Once we acknowledge that Lois is only a standard for comparison, it becomes entirely reasonable that we should design Lois's task so as to maximize its comparability with Lucinda's accomplishment. The vagueness of this desideratum is not a problem, for the desideratum merely serves to help us construct an argument. It does not pretend to exclude any other argument or counterargument.

4.2.4 The Berkeley Graduate Admissions Data

Table 4.3 shows the number of men and women who applied for admission to graduate study at the University of California at Berkeley for the fall of 1973, together with the number of each sex who were and were not admitted. These data were first published by Peter J. Bickel and colleagues in 1975.[11] These authors were concerned not only with discrimination against women but also with the shortcomings of the objectivistic models used to analyze such questions.

As Table 4.3 indicates, the rate of admission was substantially lower—almost 10 percentage points—for women than for men. Fisher's exact test produces a vanishingly small *P*-value.[12] The lower rate of admission for women is significant both substantively (10 percentage points is a lot) and statistically (the *P*-value is practically zero).

Here, as in the case of Lucinda, we can explain the statistical significance in terms of a comparison with Lois. Suppose we tell Lois that 8442 of the 12 763 applicants are men. We then give her ID numbers for the 12 763 applicants, and we ask her to try to pick out from them 5232 numbers that identify men. Since she has no way of knowing which of the numbers identify men, she uses her personal computer to choose 5232 of the 12 763 numbers at random. What is the chance that she will choose as many men as the admissions committees did? This question is answered by Fisher's exact test:

Table 4.3 Graduate admissions at Berkeley in 1973 (reproduced from Bickel *et al.* (1975) by permission of *Science*. © 1975 American Association for the Advancement of Science)

	Admitted	Not admitted	Total	% admitted
Men	3 738	4 704	8 442	44.3%
Women	1 494	2 827	4 321	34.6%
Total	5 232	7 531	12 763	41.0%

the chance is vanishingly small. So the Berkeley admissions process did much better at picking out men than we could possibly expect Lois to do. It picked out more men than could possibly happen by chance.

Though having Lois try to pick out men makes the comparison between Lois and the admissions process simple and rhetorically effective, other ways of setting up the comparison are equally valid and lead to the same conclusion. Suppose, for example, that we ask Lois to pick out 8442 numbers, trying to include as many admittees as possible. Since she knows nothing about which of the 12 763 numbers represent admittees, she will again choose the 8442 numbers at random. Suppose Amanda knows which numbers identify men and chooses them. Amanda will have 3738 admittees among her 8442 choices, and Lois has practically no chance of doing as well. In fact, her chance of doing as well is given once again by the *P*-value from Fisher's exact test. So we can conclude that being male predicts admission better than could possibly happen by chance.

An objectivistic treatment of Table 4.3 would follow the same lines as the objectivistic treatment I sketched for Lucinda and the train. We assume that there is a constant probability of admission for men and a constant probability of admission for women, and we test for equality of the two probabilities. Alternatively, we assume that there is a constant probability of an admittee being a woman and a constant probability of a non-admittee being a woman, and we test for the equality of these two probabilities. As Bickel and his co-authors and many other commentators have pointed out, none these objectivistic assumptions are plausible. As Freedman & Lane (1983b, p. 192) put it, they are known from the first "to be inadequate to describe any aspect of the physical process that generated the data."

Though the objectivistic models are useless for this example, the comparison with Lois is meaningful. It tells us that something is going on that favors men. This something may or may not be stochastic.[13] But since it has a stronger effect than could happen by chance, we can reasonably hope that further investigation will yield some insights. Bicker and his colleagues, upon undertaking such an investigation, found that the bias in favor of men was related to how the numbers of places and numbers of men and women applicants were distributed over departments. The rate of admission (number of places available per applicant) was smaller in departments where the proportion of women among applicants was higher. So we can ask why proportionally fewer places were provided in departments to which women more often applied. As it turns out, departments where proportionally more places were provided required, on the average, more mathematical preparation of their applicants. Perhaps society needed a greater fraction of those who were prepared, willing, and asking to study in these demanding fields. This can be contested, but if it is accepted, then the question of why women were being discriminated against in graduate admissions comes down to the question of why they were underrepresented among those prepared, willing, and asking to study in departments requiring more mathematical preparation.

NOTES

(1) See especially Shafer (1990a), which describes the informal picture, and Shafer (1992), which sketches one axiomatization. In both these essays, I used the phrase "ideal picture of probability" for what I am here calling the "informal story of probability". Unfortunately, the adjective "ideal" seems to have been a source of misunderstanding. One misunderstanding is that the informal story is a representation, less the rough edges, of some reality. This is not my meaning; my theme is that the informal story has many uses; its use to represent a reality we want to understand is only one of these uses. A related misunderstanding is that the merit of the informal story lies entirely in its lack of rough edges—the more ideal the better. This provides an excuse for pushing on to one of the narrower stories, where either the subjective or the objective aspects of probability are idealized away.

(2) Huygens drew an event tree in a manuscript dated 1676 (Edwards, 1987, page 146).

(3) Figures 4.1 calls for the coins to be spun rather than flipped, so that a biased coin—one that is heavier on one side than the other—can exhibit its bias by falling more often on its heavier side. Such a coin is equally likely to fall on either side when it is fairly flipped (Engel, 1992).

(4) For the subjectivist critique, see Berger & Delampady (1987) and the references therein. For a survey of other critiques, see Morrison & Henkel (1970).

(5) As Stephen Brush (1988) has noted, scientists often use the word "prediction" without regard to whether what is being predicted is already known. In many cases, at least, the credit that a scientific theory earns by predicting an effect does not seem to depend on whether the effect was known before the prediction was made.

(6) These numbers can be obtained from the normal approximation to the binomial in the usual way: $P(T \geqslant t)$ is the probability that a normal deviate with mean 50 and standard deviation 5 exceeds $t - \frac{1}{2}$.

(7) This criticism seems to go back to Harold Jeffreys. See Berger & Delampady (1987), pages 329 and 348.

(8) See Miller 1986, pages 47–48.

(9) The P-value for (4.3) is usually computed using a normal approximation. Under the null hypothesis, (4.3) is approximately normally distributed with mean zero and variance $p(1-p)\left(\frac{1}{60}+\frac{1}{40}\right)$. Since we can estimate p by $(X_1 + X_2)/100$, this implies that

$$\frac{X_1/60 - X_2/40}{\sqrt{\dfrac{X_1+X_2}{100}\left(1-\dfrac{X_1+X_2}{100}\right)\left(\dfrac{1}{60}+\dfrac{1}{40}\right)}} \tag{4.5}$$

should be approximately standard normal. Substituting 55 for X_1 and 25 for X_2, we find that (4.5) is approximately equal to 3.6. The probability of a standard normal deviate exceeding 3.6 is approximately 0.0002. As it turns out the square of (4.5) is equal to the chi-squared statistic used to approximate the sum of hypergeometric probabilities in our comparison of Lucinda with Lois. So the agreement between the two P-values does not depend on the particular numbers we have used in the example.

(10) See Fisher (1973), pages 89–92.

(11) Their article was originally published in *Science* (Bickel, Hammel & O'Connell 1975). It was reprinted, together with comments by William H. Kruskal and Petter J. Bickel, in Fairley & Mosteller (1977). The issues raised by the data were also discussed by Freedman & Lane (1983b) and Freedman, Pisani, & Purves (1978, pp. 12–15). Inspired by this example, Freedman and Lane (1983b) propose a general way of understanding tests of independence in two-way contingency tables. My discussion here is influenced by their proposal but does not follow it. The comparison I suggest with a unified stochastic story is, I think, better motivated and more persuasive than Freedman and Lane's purely "descriptive" and "nonstochastic" treatment, and it applies only to 2×2 tables.

(13) The chi-squared statistic, which has one degree of freedom, is 110.8.

(14) On our unified understanding of stochasticity, it surely was not stochastic at the beginning of the investigation by Bickel and his colleagues, for stochasticity requires an observer, and no one had been closely observing what was going on.

REFERENCES

Beaton, A.E. (1981) Interpreting least squares without sampling assumptions. Research Report RR-81-38. Educational Testing Service, Princeton, New Jersey.

Berger, J.O. & Delampady, M. (1987) Testing precise hypotheses (with discussion). *Statistical Science*. **2**, 317–52.

Bickel, P.J., Hammel, E.A. & O'Connell, J.W. (1975) Sex bias in graduate admissions: data from Berkeley. *Science*. **187**, 398–404.

Brush, S.G. (1988) Prediction and theory evaluation. *Science*, **246**, 1124–9.

Dawid, A.P. (1982) The well-calibrated Bayesian. *Journal of the American Statistical Association*. **77**, 605–13.

Dawid, A.P. (1984) Statistical theory: the prequential approach. *Journal of the Royal Statistical Society, Series A*. **147**, 278–92.

David, A.P. (1985) Calibration-based empirical probability. *The Annals of Statistics*, **4**, 1251–73.

Dawid, A.P. (1986) Probability forecasting. In S. Kotz, N.L. Johnson, and C.B. Read (eds.) *Encyclopedia of Statistical Sciences* Vol. 7. Wiley-Interscience, New York, pages 210–18.

Dawid, A.P. (1990) Fisherian inference in likelihood and prequential frames of reference. *Journal of the Royal Statistical Society, Series B*. **53**, 79–109.

Edwards, A.W.F. (1987) *Pascal's Arithmetic Triangle*. Oxford University Press, New York.

Engel, E. M.R.A. (1992) *A Road to Randomness in Physical Systems*. Lecture Notes in Statistics #71. Springer, New York.

Fairley, W.B. & Mosteller, F. (1977) *Statistics and Public Policy*. Addison-Wesley, Reading, Massachusetts.

Fisher, R.A. (1973) *Statistical Methods and Scientific Inference* (3rd edn). Macmillan, New York.

Freedman, D. & Lane, D. (1983a) A nonstochastic interpretation of reported significance levels. *Journal of Business and Economic Statistics*. **1**, 292–8.

Freedman, D. & Lane, D. (1983b) Significance testing in a nonstochastic setting. In P. Bickel *et al.* (eds), *Lehmann Festchrift*. Wadsworth, London.

Freedman, D., Pisani, R. & Purves, R. (1978) *Statistics*. Norton, New York.

Miller, R.G., Jr. (1986) *Beyond ANOVA, Basics of Applied Statistics*. Wiley, New York.

Morrison, D.E., & Henkel, R.E. (eds.) (1970). *The Significance Test Controversy.* Aldine, Chicago.
Shafer, G. (1985) Conditional probability (with discussion). *International Statistical Review.* **53**, 261–77.
Shafer, G. (1990a) The unity and diversity of probability (with discussion). *Statistical Science.* **5**, 435–62.
Shafer, G. (1990b) Perspectives on the theory and practice of belief functions. *International Journal of Approximate Reasoning.* **4**, 323–62.
Shafer, G. (1992) Can the various meanings of probability be reconciled? In G. Keren and C. Lewis (eds.), *A Handbook for Data Analysis in the Behavioral Sciences: Methodological Issues.* Lawrence Erlbaum, Hillsdale, N.J.
Vovk, V.G. (1993) The logic of probability. *Journal of the Royal Statistical Society, Series B,* **55** (2), 317–351.

Chapter 5

On the Necessity of Probability: Reasons to Believe and Grounds for Doubt

John Fox
Imperial Cancer Research Fund Laboratories

5.1 INTRODUCTION

The objective of this volume is to study "subjective probability", a concept which has been developed for use in the human decision sciences but which inherits its conceptual framework and most of its technicalities from mathematical probability theory. ("Decision scientists" are taken to include psychologists, statisticians, economists, management scientists and others.) The theory is commonly held to provide the normative standard against which the "rationality" of any judgement under uncertainty must be assessed. In this context human judgement under uncertainty is thought to be, at best, a degenerate form of that prescribed by the theory.

It is an irony that the history of probability theory, a subject whose heart is the study of uncertainty, has been surrounded by a great deal of dogma. When the modern idea of probability appeared (generally reckoned to be about 1660) Europe was torn by sectarianism and political and religious rivalry. Indeed its emergence has been linked to the rejection of the detested doctrine that uncertainty and dispute must be resolved by approved opinion (notably that of the church or some other proper authority) rather than by

Subjective Probability. Edited by G. Wright and P. Ayton.
© 1994 John Wiley & Sons Ltd.

rational debate and the marshalling of evidence. Nowadays systematic concepts of evidence and its quantitative assessment have overthrown mere opinion in many fields and mathematical probability is a well-developed and deeply understood subject. One might therefore have expected the major philosophical questions to have been settled. However, notwithstanding its many practical benefits, the use of probability mathematics continues to produce controversy about its true meaning and proper application (e.g. Cheeseman, 1987; Saffiotti, 1987). External signs of this controversy are not merely frequent (and sometimes noisy) disputes, but also, and much more interestingly, the continuing appearance of new mathematical systems.

For many, such electicism is neither desirable nor necessary, for the reason that all rival systems to probability are demonstrably mistaken. This claim is justified by the argument that if a system is not based on certain fundamental assumptions, the probability axioms, then it will be demonstrably irrational or, more technically, "incoherent" (Lindley, 1985).

The usual approach to adjudicating on competing mathematical theories is to review the different systems, examine their axioms and theorems, and rule on their correctness, completeness, universality or whatever. However, too many talented mathematicians have followed this path and ended up with firmly held but opposing positions. After deep study some conclude that probability is both necessary and sufficient as a mathematical theory of uncertainty and belief; others that probability is neither universally appropriate nor adequate for the tasks we must ask of it. The large and disputatious literature surrounding the uncertainty debate suggests that this issue is unlikely to be resolved by such adjudication.

The problem is to be resolved, I believe, with a more open position, in which we accept that (a) there is a family of distinct theories of uncertainty which can be shown to have sound mathematical foundations, (b) these theories capture different intuitions about uncertainty and belief, and (c) that a more liberal attitude will permit the development of a deeper understanding of human judgement under uncertainty and more sophisticated technologies for aiding such judgement.

The argument for my position starts in Section 5.2 with a brief presentation of some ideas taken from the history of probability concepts by the philosopher Ian Hacking (1975), together with some other issues that have been raised about the general adequacy of the probability paradigm. Hacking is especially interesting because he was the first, to my knowledge, to propose that there is a "space" of alternative probability theories, though he does not tell us how this space may be characterized. In Section 5.3 I introduce some questions from artificial intelligence (AI), and argue that they represent a fundamental challenge to the probabilists' claim to universality. In Section 5.4 I review a number of alternative uncertainty formalisms, many of which have emerged from work in AI. Finally, in Section 5.5 I attempt to address

Hacking's conjecture by offering a proposal for a general framework within which uncertainty formalisms may be understood.

5.2 THE PROBABILITY PARADIGM AND ITS DETRACTORS

5.2.1 Historical Questions

Faced with the problem of accounting for the adoption of modern probability theory against its historical rivals, Hacking arrived at the remarkable conclusion that the emergence of the modern theory was historically somewhat arbitrary. In his elegant and informative book *The emergence of probability* he identifies a puzzle: that "...theories of frequency, betting, randomness and probability appear only recently. No one knows why" (page 2). This may be contrasted with other mathematical and scientific paradigms, such as those of geometry, chronology, astronomy and navigation which, like probability, had great practical implications but appeared much earlier.

Hacking observes that "around 1660 a lot of people independently hit on the basic probability ideas" (page 11), and adds "...The time, it appears, was ripe for probability. What made it ripe?" (page 12). One answer to this is that probability could have been discovered earlier, except that a variety of cultural obstacles prevented its emergence. Among the reasons Hacking considers for this are that probability's appearance could have been blocked in cultures which (a) took a deterministic, necessitarian view of the world, (b) believed that gods settle things—fate dominates chance, (c) failed to notice or understand the idea of "equally probable alternatives" which is necessary before one can move on to more useful ideas, (d) did not face (economic) problems for which probability is the solution, or (e) lacked a sufficiently rich set of mathematical ideas to permit the development of a probability calculus. From this one might predict that if one found a culture that was somewhat impious, which took a physical rather than a fatalistic approach to causality, and had a developed trading culture and arithmetic skills, then this should be conducive to the development of probability mathematics. Hacking notes that this may have been true in India about 2000 years ago, and that there are indeed hints in the historical record of a theory of probability (page 8). Appealing as this is Hacking finds the evidence for this cultural explanation to be weak. He and others continue to raise questions about the inevitability, or necessity, of probability ideas.

5.2.2 Technical Issues

I take the paradigm of probabilistic reasoning to be: the process of assessing the relative credibility of a number of alternative hypotheses by assigning degrees of belief to the various alternatives, based on evidence, in such a way that certain axioms, the probability axioms, are satisfied. Consider the following scenario:

> A doctor is aiming to maximize the likelihood of making a correct diagnosis for a patient who is complaining of abdominal pain. She identifies all the possible causes of abdominal pain and the symptoms that are associated with each disease. An estimate is now made of the conditional probability of each symptom being caused by each disease, and the prior probability of each possible disease. After establishing the presence or absence of each symptom she is now in a position to calculate the posterior probability of each disease using Bayes' rule.

Probabilistic approaches to tasks like diagnosis have been extensively studied. The scenario describes a simple method for probabilistic diagnosis. More sophisticated methods may now be preferred (e.g. Heckerman, 1991) but the complexities would distract us unnecessarily. While probabilistic methods have yielded some striking successes, they are open to a number of technical criticisms.

Firstly, a classical probability analysis places strong requirements on the completeness of our knowledge. It requires, for example, that a doctor has exhaustively identified the possible hypotheses (e.g. the patient is suffering from gastric cancer *or* gastric ulcer *or* duodenal ulcer, only) and that a complete set of conditional probabilities representing the dependencies between hypotheses and evidence has somehow been obtained. Frequently (indeed one might argue invariably), assumptions of exhaustiveness are unrealistic. Practical diagnosis has often to be carried out in the face of high levels of ignorance; it is as much about understanding what the problem is and the accommodation of uncertainty about relevant data or hypotheses, as it is about a precise weighing of evidence for a known set of possibilities (Fox et al., 1990).

In his admirable presentation of mathematical decision theory Dennis Lindley (1971; 1985) acknowledges this: "The first task in any decision problem is to draw up a list of the possible actions that are available. ... It is almost certainly true that some successful decision-makers derive their success from their ability to think of new ideas, rather than from any ability to select among a list, so providing an example of the human element ... Such initiative and enterprise is to be encouraged [but] we can offer no scientific advice as to how it is to be developed."

Secondly, a narrowly probabilistic position does not recognize knowledge other than that expressed in probabilistic form. This is surely restrictive. In predicting the structure of a complex molecule for example, such as a protein, knowledge of the *function* of the molecule, its *evolution*, *topological* and *geometrical features*, *charges* on its components, and many other kinds of knowledge can place substantial constraints on the possible structures a molecule may have. A strong case can be made, which we shall develop, that many kinds of knowledge other than probabilistic knowledge have considerable predictive value in reasoning under uncertainty, particularly in the absence of precise quantitative data.

These two points, the frequent absence of precise quantitative data and the importance of knowledge of general principles, are powerfully illustrated in the context of risk assessment. In the conclusion of a report of the UK Department of Health's Committee on Carcinogenicity of Chemicals in Food (1990) the committee concludes that it

> ... does not support the routine use of [probabilistic] risk assessment for chemical carcinogens. This is because the present models are not validated, are often based on incomplete or inappropriate data, are derived more from mathematical assumptions than from a knowledge of biological mechanisms and, at least at present, demonstrate a disturbingly wide variation in risk estimates depending on the model adopted.

5.2.3 Doubts about "Subjective" Probability

Related questions can be raised about the adequacy of the probability paradigm, even some vaguer "subjective" version, in accounting for patterns of human reasoning under uncertainty. People are remarkably good at solving *poorly structured* problems, involving high degrees of uncertainty, which are well beyond the capabilities of current formal reasoning systems. A compelling example of this is the process of formulating scientific theories where, individually and collectively, scientists achieve explanatory order in the face of ignorance, contradictions and the appearance of challenges to the theoretical framework as well as uncertainty about what the theories imply. Most scientists will be unsurprised by Glymour's observation that "probability is a distinctly minor note in the history of science" (Glymour 1980).

The structure of the world is prodigiously complex, and this is mirrored in the heterogeneity and complexity of our knowledge of it. This is evidenced by the public language we use in talking about the way the world "works" (recall the discussion of protein structure), our private experience of it, and scholarly analysis of the complex ontology of concepts which underpins our understanding. Indeed, the subjective features of "belief" itself appear to be quite complex; the natural language vocabulary that we routinely employ appears to have an underlying semantics which is, at least, two-dimensional (Clark 1988). Insistence that the laws of objective and subjective belief must be axiomatized in the same way (via the laws of probability) blurs a distinction between what is a useful technical device and common experience.

Doctors, lawyers, scientists, and other professions who commonly have to place their opinions on the record appear to arrive at their judgements by means of processes which do not seem to resemble probability assessment. Spoken and written records suggest that debates and disputes are pursued through patterns of *argument* and *counter-argument*, in which the assumptions, structure and other properties of arguments are challenged, and not just the degree of belief warranted by an argument for a claim. Individual decision-making and judgement also seem to show reflective or "metacognitive" styles of reasoning about the validity of arguments. The probability paradigm provides few tools by which to understand such processes.

5.2.4 What May We Conclude?

These and many other doubts have been around for a long time, but they are not universally shared, and many probability theorists are clearly unimpressed by them. From their point of view Hacking's observation that probability had a difficult birth does not bring into question the manifest health of the child; if anything it merely underlines the intellectual subtlety of the achievement. Probabilists may also argue, with justice, that the theory is not static but constantly advancing, and current technical limitations merely stimulate technical advances. Finally, any inadequacies of the theory to explain human judgement are largely irrelevant; the claim for probability is that it tells us how we *ought* to make judgements under uncertainty, not how we actually make them.

In short, while such doubts may make us pause they are unlikely to lead to a radical reassessment of such a successful tradition. If there are compelling arguments against the universal appropriateness of probability we are going to have to look outside these familiar areas of debate.

I believe there are such compelling arguments. The case I shall make is grounded in the observation that mathematical probability has been developed as a *tool for people to use*; a body of concepts and techniques which helps them to analyse uncertainty and make predictions in the face of it. While the theory is highly successful in this respect, the presupposition that it is for use by, say, a human decision scientist, who brings an understanding of the world to bear in applying the theory, has profound implications for its interpretation and its limitations.

5.3 UNDERSTANDING INTELLIGENCE: A DIFFERENT CHALLENGE

Whether a decision-maker is a scientist formulating a hypothesis or a theory, a doctor diagnosing a new and complex case, a company manager developing a marketing strategy, or a lawyer designing a client's defence, the framework of probability theory gives little help in *formulating* the decision problem, *understanding* what is a relevant solution or information source—or recognizing that there is a problem in the first place. Intuitively we identify such abilities with *intelligence* (Lindsey's "human element", perhaps); any view of uncertainty which ignores the fact that it lies in a larger context of intelligent problem solving seems to me to be rather unsatisfactory.

Of course we face a serious problem; it has proved notoriously difficult to achieve an unambiguous and generally accepted definition of the concept of intelligence. Attempts to relate it to intuitive notions of ability, or objective tests of educational or other achievement, have had limited success. We struggle with such weak definitions of intelligence as "that which intelligence tests measure". Concepts like knowledge, understanding, rationality and so forth seem to be at the core of what we mean, yet seem to have a curiously marginal place in modern psychological theories. I presume that this is because psychology aspires to be an objective, empirical discipline and it is difficult to bring such abstract ideas into the realm of empirical observation. Mentalistic ideas like uncertainty, belief and rationality are implicitly present in theories of human decision-making and judgement, of course, but in rather impoverished forms. Uncertainty and belief are equated, *a priori*, with subjective assessment of probability, and the interpretation of rationality is limited to demonstrations of compliance with weak mathematical constraints on "coherent" manipulation of such probabilities (Lindley, this volume, Chapter 1).

The scientific study of artificial intelligence (AI) shares many of the concerns of psychology in that it is attempting to understand and emulate human (or at least human-like) capabilities. As with psychology AI's attempts to achieve a general definition of (artificial) intelligence have had limited success, however: "artificial intelligence is the study of how to do things which, at the moment, people do better" (Rich & Knight, 1991) or AI systems are "sophisticated electronic agents in the form of computer systems that people could regard as 'intelligent'" (Besnard, 1989). Unlike psychology, however, AI is free to sidestep difficulties like this because it is not required to justify mentalistic theories in terms of empirically observable and measurable phenomena.

In order to argue that a theory illuminates a notion of intelligence many AI researchers simply attempt to demonstrate that a computer program is *sufficient* to manifest some interesting kind of competence (Newell, 1973), where such competence may include the ability to interpret sentences in a natural language or images seen through a camera, or solve a complex medical problem and plan a suitable therapy. Lately, the criterion of sufficiency has become seen as too weak to demonstrate that a theory is principled, and it is now widely expected that a computational theory should also be *formally stated*. It should be formulated in a precise language or notation and developed in such a way that it is possible to unequivocally establish its properties and verify the claims made for it.

Not all decision scientists will be sympathetic to a computational style of theorizing. Nevertheless the freedom to investigate intuitive features of intelligence in this mathematical way has stimulated the development of entirely new theories of knowledge and reasoning, belief and intention, and what is called "common sense" understanding of the world. Although the motivation for developing such theories is not primarily psychological it is informed by intuition, and consequently the results may provide an interesting body of ideas which may in turn inform psychology. In the case of uncertainty and belief formal theories are now available which go substantially beyond that of the probability tradition.

5.3.1 The AI Paradigm

Perhaps the most significant influence on AI that distinguishes it from other mathematical and formal traditions is its goal of understanding, and eventually building, agents that can operate *autonomously* in some world. Consider, for instance, NASA's interest in constructing an autonomous vehicle which is capable of operating on the dark side of the moon or some even more exotic environment. NASA mission planners cannot be confident of anticipating all the circumstances that the vehicle might confront (they have, after all, never been there). Unanticipated threats could arise at any time. The planners may believe that they can predict familiar types of threat but not their likelihood of occurring nor their detailed manifestations, and they must assume situations will arise that they cannot predict. Since such a vehicle cannot "phone home" for instructions or reprogramming, it would benefit from capabilities reminiscent of those of a human astronaut, i.e. abilities to perceive and interpret its environment, understand when it is facing a problem, formulate possible solutions, and identify relevant sources of information which will allow it to judge its best course of action.

The challenge of an alien world is colourful but we do not in fact need to go so far afield to find problems of similar character. Much medical software, for instance, is "safety-critical", meaning that errors in operation or use can lead to death, injury and other consequences. Safety engineers have developed various methods for predicting the problems that can occur but clinical environments are so complex that all possible hazards cannot be anticipated, even for simple systems. Recent cases of software for controlling radiotherapy equipment giving incorrect dosages to patients in circumstances which were not predicted by the designers are a pointer to the increasing dangers of using even semi-autonomous equipment.

The language in which the theorems and equations of probability is expressed is ill-suited to the design of procedures which are flexible in the face of such poorly defined environments. Arithmetic operators and ordinary algebraic formalisms were not designed to speak of generalized threats or their causes, nor for formulating hypotheses about the state of the world *ab initio*. At the very least the language needs to be augmented with other formalisms in which these concepts can be expressed and manipulated. From its inception AI has been preoccupied with finding languages for coping with ill-defined problems; the resulting languages are very different from, and in many respects much more powerful than, the languages of numerical mathematics. One such family of languages is based on formal logic.

5.3.2 Computational logic

In traditional work on probabilistic reasoning the interest has centred on the properties of mathematical functions whose ranges and domains consist of numbers. When used in computer programs functions are implemented as algorithmic procedures which accept sets of numbers as input (e.g. prior and conditional probabilities) and return sets of numbers as output (posterior-probabilities). AI languages are designed to represent and manipulate more general data structures. These can include numbers but also more complex terms such as symbolic descriptions of objects; properties and interrelationships of objects, goals and actions of agents, and so forth.

For instance "factual" knowledge about a topic can be captured in a database consisting of a collection of expressions such as the following:

```
has_property(cancer, life_threatening)          (5.1)

is_more_dangerous_than(cancer,peptic_ulcer      (5.2)
```

(We could use a more familiar English-like presentation, as in *cancer is more dangerous than peptic ulcer*, but the notation avoids ambiguity about the structure, particularly in complex expressions.)

Logic programs can be thought of as procedures for proving that statements are true given some database. For example a database may contain a set of rules for processing sentences in natural language, such as:

```
is_grammatical(NLsentence) if
                          <some set of conditions>
                                             (5.3)
```

A "theorem prover" can establish whether the predicate *is_grammatical* is true or not for some sentence *NLsentence* by establishing whether the associated set of conditions is true. (*NLsentence* is a variable representing any input sentence; variables are indicated by capitalizing the first letter.)

Functional computations can also be captured by logic programs, as in:

```
has_interpretation(NLsentence,Meaning) if
                                     <conditions>   (5.4)
```

Here the function *has_interpretation* returns a meaning representation for the sentence if it can be proved to be consistent with some set of linguistic rules, or "meaning postulates", in the database.

The conditions of rules can be arbitrarily complex, made up of conjunctions, disjunctions and negations of simpler conditions. Simple conditions include facts, like (5.1) and (5.2), or conclusions of other rules. A logic program may invoke any number of rules to any depth. Example (5.5) shows (5.4) fleshed out to a single level:

```
has_interpretation(NLsentence, Meaning) if
            is_grammatical(NLsentence,Parse_tree) and
            has_meaning(Parse_tree,Meaning)          (5.5)
```

This rule should be interpreted to mean that if the theorem prover can find a valid grammatical structure, *Parse_tree*, for the sentence and assign a meaning to the structure, then it will succeed with result *Meaning*.

Variables are normally universally quantified. This means that logic programs implicitly define all possible solutions to a problem. In the case of (5.5) the program will find all the interpretations of an input sentence which are justified by the linguistic knowledge encoded in the database.

Much of the power of AI languages arises because variables can take any kind of term as a value. Consequently logic programs and other symbolic languages can not only reason with numbers and all the other datatypes of classical programming and mathematics, but also complex symbolic structures like proofs, parse trees and meaning representations. Logical formalisms can explicitly capture properties of rules, functions and programs (called "metalevel" representation) and logic programs can reason about programs as well as simply execute them. Logic programming languages, such as PROLOG (PROgramming in LOGic), provide many more computational techniques than we need to discuss here (see any text, such as Clocksin & Mellish, 1972), for details) but PROLOG provides a fairly standard notation for logic programs and we shall use it to present many of the examples that follow.

5.4 SYMBOLIC REASONING AND UNCERTAINTY

Logic languages provide a formal machinery for representing and reasoning with knowledge of the world. They can accommodate concrete concepts, as in the medical examples, but also abstract concepts like causality, hypotheses and beliefs. We shall also see that metalevel reasoning about what an agent knows or believes can play an important role in its reasoning under uncertainty. To keep the presentation short, and as widely accessible as possible, I shall discuss these ideas in a relatively non-technical style. However, most of the concepts have received considerable technical development. An up-to-date and comprehensive but accessible technical survey is Krause & Clark (1993).

5.4.1 Uncertainty and Rule-based Reasoning

The earliest experiments with rule-based reasoning systems used quite simple inference methods. For example, the following rule defines a simple piece of "diagnosis" knowledge:

```
evidence_for(Patient,cancer) if
                    known(Patient,weight_loss) and
                    known(Patient,elderly)        (5.6)
```

Which is to say: if a patient is known to be elderly and has lost weight then we are entitled to conclude that there is evidence for the patient having cancer. The rule is analogous to a conditional probability expression, $p(H \mid E1 \ \& \ E2)$ but is only qualitative; it says there is evidence for cancer but not how much. The absence of any probabilities may appear to be a weakness but it immediately confers an important freedom; since we do not have to distribute a fixed quantity of belief over a set of hypotheses we do not have to fix the set of hypotheses at the outset. In fact a simple extension to rule (5.6) allows us to introduce hypotheses progressively, as evidence is obtained:

```
hypothesis(Patient,Disease) if
               evidence_for(Patient,Disease) and
               not(excluded(Patient,Disease))  (5.7)
```

Which is to say, if we acquire any information that is evidence for a disease, and we have no reason to exclude the disease, then we are entitled to include it as a hypothesis.

The practical importance of such "open-mindedness" was once encountered by a group working on computer-based interviewing of patients. An interview program fed data directly into a diagnosis system which made the prior assumption that all patients were suffering from one of a number of gastro-intestinal diseases. Down the hall was an alcoholism clinic. Patients would occasionally stray from this clinic, be interviewed by the system, and promptly be diagnosed as suffering from an ulcer, gall-bladder disease etc.

Medical knowledge is frequently just empirical, recording that this condition and that symptoms are statistically associated, but much medical knowledge is deeper than this. For example causal knowledge (how diseases cause symptoms), taxonomic knowledge (such as the features of cancer as a class of diseases as distinct from the features of specific cancers) and knowledge of anatomical structures, physiological processes and functions etc. can all come into play in medical decision-making and judgement. Symbolic languages are well adapted to expressing such ideas. Rule (5.6) above captures a specific association between patients with cancer and those who are elderly and have lost weight. We may modify the rule to cover (an indefinite variety of) causal relationships:

```
evidence_for(Patient,Condition) if
                known(Patient,Symptom) and
                could_cause(Condition,Symptom)   (5.8)
```

Proving the predicate `could_cause(Condition,Symptom)` may require quite a complex process. For example if we have detailed knowledge of the actions of two drugs it may be necessary to demonstrate from a detailed physiological theory that they could interact to cause the observed condition. Reasoning from first principles in this way can be an important diagnostic strategy, particularly in unusual or difficult medical cases. Barahona (1993) provides a detailed analysis of causal reasoning in medicine, in terms of general knowledge of structures, functions and processes.

Medical knowledge does not consist merely of a "flat" set of diseases and their associated symptoms, but a complex network of concepts (such as ulcers of the stomach and duodenum), their classes (peptic-ulcers), classes of classes (gastrointestinal diseases) and so forth. Likewise for symptoms, treatments, tests and other medical concepts. As remarked earlier the conceptual structure, or ontology, of many domains is complicated and this has implications for how evidence is to be interpreted. Suppose we know that

```
is_a_kind_of(duodenal_ulcer, peptic_ulcer)
```

and we have evidence that a patient has a duodenal ulcer (such as pain immediately after meals in an older patient); then using the following rule:

```
evidence_for(Patient,Disease_Class) if
          evidence_for(Patient,Disease) and
          is_a_kind_of(Disease,Disease_Class)   (5.9)
```

we can directly make the inference that there is evidence for the patient having a peptic ulcer. Reasoning from the specific to the general (or vice versa) may be simple but it is important. Many peptic ulcers are treated the same way using a class of drugs called H2-antagonists. If we can obtain convincing evidence that other competing types of gastrointestinal disease are implausible (such as gall-bladder disease or cancer) then it is unnecessary to carry out a detailed differential diagnosis of the particular type of ulcer the patient is suffering from, since the treatment is the same in all cases.

We have emphasized logical aspects of rule-based reasoning but extension to include quantitative information is quite straightforward. Various systems for attaching numerical coefficients to facts and rules have been proposed.

The metalevel expressiveness of logic languages allows the expression of a quantified belief in an assertion:

```
holds( Assertion, Coeff)                              (5.10)
```

where `Coeff` represents a numerical coefficient, meaning "`Assertion` is certain to degree `Coeff`", e.g.

```
holds(known('Fred Smith',elderly),0.99)
holds(known('Fred Smith',weight_loss),0.9)
```

a conditional rule can be treated similarly:

```
holds ((Conclusion if Premisses), Coeff)              (5.11)
```

meaning "`Conclusion` can be conditionally inferred from `Premisses` with certainty `Coeff`". For example embedding rule (5.6) in (5.11) we may have

```
holds((evidence_for(Patient,cancer)if
                     (known(Patient,weight_loss) and
                     known(Patient,elderly)), 0.5)    (5.12)
```

Finally a suitable program, or "meta-interpreter" can be designed which manipulates these expressions to carry out both logical deduction and numerical calculations. If the set of hypotheses is closed and a complete set of prior probabilities and probabilistic evidence rules is available, then the meta-interpreter can derive probability-quantified conclusions by combining an appropriate probability revision procedure with the normal deductive one (we omit the details, an example program is available from the author).

Early AI methods for combining numerical uncertainty representation and revision were criticized because, among other reasons, they used *ad hoc* rather than probabilistic methods (e.g. Cheeseman, 1985), and they have been largely replaced by more established techniques. Probabilistic methods have been extended to permit propagation of probabilities over complex networks of evidence and hypotheses (Pearl, 1988; Lauritzen & Spiegelhalter, 1988). *Belief functions* can be used where evidence is to be distributed over class-structured hypotheses (Gordon & Shortliffe, 1984). *Fuzzy logic* has been used to permit vagueness in the definition of logical categories (Zadeh, 1978) and has since developed into *possibilistic logic*, a well developed alternative to the probability calculus (Dubois & Prade, 1988).

To summarize, a non-numerical calculus such as symbolic logic can provide a formally sound and well-understood inference system for capturing intuitive ideas about knowledge and for introducing and reasoning about hypotheses in the absence of quantitative uncertainty data.

The emergence of such methods raised a challenge to probability. As so often before, however, the probability community responded vigorously by extending existing techniques to address the wider range of applications that AI was addressing. One might be tempted to conclude from this that logic is a useful tool for the deductive elements of reasoning, but when uncertainty is encountered a numerical calculus is still necessary. Even this weaker position can be questioned, however, and we turn now to developments that support the opposing view.

5.4.2 Non-monotonic Reasoning

Classical logic makes the fundamental assumption that if we can validly deduce some assertion, A, then whatever other conclusions we may subsequently deduce A must remain true. If this is not the case we will be faced with a contradiction. (In ordinary logic contradictions cannot be tolerated because it is a formal property of classical logic that anything can be deduced in the presence of an inconsistency.) Consider the familiar example from elementary logic that if `X` is a man then `X` is mortal. More generally:

```
mortal(X) if
  is_biological(X)                                    (5.13)
```

Plants, animals, man, and collections of cells in culture, are generally consistent with this rule, so whenever a biological entity is encountered we may reasonably infer that it will not live for ever. Unfortunately, there is an important complication. In culture, normal cells divide for a few generations and then the "cell-line" tends to die out. However, when a cell is transformed into a tumour cell the line does not die out; it acquires a property which may be called "immortality" (although this is a slight abuse of this term, for purposes of illustration). Rule (5.12) is consequently a little too strong. We have seen two approaches to solving this problem; both involve weakening the conclusion, either by concluding that there is merely "evidence for" mortality, or by attenuating the conclusion with a probability or other numerical coefficient.

An alternative approach is to say that if I know something is biological in origin then I can reasonably assume that it is mortal unless and until I find out I am wrong, in which case I change my mind. Classical logic does not permit us to change our minds, but non-monotonic logic is designed to overcome this restriction. Example (5.13) can be rewritten in a non-monotonic form, such as:

```
mortal(X) if
  is_biological(X) and
  consistent( not(immortal(X)) )                      (5.14)
```

This is to be read as "if `X` is known to be a biological entity and it is consistent, given all that is currently known, to assume that `X` is not immortal, then `X` is mortal". If we now find that the biological entity is immortal by the following ordinary rule, say,

```
immortal(X) if
        tumour_cell_line(X)                    (5.15)
```

then it is no longer consistent to conclude `mortal (X)` and the conclusion must be withdrawn. (Speaking a little loosely, classical logic is *monotonic* because deduction can only result in increases in the collection of beliefs, while non-monotonic inference can result in either increases or decreases in our beliefs.)

Default logics are an important formalization of this idea (Besnard, 1989) which sanctions the ability to "jump to conclusions" without having to consider all the circumstances in which the conclusion could conceivably be wrong. Suppose we are planning a business trip from Paris to Greece; we choose Air France flight AF97 from Paris to Athens, and then make a hotel reservation according to the scheduled arrival time in Athens. Default logic provides a sound framework for the following kind of reasoning:

> AF97 is scheduled to arrive at 10.00 and there is no reason to believe that it will not arrive at that time so I will infer that it will arrive in Athens at 10.00.

The classical inference rule

```
will_arrive(Flight,Time) if
                       scheduled(Flight,Time)  (5.16)
```

is too strong so we introduce a "default condition" into it:

```
will_arrive(Flight,Time) if
           scheduled(Flight,Time) and
           consistent( not(delayed(Flight)) )  (5.17)
```

so long as the assertion `not (delayed (AF97,10.00))` is consistent with the other things we believe then we are entitled to assume that AF97 will arrive on time. There is any number of reasons why our flight might be delayed—strikes, fog, cancellation, bomb scares, equipment faults, loss of the aircraft in the Bermuda triangle, etc.

Of course, it would be theoretically possible to estimate the probability of one or more of these events, but would it be useful? If we hear of a strike by air traffic controllers in the Paris area then we simply retract the default assumption and rearrange our accommodation. If we do not find out in time, and arrive late, then hotels are not usually full, there are always other hotels or, at worst, an uncomfortable night at Athens airport.

A number of non-monotonic logics have been formalized. In modal logic a proposition is *possible* if we cannot prove that it is *necessarily* false (Mott,

1988). For example the autonomous vehicle referred to earlier may assume that it is possible to get from A to B if it cannot establish any reason that it cannot, such as an equipment failure or an obstacle. "Autoepistemic logic" is concerned with the formalization of agents' reasoning about what they know or believe (Moore, 1988). As scientists if we come up with an interesting idea then we may carry out a considerable amount of work on the autoepistemic argument that "I am expert in my field; if anyone else had followed up this idea I would have heard about it, so I can assume they haven't". This often works well, thought sadly not always.

Non-monotonic logics are an advance on classical logic because they have a well-grounded theory for avoiding assumptions of omniscience (knowing everything about a situation so one will never have to change one's mind). They are a practical alternative to probabilistic and other numerical uncertainty frameworks because they do not require quantification of uncertainty for reasoning to proceed. Krause & Clark (1993) provide a good review of developments in the area.

5.4.3 Argumentation

Default reasoning is a recent development but criticisms of classical mathematical logic are not new. Toulmin (1958) raises a number of questions about the role of logic in practical reasoning. In classical logic an *argument* is a sequence of inferences leading to a conclusion, which may be either true or false. The interest of the logician has traditionally been in procedures by which such arguments may be judged valid or invalid. Toulmin was more interested in the kinds of reasoning which go on in everyday debate, and in the conditions that determine whether arguments are persuasive. For example, a doctor may argue "it is possible that the patient has gastric cancer because he is elderly and has recently lost weight and I know this is a classical presentation of an advanced malignancy." Toulmin characterizes such arguments by means of the following schema:

```
Date ————————→ Claim Qualifier
          |          |
       Warrant    Rebuttal
          |
       Backing                    (5.18)
```

Data here corresponds to all the things the doctor knows about the patient (the patient is elderly and has lost weight); the *Claim* is the base sentence "the patient has cancer" but this is *Qualified* because it is said to be "possible" but not asserted to be true. The *Warrant* is the doctor's knowledge about typical presentations and relationships between pathologies, which gets its *Backing* from reference texts, research findings and so forth. Toulmin anticipated non-monotonic reasoning with his notion of *Rebuttal*; the doctor's colleague may point out "but the results of all the tests we have done seem inconsistent with cancer".

Toulmin's analysis is not formal but it is intuitively appealing. A number of workers are now investigating ways in which such ideas might be formalized (Loui, 1987; Pollock, 1992; Fox, Krause & Ambler 1992). One motivation for developing a formal theory of argumentation arises from the requirement for a theory of decision-making which is appropriate to the design of autonomous agents (Fox, 1991), but the results are quite general. In this section we introduce our formalization of argumentation, showing how it may add to the available techniques for reasoning under uncertainty. In the next section we take the ideas further by considering how the approach could provide a general framework for understanding a number of methods for reasoning under uncertainty.

A logic of argument

In propositional logic (PL) if we hold some fact to be true, say p, and also hold that p implies q ($p \rightarrow q$) then by the rule of *modus ponens* we are entitled to conclude q. Taken together, *modus ponens* and the other rules of propositional logic define a relationship between the set of sentences in a database (the antecedents), and a database extended with the set of sentences which may be validly derived by the rules of the logic (the consequences). This is summarized by:

$$\text{Antecedents} \vdash_{\text{PL}} \text{Consequents} \tag{5.19}$$

in which the "turnstile" symbol $\vdash$ represents the *consequence relation* of the logic. The subscript PL reminds us that particular consequences are only valid if we accept the inference rules of PL but may not be valid if we adopt some other logic in which *modus ponens* is not a rule of inference. (Note that "object-level" rules in the database of antecedents, such as $p \rightarrow q$, should not be confused with the inference rules or "meta-rules" of the logic, such as *modus ponens*.)

Countless logics with specialized consequence relations have been developed for particular kinds of reasoning (e.g. Haack, 1978). Some logics drop inference rules from classical logic, others add rules. For example, propositional logic includes the rule of the excluded middle, "A or not (A) but not both", while intuitionistic logic omits this rule. A logic can have an entirely different consequence relation from that of classical logic.

AI has been particularly concerned with developing specialized logics for "common sense" reasoning about space, time, belief and so forth. Default logic is an example of the latter; it copes with changing belief with a non-monotonic consequence relation. Argumentation provides another approach to reasoning under uncertainty. The logic of argument (LA) is a variant of intuitionistic logic in which two modifications have been made (Fox, Krause & Ambler 1992). First, it is assumed that an argument may not only *prove* a claim but may also, more weakly, *support* it. Second, we view arguments as contingent on the acceptance of some "view of the world". We call such a view a "theory", since it can be modelled as a collection of object-level rules and facts. The consequence relation of LA is summarized by the following meta-level schema:

$$\text{Context} \cup \text{Theory} \vdash_{\text{LA}} (\text{Claim, Grounds, Qualifier}) \qquad (5.20)$$

The schema simply says that given some set of beliefs about a particular situation (Context) together with a set of general beliefs (Theory), we are entitled by the axioms and inference rules of LA to make certain Claims. Claims are justified by some set of rules and facts drawn from the union of the Context and Theory, and are qualified in ways that we discuss in a moment.

Axioms and inference rules for LA are given in Figure 5.1. In this version of LA the qualifier is one of "supports" or "confirms" (abbreviated as + and + +). If an argument supports a claim then it *increases belief* in it (but we don't say by how much). If it confirms the claim it means that, from the point of view embodied in the theory, the claim is certain. Note that different agents may hold different theories, and in principle a single agent can adopt different and possibly inconsistent theories at different times. Consequently confirmation is not quite the same as saying the claim is logically true (or that its probability is 1.0) because there is always the possibility that even a certain claim will, in Toulmin's terms, be "rebutted" by an equally certain argument based on different theoretical assumptions.

LA has been embodied in a Prolog program called an *argumentation theorem prover* (ATP). The consequence relation of LA is embodied in the ATP which, when supplied with a proposition and a set of rules and facts (the theory) constructs all and only those arguments which support or confirm the proposition which are justified on the theory. (LA and the ATP were developed in collaboration with Paul Krause, Simon Ambler and Michael Clarke.) The theorem prover returns argument terms which we can represent in the familiar Prolog style:

```
argument(Claim,Grounds,Sign)                    (5.21)
```

For example:

```
argument(''the patient has cancer'',
         grounds(wt_loss,elderly),
         supported)                              (5.22)
```

which can be glossed as "since the patient is elderly and has lost weight this supports the claim that he has cancer." The ATP is neutral about the content

Axiom

$$\frac{\varphi : a : S}{\varphi : a : S}$$

Weakening

$$\frac{\varphi : a : ++}{\varphi : a : +}$$

Introduction Rules

$$(\wedge\,\mathrm{I})\quad \frac{\varphi : a : S \qquad \psi : b : S'}{\varphi \wedge \psi : a \cup b : \min(S,S')}$$

$$(\rightarrow\mathrm{I})\quad \frac{\begin{matrix}\cancel{\varphi : \varnothing : ++}\\ \vdots \\ \psi : a : S\end{matrix}}{\varphi \rightarrow \psi : a : S}$$

$$(\vee\,\mathrm{I})\quad \frac{\varphi : a : S}{\varphi \vee \psi : a : S} \qquad \frac{\psi : a : S}{\varphi \vee \psi : a : S}$$

$$(\neg\,\mathrm{I})\quad \frac{\begin{matrix}\cancel{\varphi : \varnothing : ++}\\ \vdots \\ \bot : a : S\end{matrix}}{\neg\varphi : a : S}$$

Elimination Rules

$$(\wedge\,\mathrm{E})\quad \frac{\varphi \wedge \psi : a : S}{\varphi : a : S} \qquad \frac{\varphi \wedge \psi : a : S}{\psi : a : S}$$

$$(\rightarrow\mathrm{E})\quad \frac{\varphi : a : S \qquad \varphi \rightarrow \psi : b : S'}{\psi : a \cup b : \min(S,S')}$$

$$(\vee\,\mathrm{E})\quad \frac{\varphi \vee \psi : a : S \qquad \begin{matrix}\cancel{\varphi : a : ++}\\ \vdots \\ \sigma : b' : S'\end{matrix} \qquad \begin{matrix}\cancel{\psi : a : ++}\\ \vdots \\ \sigma : b'' : S''\end{matrix}}{\sigma : b' \cup b'' : \min(S, S', S'')}$$

$$(\neg\,\mathrm{E})\quad \frac{\varphi : a : S \qquad \neg\varphi : b : S'}{\bot : a \cup b : \min(S,S')}$$

Figure 5.1 Summary of LA inference rules. The labelling of propositions follows the syntax *formula* : *grounds* : *qualifier*, where *qualifier* $\in \{+, ++\}$, and $\min(+, ++) = +$

of the theory; the details of this may warrant quite different kinds of argument, including appeals to *a priori* beliefs, external authoritative opinion, direct or indirect observation, inferences from knowledge of causality, structure, function, and so on (Fox and Clarke, 1991).

Practical decision-making and debate frequently do not afford the use of confirmatory arguments, so we may only have collections of supporting arguments for competing claims. What may we conclude from a number of arguments? Toulmin gives us little guidance here but clearly we need to combine or *aggregate* the arguments in some way if we are to choose among the alternatives. As discussed earlier it is unwise to assume that there will always be some basis for attaching numerical weights or confidences to individual arguments (though this can be done as we shall see later). In fact there are various ways to aggregate arguments without relying on weights, including "semi-quantitative", "meta-argumentation" and "linguistic" techniques.

One method is simply to compare the relative numbers of arguments for alternative claims; this yields a total ordering on the alternatives. Selecting the most preferred alternative on this basis amounts to applying an *improper linear decision rule with uniform weights*. It is well known that this rule gives quite a good account of much human decision-making under uncertainty (Dawes, 1979). Although apparently weak it can give results which are surprisingly similar to those given by a precise probabilistic decision procedure Fox, Barber & Bardham (1980); O'Neil and Glowinski (1990) and Chard (1991) show the effectiveness of the method in several medical applications. The linear aggregation of arguments can also be shown to be formally sound (Ambler, 1993).

The linear decision rule can be refined by taking advantage of the fact that the grounds of the arguments are explicit. Suppose we have equal numbers of arguments for a patient having gastric ulcer and having duodenal ulcer. The patient reports, among other things, that he has lost weight and we can argue that this is best explained by the former hypothesis. However, we also observe that "the patient is elderly and confused so we should question the accuracy of his memory" and hence doubt the argument. This sort of meta-level reasoning over arguments yields a more precise ordering on the alternatives; since certain arguments are weaker than others this must also weaken the claim.

Toulmin saw arguments as embedded in natural language, notably in the use of linguistic qualifiers. In standard logic qualifiers are undistinguished elements of a sentence which as a whole can only be true or false. In the argumentation framework we wish to distinguish them, in order to permit reasoning about them separately from the base proposition. In (Fox, 1986) I proposed extending logic with a set of *linguistic predicates* which are reminiscent of belief terms in natural language. For example we might wish to be able to express the decision rule that "any patient should be investigated who is possibly suffering from a disease that is life-threatening". In our usual notation:

```
should_investigate(Patient,Disease) if
                possible( hypothesis(Patient,Disease) )
                and life_threatening(Disease)          (5.23)
```

The first predicate in the conditions of the rule has the form

```
possible (Proposition).
```

This and other linguistic predicates might be interpreted in terms of the properties of the set of arguments for and against the proposition

```
possible(Claim) if
        argument(Claim,Grounds,supported) and
        complement(Claim, Complementary_claim) and
not( argument(Complementary_claim,_, confirmed) ) (5.24)
```

In other words a claim (e.g. Fred Smith has cancer) is *possibly* true if there is a supporting argument for it and the opposing claim (i.e. Fred Smith does not have cancer) cannot be confirmed. (The second variable of the negate condition, _, means that we do not require any specific grounds since the theorem prover is going to fail to find any argument.) Other definitions, covering such concepts as plausibility, suspicion, belief and doubt are discussed in Fox (1986).

Linguistic predicates offer a symbolic way of aggregating arguments, based on patterns rather than numbers of arguments. They extend the expressiveness of the knowledge representation language. This is analogous to the use of linguistic terms in reasoning about time (*A* occurs "before" or "during" *B*, etc.) and space (*A* is "above" or "inside" *B*, etc.) in that they do not depend on precise chronological or geometric quantities. Reasoning systems which employ such predicates are flexible and, arguably, a natural basis for capturing and communicating human knowledge. Of course the psychological validity of any specific linguistic scheme is open to dispute, just as the psychological validity of the notion of subjective probability is debated. As with the latter, this is an empirical question, though, as we discuss in more detail in a moment, the general approach can be given a clear mathematical interpretation.

5.5 A GENERAL FRAMEWORK FOR UNCERTAINTY

> I am inviting the reader to imagine . . . that there is a space of possible theories about probability that has been rather constant from 1660 to the present . . . perhaps an understanding of our space and its preconditions can liberate us from the cycle of probability theories that has trapped us for so long. (Hacking, 1975).

In this section we stand back from the details of particular calculi in order to consider whether different calculi share any common features. Like Hacking I believe that there is a range of possible uncertainty calculi, some of which are already recognized while others, perhaps many others, remain to be discovered. If we can understand this space of alternatives then benefits should follow. These may include a deeper understanding of rival systems; better criteria for selecting the appropriate calculus for solving different kinds of problem, and a more informed view of the strengths as well as the weaknesses of human reasoning under uncertainty.

I wish to suggest that various proposals for specific uncertainty calculi are specializations of a "generic" though relatively weak calculus. I shall suggest that the calculus is a logic of argument (such as LA, but there may be other candidates) augmented with different systems of qualifiers and their corresponding aggregation functions. For brevity I shall only present the essential ideas, technical details can be found in (Fox, Krause & Elvang Goransson, 1993; Krause, Ambler & Fox, 1993). For clarity I shall use the following simple notation for arguments in this section:

$$S : G : Q$$

meaning "the sentence S is qualified by Q given grounds G". An expression of the form $\{S : G : Q\}$ represents a set of arguments about S.

Assume we have some method for constructing sets of arguments, such as LA. As presented earlier LA has only two qualifying symbols, $\{+, ++\}$ representing supporting and confirming arguments. We generalize this set to the idea of a "dictionary" of "signs" which can be used to label any (proposition, grounds) pair. The definition of a dictionary has the general form:

$$dict(D) =_{\text{def}} \{S1, \ldots, Sn\}$$

meaning dictionary D consists of the set of symbols $S1, \ldots, Sn$. There is no reason to restrict dictionaries to finite sets of symbolic signs. Dictionaries for probabilities, possibilities, certainty factors, belief functions etc. can be similarly defined, e.g.

$$dict(prob) =_{\text{def}} [0, 1]$$

Given some set of arguments for P whose signs are drawn from a dictionary D we wish to aggregate these arguments to yield a new qualifier in D (or possibly in some other dictionary) representing our overall confidence in P. Suppose the propositions of interest are formulae of a language L, the grounds are constructed from a class of sentences G, and signs drawn from the dictionary D. Then an *aggregation function* has the general form:

$$AGG_D : \mathbf{P}\,(L \times G \times D) \longrightarrow L \times D$$

meaning that that the function maps from the power set (the set of all subsets) of arguments into the set of possible claims with their associated qualifiers. This general formalization summarizes a structure which is common to a number of calculi.

5.5.1 The Generic Calculus

In standard logical inference we assign the value "true" to a sentence on the basis that we can construct a proof from other sentences which are held to be true. In argumentation we assume that while an inference procedure may be valid the sentences that it assumes may not be true, and consequently any conclusions that can be deduced using them may be in error. We therefore substitute the sign + for "true" to acknowledge this possibility. This gives us the simplest uncertainty calculus:

$$dict(generic) =_{\text{def}} \{+\}$$

the argument ca:lw: + simply says "the fact that the patient has lost weight increases my confidence in her having cancer, but I cannot say by how much". Suppose we have two arguments in the generic calculus:

$$P:G1:+ \ \& \ P:G2:+$$

how may we aggregate them? In general the more arguments we have for P the greater may be our confidence in P. More formally we can characterize the *force* of a set of arguments (represented by $|Args|$) with the constraint:

$$\text{Let } Args = \{P:G1:+, \ldots, P:Gn:+\} \text{ then}$$
$$|Args'| \geqslant |Args| \text{ if } Args' = \{P:Gn+1:+\} \cup Args \qquad (5.25)$$

This captures the simplest aggregation procedure, the improper linear model in which we just count arguments for a proposition to assess our relative confidence in alternative claims.

5.5.2 The Bounded Generic Calculus

There is no limit to the number of arguments that can be constructed for a proposition. Intuitively, however, some arguments are conclusive; to represent this we may define a more specialized calculus, introducing a dictionary with an additional sign ++, i.e.:

$$dict\ (bounded) =_{\text{def}} \{++, +\}$$

which is the dictionary on which we defined LA. We now require an extended aggregation function. Informally, if we have a conclusive argument for some proposition, then this argument will dominate the aggregation procedure, that is the aggregation function is restricted by (5.25) and the additional constraint:

$$|\{P:G1:++\}| = |\{P:G1:++, P:G2:+\}| \qquad (5.26)$$

(5.26) applies to many quantitative calculi as well as this symbolic one, such as probability. If we know the precise probability of a hypothesis H given evidence E is 1.0 and E is observed (with certainty) then the probability of H is certain, notwithstanding any other pieces of marginal evidence against H.

5.5.3 The Delta Calculus

In some situations we want to be able to argue *against* as well as *in support of* propositions, i.e. we would like to be able to write arguments of the form:

$$P:G:-$$

which means that G justifies a *reduction* in belief in P. To do this we introduce the extended dictionary:

$$dict(delta) =_{\text{def}} \{+,\ -\}$$

As in intuitionistic logic an argument for P is not necessarily an argument against $not(P)$; we may have an argument in favour of some scientific theory, say, but if we cannot be sure we have exhaustively enumerated all possible theories we cannot be certain that some alternative theory will not also be supported by the same argument. However, for cases where we have exhaustively identified the alternatives (e.g. we hold that the patient has cancer or does not have cancer) we can strengthen the generic calculus by explicitly defining an exclusiveness constraint on the mutually exclusive alternatives:

$$P:G:- \text{ if and only if } Q:G:+ \tag{5.27}$$

An aggregation procedure for the delta dictionary honours the generic constraint (A1), but in addition we add a complementary constraint which takes account of arguments against propositions. As in (A1) let *Args* be some set of arguments about a proposition P, then

$$|Args'| \leqslant |Args| \text{ if } Args' = \{P:G:-\} \cup Args \tag{5.28}$$

5.5.4 The Bounded Delta Calculus

Finally, we specialize the delta calculus with an extended dictionary including symbols for upper and lower confidence bounds:

$$dict(bounded\_delta) =_{\text{def}} \{++,\ +,\ -,\ --\}$$

The constraints (A1–A3) all hold on aggregation, but we have the additional bounded-complement constraint analogous to (5.27)

$$P:G:++ \text{ if and only if } Q:G:-- \tag{5.29}$$

5.5.5 Argumentation and Numerical Calculi

In conventional quantitative calculi the logical arguments underlying inference receive little attention. These calculi must satisfy the aggregation constraints (5.25), (5.26), (5.28), (5.29) , (the complementation constraint (5.27) is only necessarily obeyed by classical probability). For example, if we can logically construct an argument then the belief coefficient (probability, possibility) associated with the conclusion of the argument must increase or decrease consistently with the constraints. In other words if we can construct an argument in support of a claim, but the overall belief in the claim as determined by the aggregation function of the calculus is reduced, then the function is incoherent. The probability calculus is proveably coherent, but it is not alone in having this property.

My basic proposal is that the most elementary form of reasoning under uncertainty consists of constructing arguments *in support of* propositions and aggregating the signs of those arguments within constraints determined by the dictionaries from which the signs are drawn. The ATP provides the machinery for the first step; we now need to demonstrate that it is compatible with different dictionaries and their characteristic aggregation functions. The necessary demonstration is only summarized here; technical details are provided in Krause, Ambler & Fox (1993).

(1) The simplest aggregation function has "confirming" arguments outweighing "supporting" arguments, which in turn outweigh "vacuous" arguments.
(2) The next refinement is the improper-linear aggregation rule: The more supporting arguments there are for a claim the greater confidence we may have in that claim.
(3) "Weakest link" aggregation. Each sentence in a database is labelled with a value drawn from [0,1]. Confidence in an argument is the minimum value of the confidences assigned to the individual grounds of the argument. Overall confidence in a claim is the maximum value of the confidences of the arguments which support it. This is the aggregation function of possibilistic logic (Dubois & Prade, 1988).
(4) "Shortest path" aggregation. Similar to (3) but the confidences of the grounds of an argument are multiplied together. If the grounds have confidences less than 1, then the greater the number of axioms used in constructing an argument, the lower will be the resulting confidence. Again, the confidence in a claim is the maximum value of the confidences of the arguments which support it.
(5) The final aggregation procedure has certain similarities to belief functions. In the simplest case, the confidence in a single argument is just the product of the confidences in the axioms used to construct that argument. However, if two arguments are aggregated, the confidence of the combined arguments is given by the formula

$$c(A) + c(B) - c(A \wedge B)$$

which is the formula for the combination of "pure arguments" given by Bernoulli (Shafer, 1978) and is essentially a probabilistic method without normalization.

5.5.6 Argumentation and Non-monotonic Reasoning

Standard default logic can be interpreted in terms of the bounded delta calculus. Suppose we can construct an argument for P on the basis of a default rule. By definition a default is not guaranteed to be correct, so in this calculus the argument has the form

$$P : \text{default} : +$$

If we subsequently identify arguments for rejecting P (i.e. not(P) : G : ++) then aggregation will yield the conclusion not(P) by constraints (5.26) and (5.29). Argumentation therefore permits behaviour much like that of default logic, but it may also illuminate the relationship between default reasoning and quantitative uncertainty. Suppose we have a reason to doubt P but not to reject it (i.e. we can construct the argument not(P) : G : +), then this balances the default argument, and we are equivocal about P and not(P). If we have further arguments against P then the balance of argument turns against it (constraint (A4)) but we can still hold both P and not(P) as possibilities, a behaviour similar to the normal behaviour of probabilistic, possibilistic, belief function and other quantitative calculi.

5.5.7 Argumentation and Natural Language

As remarked earlier the presence of an inconsistency in a logical theory is pathological for classical logic; everything can be deduced from a contradiction. This seems counter-intuitive since we constantly encounter contradictions in everyday life, yet in some way manage to cope with them. Elvang-Goransson, Krause & Fox (1993) attempt to explain how arguments constructed from an inconsistent database using classical logic can be assigned different levels of "acceptability" rather than leading to general logical incoherence. They view reasoning, as here, in terms of the construction and evaluation of arguments, but rather than merely deciding from the arguments which claims to accept and which to reject they define a hierarchy of "acceptability classes". The weakest class of claims, $c1$, is the set for which there is, simply, some argument. A stronger class of claim, $c2$, consists of the subset of $c1$ for which the argument is consistent. An even more acceptable set of claims, $c3$, is that subset of $c2$ for which there are no rebutting arguments, while class $c4$ is the set of claims in $c3$ whose *grounds* are not subject to a rebutting argument. Class $c5$ contains the claims which are tautologies.

This is a purely formal system which attempts to capture the idea of *relative uncertainty* in purely logical terms. However, the authors also conjecture a relationship between various properties of these acceptability classes and linguistic uncertainty terms discussed earlier and informally defined in terms of *patterns* of argument. Elvang-Goransson, Krause & Fox (1993) indicate how a substantial set of English descriptors, such as supported, doubted, probable and plausible (and their lexical and affixal negations) might be derived. Any claim for psychological validity of their interpretations is of course subject to the earlier caveat that it requires empirical validation. However, the proposal is attractive in that it offers a formal framework in which intuitive ideas about relative uncertainty could be understood without requiring a quantitative interpretation.

5.6 CONCLUSIONS

Probability theory is an extremely successful branch of applied mathematics. It has never, however, entirely shaken off disputes about its claim to provide the only proper framework within which to reason under uncertainty. Since Hacking's careful historical analysis of the emergence of probabilistic ideas, there is reason to be somewhat concerned that these ideas have "tended to constrain the space within which discussions ... are conducted". Attempts to escape this space have been repeatedly frustrated by the impressive ability of probability theorists (or, they might say, the extraordinary power of the theory) to achieve technical advances which appear to rebut the arguments of their critics.

Faced with such implacable sophistication it is easy to understand that many decision scientists should have accepted probability as the "correct" theory against which human judgement under uncertainty must be assessed. The consequences, it seems to me, have been that discussions have been forced into an overly restrictive framework.

AI is making significant advances in foundational theories of mental concepts. One of the most striking areas of development is in work aimed at formalizing intuitive concepts of common sense, or "everything everyone knows" as the AI theorist John McCarthy put it. Having beliefs seems to have something to do with having common sense, and consequently AI has become the latest stage on which the probability debate is being conducted. However, AI brings an entirely new set of questions and techniques to bear. It is less concerned with how to refine the current set of tools for addressing such human tasks as prediction and forecasting than with the question of what a rational, adaptive agent needs to do to cope in a complex and often poorly understood world. Its answers are far from complete, but its concepts and languages are producing many new ideas. The willingness to confront the problem of autonomy is yielding advances in our understanding of such notions as knowledge and, as I hope to have shown here, uncertainty and belief. The availability of good ideas about such concepts is fundamental to developing strong theories of intelligence, whether natural or artificial.

The technical adequacy of non-probabilistic inference, non-monotonic logic and argumentation, seem to me to be persuasive. Somewhat less convincing, at this stage of the investigation, is the utility of these ideas for understanding human judgement. I believe there is a strong case for their consideration, but perhaps more important than the specific proposals are the new freedoms conferred by the paradigm and languages of AI. Whatever the fate of the techniques discussed here decision scientists have no excuse for ignoring the opportunity to think about uncertainty and belief outside the framework of probability.

I think that people make a lot of judgements under uncertainty rather well, despite the laboratory evidence. Most of the traditions, artifacts and intellectual triumphs of humanity—including probability theory itself—have been achieved in the face of uncertainty. Their emergence is owed to intellectual processes that cannot be matched by procedures based on any current theory or technology. However, this is not an argument for the nobility of man. People make mistakes and are probably often irrational. My claim is that probability theory does not have a monopoly as a source of insight into the frailties of our judgement. The new thinking which I have credited to AI has considerable promise for releasing us from "the cycle of probability theories that has trapped us for so long".

REFERENCES

Ambler, S. (1992) A categorical approach to the semantics of argumentation. Queen Mary and Westfield College, Department of Computer Science.

Ambler, S. (1993) A categorical approach to the semantics of argumentation. Technical report 255, Imperial Cancer Research Fund, London.

Barahona, P. (1993) A causal-functional model for medical knowledge representation. In E. Keravnou (ed.) *Deep Models for Medical Knowledge Engineering*. Elsevier, Amsterdam.

Bell, J. Nonmonotonic reasoning, nonmonotonic logic and reasoning about change. *Artificial Intelligence Review*, **4**, 79–108.

Besnard, P. (1989) *An Introduction to Default Logic*. Springer, Berlin.

Carcinogenicity of Chemicals in Food (1990). Report of UK Department of Health Committee.

Chard, T. (1991) Qualitative probability versus quantitative probability in clinical diagnosis: a study using computer simulation. *Medical Decision Making*, **11**, 38–41.

Cheeseman, P. (1985) In defense of probability. In *Proceedings of the International Joint Conference on Artificial Intelligence*, pp. 1002–1009: Kaufman, Los Angeles.

Cheeseman, P. (1987) Probabilistic versus fuzzy reasoning. In L. Kanal and J. Lemmer (eds.), *Uncertainty in Artificial Intelligence*. North Holland, Amsterdam.

Clark, D.A. (1988) Psychological aspects of Uncertainty and their Implications for Artificial Intelligence (Thesis). University of Wales.

Clark, D.A. (1990) Numerical and symbolic approaches to uncertainty management in AI. *Artificial Intelligence Review*, **4**, 109–46.

Clocksin, W. & Mellish C. (1981) *Programming in Prolog* (3rd edn). Springer, Berlin.

Dawes, R.M. (1979) The robust beauty of improper linear models. *American Psychologist*, **34**, 571–582.

Dawes, R.M. (1982) The robust beauty of improper linear models in decision making. In D. Kahneman, P. Slovic & A. Tversky (eds.), *Judgement Under Uncertainty: Heuristics and Biases*. Cambridge University Press, Cambridge, pages 391–407.

Dubois, D. & Prade, H. (1988) An introduction to fuzzy and possibilistic logics. In P. Smets, E.H. Mamdani, D. Dubois & H. Prade (eds.) *Non-standard Logics for Automated Reasoning*. Academic Press, London.

Elvang-Goransson, M., Krause, P.J. & Fox, J. (1993) Dialectic reasoning with inconsistent information. *Uncertainty in Artificial Intelligence*, **8**. Morgan Kaufman, San Mateo.

Fox, J. (1986) Three arguments for extending the framework of probability. In L. Kanal & J. Lemmer (eds.) *Uncertainty in Artificial INtelligence*. North Holland, Amsterdam.

Fox, J., Barber, D.C., & Bardhan, K.D. (1980) Alternatives to Bayes: a quantitative comparison with rule-based diagnosis. *Methods of Information in Medicine*.

Fox, J. (1991) Decision making and autonomous systems. In M. Singh and L. Trave-Massuyes (eds.), *Qualitative Reasoning and Decision Support Systems*, North Holland, Amsterdam.

Fox, J. & Clark, M. (1991) Towards a formalisation of arguments in decision making. *Proc. Stanford Symposium on Argumentation and Belief*, 92–99, AAAAI.

Fox, J., Krause, P.J. & Ambler, S. (1992) Arguments, contradictions and practical reasoning. In *Proceedings of 10th European Conference on Artificial Intelligence*, Vienna.

Fox, J., Krause, P.J. & Elvang Goransson, M. (1993) Argumentation as a general framework for uncertain reasoning. *Uncertainty in Artificial Intelligence*, **8**. Morgan Kaufmann, San Mateo.

Fox, J., Clark, D.A., Glowinksi, A. & O'Neil, M. (1990) Using predicate logic to integrate qualitative reasoning and classical decision theory. *IEEE Tran. Systems, Man and Cybernetics*, **20**, 347–57.

Glymour, C. (1980) *Theory and Evidence*. Princeton University Press, Princeton.

Gordon, J. & Shortliffe, E.H. (1984) The Dempster–Shafer theory of evidence. In B.G. Buchanan & E.H. Shortliffe (eds.) *Rule-based Expert Systems: the MYCIN Experiments of the Stanford Heuristic Programming Project*. Addison-Wesley, Reading, MA.

Haack, S. (1978) *Philosophy of Logics*. Cambridge University Press, Cambridge.

Hacking, I. (1975) *The Emergence of Probability*. Cambridge University Press, Cambridge.

Heckerman, D. (1991) *Probabilistic Similarity Networks*. MIT Press, Cambridge, MA.

Hopkins, M. & Clarke, M. (1988) Non-monotonic and counterfactual reasoning: some experiments with a practical system. In B. Smith and G. Kelleher (eds.), *Reason Maintenance Systems and Their Applications*. Ellis Horwood, Chichester.

Krause, P.J., Ambler, S. & Fox, J. (1993) The development of a "logic of Argumentation". In B. Bouchon-Meunier, L. Valverde and R. Yager (eds.), *Advanced Methods in Artificial Intelligence*, Springer-Verlag, Berlin.

Krause, P.J. & Clark, D.A. (1993) *Representing Uncertainty: AI Perspectives*, Intellect Press, Oxford.

Kruse, R. & Siegel, P. (eds) (1991) *Symbolic and Quantitative Approaches to Uncertainty* (Proceedings of European Conference ECSQAU). Springer-Verlag, Berlin.

Lauritzen, S.L. & Spiegelhalter, D.J. (1988) Local computations on graphical structures and their applications to expert systems. *Journal of the Royal Statistical Society*, B. 50.

Lindley, D.V. (1971) *Making Decisions*. Wiley, London.

Lindley, D.V. (1985) *Making Decisions* (2nd edn). Wiley, Chichester (reprinted 1994).

Loui, R.P. (1987) Defeat among arguments: a system of defeasible inference. *Computational Intelligence*, **3**, 100–106.

Moore, R.C. (1988) Autoepistemic logic. In Ph. Smets, E.H. Mamdani, D. Dubois & H. Prade (eds.) *Non-standard Logics for Automated Reasoning*, Academic Press, London.

Mott, P. (1988) Default non-monotonic logic. *The Knowledge Engineering Review*, **3**(4), 265–84.

Newell, A. (1973) You can't play twenty questions in nature and win. In W.G. Chase (ed.) *Visual Information Processing*. Academic Press, New York.

O'Neil, M. & Glowinski, A.J. (1990) Evaluating and validating very large knowledge based systems. *Medical Informatics*, **15**, 237–51.

Parsons, S. (1992a) Qualitative belief networks. *Proceedings of the 10th European Conference on Artificial Intelligence*, Vienna.

Parsons, S. (1992b) Qualitative possibilistic networks. *Proceedings of the International Conference on Information Processing and Management of Uncertainty*. Palma de Mallorca.

Parsons, S. & Fox, J. (1991) Qualitative and interval algebras for robust decision making under uncertainty. In M.G. Singh and L. Trave-Massuyes (eds.) *Decision Support Systems and Qualitative Reasoning*. North Holland, Amsterdam.

Pearl, J. (1988) Probabilistic reasoning in intelligent systems: networks of plausible inference. Morgan Kaufman, San Mateo, CA.

Pollock, J.L. (1992) How to reason defeasibly. *Artificial Intelligence*, **57**, 1–42.

Rich, E. and Knight, K. *Artificial Intelligence* (2nd edn.), McGraw-Hill, New York.

Safiotti, A. (1987) An AI view of the treatment of uncertainty *The Knowledge Engineering Review*, **2**, 75–97.

Shafer, G. (1978) Non-additive probabilities in the work of Bernoulli and Lambert. *Archive for History of Exact Sciences*, **19**, 309–70.

Shortliffe, E.H. (1976) Computer based medical consultations: MYCIN. Elsevier, New York.

Smets, Ph. (1990) The transferable belief model and other interpretations of Dempster–Shafer's model. *Proceedings of 6th conference on Uncertainty in Artificial Intelligence*, Boston.

Toulmin, S. (1958) *The Uses of Argument*. Cambridge University Press, Cambridge.

Wellman, M.P. (1990) Fundamental concepts of qualitative probabilistic networks. *Artificial Intelligence*, **44**, 257–303.

Wilson, N. (1990) Rules, belief functions and default logic. In *Proceedings of the 6th Conference on Uncertainty in AI*, Boston.

Zadeh, L.A. (1978) Fuzzy sets as a basis for a theory of possibility. *Fuzzy Sets and Systems*, **1**, 3–28.

Part Two

Studies in the Psychological Laboratory

Chapter 6

Laboratory Studies of Subjective Probability: a Status Report

Lee Roy Beach and Gary P. Braun
University of Arizona

The year 1967 marks the beginning of general interest in subjective probability as an area of research. In that year, Edwards and Tversky published a collection of theoretical and empirical papers in a book called *Decision Making*, parts of which examined subjective probability. In that year also, Peterson and Beach published a *Psychological Bulletin* article called "Man as an intuitive statistician", sections of which reviewed the relatively few existing studies of subjective probability. Together, these publications identified a small, widely dispersed literature, and at about the time that they were published, psychology emerged as the official home for research on subjective probability.

The literature was small because not many researchers had yet become interested in studying subjective probability. It was diverse because the few researchers who were interested in it came from diverse disciplines. The research found a home in psychology because, as the 1960s progressed, the editors of psychology journals became increasingly willing to publish work on subjective probability. The new journal, *Organizational Behavior and Human Performance*, became the primary outlet for this research.

In 1967 the idea that probability theory might prove to be a formal and precise model of human uncertainty was very exciting. Because of their training in statistics, most behavioral scientists were familiar with probability

Subjective Probability. Edited by G. Wright and P. Ayton.
© 1994 John Wiley & Sons Ltd.

theory. Many were intrigued by the possibility that it could serve both as a tool for data analysis and as a behavioral model. The research paradigm seemed fairly straightforward: present subjects with scenarios containing events that could be described in probability terms, not unlike the problems at the end of the chapters in a statistics book. Have the subjects make judgments about the probabilities of various events in the scenarios or of events that might follow from the described events. Finally, compare the judgments to the probabilities that a statistician would derive using probability theory—that is, compare the subjects' subjective probabilities to the "correct" probabilities.

The rationale for all of this, in so far as one was required, was borrowed from Brunswik (1956). First, if it is assumed that the world is not wholly predictable, then people must, of necessity, be uncertain about future events. In order to cope effectively, they must learn to accurately evaluate that unpredictability.

The second assumption was that if people have to evaluate unpredictability, then the best way to do it is to behave like a statistician using the logic of probability theory, hence the idea of "man as an intuitive statistician". Probability theory was regarded as the "normative" (correct) way of evaluating and characterizing unpredictability and uncertainty.

Much of the interest in subjective probability came from the fact that probability is a component of the expected value model of choice. Given the assumptions outlined above, it seemed reasonable to view attempts to cope with an unpredictable environment as, in effect, attempts to gamble successfully. Therefore, maximization of expected value provided the seemingly appropriate model for coping because it was the optimal model for gambling (Bernoulli, 1738). Moreover, the psychological nature of subjective probability meshed nicely with earlier work by Savage (1954) on personal probabilities in economic theories of decision-making. Thus, by assuming that humans deal with unpredictability and uncertainty in the same way that statisticians do, one could rationalize the use of probability theory to summarize personal uncertainty, and in doing so provide a link between psychology and economics. It was intriguing. Not logically flawless, perhaps, but nevertheless intriguing.

From the beginning, research focused on examining the degree to which subjective probabilities conform to probability theory. This was complicated by the existence of two ways of viewing such conformity. One was an "objectivist" view, which follows from the assumption that humans must learn about unpredictability in the environment. Work in this vein used relative frequency as the criterion for evaluating subjective probabilities, and the subjects' problems involved things like draws from urns of red and blue poker chips, rolls of dice, draws from decks of cards, or throws of roulette wheels. The dependent variable was *accuracy*, the difference between the subjects' judgments (or subjective probabilities inferred from choices among bets, selling or buying prices, etc.) and the relative frequencies defined by the presented problem.

In contrast, the "subjectivist" view follows from the assumption that even though subjective probabilities may not mirror the environment, something like cognitive consistency should force them to be coherent. Coherent means that relationships among a subject's judgments for a set of mutually exclusive, exhaustive events should be constrained in the ways required by probability theory. For example, the judged probability of the union of two independent events should be equal to the sum of the judged probabilities of the two individual events. Subjects were asked for judgments of the probability of the union and for each of the two individual events. Then the experimenter would sum the two judgments for the individual events and compare it to the judgment for the intersection. The similarity (difference), called "*coherence*", was the dependent variable.

The most celebrated research on coherence was the "conservatism" studies by Edwards and his associates (e.g., Phillips and Edwards, 1966). This line of research was in large part responsible for the surge of interest in subjective probability. Subjects were presented with problems that required them to revise their judgments about the truth of two or more hypotheses as they observed a sequence of data that were pertinent to the hypotheses. Bayes's theorem was the normative model from probability theory and the question was whether subjects revised as much and in the same direction as prescribed by the theorem. The general finding was that revisions were in the prescribed direction but they were smaller than the prescribed amount—they were conservative. This finding, which was frequently replicated, means that the judgments are inaccurate but it does not prove that they are incoherent. Follow-up studies (e.g. Beach 1966; Wheeler & Beach, 1968) suggested that the inaccurate, conservative judgments often approximated coherence, but how close is close enough and how often is often enough? At any rate, conservatism became a "phenomenon", and was of sufficient note to merit mention in a few introductory psychology texts.

Interest in Edwards's findings soon was superseded by interest in a more general set of findings. Promoted by Kahneman and Tversky (1973), research on heuristics and biases was aimed at demonstrating that probability theory could not be regarded as the calculus of uncertainty. Its intended audience was as much economists as it was psychologists, and its message was that no theory of choice can justifiably represent human uncertainty in terms of probability theory. This was a profound statement and it had profound effects on both economic and behavioral decision theory.

The second message of the heuristics and biases results soon eclipsed the importance of the first message for many researchers. The results were broadly interpreted as an indictment of human reasoning. In its most benign form this indictment simply states that humans use heuristics to make judgments about probabilities—simple rules of thumb that make the task easier. To the degree that the heuristics fall short of the normative procedures, the resulting judgments also fall short, called bias. Research consists of comparing judgments to normative prescriptions from probability theory and inferring the heuristic that might account for the bias. In the ideal case, this is followed by studies that set up conditions that might reasonably promote use of the inferred heuristic in order to demonstrate that the biased result is in fact obtained.

In its least benign form the indictment states that human reasoning is seriously flawed, that humans are irrational, intellectual cripples. Of course, this statement assumes that probability theory equates with rationality, a colossal assumption.

The heuristics and biases research appears to have settled the question of whether probability theory can serve as the calculus of uncertainty: it apparently cannot. However, it leaves unsettled the question of what to do if one is convinced that probability theory *ought* to be the calculus of uncertainty. The problem then becomes one of knowing precisely how subjective probability differs from probability theory in order to devise methods for making the former congruent with the latter. Such methods might involve training or they might involve application of transformations to subjective probabilities so that they satisfy the requirements of probability theory. Although there has been research along these lines, there has been far less than might be expected. In fact, a surprising number of researchers who develop systems for aiding decision making simply take subjective probability judgments for granted, treating them as though they conform to probability theory. These researchers seem much more concerned about careful measurement of utilities, especially for complex multi-attributed options. This is curious because these carefully measured utilities end up being combined with highly suspect probability judgments to determine the decision that the system will recommend.

6.1 METHODOLOGICAL CRITIQUE

Because of its visibility, the heuristics and biases research necessarily becomes the focus of any examination of subjective probability research. Whether the emphasis is upon the conclusion that probability cannot serve as the calculus of uncertainty or upon the conclusion that humans are irrational, the examination tends to turn on methodology. First, and most fundamental, is the question of whether probability theory is, in fact, the appropriate standard for evaluating the adequacy of subjective probabilities, and of human reasoning. Second is the question of the degree to which the research results generalize to the cognitive processes of knowledgeable, motivated judges working at familiar, important tasks that involve uncertainty. And, third, there is the question of what new views (and old data) about subjective probability imply for future theory and research. Let us look at each of these three questions in turn.

6.1.1 Is Probability Theory an Appropriate Standard?

This question is fundamental because probability theory is a unique kind of logic and it is not at all clear that it is as broadly applicable as is sometimes assumed. To apply probability logic one must treat all events as faceless members of categories and ignore any unique information one might have about them as individual events, i.e. the events in a category must be treated as intersubstitutable (Suppes, 1957). The result is that a probability theory description of a set of events may use only part of the available information about those events. This raises a question about the conditions under which it is reasonable to ignore such additional information in order to apply probability theory to make predictions about events.

Consider the following thought experiment (Beach, Christensen-Szalanski & Barnes, 1987): Suppose that an experimenter were to *randomly* select a church in the United States and visit it one Sunday afternoon in June, the traditional month for American weddings. She stands outside and waits for the wedding party to emerge. Then she approaches the Best Man and asks, "If I were to randomly select an American couple getting married this afternoon, what is the probability that they will still be married to each other ten years from now?"

Assuming that he knows the American divorce rate, the Best Man probably would give it as his answer. However, what if the experimenter asked, "What is the probability that the newly married couple for whom you just were Best Man will still be married to each other ten years from now?"

For the experimenter, the change in question does not change the problem—she randomly selected this couple and she knows absolutely nothing about them as individuals; for her they are intersubstitutable with any other newly married American couple. But, for the Best Man, the subject of this experiment, the change of question constitutes a substantial change in the problem. The divorce rate in the population as a whole may influence his answer, but only if he is particularly cynical. (Indeed, if he really thought the divorce rate accurately described his friends' chances of success he might well have declined to serve as Best Man on the grounds that it would be a poor investment of his time.) Rather, his answer to the second question, the one that is specific to his friends as individuals, properly is based upon his specific knowledge about them and his private theory about what *causes* successful and unsuccessful marriages.

Superficially, the Best Man's answer to the second question will in some way resemble the experimenter's answer, because they will both be numbers on a scale from 0.00 to 1.00. However, they will have derived from very different processes, and it is presumptuous to condemn the Best Man's process in favor of the experimenter's. The two processes simply are different, although the Best Man's is more richly informed. More important, for both the Best Man and the experimenter the resulting probabilities reflect their uncertainty about the newlyweds' chances of marital success. In short, two very different processes provide answers to the same question and there is no reason that the answers must be the same. One cannot say that the Best Man's process or answer is wrong merely because it is different from the experimenter's.

Thought experiments are all very well but the empirical issue is whether subjects actually use knowledge-based reasoning instead of statistical (probability) reasoning in making subjective probability judgments. The empirical answer is that they can use both kinds of reasoning and which kind they use in a specific instance depends upon the circumstances.

Building upon a suggestion by Kahneman and Tversky (1982), Barnes conducted an experiment using 10 undergraduate students as subjects; subjects similar to those used in the majority of laboratory experiments on subjective probability (Barnes, 1984). She presented these subjects with 15 word problems from the subjective probability literature. Five of the problems had previously been used to demonstrate that subjects ignore sample sizes (the law of small numbers, Tversky & Kahneman, 1971), five to demonstrate that they ignore or underuse base rates (the base rate bias, Kahneman & Tversky, 1973), and five to demonstrate that they judge the probability of the conjunction of events as being higher than the probabilities of the constituent events (the conjunction fallacy, Tversky & Kahneman, 1983).

The subjects read the 15 problems and made a probability judgment while "thinking aloud" about what they were doing. The transcriptions of the 'thinking aloud' were then classified according to whether they reflected statistical reasoning, knowledge-based reasoning, a mixture of the two, or were unclassifiable. It was found that:

- For the sample-size problems, 74% of the judgments were based on statistical reasoning, 22% on knowledge-based reasoning, and 4% on a mixture or were unclassifiable.
- For the base-rate problems, 50% were based on statistical reasoning, 46% on knowledge-based reasoning, and 4% on a mixture or were unclassifiable.
- For the conjunction problems, 28% were based on statistical reasoning and 72% were based on knowledge-based reasoning.

That is, most subjects used statistical reasoning on sample-size problems, most used knowledge-based reasoning on conjunction problems, and about equal numbers used statistical and knowledge-based reasoning on the base-rate problems.

However, subjects were not rigid in their choice of reasoning. For example, on one of the five conjunction problems all of the subjects used knowledge-based reasoning while for another conjunction problem most of them used statistical reasoning. Indeed, Barnes' results conform to the conclusions reached by Nisbett *et al.* (1983) that problems that encourage recognition of chance, repeatability and the like tend to evoke statistical reasoning, and those that involve individual persons tend to evoke knowledge-based logic. For example, two of Barnes' five base-rate problems were about diseases in groups of people, and virtually all subjects used statistical reasoning for them. The other three base-rate problems involved individual persons and most of the subjects used knowledge-based reasoning.

On the basis of these results, Beach, Barnes & Christensen-Szalanski (1986) proposed the contingency model of subjective probability judgment that is described in Figure 6.1. The general idea is that judges possess a repertory of strategies for making subjective probability judgments, the primary categories being strategies based upon statistical reasoning and strategies based upon knowledge-based (epistemic) reasoning, although other kinds of strategies probably exist (soothsayers, examining the entrails of chickens, repeating judgments made by others of high-status etc.). Choice of one or another of these strategies for use in a particular task is contingent upon the characteristics of the task, primarily upon the type of judgment that is required by the context or by some other person (e.g. the experimenter). The selected strategy is then applied either more or less rigorously, depending upon the judge's motivation to produce an accurate judgment. Motivation for accuracy is contingent upon the characteristics of the environment in which the judgment task is embedded. Thus, if there is little benefit from being accurate, if the judgment can be revised easily without penalty, if the judge is not expected to be particularly competent in this sphere, or if the available information is poor, there is little motivation to be accurate and little motivation to be rigorous in application of the selected strategy. If the opposite conditions obtain, motivation to be accurate is high and the judge should be more

Strategy repertory	→	Strategy selection	→	Motivation for accuracy	→	Strategy implentation	→	Offered judgment
↑		↑		↑				
Judge's strategies: statistical & knowledge-based		Task: type of judgment required		Environment: benefit, revisibility, assumed competence, quality and amount of Information				

Figure 6.1 The Beach, Barnes and Christensen-Szalanski contingency model of subjective probability judgment

rigorous in application of the selected strategy. Different combinations of high and low on these variables results in different levels of motivation.

This model has never been tested and no comparable model has been proposed for non-laboratory subjective probability tasks. However, two recent studies speak to the third stage in the model, motivation for accuracy. Ashton (1992), using experienced auditors, found that requiring an explicit, written justification for judgments about bond ratings for 16 industrial corporations increased judgment accuracy using Moody's Investors Service bond ratings as the criterion. (Note that this criterion is empirical, not a normative model's prescription, but this is not different from using empirical relative frequencies as the criterion in earlier experiments.) The increase in accuracy came from greater consistency in the judge's processes, and it led to greater consensus among the judges. In a related study, Johnson and Kaplan (1991) had auditors assess the risk of obsolescence of 20 items in an inventory. An experimental group was instructed that their judgments would be reviewed and that they would be asked to explain them; a control group was given no such instruction. The effect was that the experimental group showed more consensus than the control group, but both groups were very consistent in the processes they used to make the judgments. It would appear that, as suggested by the model in Figure 6.1, motivation can be increased by emphasizing the negative consequences of inaccuracy (a condition, by the way, that is central to auditors' jobs).

Although other aspects of the model have not yet been tested, the model is at least an attempt to address the fact that judges clearly use different kinds of reasoning (strategies) to make subjective probability judgments and that characteristics of the task and environment appear to systematically influence which kinds of reasoning they elect to use.

All of this leaves open the question about how accurate and how coherent subjective probability judgments really are. The fact is, the question really has never been properly addressed. Let us assume that in some situation the experimenter wants the judgments to agree with probability theory. Barnes' (1984) results mean that the experimenter first must make sure that the judges are using statistical reasoning, otherwise it does not make much sense to assess accuracy by comparing their judgments to the prescriptions of probability theory. This precaution was never taken in the studies published prior to the Barnes study, and we know of no studies published since that have done so. It seems safe to assume that the subjects in most if not all studies have used a variety of strategies; different strategies by different subjects and different strategies by the same subject on different problems. However, because the data are almost always pooled for analyses, it is impossible to tell from the published results just what was going on. Perhaps those subjects who were using statistical reasoning were giving judgments very like those prescribed by probability theory. Perhaps those who were using knowledge-based reasoning were giving completely different judgments, but judgments that followed rigorously from the underlying knowledge-based logic. Group averages of the two kinds of judgment will, quite likely, reveal an apparent bias. But the source of the so-called bias is in the kind of strategies that different subjects selected (and in the experimenter's method of grouping the data), not necessarily in the judgments produced by those subjects who selected strategies that used statistical reasoning. The result is that one cannot go back to the literature and untangle things. In fact, to a large degree most if not all of those studies are useless for deciding about the very hypotheses that they were designed to test.

The Barnes research (Barnes, 1984) grouped all forms of nonstatistical reasoning into one category, called knowledge-based reasoning. It is possible however to be more specific. In a series of studies, Hendrickx (1991: see also Hendrickx, Vlek & Oppenwal, 1989, and Hendrickx, Vlek & Caljé, 1992) contrasted subjects' use of frequencies and knowledge in the generation of subjective probability assessments. The studies were done in the context of risk taking, and Hendrickx first established that subjective probability plays an important role in subjects' judgments of the riskiness of various situations. Then he examined the ways in which information is used to judge accident or loss probabilities. He concluded that,

> . . . people are sensitive to different types of risk information and . . . they use different cognitive strategies when judging accident or loss probabilities. Such judgments may be either: (a) based on information about past outcome frequencies, or (b) derived through mental simulation, i.e., the construction and evaluation of possible event scenarios, or (c) logically deduced from knowledge about relevant characteristics of the outcome-generating mechanism (page 121).

Hendrickx's results showed that when using scenarios, people judged an event to be more likely if they had available more scenarios that described how the event might occur, and that extensive, concrete scenarios had a greater effect on increased probability judgments (and decreased risk-taking behavior) than did brief, abstract scenarios. Moreover, the less personal control afforded by a scenario the more risky the situation was judged to be. Finally frequency information was dominated by scenario information, and frequency information was ignored altogether when subjects could logically infer the relevant probabilities through their understanding of the process that would generate the event in question.

Together, the Barnes results and the Hendrickx results demonstrate that subjects have various strategies for assessing probabilities. In Hendrickx's words,

> Our findings underscore the constructive nature of human probability assessment... Apparently, people do not possess some standardized 'mental

> algorithm' for probability assessment, which is blindly applied every time a probability assessment is required. Instead, probability judgments appear to result from several, highly divergent mental processes ("cognitive strategies") which people apply in a flexible manner; dependent on the informational conditions—and probably also on other task-specific factors...—different strategies are applied to the task (page 122).

He further concludes that probability assessment deriving from different cognitive strategies may have different relevance and different meaning to people. These conclusions, as well as those that follow from Barnes' results, point out the necessity for a more thorough description of the various strategies people use and for the identification of the conditions that promote or suppress the use of the strategies. "The latter point is crucially relevant if we want to overcome the low predictive value of the results of the 'heuristic approach'..." (Hendrickx, 1991, page 122).

6.1.2 How Generalizable are the Heuristics and Biases Results?

The second question arising from an examination of the methodology used in subjective probability research is about the degree to which the research results generalize to the cognitive processes of knowledgeable, motivated judges working at familiar, important tasks that involve uncertainty.

A great deal has been written of late about expert judgment, some of which focuses on subjective probability judgments. This is not the place to recapitulate all that has been said (see the special issue of *Organizational Behavior and Human Decision Processes*, November, 1992). Instead, we will examine one area of expertise, financial auditing, that has been the subject of a recent exhaustive review (Smith & Kida, 1991). In doing so we will explore both the empirical results on subjective probability in the auditing literature and we will return to our first question, whether probability theory is an appropriate evaluative standard in this area of expert functioning.

There is a growing body of empirical research on auditor judgment that is inspired by the psychological literature on subjective probability, particularly the heuristics and biases literature. Moreover, because the subjects in the auditing research often are practising auditors performing familiar tasks, some of the studies address the question of the generalizability of the results of the psychology studies.

We will begin by describing what it is that auditors do and why this is such an apt area for subjective probability research. Then we will summarize the results of the Smith and Kida (1991) review of the empirical literature. Finally we will discuss the appropriateness of assuming that auditors should be sensitive to base rates as an example of how real-life judgment may involve more subtle matters than can be dealt with using probability theory. The latter will bring us back to the issue of whether probability theory is always the appropriate standard for evaluating how people deal with uncertainty.

What auditors do. Most organizations are required to make periodic reports about their financial condition, called financial statements. Errors in these statements can occur because those who prepare them lack valid or pertinent information or because it is in some way beneficial to them to misrepresent the facts. To reduce the risk of material error in the statements, auditors are engaged by the organization itself (the client) to examine its statements, to review the information upon which they are based, and to evaluate the processes through which the information was obtained and compiled to arrive at the statements. The auditors then issue one of three opinions: an unqualified opinion that material error is absent, a qualified opinion (which notes possible sources of difficulty), or an adverse opinion. Or, the auditor can disclaim an opinion altogether. The client appends the opinion to its financial statements prior to distributing them to interested users, e.g., stockholders, regulatory agencies.

Waller and Felix (1984) divide the audit process into four steps:

(1) deciding to perform the audit for the client organization;
(2) gaining an understanding of the client and making a preliminary evaluation of the client's internal accounting controls;
(3) planning and execution of the audit activities;
(4) forming an opinion.

Research on subjective probability in auditing has focused on the fourth step because it is here that the auditor looks at information and revises his or her opinion about its implications for the financial statements that the client has made. The predominant research metaphor has been that of the auditor as a Bayesian statistician revising his or her prior probabilities about the presence or absence of material error in the statements in light of data about the client's internal controls and accounting procedures. The final audit opinion is seen as being determined by the posterior probabilities at the end of the audit process.

Uncertainty in auditing. Contrary to general opinion, it is not the auditors' task to decide whether the client's financial statements are or are not true. Rather, the task is to express an opinion about whether the financial statements are a fair representation of the client's financial condition. This definition of the task recognizes that there is uncertainty in the process leading to formation of the financial statements as well as of the audit opinion, and that uncertainty also exists for the user of the financial statements. Within the professional standards that guide auditors' activities and that govern the opinions they issue, the auditors' uncertainty is explicitly recognized. It is called *audit risk* and is identified as the risk that the auditor will unknowingly express an unqualified opinion about financial statements that, in fact, are not fairly presented.

Audit risk has two components, the risk that the statements actually are unfairly presented and the risk that the auditor will not detect this fact. When both components obtain, the auditors will incorrectly give an unqualified opinion. From the auditors' point of view, the detection component is of primary interest—the task is to devise tests that will detect unfair presentation if, in fact, it is present. In this the auditor is much like an experimenter whose task is to detect significant differences if they in fact exist. Because of this parallel, the use of probability logic seems to be a natural fit to the audit process.

Just as researchers seldom exhaustively sample the populations to which they wish to generalize, either because it would be too expensive or because it would be virtually impossible to obtain complete data, auditors must rely upon partial information. The sheer volume of transactions in which the client organization engages, together with the particular accounting procedures used by the client, would make any effort to obtain complete knowledge about the financial affairs of the client much too expensive. As a result, auditors must restrict their information procurement and accept the consequent uncertainty. The question is whether in learning to deal with this uncertainty auditors come to behave like statisticians.

Empirical studies of auditors. Studies of auditors' ability to deal with uncertainty have adopted the strategy of replicating heuristics and biases studies in "audit contexts". This means that auditors or accounting students have served as subjects and that the problems have been adapted to include features that relate to auditing or accounting. The bulk of the studies have examined the anchoring and adjustment heuristic or the representativeness heuristic, with a few studies examining other heuristics.

Smith and Kida (1991) report six tests of auditor use of the anchoring and adjustment heuristic. Results suggest that auditors' answers were biased in the expected (anchored) direction, suggesting use of the heuristic, but in most cases the degree of bias was less than that observed in studies that do not use experts or that do not use tasks that are highly familiar to the subjects (Butler, 1986; Joyce & Biddle, 1981a; Kinney & Uecker, 1982).

Tests of the representativeness heuristic, as reflected in an underuse of base rates, provide similar results. When the subjects were business students performing unfamiliar tasks (Johnson, 1983; Swieringa *et al.* 1976), results showed that the subjects ignored base rates, especially when unique information was available about individual events. In contrast, when auditors, rather than students, are asked to make judgments in an unfamiliar task (Joyce & Biddle, 1981b) the results showed under-use of the base rates, but they were not completely ignored. Finally, with highly experienced auditors and a highly familiar task, Kida (1984a) found that base rates were used more when they were accompanied by an explanation than when they were not. The explanation appeared to encourage the auditors to draw a causal link between the explanation (cash flows) and the base rate (failure rate in the industry) in judging the probability that a particular firm would fail. Just presenting the base rate without the causal link resulted in its not being used by the auditors. The conclusion implied by these studies is that base rates are not much used by subjects in general, but that if the base-rate information can be related to the subject's knowledge base (i.e. cash flows for auditors' theories of what causes business failures), the base rate influences the subsequent judgment.

Returning to the remainder of the Smith and Kida (1991) review: two studies showed auditors' judgments reflected greater sensitivity to sample size than is found using non-expert judges and unfamiliar tasks (Biddle & Joyce, 1979; Uecker & Kinney, 1977). Three studies found that auditors are sensitive to the reliability of information sources (Bamber, 1983; Cohen & Kida, 1989; Joyce & Biddle, 1981b). Five studies showed auditors not to be subject to confirmatory biases in information search (Anderson, 1988; Anderson & Kida, 1989; Butt & Campbell, 1989; Kida, 1984b; Trotman & Sng, 1989). And, one study (Tomassini *et al.*, 1982) found auditors to be better calibrated than are student subjects.

With the possible exception of the anchoring and adjustment results, these studies suggest that auditors' judgments deviate from normative prescriptions, but not as much as is usually observed in studies that use nonexpert subjects and relatively unfamiliar tasks. However, the exact source of the reduced mean "error" is not at all clear. That is, do these results mean that fewer of the auditors used the heuristic or do they mean that the auditors all used it less, resulting in less bias, or just what do they mean? (Indeed, what does "used it less" actually mean?) The Barnes (1984) results would suggest that some auditors were using statistical reasoning and some were using something else, perhaps the heuristic of interest. Whatever was happening, the fact remains that performance *vis* normative prescriptions is superior for auditors (experts) making judgments about events with which they are familiar.

Base rates in auditing. Kida's (1984a) finding that highly experienced auditors who were performing a familiar judgment task only took the base rate into account when it could be linked to a cause for its occurrence suggests that base rates may not normally be used in auditing in the way they are used in probability theory. The question is whether they should be used in the normatively prescribed manner or whether a more complicated use involving causal reasoning is appropriate. In what follows we will argue that normative, noncausal, reasoning is not appropriate. In so far as this is true for base rates, it may also be true for other aspects of the normative model—at least it raises that possibility.

One way of examining the appropriateness of statistical use of base rates in auditing is to contrast the auditing task with the task performed by an insurance actuary, for whom the statistical use of base rates clearly is appropriate. An actuary uses base rate information to categorize customers for the purpose of setting a premium. By examining the base rate for claims for persons or property in various categories, a premium can be devised that assures that the insurance company makes a profit in the long run. It is not unreasonable to think that an auditor might do the same thing for assaying the potential for material errors in financial statements by defining various categories of clients, kinds of internal controls, etc. However, it can be argued that doing so would be inappropriate and, to return to the heuristics and biases literature, because they have no practice, expecting auditors to be more adept at base-rate problems in experiments is equally inappropriate.

First, actuaries make judgments about the future, auditors make judgments about the past. That is, insurance is a gamble about a future event, with the insurance company betting that a claim will not be made and the insured betting that it will. In contrast, an auditor's task is to make correct judgments about past events, issuing a correct opinion about the financial statements made by the client. In light of this, it seems rational for the actuary to heed base rates and for the auditor to search for information that justifies and strengthens his or her opinion. Indeed, in the face of perceived risk the actuary can merely raise premiums to cover the potential losses. The auditor has less flexibility if only because fees sufficient to cover all possible lawsuits by those who use the financial statements become too high and clients would defect. Therefore, the auditor's major protection is provided by performance of additional substantive tests of details of the client's financial records.

Suppose for the moment that the auditor actually would like to use statistical reasoning in the audit, particularly base rates. However, as we have seen, he or she is motivated to reduce audit uncertainty in order to reduce the possibility of an incorrect opinion. In the course of performing additional substantive tests in order to accomplish this, the auditor learns a great deal that is specific to the individual client. Unless one carefully ignores part of what is learned, information of this kind acts to reduce the size of the category to which the client belongs because fewer and fewer past clients are similar enough to this client to be regarded as intersubstitutable. Indeed, the more that is learned the smaller the category becomes, until it is unique to the client alone. The base rate has no meaning when the category has only one member, because the probabilities are 1.0 or 0.0. Thus, even if statistical reasoning were preferred, the knowledge gained in the interest of raising accuracy actually serves to reduce the applicability of statistical reasoning.

Actuaries differ from auditors in yet another way. Actuaries know when they bet wrong, auditors do not. That is, a verified claim is a clear indication of having lost the bet with the insured, and it is in the interests of the insured to notify the insurance company when the company has lost the bet. Each loss is added to the records kept by the insurance industry, and these records are consulted when future bets are made. In contrast, it often is not in the best interests of auditors' clients to inform the auditing firm when an unqualified opinion is in fact wrong. Quite the opposite. Moreover, records are not kept, so they cannot be consulted for future audits. Generally, the only feedback auditors get about their errors is from lawsuits brought by disgruntled users of the financial statements. Even lawsuits are not infallible feedback because the verdict is based on which side presents the best argument rather than upon whether or not the auditor's opinion actually was correct—if the auditor can defend the opinion, he or she may win the lawsuit even if the opinion was in fact wrong. In view of this, it is not at all clear what base rates even mean in an auditing context and how an auditor would go about assessing them if he or she wanted to use them.

Lawsuits, as feedback or otherwise, are a prominent fact of life for auditors. Insurance companies may have to justify their categorizations (proving that they do not discriminate against this or that group), but auditors must justify their opinion in court. Causality, not probability, is the language of the courtroom. Auditors must keep in mind the possibility that they will have to justify their procedures and conclusions, and justification will have to be in terms of why they did what and what the results meant. It is doubtful that a defence couched in terms of Bayes's theorem would sway a jury, or be admitted as a legitimate form of argument.

Experience in laboratory settings. Experience can be acquired in many ways. We assume that the auditors in the studies reviewed by Smith and Kida (1991) acquired theirs by participating in numerous audits. Following Hendrickx (1991), perhaps their experience was reflected in mental records of the frequencies of various features of audits. Perhaps it was reflected in scenarios about how clients behave, constructed as a result both of observation and of sharing stories with fellow auditors. Perhaps it was represented as a mental model of the processes by which various kinds of audit information are generated. Perhaps its representation contained elements of all three, but for the moment, let us focus on the first—mental records of frequencies as a reflection of experience.

Much of the research on subjective probability has involved presenting subjects with rather complicated problems. Conservatism usually was studied by telling subjects that there were urns filled with various proportions of red and blue balls; the task was to estimate the probability of a particular urn being the one from which a displayed sample was drawn. Heuristics and biases were studied by presenting subjects with a written paragraph in which the various probabilities were given; the task was to estimate the probability of some event deriving from the information in the paragraph. The point is, the relevant information was given to the subjects—they did not experience it themselves.

It should not be thought that the results of studies of "man as an intuitive statistician" have uniformly shown people to be inept. As Christiansen-Szalanski and Beach (1984) found, the results are more evenly split than one might imagine—it is just that poor performance is cited more frequently than good performance. The question is, of course, what differentiates between the two kinds of results? Space does not permit a literature review that might answer this question. However, in the interest of prompting such a review (by someone) in the future, we submit the following nonrandom sample of studies that may suggest an answer.

Early in his career, the senior author (Beach) found himself repeatedly embarrassed by the inability to find compelling evidence that subjects were rotten intuitive statisticians. Study after study obtained results that were at variance with everything else in the then-growing literature. But, in retrospect, these studies can be seen to have one feature that makes them different from the majority of other studies, a feature that was quite overlooked at the time. In these studies, subjects were given experience with the events before they were asked to assess probabilities.

For example, Beach and Peterson (1966) presented a random sequence of flashing lights—seven different lights that each flashed a prespecified proportion of the time in a sequence of flashes. Then the subjects assessed the probability for each light and for unions of pairs of lights. The assessments for the single lights were compared to the true proportions, and because the sequence was fairly short they were not particularly accurate (although they were not far off). Then the subjects' assessments for the unions were predicted using their assessments for the individual lights, and there was very high agreement; subjects evidenced coherence among their subjective probabilities.

Using the same seven lights, Beach and Phillips (1967) found that subjects' assessments of the probabilities correlated 0.92 (slope = 0.918) with the lights' proportions after 300 flashes. Moreover, when subjects made bets about which of two specified lights would occur first if the sequence of lights continued, the subjective probabilities inferred from the bets for each light correlated 0.93 (slope = 1.06) with the previously assessed probabilities for the individual lights and 0.90 (slope 0.722) with the lights' proportions. Again, the subjects' assessed subjective probabilities were highly coherent.

Beach (1966) presented subjects relative frequencies via cards in decks and had them revise their subjective probabilities about which deck a sample was drawn from. The experiment was complex, but the upshot was that even when the revised probabilities were not accurate in relation to the deck compositions, they were internally coherent and suggested that revisions were not conservative. In a similar vein, Wheeler and Beach (1968) taught subjects about sampling distributions for urns of different compositions (0.40/0.60 and 0.80/0.20) by showing them sequences of samples from the urns. A subsequent revision task found that conservatism was almost eliminated as a result of the training.

More recently, Christensen-Szalanski and Beach (1982) taught subjects frequencies for a problem that was drawn from the heuristics and biases literature. Subjects saw a series of 35 mm slides which represented suspected cases of disease. Some slides indicated that the patient had the disease, some indicated that he or she did not. This was crossed with either positive or negative test results. After seeing the slides the subjects were told that in a city of 100 000 people there are 7000 who have contracted the disease. A test for the disease is positive for 80% of the people who have the disease and is negative for 80% of the people who do not have the disease. If the test were given to all the people in the city, what is the probability that a person with a positive test has the disease? In this experiment, subjects' answers were very close to the correct answer, 0.23, in contrast to the usual finding (0.80) when subjects had not seen the frequencies.

Finally, Barclay and Beach (1972) examined the combinatorial properties of subjective probabilities for problems that college students were likely to understand easily. Thus, for example, subjects were asked for the probability that the big fountain on campus would be turned on if they were to walk over to the plaza. And they were asked the probability that Mt Rainier would be visible from the plaza if they were to go look. Then they were asked the probability that *both* the fountain would be on and the mountain would be visible. Data analysis consisted of summing the two probabilities for the single events and comparing them with the probability for the union. This strategy was used to examine various problems involving (1) unions of mutually exclusive events, (2) unions of nonexclusive events, (3) intersections of independent events, and (4) intersections of nonindependent events. The respective mean correlations, slopes, and intersections across individual subjects were:

(1) $r = 0.94$, slope = 0.94, intercept = 0.01;
(2) $r = 0.85$, slope = 0.98, intercept = 0.05;
(3) $r = 0.85$, slope = 1.00, intercept = 0.08;
(4) $r = 0.86$, slope = 1.02, intercept = 0.05.

The correlations are high, but the most surprising results are how close the slopes are to 1.00 and how close the intercepts are to 0.00, both of which are necessary conditions for establishing identity between the combinations of the assessed probabilities of the elementary events and the assessed probabilities of the union or intersection of the events. In short, subjects' probability assessments were coherent. Moreover, there was no evidence of the so-called "conjunction fallacy" in the assessments of probabilities for intersections of events.

6.2 CONCLUSIONS

Our conclusions can best be presented by reviewing the questions that guided the discussion.

Is probability theory an appropriate standard for evaluating the adequacy of subjective probabilities, and of human reasoning? The answer to this question depends upon the domain to which it is applied. If the domain is mutually agreed upon by the experimenter and judge to be one that is properly addressed by probability theory, then probability theory is indeed the appropriate standard. Unfortunately, mutual agreement seldom has obtained because no effort was made on either side to assure that it did. The result is that while experimenters generally have assumed that the domain covered by the problems they gave judges was properly (and, usually, solely) addressed by probability theory, the judges frequently have assumed otherwise. This means that much of the empirical "evidence" that is claimed to show that subjective probabilities are biased and that judges are irrational simply misses the point. Moreover, it probably is impossible to go back and sort things out because the data are a hopeless tangle; some subjects assumed the tasks to require statistical reasoning and others assumed them to require knowledge-based reasoning, and one cannot tell in retrospect which are which.

The results of studies that have examined judgment strategies have, however, provided a new and potentially important way of thinking about subjective probability. Barnes (1984) and Hendrickx (1991), and others no doubt, provide evidence that subjective probability assessments can derive from knowledge-based reasoning. Hendrickx (1991) showed that subjects' stated willingness to take risks was related to these knowledge-based subjective probabilities, implying that they in fact reflect perceived riskiness, confidence and similar varieties of subjective uncertainty. Indeed, the research agenda for the future might profitably focus on judges' use of casual logic in the context of scenarios as a way of generating predictions of future events and as a way of evaluating uncertainty.

The second question is about the degree to which the results of subjective probability research generalizes to the cognitive processes of knowledgeable, motivated judges working at familiar, important tasks that involve uncertainty. There is a considerable body of research on this question, but we have focused on the performance of experienced auditors who are presented problems that are similar to the problems they face in the course of their occupation (Smith & Kida, 1991). The results strongly suggest that the "heuristics and biases" results are greatly attenuated by experience and task familiarity. Moreover, in laboratory experiments, naive subjects can be given experience with the frequencies of the events in question, with the result that their performance is close to the prescriptions of normative theory.

It is important to note in this latter case that when subjects are taught frequencies, their answers to subsequent questions are based upon repartitioning the set of observations rather than upon a cognitive counterpart of the normative computations. Of course, partitioning also is what the calculus of probability does, but there is a meaningful difference in the operations involved. Imagine that you have a set of marbles, each of which has a label. If the experimenter asks you what proportion of the marbles are labeled B or Y or G, it is a simple matter to segregate the marbles into the required classes and estimate the proportions. Conditional probabilities (which is what Bayes' theorem is all about) would simply involve repartitioning the classes of marbles according to the labels and order of precedence dictated by the statement of conditionality. In neither case need you have recourse to the mathematical operations prescribed by normative probability theory.

Similarly, partitioning and repartitioning of frequentistic events in memory is different from the abstract logical task implied by the processes involved in deriving probabilities and conditionalities using the equations of probability theory. The former uses the judge's experience, the latter uses his or her knowledge of probability theory, and the two may not be applied with equal ease to probability assessment. Recent research by Gigerenzer (1991), as well as the frequentistic laboratory studies cited above, suggests that when the familiar subjective probability assessment problems are rephrased as partitioning of frequentistic events with which the judge has some familiarity, the results look quite different from those that generally are reported. Indeed, Gigerenzer's (1991) results indicate that the biases that supposedly arise from the celebrated heuristics disappear or be greatly diminished when the problems are posed in terms of frequencies.

Many researchers have lost interest in subjective probability as a research topic because they regard the major issues as settled by the heuristics and biases studies; probability is not the calculus of uncertainty and judges' rationality is flawed. However, as we have seen, things are not quite so simple. Subjective probability may not necessarily follow probability theory, but it still is interesting for its own sake. Knowledge-based reasoning as well as statistical reasoning may give rise to subjective probability, and those probabilities may well account for some of the variance in human behavior, particularly in risk assessment (e.g. Hendrickx, 1991). Moreover, it may well turn out that judges are frequentists; that they are able to use frequentistic information in ways that have yet to be fully explored (e.g. Gigerenzer, 1991). In short, a non-normative approach to the study of uncertainty and subjective probability may well prove to be profitable, broadening and enriching our understanding of judgment and decision making.

REFERENCES

Anderson, B. (1988) An examination of factors affecting auditors' hypothesis testing strategies. Unpublished doctoral dissertation, University of Massachusetts, Amherst.

Anderson, B. & Kida, T. (1989) An examinations of the strategies used by auditors to test inherited hypotheses. Working paper, Boston University.

Ashton, R.H. (1992) Effects of justification and a mechanical aid on judgment performance. *Organizational Behavior and Human Decision Processes*. **52**, 292–306.
Bamber, E.M. (1983) Expert judgment in the audit team: A source reliability approach. *Journal of Accounting Research*, **21**, 396–412.
Barclay, S. & Beach, L.R. (1972) Combinatorial properties of personal probabilities. *Organizational Behavior and Human Performance*, **8**, 176–83.
Barnes, V.E. (1984) *The quality of human judgment: An alternative prospective*. Unpublished Ph.D. Dissertation, University of Washington, Seattle.
Beach, L.R. (1966) Accuracy and consistency in the revision of subjective probabilities. *IEEE Transactions on Human Factors in Electronics*. **HFE-7**, 29–37.
Beach, L.R., Barnes, V.E. & Christensen-Szalanski, J.J.J. (1986) Beyond heuristics and biases: A contingency model of judgmental forecasting. *Journal of Forecasting*, **5**, 143–57.
Beach, L.R., Christensen-Szalanski, J.J.J. & Barnes, V.E. (1987) Assessing human judgment: Has it been done, can it be done, should it be done? In G. Wright & P. Ayton (eds.), *Judgmental Forecasting*. Wiley, Chichester.
Beach, L.R. & Peterson, C.R. (1966) Subjective probabilities for unions of events. *Psychonomic Science*, **5**, 307–8.
Beach, L.R. & Phillips, L.D. (1967) Subjective probabilities inferred from estimates and bets. *Journal of Experimental Psychology*, **75**, 354–9.
Bernoulli, D. (1738) Specimen theoriae novae de mensura sortis. *Commentarii Academiae Scientarum Imperialis Petropolitanae*, **5**, 175–92.
Biddle, G. & Joyce, E. (1979) The role of sample size in probabilistic inference in auditing. Unpublished manuscript, University of Chicago.
Brunswick, E. (1956) *Perception and the Representative Design of Experiments*. University of California Press, Berkeley, CA.
Butler, S. (1986) Anchoring in the judgmental evaluation of audit samples. *The Accounting Review*, **61**, 101–11.
Butt, J. & Campbell, T. (1989) The effects of information order and hypothesis-testing strategies on auditors' judgments. *Accounting, Organizations and Society*, **14**, 471–9.
Christensen-Szalanski, J.J.J. & Beach, L.R. (1982) Experience and the base-rate fallacy. *Organizational Behavior and Human Performance*, **29**, 270–8.
Christensen-Szalanski, J.J.J. & Beach, L.R. (1984) The citation bias: Fad and fashion in the judgment and decision literature. *American Psychologist*, **39**, 75–8.
Cohen, J. & Kida, T. (1989) The impact of analytical review results, internal control reliability and experience on auditors' use of analytical review. *Journal of Accounting Research*, **27**, 263–76.
Edwards, W. & Tversky, A. (1967) *Decision Making*. Penguin, Harmondsworth.
Gigerenzer, G. (1991) How to make cognitive illusions disappear: Beyond "heuristics and biases." In W. Stroebe & M. Hewstone (eds.), *European review of social psychology (vol. 2)*. Wiley, Chichester.
Hendrickx, L. (1991) How versus how often. Doctoral dissertation, University of Groningen, The Netherlands.
Hendrickx, L., Vlek, C. & Caljé, H. (1992) Effects of frequency and scenario information on the evaluation of large-scale risks. *Organizational Behavior and Human Decision Processes*, **52**, 256–75.
Hendrickx, L., Vlek, C. & Oppenwal, H. (1989) Relative importance of scenario information and frequency information in the judgment of risk. *Acta Psychologica*, **72**, 41–63.
Johnson, V.E. & Kaplan, S.E. (1991) Experimental evidence on the effects of accountability on auditor judgments. *Auditing*, **10**, 96–107.

Johnson, W. B. (1983) "Representativeness" in judgmental predictions of corporate bankruptcy. *The Accounting Review*, **63**, 78–97.
Joyce, E. & Biddle, G. (1981a) Anchoring and adjustment in probabilistic inference in auditing. *Journal of Accounting Research*, **19**, 120–45.
Joyce, E. & Biddle, G. (1981b) Are auditors' judgments sufficiently regressive? *Journal of Accounting Research*, **19**, 323–49.
Kahneman, D. & Tversky, A. (1972) Subjective probability: A judgment of representativeness. *Cognitive Psychology*, **3**, 430–54.
Kahneman, D. & Tversky, A. (1973) On the psychology of prediction. *Psychological Review*, **80**, 237–51.
Kahneman, D. & Tversky, A. (1982) On the study of statistical intuitions. *Cognition*, **11**, 123–141.
Kida, T. (1984a). The effect of causality and specificity on data use. *Journal of Accounting Research*, **22**, 145–52.
Kida, T. (1984b) The impact of hypothesis-testing strategies on auditors' use of judgment data. *Journal of Accounting Research*, **22**, 332–40.
Kinney, W.R., Jr. & Uecker, W. (1982). Mitigating the consequences of anchoring in auditor judgments. *The Accounting Review*, **63**, 55–69.
Nisbett, R., Krantz, D., Jepson, C. & Kunda, A. (1983) The use of statistical heuristics in everyday inductive reasoning. *Psychological Review*, **90**, 339–363.
Peterson, C.R. & Beach, L.R. (1967). Man as an intuitive statistician. *Psychological Bulletin*, **68**, 29–46.
Phillips, L. & Edwards, W. (1966). Conservatism in a simple probability inference task. *Journal of Experimental Psychology*, **72**, 346–54.
Savage, L.J. (1954). *The Foundations of Statistics*. Wiley, New York.
Smith, J.F. & Kida, T. (1991). Heuristics and biases: Expertise and task realism in auditing. *Psychological Bulletin*, **109**, 472–85.
Suppes, P. (1957). *Introduction to Logic*. Von Nostrand, Princeton, NJ.
Swieringa, R., Gibbins, M., Larsson, L. & Sweeney, J.L. (1976). Experiments in the heuristics of human information processing. *Journal of Accounting Research*, **4** (Suppl.), 159–87.
Tomassini, L., Solomon, I., Romney, M. & Krogstad, J. (1982) Calibration of auditors' probabilistic judgments: Some empirical evidence. *Organizational Behavior and Human Performance*, **30**, 391–406.
Trotman, K. & Sng, J. (1989). The effects of hypothesis framing, prior expectations and cue diagnosticity on auditors' information choice. *Accounting, Organizations and Society*, **14**, 565–76.
Tversky, A. & Kahneman, D. (1971). The belief in the law of small numbers. *Psychological Bulletin*, **76**, 105–10.
Tversky, A. & Kahneman, D. (1983). Extensional versus intuitive reasoning: The conjunction fallacy in probability judgment. *Psychological Review*, **90**, 293–315.
Uecker, W. & Kinney, W. (1977). Judgmental evaluation of sample results: A study of the type and severity of errors made by practicing CPAs. *Accounting, Organizations and Society*, **2**, 269–75.
Waller, W.S. & Felix, W.L. (1984). Cognition and the auditor's opinion formulation process: A schematic model of interactions between memory and current audit evidence. In S. Moriarity & E. Joyce (eds.), *Decision Making and Accounting: Current Research*. University of Oklahoma Press, Norman, OK.
Wheeler, G. & Beach, L.R. (1968). Subjective sampling distributions and conservatism. *Organizational Behavior and Human Performance*, **3**, 36–46.

Chapter 7

Why the Distinction between Single-event Probabilities and Frequencies is Important for Psychology (and Vice Versa)

Gerd Gigerenzer
University of Chicago

Some years ago in Stanford I was lunching with a motley group of colleagues, mostly psychologists and economists, all interested in judgment under uncertainty. We gnawed our way through our sandwiches and through the latest embellishments of the prisoners' dilemma, trading stories of this or that paradox or stubborn irrationality. Finally, one economist from Princeton concluded the discussion with the following dictum: "Look," he said with conviction, "either reasoning is rational *or* it's psychological."

This forked opposition between the rational and the psychological has haunted me ever since. Frege scholars will hear in it an echo of the nineteenth-century debate between the logician Frege and the psychologist Wundt over the status of the "laws of thought"; the economists and psychologists seated at the picnic table with me that afternoon had in mind the more recent findings of the "heuristics and biases" research program in cognitive psychology (e.g. Tversky & Kahneman 1974, 1983). Certainly anyone acquainted with only this aspect of contemporary psychology—and it remains among the best publicized, both to colleagues in other disciplines and to the public at large—could

Subjective Probability. Edited by G. Wright and P. Ayton.
© 1994 John Wiley & Sons Ltd.

easily have come to think that psychology is about revealing and explaining human irrationality. The *conjunction fallacy*, the *base-rate fallacy*, the *overconfidence bias*—this was the gloomy litany of sins people seemed to commit routinely and incorrigibly against reason. According to the exponents of the "heuristics and biases" program, human beings were programmed to be systematically, stubbornly *irrational* when making judgments under uncertainty—at least, most of the time. (Experimental subjects were not dazzling at logical thinking either, but that is another story and another research program.) No wonder the psychology of reasoning had become nearly synonymous with the investigation of the irrational (you get a taste from Table 7.1).

What exactly did it mean to be irrational, according to the psychologists of the heuristics and biases program? Let me use a well-known example, the "Linda problem". Assume you are a subject in a psychological experiment. In front of you is a text problem and you begin to read:

> Linda is 31 years old, single, outspoken and very bright. She majored in philosophy. As a student, she was deeply concerned with issues of discrimination

Table 7.1 A sample of conclusions from the *heuristics and biases* program

In making predictions and judgments under uncertainty, people do not appear to follow the calculus of chance or the statistical theory of prediction. Instead, they rely on a limited number of heuristics which sometimes yield reasonable judgments and sometimes lead to severe and systematic errors.

Daniel Kahneman & Amos Tversky (1973, page 237)

It appears that people lack the correct programs for many important judgmental tasks. ...we have not had the opportunity to evolve an intellect capable of dealing conceptually with uncertainty.

Paul Slovic, Baruch Fischhoff & Sarah Lichtenstein (1976, page 174)

The genuineness, the robustness, and the generality of the base-rate fallacy are matters of established fact.

Maya Bar-Hillel (1980, page 215)

The biases of framing and overconfidence just presented suggest that individuals are generally affected by systematic deviations from rationality.

Max Bazerman & M.A. Neale (1986, page 317)

[Overconfidence bias] has proved so robust that it is hard to acquire much insight into the psychological processes producing it.

Baruch Fischhoff (1988, page 172)

[We are] a species that is uniformly probability-blind, from the humble janitor to the Surgeon General ... We should not wait until A. Tversky and D. Kahneman receive a Nobel prize for economics. Our self-deliberation from cognitive illusions ought to start even sooner.

Massimo Piattelli-Palmarini (1991, page 35)

> and social justice, and also participated in antinuclear demonstrations. Which of these two alternatives is more probable?
>
> (a) Linda is a bank teller
> (b) Linda is a bank teller and active in the feminist movement.

Which alternative would you choose? Assume you chose (b), just as most subjects—80% to 90%—in previous experiments did. Tversky and Kahneman (1983) argue: (b) is the conjunction of two facts, namely that Linda is a bank teller and is active in the feminist movement, whereas (a) is one of the conjuncts. Because the probability of a conjunction cannot be greater than that of one of its conjuncts, the correct answer is (a), not (b). Therefore, your judgment is recorded as an instance of a celebrated reasoning error, known as the *conjunction fallacy*. Tversky, Kahneman, and others have shown that this type of judgment is highly stable across experimental manipulations. By analogy to stable visual illusions, stable reasoning errors, such as the conjunction fallacy, have been labeled *cognitive illusions*. The standard conclusion is that the mind does not possess the proper statistical algorithms, but relies on non-statistical quick-and-dirty algorithms, such as the *representativeness heuristic*. That is, the mind assesses the probability by calculating the similarity between the description of Linda and each of the alternatives, and chooses the alternative with the highest similarity. Judging probability by similarity has been termed the *representativeness heuristic*.

This alleged demonstration of human irrationality in the Linda Problem has been widely publicized in psychology, philosophy, economics and beyond. Stephen J. Gould (1992, page 469) puts the message clearly:

> I am particularly fond of [the Linda] example, because I know that the [conjunction] is least probable, yet a little homunculus in my head continues to jump up and down, shouting at me—"but she can't just be a bank teller; read the description." ... Why do we consistently make this simple logical error? Tversky and Kahneman argue, correctly I think, that our minds are not built (for whatever reason) to work by the rules of probability.

In what follows I will argue that Gould should have had more trust in the intuition of his homunculus.

Two aspects of the standards of rationality versus irrationality assumed by this and other celebrated experiments cry out for closer inspection.

(1) *The distinction between single-event probabilities and frequencies.* In the supposed demonstrations of the conjunction fallacy, the base rate fallacy, and the overconfidence bias, rationality is characterized not simply by probability theory, but some particular *interpretation* of probability theory, often a narrow version of Bayesianism. In the Linda problem, probability theory

is applied to a single event—that Linda is a bank teller—rather than to frequencies. Does probability theory apply to single events?

This is a controversial matter amongst probabilists, who have long and heatedly debated the merits of subjective Bayesian versus objective frequentist interpretations of probability. The influential Bayesian Leonard J. Savage (1954), for instance, introduced his notion of personal probability with everyday examples of reasoning about *singular* events: "I personally consider it more probable that a Republican president will be elected in 1996 than that it will snow in Chicago sometime in the month of May, 1994. But even this late spring snow seems to me more probable than that Adolf Hitler is still alive" (page 27). Savage's proposal challenged the frequentist schools which were then dominant, as they are in most statistics departments today. Savage was quite explicit about the deviant character of his proposal, when he added, "Many, after careful consideration, are convinced that such statements about probability to a person mean precisely nothing, or at any rate that they mean nothing precisely" (page 27).

The mathematician Richard von Mises (1957) was one of those many. In his view, a reference class (collective) has to be defined first, and then the probability of a *repetitive* event is the relative frequency of this event in its class. One of his examples is the probability of death at age 40, as determined from the data of insurance companies. The class is "all men insured before reaching the age of forty after complete medical examination and with the normal premium". The number of deaths at age 40 was 940 out of 85 020, which corresponds to a relative frequency of about 0.011. This probability is attached to a class, but not to a particular person or a single event. Every particular person is always a member of many different classes, whose relative frequencies of death may have different values. Therefore, von Mises concluded, "It is utter nonsense to say, for instance, that Mr. X, now aged forty, has the probability 0.011 of dying in the course of the next year." (pages 17–18).

By now it should be clear that according to a strong frequency view of probability (e.g. Neyman, 1977; von Mises, 1957), what has been labeled the conjunction fallacy is *not* an error in probabilistic reasoning. In this view, probability theory is about frequencies and simply doesn't apply to single events.

(2) *Content-independent rationality*. There is a peculiar indifference in this standard of rationality to background knowledge: rationality here means the deployment of formal algorithms (or rules, such as the conjunction rule) which are content-independent. That is, it is assumed that they can and should be applied to tasks with different specific contents, provided the formal structure remains constant. From this point of view, rationality is not bound to any specific domain, and knowledge ideally is irrelevant to proper reasoning.

In the Linda problem, for instance, whatever you know about bank tellers and feminists is assumed to be entirely irrelevant; indeed, you need not read the description of Linda at all—it is irrelevant to content-independent rationality. Hence there is little analysis of how the content of a problem cues the understanding of the term "probable" in this research tradition. "Probable" can refer to *typical*, *prototypical*, *frequent*, *credible*, to the *weight of evidence*, to a *plausible causal story*, or to what "may in view of present evidence be reasonably expected to happen," as the *Oxford English Dictionary* informs us. Most of these uses do not obey the laws of probability. For instance, judgments of typicality do *not* follow the conjunction rule. Betty Friedan may count as a typical *feminist writer*, but not as a typical *writer*. Such psychological considerations, however, are not part of the content-independent rationality that defines right and wrong reasoning in the heuristics and biases program.

These two issues are not independent. For instance, suppose one insists that every single-event statement involving the term "probable," as in the Linda problem, must obey the laws of probability theory rather than, say, the guidelines of the *Oxford English Dictionary*. (I take the statements in Table 7.1 to epitomize this conviction.) One would then be uninterested in how content (add physical and social context, goals, if you want) determines what is reasonable in a given situation.

I will focus in this chapter on the distinction between single-event probabilities and frequencies, and will say little about the role of content in understanding what is rational (on the latter see Cosmides & Tooby, 1992; Gigerenzer, 1991; Gigerenzer & Hug, 1992).

In the first part, drawing on recent work in the history of probability, I will show that the distinction between single-event probabilities and frequencies was dependent on theories of mind: the meaning of probability changed when theories of mind changed. In the second part, drawing on recent experimental work, I will show that apparently stable cognitive illusions are dependent on the distinction between single-event probabilities and frequencies: cognitive illusions tend to disappear when single-event probabilities are changed into frequencies. Thus, I argue that the conceptual distinction between single-event probabilities and frequencies is of direct relevance for psychology, and *vice versa*.

7.1 HOW THEORIES OF PSYCHOLOGY SHAPED THE MEANING OF PROBABILITY

According to legend, probability is one of the few seminal ideas that has an exact birthday. In 1654, precisely three hundred years before Savage's treatise, the now famous correspondence between Blaise Pascal and Pierre Fermat first cast the calculus of probability in mathematical form. Ian Hacking (1975) argued that the notion of probability that emerged so suddenly was

Janus-faced from the very beginning. One face was aleatory, concerned with observed frequencies (e.g. co-occurrences between fever and disease, comets and deaths of kings); the other face was epistemic, concerned with degrees of belief or opinion warranted by authority. In his view, the twentieth-century duality between objective frequencies and subjective probabilities existed then as it does now. Barbara Shapiro (1983) and Lorraine Daston (1988), however, have argued that probability in the seventeenth and eighteenth centuries had more than Janus's two faces. It included physical symmetry (e.g. the physical construction of dice, now called "propensity"); frequency (e.g. how many people of a given age died annually); strength of argument (e.g. evidence for or against a judicial verdict); intensity of belief (e.g. the firmness of a judge's conviction in the guilt of the accused); verisimilitude; and epistemological modesty, among others. Over the centuries, probability also conquered new territory and created further meanings, such as in quantum physics, and lost old territory, such as the probability of causes (Daston, 1988). Rather than Janus's two faces, probability seems more like a group of visages loosely assembled in a family portrait, with some members joining over time and others dropping out.

7.1.1 The Unity: Frequencies *and* Subjective Beliefs

The puzzling fact about the Enlightenment probabilists is *the ease with which they slid from one meaning of probability to the next*—and this holds independently of whether you see probability as Janus-faced or more like a family portrait. This ease created the apparent paradox that competing present-day interpretations of probability could claim the same work as their ancestor. Jakob Bernoulli's *Ars conjectandi* (1713), for instance, has been variously claimed as anticipating the 20th century's subjective interpretation, Rudolf Carnap's logical interpretation, and the extreme frequentist interpretation of Jerzy Neyman and Richard von Mises (Hacking, 1975, pages 15–16).

The solution to this puzzle lies in the intimate link between psychology and probability. Daston (1988, Chapter 4) argues that only with hindsight does it seem that Bernoulli and other classical probabilists vacillate between objective and subjective interpretations. Whereas today these interpretations look incompatible to many, the classical probabilists were able to reconcile the subjective and objective facets of probability on the basis of the theories of mind advanced by John Locke, David Hartley, and David Hume. The following account of how associationist psychology shaped and incorporated ideas of probability is a condensed version of Daston's (1988) detailed study.

Philosophers such as Hartley and Hume, and mathematicians like Condorcet and Laplace, treated associationist psychology and mathematical probability as kindred topics. Following Locke's associationism, Hume held that the mind unconsciously and automatically tallied frequencies and proportionated degree of belief (for Hume, the *vivacity* of an idea). And Hume insisted that the psychological mechanism that converted frequency into belief was finely tuned: "When the chances or experiments on one side amount to ten thousand, and on the other to ten thousand and one, the judgment gives the preference to the latter, upon account of that superiority." (Hume, 1739/1975, page 141). Despite his reservation about the validity of induction, Hume made probabilistic thinking the *de facto* standard of reasonableness. Hume linked frequency with belief, but his account contained almost no reference to the mathematical theory of probability. David Hartley's (1749) work did. He combined elements from Locke's sketch of associationism and Newton's physiological speculations concerning the vibratory basis of sensations, and worked it into a full-blown associationism that connected the laws of mind with the laws of probability. Repeated associations created cerebral vibrations until grooves of mental habit were etched in the brain. Through this physiological mechanism, human judgment, when undeflected by strong emotion or passion, imitated the law of large numbers.

The list of psychological mechanisms underlying the mapping of objective frequencies into subjective belief, postulated from Locke to Hartley to Laplace, seems surprisingly familiar to a contemporary psychologist: observed frequencies are transformed into degrees of belief through "traces", "vibrations", "interior images", and "impressions". All these mechanisms assumed the passive, automatic and unconscious mapping of experienced frequencies into subjective probabilities. Being built up from frequencies, degrees of belief were considered to be rational. The Enlightenment empiricists had taken due notice of the distortion of rational belief through passion and interest, but they believed these were corrigible aberrations.

These psychological theories were the backbone of what is now known as the *classical interpretation of probability* (from 1660 to c. 1840) and they explain some of its central features. First, classical probability conflated subjective belief and objective frequencies, based on associationist psychology. Second, probabilities were epistemic, a figment of human ignorance and therefore subjective, not part of the physical world. Classical probabilists, from Jakob Bernoulli through Laplace, were arch determinists (Daston, 1992). God, or Laplace's secularized demon, could dispense fully with probability. However, we humans are, as John Locke put it, most of the time condemned to live in the twilight of probability rather than the noonday sun of certainty. Although the world itself is deterministic, human cognition is inherently probabilistic and empirical in its working—a view that was revived, among others, in Egon Brunswik's (1955) functional probabilism. Third, the mapping of frequencies into subjective probabilities was considered to be rational, as were subjective probabilities, unless disrupted by passion or interest. The Enlightenment probabilists cherished the fiction of the *hommes éclairés*, an elite of educated people who could prevent such disruptions from affecting their beliefs. Probability theory mirrored their reasoning and provided a tool for those unfortunates needing help to stay clear of these disruptions. Human reasoning and probability theory were two sides of the same coin. In Laplace's famous phrase, probability theory was nothing more than "good sense reduced to a calculus".

By the time Siméon-Denis Poisson (1837) published his major work on probability, the classical interpretation was under attack on several fronts. The psychological theories postulating mechanisms that guaranteed the proportioning of belief to frequencies had given way to those that emphasized the illusionary nature of human belief. Etienne de Condillac (1754) was one of the first to express misgivings about the reliability of the link between frequency and belief. In his psychology, wishful thinking became the rule rather than the exception. Condillac was preoccupied with pathological associations caused by experiences early in life, by prejudice, or by brain consistency. He held, for instance, that young girls were prone to confuse chimeras for realities, because their brains were soft and even faint associations left permanent impressions in a soft medium. Condillac and his followers shifted the associationist psychology of Hume and Hartley to a psychology in which needs, wants, and temperaments (and other sources of pathologies) determined how the mind distributed attention, which in turn organized experience (Daston, 1988). The unity between frequency and belief was slowly eroded. What psychology had given to probability, it now took away. Poisson was the first to distinguish clearly in print, in 1837, between the *subjective* and the *objective* meaning of probability.

There is a broader intellectual and social context in which the rise and the fall of the classical interpretation of probability is embedded. The French revolution and its aftermath seems to have shaken the confidence of the probabilists in the existence of a single shared standard of reasonableness. What constituted "good sense" was no longer self-evident. The consensus and the values of the intellectual and political élites fragmented and disappeared, as did *l'homme éclairé*, the fiction of the reasonable man who embodied this consensus (Daston, 1988; Gigerenzer *et al.*, 1989, Chapter 1).

7.1.2 The Divorce: Frequencies *versus* Subjective Belief

Subjective belief and objective frequencies began as equivalents and ended up as diametric opposites. Poisson had distinguished the two, and the political economist and philosopher Antoine Cournot (1843) seems to have been the first who went one step further and eliminated subjective belief from the realm of mathematical probability: mathematical probability was not a measure of belief. Only then did it become evident that the classical interpretation of probability had been an *interpretation*. Classical probability was a form of "mixed mathematics", a term stemming from Aristotle's explanation of how optics and harmonics mixed the forms of mathematics with the matter of light and sound. Classical probability theory had no existence independent of its subject matter—the beliefs of reasonable men. The modern view that a mathematical theory might indeed exist independently of a particular subject matter—the distinction between formal theory and application—was foreign to mixed mathematics. Arguably, mathematical probability did not free itself from its particular applications until very recently, when in 1933 A.N. Kolmogoroff presented his axiomatization of probability.

The new associationist psychology which focused on illusions had, by the early nineteenth century, provided the arguments for severing subjective probabilities from objective frequencies, and, ironically, associationist psychology from probability theory. By about 1840, *l'homme éclairé* had given way to *l'homme moyen*. Probability was no longer about mechanical rules of rational belief embodied in an élite of reasonable men, but about the properties of the average man (*l'homme moyen*), the embodiment of mass society if not mediocrity. Adolph Quetelet's (1835) *social physics* determined the statistical distributions of suicide, murder, marriage, prostitution, height, weight, education, and almost everything else in Paris, and compared these with the distributions in London or Brussels. The means of these distributions defined the fictional *average man* in each society. The means and rates of moral behaviors, such as suicides and crimes in Paris or in London, proved to be strikingly stable over the years; this was cited as evidence that moral phenomena are governed by the laws of a society rather than by the free decisions of its individuals. In nineteenth-century France, statistics became known as "moral science".

This new focus on mass phenomena had a tremendous impact on pioneer sociologists such as Herbert Spencer and Emile Durkheim, and shaped demography, insurance, epidemiology, Prussian bureaucracy, the debates on free will, Francis Galton's enthusiasm for the normal curve and Gustav Theodor Fechner's statistical aesthetics, *inter alia* (Hacking, 1990; Stigler, 1986). Quetelet's model of human behavior as erratic and unpredictable at the individual level, but governed by statistical laws and predictable at the level of society, was independently adopted by James Clerk Maxwell and Ludwig Boltzmann to justify, by analogy, their statistical interpretation of the behavior of gas molecules (Porter, 1986). By this strange route, physics became revolutionized through the analogy with statistical laws of society.

Throughout most of the nineteenth and twentieth centuries, the "probabilistic revolution" (Krüger, Daston & Heidelberger, 1987; Krüger, Gigerenzer & Morgan, 1987) was about frequencies: from the kinetic theory of gas to quantum statistics, and from population genetics to the Neyman–Pearson theory of hypothesis testing. The urn model of classical probability was now concerned with these mass phenomena, excluding subjective degrees of belief such as single-event probabilities. Joseph Bertrand in his *Calcul des probabilités* (1889), for instance, criticized Laplace's applications of Bayes' theorem to calculate degrees of belief: we believe the sun will rise tomorrow because of "the discovery of astronomical laws and not by renewed success in the game of chance" (page xliv).

As is well known, subjective probability has regained acceptance in the second half of this century with the pioneering work of Bruno de Finetti and Frank Ramsay in the 1920s and 1930s and Leonard Savage in the 1950s. The reasonable man, once exiled from probability theory, had his comeback. Economists, psychologists, and philosophers now struggle again with the issue of how to codify "reasonableness" in mathematical form—the same issue once abandoned by mathematicians as a thankless task. Before the 1970s, the return of subjective probability still provoked a particularly lively debate between frequentists and subjectivists (now called "Bayesians"). Today, both sides pretend to know each other's arguments all too well and seem to have stopped listening. Frequentists dominate statistics and the experimental sciences; subjectivists the areas of theoretical economics and artificial intelligence. The territory has been divided up. As Glenn Shafer (1989) complained, "conceptually and institutionally, probability has been balkanized" (page 15).

To summarize: theories of psychology have been important in shaping the meaning of probability, and therewith the subject matter of probability theory. In particular, the associationist psychology of Locke, Hume and Hartley provided the grounds for *not* distinguishing objective frequencies and subjective degrees of belief—from the inception of probability theory *circa* 1650 to roughly 1840. The turn of associationist psychology towards illusions dethroned the reasonable man of classical probability theory and made the distinction between degrees of reasonable belief and frequencies obvious. After this conceptual transformation, psychology found itself dissociated from probability theory, too.

7.2 HOW THE DISTINCTION BETWEEN SINGLE EVENTS AND FREQUENCIES AFFECTS COGNITIVE ILLUSIONS

Psychologists like precise birthdays too. Textbooks celebrate 1879 as the beginning of what is referred to as *scientific psychology*, when Wilhelm Wundt devoted some space at the University of Leipzig for conducting experiments. For Wundt, the experimental method was a means to study elementary cognitive processes, such as attention and perceptual thresholds, but not (what he believed to be) deeply culture-bound processes such as thinking (Danziger, 1990). For these and other reasons, such as the dominance of American behaviorism, probabilistic reasoning was only occasionally a topic for psychologists in the first half of this century.

The classical probabilists would have felt a strong sense of *déjà vu* upon learning about some of the theoretical developments in the second half of the twentieth century. Around 1950, Jean Piaget in Geneva revived the reasonable man of classical probability theory. In Piaget and Inhelder's (1951/1975) experimental work, the formal laws of probability are the laws of the adolescent and adult mind. Errors in probabilistic reasoning were characteristic only during ontogenetic development, until the age of fourteen or so, when formal probabilistic reasoning emerges. Take for instance the law of large numbers. In 1703 Jacob Bernoulli had written in a letter to Leibniz that the law of large numbers is a rule that "even the stupidest man knows by some instinct of nature *per se* and by no previous instruction" (see Gigerenzer et al., 1989, page 29). More than two centuries later, Piaget and Inhelder concluded that even twelve to thirteen year olds intuitively apply the law of large numbers and understand the reasons for the law (page 207).

Locke, Hartley, and Hume had assumed that the mind unconsciously tallies frequencies and converts them into degrees of belief. Hasher and Zacks (1979) concluded from their experiments that frequencies are one of the few kinds of information (the others being word meaning and spatial and temporal location) that are monitored *automatically*—that is, without intention or much attention, and without interfering with other tasks. Moreover, what is now called *automatic frequency processing* seems to be generally accurate, a conclusion independently arrived at by others (e.g. Brehmer & Joyce, 1988). The thesis that objective frequencies eventually shape degrees of belief has now been experimentally demonstrated (Hasher, Goldstein & Toppino, 1977). Locke, Hartley, and Hume would have been enthusiastic about these experimental findings. The reasonable man is back, dressed in modern fashion: less élite (everyone is a reasonable intuitive statistician) and confirmed by numerous experimental results.

The *déjà vu*, however, goes beyond the recreation of the reasonable man. Around 1970, much of cognitive and social psychology turned away from the rational intuitive statistician and focused on illusions (Kahneman, Slovic & Tversky, 1982; Nisbett & Ross, 1980). One and a half centuries earlier, associationist psychology had turned to illusions, and the reasonable man had crumbled along with the classical interpretation of probability. Now illusions were again being used to destroy belief in the rational *homo sapiens*, and to challenge economists' rational *homo economicus*. Now, as then, illusions were no longer the exception, but the rule.

Here the historical parallels end. The old challenge was that passion and wishful thinking almost always interfere with the rational laws of thought. Freud's attack on human rationality is a well-known variation on that theme. The unconscious wishes and desires of the Id are a steady source of intrapsychic conflict that manifests itself in all kinds of irrational beliefs, fears, and behaviors. The new challenge, however, does not invoke passion or wishful thinking as interfering with otherwise rational reasoning. This challenge is stronger: the human mind does not possess the proper statistical algorithms. Poor reasoning is seen as a straightforward consequence of the laws of human reasoning, which are non-statistical "rule-of-thumb" heuristics. The mind is a poor intuitive statistician whether or not passion and wishful thinking compound this state of affairs.

Ironically, the departure point of the *unreasonable man* that emerged two decades after Piaget's revival of the *reasonable man* was Savage's neo-Bayesianism. In the 1960s, Ward Edwards and his colleagues at the University of Michigan made two related proposals. First, Edwards, Lindman & Savage (1963) attempted to persuade experimental psychologists to turn Bayesian and to dispense with frequentist hypothesis testing. Second, Edwards (1968) proposed to study empirically whether intuitive reasoning follows Bayesian statistics. The first proposal fell still-born from the press; the second became a raging success.

Experimenters already had their frequentist statistics, a curious and confused mishmash of Fisher's significance testing and Neyman–Pearson's hypotheses testing (Gigerenzer, 1993). This was generally presented as the *sine qua non* of scientific method. Textbooks did not tell their readers that they were teaching a shotgun marriage between Fisher and Neyman–Pearson. Rather, the textbooks created the illusion that "statistics is statistics is statistics". Since the 1950s, statistical inference had become a mechanical ritual in psychology and beyond, enforced by journal editors and internalized by researchers as the guardian of objectivity and scholarly morality. Bayesianism, by contrast, looked subjective and, above all, unnecessary.

Thus, in the 1970s and 1980s, Bayesianism became a rational yardstick for the *subjects* in psychological experiments, but not for the *experimenters* who analyzed them. Subjects were judged rational if their inferences from data to hypotheses followed Bayes' theorem; otherwise their judgments were recorded as errors in reasoning, such as the *base rate fallacy* (see below). However, when experimenters made inferences from data to hypotheses—here, whether subjects are Bayesians—they did *not* use Bayes' theorem. They used, as they had been taught for two decades before Edwards' proposal, frequentist statistics. But the most commonly used kind of frequentist statistics, R.A. Fisher's significance testing, does not use prior probabilities or base rates. This neglect of base rates by experimenters was not recorded as an error in reasoning, although it had all the characteristics of the base rate fallacy. Nor do I know of a single experimenter who noticed and remarked on that amazing double standard. The split between Bayesians and frequentists not only divides disciplines today, but can also go right through a single individual (Gigerenzer, 1993).

Edwards seems to have soon become dissatisfied with pointing out discrepancies between subjects' reasoning and Bayes' formula, and no interesting and rich theory of how subjects actually do reason had emerged. He turned to the task of designing tools that help people reason the Bayesian way. In the 1970s, Amos Tversky and Daniel Kahneman took over Edwards' second proposal and turned it into what is now known as the *heuristics and biases* program.

The heuristics and biases program arrived at a view of human rationality (Table 7.1) diametrically opposed to that of classical probability theory. Yet this modern program neglects the distinction between single-event probabilities and frequencies just as the classical probabilists ignored the distinction between subjective degrees of certainty and objective chances.

I will now use the distinction between single-event probabilities and frequencies to unearth the reasonableness hidden by the perspective of the heuristics and biases program.

7.2.1 Representation of Information: Single-event Probabilities versus Frequencies

My point here is precisely *not* to champion one side over another—frequentism over Bayesianism, or vice versa—but to point out a connection between the single-event probabilities/frequencies distinction and a second distinction, that between algorithms and information representation.

Much ink flowed in debates about mental *algorithms*: is the mind equipped with the right statistical algorithms or only with suboptimal algorithms based on rules of thumb such as the representativeness heuristic? These two alternatives, however, are not sufficient for a theory of cognitive processes underlying judgment under uncertainty, because they only deal with the level of algorithms. Algorithms need *information*, and information needs *representation*. This distinction between algorithms and information representation is central to David Marr's (1982) analysis of visual information-processing systems.

For example, consider *numerical* information. This information can be represented by the Arabic numeral system, the binary numeral system, Roman numerals, and other symbol systems. These different representations can be mapped *one-to-one onto each other*, and are in this sense formally equivalent representations. But they are not necessarily equivalent for calculating algorithms. The algorithms programmed into my pocket calculator work well when I feed them Arabic numerals, but not when I feed them binary numbers. The human mind seems to acquire analogous preferences for one form of representation: contemplate for a moment long division in Roman numerals.

Let me now return to the distinction between single-event probabilities and frequencies. Instead of squabbling over which captures the "real" meaning of probability, let us instead regard them as two different representations of probability information. Finer distinctions can be made, but this will suffice to start.

An evolutionary speculation links these two distinctions. Assume that some capacity or algorithm for statistical reasoning has been built up through evolution by natural selection. For what information representation would such an algorithm be designed? Certainly not for percentages and single-event probabilities (as is assumed in many experiments on human reasoning), since these took millennia of literacy and numeracy to evolve as tools for communication. Rather, in an illiterate and innumerate world, the representation would be frequencies of events, sequentially encoded as experienced—for example, *3 out of 20* as opposed to 15% or $p = 0.15$. Such a representation is couched in terms of discrete cases, that is, natural numbers.

Note that bumblebees, birds, rats, and ants all seem to be good intuitive statisticians, highly sensitive to changes in frequency distributions in their environments, as recent research in foraging behavior indicates (Gallistel, 1990; Real & Caraco, 1986). One wonders, reading that literature, why birds and bees seem to do so much better than humans.

In short, the proper functioning of a mental algorithm depends on the way in which information is represented. So, to analyze probabilistic reasoning, we must attend to the difference between, at least, the frequency and the single-event representation of probability. If evolution has favored one of these forms of representation, then it would be frequencies, which prelinguistic organisms could observe and act on.

Attending to this distinction suffices to make several apparently stable cognitive illusions disappear.

7.2.2 How to Make the Conjunction Fallacy Disappear

Now we apply the distinction between single-event and frequency information representation to the Linda problem. We only change the format from single event to a frequency representation, leaving everything else as it was.

> Linda is 31 years old, single, outspoken and very bright. She majored in philosophy. As a student, she was deeply concerned with issues of discrimination and social justice, and also participated in antinuclear demonstrations.
>
> *There are 100 people who fit the description above. How many of them are*:
>
> (a) bank tellers
> (b) bank tellers and active in the feminist movement.

Subjects are now asked for frequency judgments rather than for single-event probabilities. If the mind solves the Linda problem by using a representativeness heuristic, changes in information representation should not matter, because they do not change the degree of similarity. The description of Linda is still more representative of (or similar to) the conjunction "teller and feminist" than of "teller." Subjects therefore should still commit the conjunction fallacy.

However, if there is some statistical algorithm in the mind *that is adapted to frequencies* as information representation, then something striking should happen to this stable cognitive illusion. Violations of the conjunction rule should largely disappear.

Table 7.2 How to make the *conjunction fallacy* disappear

Linda problem	Conjunction violations (%)
Single-event versions	
Tversky & Kahneman (1983)	
Which is more probable?	85
Probability ratings	82
Probability ratings T^*	57
Betting	56
Fiedler(1988)	
Probability ranking, Exp. 1	91
Probability ranking, Exp. 2	83
Hertwig & Gigerenzer (1993)	
Probability ranking	88
Frequency versions	
Fiedler (1988)	
How many out of 100?	22
How many out of X?	17
Hertwig & Gigerenzer (1994)	
How many out of 200?	13
How many	16

Note: The various versions of the Linda problem are (i) which is more probable (see text, $n = 142$), (ii) probability ratings on a nine-point scale ($n = 119$), (iii) probability ratings using the alternative "Linda is a bank teller whether or not she is active in the feminist movement" (T^*) instead of "Linda is a bank teller" (T) ($n = 75$), (iv) hypothetical betting, i.e. subjects were asked "if you could win $10 by betting on an event, which of the following would you choose to bet on?" ($n = 60$). Fiedler asked subjects to rank order T, T&F, and other alternatives with respect to their probability. In his first frequency version the population size was always 100, in the second it varied ($n = 44$ and 23, in Experiments 1 and 2, respectively). Hertwig & Gigerenzer asked subjects to rank order T, T&F, and F, with respect to their probability (single-event version, $n = 24$), or estimate the frequency of T, T&F, and F (in the two frequency versions, each $n = 25$). In one of the frequency versions, the number of women was specified (200); in the other, this number was not specified.

The available experimental evidence confirms this prediction. Klaus Fiedler (1988) reported that the number of conjunction violations in the Linda problem dropped from 91% in the original, single-event representation to 22% in the frequency representation. A similar result was found when he replaced "there are 100 people" by some odd number such as "there are 168 people." The drop in the number of conjunction violations here was from 83% to 17%. Hertwig and Gigerenzer (1994) used three alternatives: F (Linda is active in the feminist movement), T&F (Linda is a bank teller and active in the feminist movement) and T (Linda is a bank teller). In the single-event task, subjects rank-ordered F, T&F and T with respect to their probability; in the frequency task, they estimated the frequency of T, T&F and F ("how many out of 200?"). The percentage of conjunction violations dropped from 88% in the single-event task to 13% and 16%, respectively, in two frequency tasks.

Hertwig and Gigerenzer as well as Fiedler reported similar results for other reasoning tasks from which the conjunction fallacy has been inferred as a stable cognitive illusion. Tversky and Kahneman (1983) had reported a similar case in their original paper, but maintained the claim that people commit a fallacy when choosing the conjunction in the single-event case.

To summarize: The philosophical and statistical distinction between single events and frequencies clarifies that judgments hitherto labeled instances of the "conjunction fallacy"' cannot be properly called reasoning errors in the sense of violations of the laws of probability. The conceptual distinction between single-event and frequency *representations* suffices to make this allegedly stable cognitive illusion largely disappear. The conjunction fallacy is not the only cognitive illusion that is subject to this argument.

7.2.3 How to Make the Base-rate Fallacy Disappear

In the 1960s, Ward Edwards and his colleagues designed probability revision problems to find out whether their subjects were Bayesians. Many of these problems used the tried-and-true urns-and-balls problems, and the major finding was that subjects exhibited *conservatism*—that is, that they seemed to give too much weight to the base rates. From the 1970s on, however, Tversky, Kahneman, and many of their followers claimed that reasoning deviates from Bayes' rule in the opposite direction, that subjects in fact ignore base rates—the so-called *base-rate fallacy*.

Recently, some researchers have weakened their claims about the generality and robustness of the base-rate fallacy, but some of the fundamental confusions with which this stimulating research was burdened from the very start have survived (Gigerenzer & Murray, 1987, ch.5).

The two confusions I will point out are both instances of blurring single-event probabilities and frequencies. The first confusion is between the Bayesian notion of a person's *prior probability* and the frequentist concept of a *base rate*. Tversky and Kahneman (e.g. 1974; Kahneman & Tversky, 1973) started out using the expressions "neglect of base rates" or "insensitivity to base rates" interchangeably with those of "neglect of prior probabilities" or "insensitivity to prior probabilities." However, priors and base rates are different things. Priors are subjective degrees of belief that may be informed by objective base rates, but need not be identical. (Similarly, the subjective likelihoods that enter Bayes' theorem and the "individuating" information presented by the experimenter need not be identical; see Birnbaum, 1983; Schum, 1990). This confusion, however, was necessary to argue that if a subject does not give much weight to whatever base rate information the experimenter has presented, this counts as a demonstration of a fallacy, i.e., that the subject does not reason by Bayesian principles. Whether or not the mind actually reasons by Bayesian principles, this confusion between a *base rate* and a *subjective prior* has prevented us from drawing adequate conclusions from experimental work.

The second and related confusion is between normative theories of the subjective and frequentist varieties. For instance, when subjects seemed not to pay much attention to base rate information, Kahneman and Tversky (1973, p. 243) asserted: "The failure to appreciate the relevance of prior probability in the presence of specific evidence is perhaps one of the most significant departures of intuition from the normative theory of prediction." But *which* normative theory? They seem to have had Bayesianism in mind, and at that, a narrow version thereof—e.g., one that conflated base rates with priors (Gigerenzer, Hell & Blank, 1988). But what if intuition were measured against the frequency view?

I will now apply the distinction between single-event and frequency information representation to the base-rate fallacy . Here is an observation to start with.

Some researchers tend to *change* the representation of a problem from single-event probabilities to frequencies when they turn away from their subjects and explain the correct solution to their readers. An early example is Hammerton (1973, p. 252) who used *single-event probabilities* to communicate information to his subjects:

> 1. A device has been invented for screening a population for a disease known as psylicrapitis. 2. The device is a very good one, but not perfect. 3. If someone is a sufferer, there is a 90% chance that he will be recorded positively. 4. If he is *not* a sufferer, there is still a 1% chance that he will be recorded positively. 5. Roughly 1% of the population has the disease. 6. Mr. Smith has been tested, and the result is positive. The chance that he is in fact a sufferer is:

Hammerton seems to have been surprised that his subjects gave a median response of 85% (which is close to the 90% hit rate) despite the 1% base rate.

Such judgments have been labeled by others as the base-rate fallacy. When the author explained the correct answer to his readers, he switched however, without comment into a *frequency* representation (p. 252):

> Out of every 100 persons tested, we expect 1 to have the disease; and the device is nearly certain to say that he has. Also, out of that 100, we expect the machine to say that 1 healthy person has the disease. Thus, in the long run, out of every 100 persons tested, we expect 2 positive results, one of which will be correct and the other incorrect. Therefore the odds on any positive result being valid are roughly even.

The frequency format could be easily digested by Hammerton's readers. You can "see" that the relative frequency is one out of two (i.e. 50%), and not 85%. Hammerton's subjects, however, were tested and failed on a single-event representation.

Here is a second example. In a fascinating article on mammography, Eddy (1982) reports that he asked 100 physicians questions of the following kind:

> The prevalence of breast cancer is 1% (in a specified population). The probability that a mammography is positive if a woman has breast cancer is 79%, and 9.6% if she does not. What is the probability that a woman who tests positive actually has breast cancer?____%

Eddy (1982) reports that 95 out of 100 physicians estimated the probability $p(cancer \mid positive)$ to be about 75%. However, if one applies Bayes' theorem to the information given, $p(cancer \mid positive)$ is only about 0.08 (or 8%). The judgment of these 95 physicians once more looks like an instance of the base-rate fallacy. College students, physicians, writers of medical textbooks (Eddy, 1982), and staff at the Harvard Medical School (Casscells, Schoenberger & Grayboys 1978) all seem to have equally great difficulties with problems of this kind. Reasoning about single-event probabilities (or percentages) does not seem to come naturally to them.

Let us now perform a thought-experiment with the mammography problem. Change the information representation in the mammography problem from single-event probabilities to frequencies:

> Imagine 100 people (think of a 10×10 grid). We expect that one woman has cancer and a positive mammography. Also, we expect that there are 10 more women with positive mammographies but no cancer. Thus we expect 11 people with positive mammographies. How many women with positive mammographies will actually have breast cancer?

With frequencies, you immediately "see" that only about *1 out of 11* women who test positive will have cancer. The base-rate fallacy disappears if the information is represented in frequencies. Note that by "frequencies", I mean natural numbers. Let us now turn from the thought-experiment to real experiments.

Casscells et al. (1978) gave 60 staff and students of the Harvard Medical School the following problem, cast in *single-event probabilities* (except for the base rate):

> If a test to detect a disease whose prevalence is 1/1000 has a false positive rate of 5%, what is the chance that a person found to have a positive result actually has the disease, assuming you know nothing about the person's symptoms or signs?

If one inserts these numbers into Bayes' theorem, the posterior probability that the person actually has the disease is 0.02 (assuming the test correctly diagnoses every person who has the disease—a piece of missing information).

Most of the staff and students at Harvard Medical School were hopelessly lost—almost half estimated this probability as 0.95 not 0.02. Only 11 participants answered 0.02. Note the amount of variability in the physicians' judgments about the probability of the disease! The modal answer of 0.95 was taken to be another instance of the *base-rate fallacy*, or *base-rate neglect*, as Tversky and Kahneman (1982) called it. The base rate of the disease (1/1000) is neglected, and judgment is based only (or mainly) on the characteristics of the test (here: the false positive rate). This seemed yet more proof of the stability of the base-rate fallacy.

But I will now apply to the Harvard Medical School problem the same frequency-representation procedure I applied to the preceding problems. If there is some kind of algorithm for statistical reasoning that works on frequency representations, changing the information representation in the Harvard Medical School problem from single-event probabilities and percentages to frequencies should make the base-rate fallacy disappear. Consequently, the large variability in judgments should also disappear.

Cosmides and Tooby (in press) have tested this prediction in a series of experiments with more than 400 Stanford undergraduates. They constructed a dozen or so variations of this medical problem, substituting step-by-step frequencies for single-event probabilities. In the original single-event version, the Stanford undergraduates gave almost the same low percentage of 0.02 answers as the staff and students at Harvard Medical School, 12% compared to 18% (Table 7.3). The original single-event version was somewhat ambiguous, because the true positive rate was not specified and Stanford undergraduates might not know what the term "false positive rate" means. Therefore, Cosmides and Tooby constructed a purified single-event version in which these ambiguities were eliminated:

> The prevalence of disease *X* is 1/1000. A test has been developed to detect when a person has disease *X*. Every time the test is given to a person who has the

THE LIBRARY
BOLTON INSTITUTE OF HIGHER
EDUCATION - DEANE CAMPUS

> disease, the test comes out positive. But sometimes the test also comes out positive when it is given to a person who is completely healthy. Specifically, 5% of all people who are perfectly healthy test positive for the disease.
> What is the chance that a person found to have a positive result actually has the disease, assuming that you know nothing about the person's symptoms or signs?____%

This made the percentage of 0.02 answers go up, but only to 36%. Rather dramatic effects were obtained, however, when the single-event format was changed into a frequency format. There were two major changes, the format of the *information* (first paragraph of the single-event version) and that of the task (second paragraph). To change the format of the information from single events to frequencies, (1) all probability information was expressed in frequencies such as "50 out of 1000" instead of 5%, and (2) a reference class ("Americans") was added on which these frequencies are defined. Nothing else was changed:

> One out of 1000 Americans has disease *X*. A test has been developed to detect when a person has disease *X*. Every time the test is given to a person who has the disease, the test comes out positive. But sometimes the test also comes out positive when it is given to a person who is completely healthy. Specifically, out of every 1000 people who are perfectly healthy, 50 of them test positive for the disease.

To transform the *task* from estimating a single-event probability to estimating a frequency, a new sample of Americans was introduced, and the second paragraph of the single-event version was replaced by the following question:

> How many people who test positive for the disease will actually have the disease? ____out of____.

If our minds were not built to reason statistically, but only equipped with crude heuristics (consult Table 7.1), then the distinction between single events and frequencies should not matter. But it does. Table 7.3 shows how one can make almost everybody (or almost nobody, or any proportion inbetween) find the answer that corresponds exactly with the result of applying Bayes' theorem to the information given—that is, a probability of 0.02 or a frequency "1 out of 51," respectively. If both the information and the task were in terms of frequencies, this percentage was over seventy; if only one of the two was represented by frequencies, the percentage was inbetween the single-event and the frequency versions. If the frequency format was combined with asking the subjects to construct a pictorial frequency representation (i.e. to represent each person by a square, and mark those who do have the disease and those who test positive), then the percentage reached 92%.

Table 7.3 How to make the *base-rate fallacy* disappear: The Harvard Medical School problem (see Cosmides & Tooby, in press)

Representation of the problem	*n*	Answers in accordance with Bayes' rule (in %)
Original single-event format (Casscells et al., 1978)	60	18
Single-event format, replication	25	12
Information in frequency format, task in single-event format	25	56
Information in single-event format, task in frequency format	75	59
Information and task in frequency format	75	73
Information and task in frequency format, pictorial representation	25	92

Cosmides and Tooby's experimental variations, both in number and in detail, go beyond what I have described here. But my summary suffices to make the same point as with the conjunction fallacy. The conceptual distinction between single-event probabilities and frequencies seems to be as important for the untutored mind as it is for probability theory. It can make apparently stable cognitive illusions disappear.

These results have direct implications for teaching statistical reasoning.

7.2.4 Natural Sampling of Frequency Information

So far I have dealt with situations in which frequency information comes in one package, as in textbook problems or in newspapers. In many natural environments, and for animals or people in an illiterate world, however, frequencies must be *sequentially* learned through experience. How does an algorithm vary if we move from the standard single-event probability textbook problem to a corresponding ecological situation, in which the structure of the environment is sequentially learned through experience? Here is another thought experiment.

Let us transpose the above medical diagnosis problem to a non-literate society where physicians have to rely on their experience alone. Assume you are a physician. Your tribe has been afflicted for one year by a previously unknown and fatal disease. Everyone suspected of having the new disease is sent to you. You were lucky to discover one symptom that seems to signal the outbreak of the disease. What would it mean to be a Bayesian physician in this non-literate society?

You would encounter all information sequentially, as discrete cases that add up to frequencies. This information gathering is sometimes called *natural sampling* (Kleiter, 1993), a concept corresponding to Brunswik's (1955) *representative sampling*. So far you have seen 30 people suspected of having the disease. Ten of these turned out to have the disease, 20 did not. Of the 10 persons afflicted, 8 showed the symptom; of the 20 persons not afflicted, only 4 had the symptom. Now they bring in number 31. She has the symptom. What mental algorithm do you need in order to calculate the Bayesian posterior probability that she actually has the disease?

It turns out that in natural sampling this algorithm is quite simple—indeed, much simpler than required in those studies from which the base rate fallacy has been concluded. The algorithm needs only two absolute frequencies: the number a of people with symptom and disease, and the number b of people with symptom and no disease. These frequencies are $a = 8$ and $b = 4$, respectively. The algorithm to calculate the relative frequency $f(D \mid S)$ of people with disease D among those who have the symptom S is:

$$f(D \mid S) = \frac{a}{a+b} = \frac{8}{8+4}$$

If you are a Bayesian and want to calculate from the frequencies monitored so far the *posterior probability* $p(D \mid S)$ that patient number 31 has the disease, your mental algorithm is just as simple:

$$p(D \mid S) = \frac{8}{8+4}$$

Compare now this algorithm to that needed in the standard probability revision tasks of the heuristics and biases program. In the latter, the information is presented in terms of three single-event probabilities (forget for a moment the confusion between base rates and subjective priors): the prior probability $p(D)$, and the likelihoods $p(S \mid D)$ and $p(S \mid -D)$. For this representation of information, Bayes' theorem is:

$$p(D \mid S) = \frac{p(D)p(S \mid D)}{p(D)p(S \mid D) + p(-D)p(S \mid -D)}$$

The information (corresponding to the natural sampling condition) would be represented as $p(D) = 0.33$, $p(S \mid D) = 0.80$, and $p(S \mid -D) = 0.20$. Inserting these numbers into Bayes' theorem results in the following calculation:

$$p(D \mid S) = 0.33 \times 0.80/(0.33 \times 0.80 + 0.67 \times 0.20) = 0.67$$

The result is the same as in natural sampling, but the calculation is much more difficult.

The general point I want to make is that the way information is represented in an experiment, versus encountered in a natural environment, can require reasoning algorithms of differing complexities. Even if these algorithms are mathematically equivalent, as they are in the thought experiment just presented, they can be computationally and psychologically different. Specifically, if information is encoded through natural sampling of frequencies—as opposed to laboratory studies which present three single-event probabilities—the following differences arise:

(1) In natural sampling, memory needs to monitor only *two* kinds of information, the frequencies a and b. *No* attention need be paid to the base rates themselves.
(2) In natural sampling, Bayes' rule reduces to a simple algorithm.
(3) Frequency information, naturally sampled, carries more information than single-event probabilities. Absolute frequencies contain information about the sample size (e.g. "3 out of 20", as opposed to $p = 0.15$), which allows for computing the precision (so-called second-order probabilities) of the information (Kleiter, 1993).

I know of very few studies that have used natural sampling instead of displaying three single-event probabilities. Christensen-Szalanski and Beach (1982) represented the information in a medical diagnosis problem (similar to those described earlier) both in the single-event probability format, as usual, and by natural sampling. In the single-event version the usual results were obtained, from which the base-rate fallacy has been concluded. In the natural sampling condition, subjects were shown 100 slides, one by one. Each slide contained information about one patient: whether or not the patient had pneumonia, and whether or not the test result was positive. As in the single-event version, the task was to estimate $p(\text{pneumonia} \mid \text{positive})$. The mean estimate in the natural sampling condition was 0.22, almost identical with the actual relative frequency $f(\text{pneumonia} \mid \text{positive}) = 6/(6 + 19) = 0.24$. (Although the means were very close, there was still considerable individual variability in estimates, perhaps due in part to individual differences in monitoring the actual frequencies.)

Here is a second study. One of the best publicized demonstrations of the base-rate fallacy outside of the realm of medical diagnosis problems is Tversky and Kahneman's (1982) *Cab problem*:

> A cab was involved in a hit-and-run accident at night. Two cab companies, the Green and the Blue, operate in the city. You are given the following data:
>
> (i) 85% of the cabs in the city are Green and 15% are Blue.
> (ii) a witness identified the cab as Blue. The court tested the reliability of the witness under the same circumstances that existed on the night of the accident and concluded that the witness correctly identified each one of the two colors 80% of the time and failed 20% of the time.
>
> What is the probability that the cab involved in the accident was Blue rather than Green?

Tversky and Kahneman reported that the modal and median response of several hundred subjects was 0.80, whereas Bayes' theorem gives only 0.41. The median response is identical with the witness' hit rate—just as in some of the medical diagnosis problems—and this has been interpreted to mean that subjects neglect base rates. Bar-Hillel (1980; 1983) has tried many variations, such as presenting the base rates before or after the other information, and concluded that the base-rate fallacy was a robust phenomenon. Tversky and Kahneman (1980) suggested that the reason for this is that base rates tend not to be used unless they are seen as causal: "The proportions of Blue and Green cabs does not induce a differential propensity to be involved in accidents and this information is therefore neglected." (page 70).

Note that the problem, like those described earlier, is presented in terms of single-event probabilities and percentages. If base rates are neglected because they are not "causal", then the distinction between single-event probabilities and frequencies should not matter, since frequencies do not induce a "differential propensity to be involved in accidents", either.

Schlotterbek (1992) displayed the information in the Cab problem by means of natural sampling. In an analogy to the study by Christensen-Szalanski and Beach, 100 incidents of hit-and-run accidents were shown, one by one, using a computer display. In each case the subjects could see whether the cab was blue or green, and what the witness reported. After they had seen all 100 incidents, subjects estimated either the probability that a cab reported as "blue" is actually blue, in a new case, or the corresponding frequency f(blue | "blue"). Subjects were also asked (after they had seen all 100 cases) for their perceived four conjoint frequencies (blue cabs and report "blue", blue cabs and report "green", and so on). This allowed one to control for individual differences in perceived frequencies. Each subject's response to the frequency task was compared with the actual frequency, which was 12 out of 29, or 0.41, and with the corresponding individual frequency, calculated from the subject's reported conjoint frequencies.

Probability judgments corresponded well to the actual frequency (as in the Christensen-Szalanski & Beach study), and frequency judgments still better (median = .42; mean = .45). Half of the frequency judgments hit exactly either the actual frequency (12 out of 29) or the corresponding number calculated from subjects' perceived conjoint frequencies. Perceived conjoint frequencies were in very good correspondence with actual frequencies, with a slight overestimation of the smallest frequency (blue cabs and report "green") and underestimation for the largest (green cabs and report "green"). Neither of these two frequencies, however, is needed to solve the frequency task.

To summarize: the examples given, including sequential frequency processing, show that the distinction between single-event probabilities and frequencies is relevant to understanding how the mind reasons about a class of problems that are often termed *Bayesian probability revision* problems.

Thus far, we have seen how to make two cognitive illusions, the conjunction fallacy and the base-rate fallacy, largely disappear. I will now turn to a third prominent illusion.

7.2.5 How to Make Overconfidence Bias Disappear

Confidence in one's knowledge has been typically studied with questions of the following kind:

> Which city has more inhabitants?
>
> (a) Hyderabad
> (b) Islamabad
>
> How confident are you that your answer is correct?
> 50%, 60%, 70%, 80%, 90%, 100%

Imagine you are an experimental subject. Your task is to choose one of these two alternatives. Suppose you choose Islamabad, as most subjects in previous studies did. Then you are asked to state your confidence, or subjective probability, that your answer "Islamabad" is correct. 50% confident means guessing; 100% confident means that you are absolutely sure that Islamabad is the larger city. From a large sample of questions, the experimenter counts how many answers in each of the confidence categories were actually correct.

The major finding of some two decades of research is the following (Lichtenstein, Fischhoff & Phillips, 1982): In all the cases where subjects said, "I am 100% confident that my answer is correct", the relative frequency of correct answers was only about 80%; in all the cases where subjects said, "I am 90% confident", the relative frequency of correct answers was only about 75%; when subjects said "I am 80% confident", the relative frequency of correct answers was only about 65%, and so on. Values for confidence were systematically *higher* than relative frequencies. This systematic discrepancy has been interpreted as an error in reasoning and has been named *overconfidence bias*. Quantitatively, overconfidence bias is defined as the difference between mean confidence and percentage correct.

Is overconfidence bias really a "bias" in the sense of a violation of probability theory? Let me rephrase the question: has probability theory been violated if *one's degree of belief (confidence) in a single event* (i.e. that a particular answer is correct) is different from the relative frequency of correct answers in the long run? From the point of view of the frequency interpretation, the answer is "no". In this view, probability theory is about frequencies; it does not apply to single-event judgments such as confidences. Therefore, no statement about confidences can violate the laws of probability. Even for Bayesians, however, the answer is not "yes". The issue here is not internal consistency, but the relation between subjective probability and external (objective) frequencies, which is a more complicated issue and depends on conditions such as independence. In particular, if there is no feedback after each answer, as in this research, and if the true answers for a series of questions are dependent, one cannot expect that one's average degree of belief matches the relative frequency of correct answers. Consider, for instance, predictions of the following type: "Will there be snowfall on December 24, 1999, in downtown Boston? Yes/No." "Will there be snowfall on December 24, 1999, at Logan (Boston) airport? Yes/No." "Will there be snowfall on December 24, 1999, in Cambridge, Mass.? Yes/No." And so on. Assume, after careful consideration, that your probability that there will be snow is 0.7 in each case. Nevertheless, you cannot expect that your single-event confidences match the relative frequencies in the long run, because the outcomes are dependent. If it snows in downtown Boston, it will most likely snow in all places, and you appear to be underconfident; otherwise you will appear overconfident.

For these various reasons, a discrepancy between confidence in single events and relative frequencies in the long run should not be labeled simply an "error" in statistical and probabilistic reasoning, contrary to the claims in the heuristics-and-biases literature. It only looks that way from the perspective of a narrow interpretation of probability theory that blurs the fundamental distinction between single events and frequencies.

However, for the last two decades, many researchers have taken it for granted that any systematic difference between confidence and frequency is a reasoning error, a regrettable deviation from rationality. And they assumed that their task is to explain this discrepancy by some deficiency in our mental or motivational programming, such as a "confirmation bias" (Koriat, Lichtenstein & Fischhoff, 1980), "insensitivity to item difficulty" (von Winterfeldt & Edwards, 1986, page 128), and the tendency of humans in the Western world to overestimate their intellectual powers (Dawes, 1980). Similar to other "cognitive illusions", overconfidence bias has been suggested as an explanation for human disasters of many kinds, including deadly accidents in industry (Spettell & Liebert, 1986), errors in the legal process (Saks & Kidd, 1980) and systematic deviations from rationality in negotiation and management (Bazerman & Neale, 1986).

Many experiments have demonstrated the stability of the overconfidence phenomenon despite various "debiasing methods", such as warning subjects about overconfidence prior to the experiment or providing monetary incentives. We even used a bottle of French champagne as an incentive, but to no avail. Edwards and von Winterfeldt (1986, page 656) concluded in a tone of regret: "Can anything be done? Not much."

I will now apply the distinction between single-event probabilities and frequencies to the overconfidence bias. Take the same kind of general-knowledge questions that have been used before to demonstrate the overconfidence bias. But now let our subjects make *frequency judgments*. After our subjects answered 50 general-knowledge questions of the Hyderabad–Islamabad type, in the usual format, they also had the opportunity to judge frequencies: "How many of these 50 questions do you think you answered correctly?"

If confidence in one's knowledge were truly biased due to confirmation bias, wishful thinking, or other deficits in cognition, motivation, or personality, then the difference between a single-event and a frequency representation should not matter. Overestimation should remain stable, as it does with warnings and bribes.

Table 7.4 shows the results of two experiments (Gigerenzer, Hoffrage & Kleinbölting, 1991). If one calculates, for each subject, the difference between mean confidence (averaged over 50 questions) and the relative frequency of correct answers, one finds as usual a stable positive difference that has been called the overconfidence bias: the value +13.8 is such a difference (multiplied by 100, and averaged across the 80 subjects in the first experiment). But the interesting issue is how the frequency estimates compare with the actual frequencies.

When we compared subjects' estimated frequencies with their true frequencies, there was *no* overestimation. Frequency judgments were quite accurate. In both experiments, the mean differences were even slightly negative, indicating a tendency towards underestimation. For instance, the figure −2.4 means that in the actual set of 50 questions, the estimated frequency of correct answers was, on the average, 1.2 lower than the true frequency. Subjects missed the true frequency by an average of only about 1 correct answer in a set of 50 questions.

Note that the very same subjects appear to be overestimating their knowledge, if one blurs the distinction between single-event probabilities and

Table 7.4 How to make the *overconfidence bias* disappear (see Gigerenzer, Hoffrage & Kleinbölting, 1991)

Difference between	Experiment 1 ($n = 80$)	Experiment 2 ($n = 97$)
Mean confidence and relative frequency of correct answers ("overconfidence bias")	+13.8	+15.4
Estimated frequency and frequency of correct answers	−2.4	−4.2

Note: To make values for frequency and confidence judgments comparable, all frequencies were transformed to relative frequencies. Values shown are differences multiplied by a factor of 100.

frequencies. You may think that this difference between single-event and frequency judgment is simply due to subjects having second thoughts about their performance at the end of the experiment. We have checked this. When the sequence "confidence judgments–frequency judgment" was repeated again and again (by presenting several sets of 50 questions in a sequence), subjects consistently gave different values for confidence and frequency.

This chapter is not the place to pursue the question of how to model these striking judgments. We have developed the theory of *probabilistic mental models* (Gigerenzer, Hoffrage & Kleinbölting, 1991), which explains this and related phenomena with an algorithm that infers both confidences and frequencies from frequency information—that is, from frequency information based on different reference classes.

To summarize: I have argued that the discrepancy between mean confidence and relative frequency of correct answers, known as "overconfidence bias", is not an error in probabilistic reasoning. It only seems that from a narrow normative perspective, in which the distinction between single-event probabilities and frequencies is blurred. If we ask our subjects about frequencies instead of single-event probabilities, we can make this stable phenomenon disappear. The conceptual distinction is much more effective with our subjects than money or French champagne.

The striking effect of frequency representations on apparent violations of probability theory, as reported in this chapter, seems to generalize to so-called violations of utility theory as well. For instance, Keren and Wagenaar (1987) showed that standard violations such as the "certainty effect" and the "possibility effect" (Kahneman & Tversky, 1979) largely disappear when a single gamble is changed into a repeated gamble (see also Keren, 1991; Lopes, 1981; Montgomery & Adelbratt, 1982). It also seems to generalize to a class of phenomena known as the "illusion of control" (Langer, 1975), which largely disappears if single-event estimates are replaced by judgments about a series of events (Koehler, Gibbs & Hogarth, in press).

7.3 CONCLUSIONS

Probability theory and psychology have historically been intertwined since the Enlightenment. The psychological theories of Locke, Hume, and Hartley provided the grounds for the classical interpretation of probability, in particular for the assumption that the mind unconsciously tallies frequencies and converts them into rational degrees of belief. This created the fiction of the reasonable man (*l'homme éclairé*) and the blurring of the distinction between objective frequencies and subjective probabilities. When, by the early nineteenth century, psychological theories had shifted to illusions, the reasonable man dissolved and the difference between frequencies and degrees of belief became clear. The reasonable man gave way to the average man (*l'homme moyen*), and the frequency interpretation of probability emerged and became dominant. When the subjective interpretation—Bayesianism and subjective utility theory—regained influence in the second half of this century, these modern versions of Enlightenment rationality often did not distinguish between single-event probabilities and frequencies, nor between single and repeated gambles—just as classical probability theory had not. And many psychologists, following in these footsteps, also failed to make this distinction and found the human mind overflowing with cognitive illusions. Conflating single-event probabilities and frequencies now served the fiction of the irrational man.

Much of the current view is condensed in my economist colleague's dictum, "either reasoning is rational or it's psychological." Rationality is now defined in terms of formal algorithms or axioms, and psychology is called upon to explain the irrational. However, algorithms work on information, and information needs representation. To discuss rationality in terms of algorithms alone, good or bad ones, is incomplete if one does not pay attention to the kind of information representation that these algorithms were designed to work upon. Consequently, one cannot simply conclude from what looks like bad performance, or cognitive illusion, that there are poor algorithms. This *non sequitur* has been a basic flaw in the heuristics and biases program. When information is represented in terms of frequencies rather than single-event probabilities, apparently stable cognitive illusions tend to disappear.

ACKNOWLEDGEMENTS

This chapter is based in part on work supported by the *Fonds zur Förderung der wissenschaftlichen Forschung (P 8842MED), Austria*. I am grateful to Lorraine Daston, Berna Eden, Dan Goldstein, Ralph Hertwig, Cheryce Kremer, Jim Magnuson, and Peter Sedlmeier for many helpful comments.

REFERENCES

Bar-Hillel, M. (1980) The base rate fallacy in probability judgments. *Acta Psychologica*, **44**, 211–33.

Bar-Hillel, M. (1983) The base-rate fallacy controversy. In R.W. Scholz (ed.), *Decision making under uncertainty*. North-Holland, Amsterdam.

Bazerman, M.H. & Neale, M.A. (1986) Heuristics in negotiation. In H.R. Arkes & K.R. Hammond (eds.), *Judgment and decision making: an interdisciplinary reader*. Cambridge University Press, Cambridge, pages 311–21.

Bernoulli, J. (1713) *Ars conjectandi*. Basel.

Birnbaum, M.H. (1983) Base rates in Bayesian inference: Signal detection analysis of the cab problem. *American Journal of Psychology*, **96**, 85–94.

Brehmer, B. & Joyce, C.R.B. (eds.) (1988) *Human judgment: The SJT view*. North-Holland, Amsterdam.
Brunswik, E. (1955) Representative design and probabilistic theory in a functional psychology. *Psychological Review*, **62**, 193–217.
Casscells, W., Schoenberger, A. & Grayboys, T. (1978) Interpretation by physicians of clinical laboratory results. *New England Journal of Medicine*, **299**, 999–1000.
Christensen-Szalanski, J.J.J. & Beach, L.R. (1982) Experience and the base-rate fallacy. *Organizational Behavior and Human Performance*, **29**, 270–8.
Condillac, E. (1754/1793) *Traité des sensations*. In *Oeuvres*, Vol. 3, Paris.
Cosmides, L. & Tooby, J. (1992) Cognitive adaptations for social exchange. In J.H. Barkow, L. Cosmides & J. Tooby (eds), *The Adapted Mind*. Oxford, Oxford University Press, pp. 163–228.
Cosmides, L. & Tooby, J. (in press) Are humans good intuitive statisticians after all? Rethinking some conclusions from the literature on judgment under uncertainty. *Cognition*.
Cournot, A.A. (1843) *Exposition de la théorie des chances et des probabilités*. Paris.
Danziger, K. (1990) *Constructing the subject: Historical origins of psychological research*. Cambridge University Press, Cambridge.
Daston, L.J. (1988) *Classical probability in the Enlightenment*. Princeton University Press, Princeton, NJ.
Daston, L.J. (1992) The doctrine of chances without chance: Determinism, mathematical probability, and quantification in the seventeenth century. In M.J. Nye et al. (eds.), *The Invention of Physical Science*. Kluwer Academic Publishers, Netherlands.
Dawes, R.M. (1980) Confidence in intellectual judgments vs. confidence in perceptual judgments. In E.D. Lantermann & H. Feger (Eds.), *Similarity and choice: Papers in honor of Clyde Coombs*. Huber, Berne, pages 327–45.
Eddy, D.M. (1982) Probabilistic reasoning in clinical medicine: Problems and opportunities. In D. Kahneman, P. Slovic & A. Tversky (eds.), *Judgment Under Uncertainty: Heuristics and Biases*. Cambridge University Press, Cambridge, pages 249–67.
Edwards, W. (1968) Conservatism in human information processing. In B. Kleinmuntz (ed.), *Formal Representation of Human Judgment*. Wiley, New York.
Edwards, W., Lindman, H. & Savage, L.J. (1963) Bayesian statistical inference for psychological research. *Psychological Review*, **70**, 193–242.
Edwards, W. & von Winterfeldt, D. (1986) On cognitive illusions and their implications. In H.R. Arkes & R.R. Hammond (eds.), *Judgment and decision making: An interdisciplinary reader*. Cambridge University Press, Cambridge, pages 642–79.
Fiedler, K. (1988) The dependence of the conjunction fallacy on subtle linguistic factors. *Psychological Research*, **50**, 123–9.
Fischhoff, B. (1988) Judgment and decision making. In R.J. Sternberg & E.E. Smith (eds.), *The Psychology of Human Thought*. Cambridge University Press, Cambridge, pages 153–87.
Gallistel, C.R. (1990) *The Organization of Learning*. MIT Press, Cambridge, MA.
Gigerenzer, G. (1991) How to make cognitive illusions disappear: Beyond "heuristics and biases". *European Review of Social Psychology*, **2**, 83–115.
Gigerenzer, G. (1993) The Superego, the Ego, and the Id in statistical reasoning. In G. Keren & C. Lewis (eds.), *A Handbook for Data Analysis in the Behavioral Sciences: Methodological Issues*, Erlbaum, Hillsdale, NJ, pages 313–39.
Gigerenzer, G., Hell, W. & Blank, H. (1988) Presentation and content: The use of base rates as a continuous variable. *Journal of Experimental Psychology: Human Perception and Performance*, **14**, 513–25.

Gigerenzer, G. Hoffrage, U. & Kleinbölting, H. (1991) Probabilistic mental models: A Brunswikian theory of confidence. *Psychological Review*, **98**, 506–28.

Gigerenzer, G. & Hug, R. (1992) Domain-specific reasoning: Social contracts, cheating, and perspective change. *Cognition*, **43**, 127–71.

Gigerenzer, G. & Murray, D.J. (1987) *Cognition as Intuitive Statistics*. Erlbaum, Hillsdale, NJ.

Gigerenzer, G., Swijtink, Z., Porter, T., Daston, L., Beatty, J. & Krüger, L. (1989) *The Empire of Chance: How Probability Changed Science and Everyday Life*. Cambridge University Press, Cambridge.

Gould, S.J. (1992) *Bully for brontosaurus. Further Reflections in Natural History*, Penguin Books, London.

Hacking, I. (1975) *The Emergence of Probability*. Cambridge University Press, Cambridge.

Hacking, I. (1990) *The Taming of Chance*. Cambridge University Press, Cambridge.

Hammerton, M. (1973) A case of radical probability estimation. *Journal of Experimental Psychology*, **101**, 252–4.

Hartley, D. (1749) *Observations on Man, His Frame, His Duty, and His Expectations*. London.

Hasher, L., Goldstein, D. & Toppino, T. (1977) Frequency and the conference of referential validity. *Journal of Verbal Learning and Verbal Behavior*, **16**, 107–112.

Hasher, L. & Zacks, R.T. (1979) Automatic and effortful processes in memory. *Journal of Experimental Psychology: General*, **108**, 356–88.

Hertwig, R. & Gigerenzer, G. (1994) The chain of reasoning in the conjunction task. Unpublished manuscript.

Hume, D. (1739/1975) *A Treatise of Human Nature*. Clarendon Press, Oxford.

Kahneman, D., Slovic, P. & Tversky. A. (Eds.). (1982) *Judgment under Uncertainty: Heuristics and Biases*. Cambridge University Press, Cambridge.

Kahneman, D., & Tversky, A. (1973) On the psychology of prediction. *Psychological Review*, **80**, 237–51.

Kahneman, D., & Tversky, A. (1979) Prospect theory: An analysis of decision under risk. *Econometrica*, **47**, 263–91.

Keren, G. (1991) Additional tests of utility theory under unique and repeated conditions. *Journal of Behavioral Decision Making*, **4**, 297–304.

Keren, G. & Wagenaar, W. A. (1987) Violation of utility theory in unique and repeated gambles. *Journal of Experimental Psychology: Learning, Memory and Cognition*, **13**, 387–91.

Kleiter, G.D. (1993) Natural sampling: Rationality without base rates. In G.H. Fischer & D. Lanning (eds.), *Contributions to Mathematical Psychology, Psychometrics, and Methodology*. Springer, New York, pages 369–82.

Koehler, J.J., Gibbs, B.J. & Hogarth, R.M. (in press) Shattering the illusion of control: Multi-shot versus single-shot gambles. *Journal of Behavioural Decision Making*.

Kolmogoroff, A.N. (1933) Grundbegriffe der Wahrscheinlichkeitsrechnung. *Ergebnisse der Mathematik*, **2**(3), 196–262.

Koriat, A., Lichtenstein, S. & Fischhoff, B. (1980) Reasons for confidence. *Journal of Experimental Psychology: Human Learning and Memory*, **6**, 107–18.

Krüger, L., Daston, L. & Heidelberger, M. (eds.) (1987) *The Probabilistic Revolution. Vol. I: Ideas in History*. MIT Press, Cambridge, MA.

Krüger, L., Gigerenzer, G. & Morgan, M.S. (eds.) (1987) *The Probabilistic Revolution. Vol. II: Ideas in the Sciences*, MIT Press, Cambridge, MA.

Langer, E.J. (1975) The illusion of control. *Journal of Personality and Social Psychology*, **32**, 311–28.

Lichtenstein, S., Fischhoff, B. & Phillips, L.D. (1982) Calibration of probabilities: The state of the art to 1980. In D. Kahneman, P. Slovic, & A. Tversky (eds.), *Judgment under uncertainty: Heuristics and Biases*. Cambridge University Press, Cambridge, pages 306–34.

Lopes, L.L. (1981) Decision making in the short run. *Journal of Experimental Psychology: Human Learning and Memory*. **7**, 377–85.

Marr, D. (1982) *Vision: A Computational Investigation into the Human Representation and Processing of Visual Information*. Freeman, San Francisco.

Mises, R.v. (1957) *Probability, Statistics, and Truth*, Allen and Unwin, London.

Montgomery, H. & Adelbratt, T. (1982) Gambling decisions and information about expected value. *Organizational Behaviour and Human Performance*, **29**, 39–57.

Neyman, J. (1977) Frequentist probability and frequentist statistics. *Synthese*, **36**, 97–131.

Nisbett, R.E. & Ross, L. (1980) *Human Inference: Strategies and Shortcomings of Social Judgment*. Prentice-Hall, Englewood Cliffs, NJ.

Piaget, J. & Inhelder, B. (1975) *The Origin of the Idea of Chance in Children*, Norton, New York, (original work published in 1951).

Piattelli-Palmarini, M. (1991) Probability blindness. Neither rational nor capricious. *Bostonia*, March/April, pages 28–35.

Poisson, S-D. (1837) *Recherches sur la probabilité des jugements en matière criminelle et en matière civile*. Paris.

Porter, T.M. (1986) *The rise of statistical thinking, 1820–1900*. Princeton University Press, Princeton.

Quetelet, L.A.J. (1835) *Sur l'homme et le développement de ses facultés, ou essai de physique sociale*. Bachelier, Paris.

Real, L.A. & Caraco, T. (1986) Risk and foraging in stochastic environments: Theory and evidence. *Annual Review of Ecological Systems*, **17**, 371–90.

Saks, M., & Kidd, R.F. (1980) Human information processing and adjudication: Trial by heuristics. *Law and Society Review*, **15**, 123–60.

Savage, L.J. (1954) *The Foundations of Statistics*. Wiley, New York.

Schlotterbek, M. (1992) *Bayesian Inference Without Base Rates*. Diploma thesis, University of Tübingen, Germany.

Schum, D. (1990) Discussion. In R.M. Hogarth (ed.), *Insights in Decision Making* University of Chicago Press, Chicago, pages 217–33.

Shafer, G. (1989) *The Unity and Diversity of Probability*, Inaugural lecture, 20 November 1989, University of Kansas.

Shapiro, B.J. (1983) *Probability and Certainty in Seventeenth-century England*. Princeton University Press, Princeton.

Slovic, P., Fischhoff, B. & Lichtenstein, S. (1976) Cognitive processes and societal risk taking. In J.S. Carrol & J.W. Payne (eds.), *Cognition and Social Behavior*. Erlbaum, Hillsdale, NJ.

Spettell, C.M. & Liebert, R.M. (1986) Training for safety in automated person-machine systems. *American Psychologist*, **41**, 545–550.

Stigler, S.M. (1986) *The History of Statistics: The Measurement of Uncertainty before 1900*. Belknap Press of Harvard University Press, Cambridge.

Tversky, A. & Kahneman, D. (1974) Judgment under uncertainty: Heuristics and biases. *Science*, **185**, 1124–31.

Tversky, A. & Kahneman, D. (1980) Causal schemas in judgments under uncertainty. In M. Fishbein (eds.), *Progress in Social Psychology*. Erlbaum, Hillsdale, NJ, pages 49–72.

Tversky, A. & Kahneman, D. (1982) Evidential impact of base rates. In D. Kahneman, P. Slovic, & A. Tversky (eds.), *Judgment under Uncertainty: Heuristics and Biases*, Cambridge University Press, Cambridge, pages 153–60.

Tversky, A. & Kahneman, D. (1983) Extensional versus intuitive reasoning: The conjunction fallacy in probability judgment. *Psychological Review*, **90**, 293–315.

Winterfeldt, D. von, & Edwards, W. (1986) *Decision Analysis and Behavioral Research*. Cambridge University Press, Cambridge.

Chapter 8

Subjective Probability: What Should We Believe?

Peter Ayton
City University

and

George Wright
University of Strathclyde

"Everybody complains about their memory but no-one complains about their judgement."

La Rochefoucauld

When introductory psychology students are shown visual illusions for the first time they are often quite fascinated. They usually want to see more. They express delight when they "see" the effects and some grumble their disappointment when they fail to experience certain effects that are rather dependent on specific lighting or viewing conditions. That this is not just because the demonstrations are a welcome break from the usual drudge of lectures seems borne out by their quite different reactions to demonstrations of reasoning problems, or what some, though for different reasons (e.g. Cohen, 1981; von Winterfeldt & Edwards, 1986) call "cognitive illusions". In our experience, those that fall prey to base-rate neglect (Kahneman & Tversky, 1973) or the conjunction effect (Kahneman & Tversky, 1982a; Tversky & Kahneman, 1983) will be more likely to complain. They sometimes argue that they were misled, didn't

Subjective Probability. Edited by G. Wright and P. Ayton.
© 1994 John Wiley & Sons Ltd.

understand the problem properly or that it was all just some silly trick. The stark contrast with the look of smug glee on the faces of the few that didn't give the wrong answer is quite inverse to the reactions of those confronted with visual illusions.

The reason for this difference in reactions presumably has something to do with how people view their psychological faculties. To students on an introductory psychology course it is not always obvious that there is such a thing as perception—or that if there is, it must be both immediate and veridical. The experiencing of visual illusions serves as a vivid demonstration that there is some sort of evidently fallible process mediating between reality and their apprehension of it. However when it comes to reasoning and solving problems, students, perhaps more than most, are self-consciously aware that they have limitations. After all, they know that at the end of the course they will face a test that will be widely interpreted as a measure of each individual's intellectual calibre. A rather defensive attitude to their own rational competency seems to be behind the motivation to explain away any apparent failings. In comparison the existence of visual illusions does not threaten anyone's sense of their own "visual competence". In fact, the opposite seems to be true; visual illusions serve to impress upon people how clever the brain is to be able to make sense of the world.

The quotation printed at the head of this article gives a further example of the different reactions people have to the fallibility of different aspects of their psychological functioning. Both of our examples suggest that people seem to hold their judgement in high regard—though with each there is a hint that this may be rather unwarranted. Much of the psychological research focusing on subjective probability can be construed as input to a long running debate about whether or not human judgement is reasonable according to the various standards employed to judge it. Indeed the debate is now more often about the standards used to judge judgement than the judgements themselves (cf. Chapters 6, 7 & 10). In some respects the reactions of students confronted with their own fallibilities rather parodies the debate. There are those who argue that there is nothing really so very wrong with the human judgements underlying subjective probabilities (e.g. Gigerenzer, Chapter 7) while others claim that there are observable inadequacies (e.g. Harvey, Chapter 14). What can be concluded from these arguments? In this chapter we consider some of the evidence concerning the quality of subjective probabilities. In doing so we consider the role of the standards that have been employed to study subjective probability.

8.1 EVALUATING SUBJECTIVE PROBABILITIES

Subjective probabilities represent degrees of belief in the truth of particular propositions. For example I may be able to say that I feel 50% sure that I have an appointment at the dentist's this afternoon, or 90% sure that the Suez canal is the longest in the world or 99% sure that a person tried in court for an offence is guilty. Such probabilities are subjective in the sense that they reflect an individual's assessment based on his or her knowledge and opinions.[1] People with different knowledge and beliefs will be perfectly entitled to offer different judgements of the likelihoods attached to the same propositions. For example, it is not possible using probability theory to calculate the probability that the Suez canal is the longest in the world. Indeed, unless we are dealing with random sampling from known populations, probability theory does not offer anyone the means to compute *the* probability of uncertain events. It follows that for any particular statement there is no "correct" subjective probability.

At first it might be supposed that this would pose an insurmountable problem for anyone interested in evaluating judgements of subjective probability, for if any value is as "correct" as any other who can say what a good judgement is? However there are ways by which *sets* of subjective probabilities are constrained by axioms of probability theory and also external correspondence to facts about the world. For example the additivity axiom states that the probabilities for a mutually exclusive and exhaustive set of events (e.g. the possible winners of a horse race) must add to one. Although, insofar as probability theory is concerned, an individual is entitled to believe anything he likes about the chances of each horse winning, the total of all the probabilities he estimates for each horse winning must add to one.[2] If the subjective probabilities produced by an individual conform to an axiom of probability theory then they are said to be coherent with respect to that axiom. The state of the world can also be referred to in order to measure how well calibrated a set of subjective probabilities are (cf. McClelland & Bolger, this volume). If the statements concern some verifiable aspect of the world then they can be checked for external correspondence (cf. Yates, 1982). For example, if I claim to be 70% sure about each of a whole set of statements being true then, to be well calibrated, 70% of them must in fact be true.

We have likened the relationship between the coherence and calibration properties of subjective probabilities to the relationship between reliability and validity in psychometric test design (Ayton & Wright, 1987). For example, a personality test designed to measure, say, intelligence is said to be reliable if on different occasions it gives the same assessment of a group of individuals. Reliability is of interest as it is a prerequisite for validity; a test that proved to be unreliable couldn't systematically be providing a valid measure of intelligence. However, as many critics of intelligence tests have pointed out, of itself, reliability is no guarantee that the test really is measuring intelligence—it might be measuring some other stable trait. The validity of an intelligence test has to be determined by comparing the test results with some other measure of intelligence. In a similar way coherence and calibration of subjective probabilities are interdependent. Subjective probabilities that are incoherent with respect to an axiom of probability theory cannot systematically be well calibrated; they therefore cannot be taken as a guide to the relative likelihood of the events that they describe. However, subjective probabilities that are shown to be coherent are not necessarily well calibrated.

Why should the quality of subjective probabilities be of any concern? In the classic decision analytic framework (see von Winterfeldt & Edwards, 1986; Goodwin & Wright, 1991) numerical probabilities are ascribed to all the different events identified in a decision tree. The best alternative is selected by combining the probabilities and the utilities corresponding to the possible outcomes associated with each of the possible alternatives. *Subjective* probabilities are thus one of the prime numerical inputs into decision analysis (Raiffa, 1968), cross-impact analysis (Dalkey, 1972), fault-tree analysis (Fischhoff, Slovic & Lichtenstein, 1978) and many other management technologies. This is because actuarial statistics concerning relative frequencies of the pertinent future events will often be unavailable or may be believed to be inappropriate for current circumstances. A decision-maker may realize for example that there have been changes in the world which have some causal impact on the events being judged. Such changes would invalidate statistical methods for calculating probabilities, for example by using regression or time-series methods based on averaging techniques. And, of course, one might take the decision analytic framework as a psychological model of (unaided) human choices. Perhaps people intuitively attempt some sort of expected utility analysis in order to make their choices; plainly, subjective expectations in this context would be crucial. For example, our decision about whether or not to go on a picnic might be strongly influenced by our estimate of the likelihood of good weather.

8.2 THE APPEARANCE OF CONSERVATISM

So how good are judgemental probabilities? One early benchmark used for comparison was Bayes' theorem. Bayes' theorem defines mathematically how probabilities should be combined and can be used as a normative theory of the way in which subjective probabilities representing degrees of belief attached to the truth of hypotheses should be revised in the light of new information. In the 1960s Ward Edwards and his colleagues conducted a number of studies using the book-bag and poker-chip paradigm. A typical experiment might involve two opaque bags. Each bag contained one hundred coloured poker-chips in different but stated proportions of red to blue. One contains 70 red chips and 30 blue while the second contains 30 red chips and 70 blue. The experimenter chooses one bag at random and draws a series of chips from it. After each draw, the poker-chip is replaced and the bag well shaken before the next chip is drawn. The subject's task is to say how confident they are—in probability terms—that the chosen bag is bag 1 or bag 2.

A crucial aspect of the logic of these studies is that the experimenter is able to say what the correct subjective probabilities should be for the subjects by the simple expedient of calculating them using Bayes' theorem. All of the information required as inputs to Bayes' theorem is explicit and unambiguous. Ironically enough though this meant that the *subjectivity* of probability was not a part of the studies in the sense that the experimenters assumed that they could *objectively* compute the correct answer—which they would be able to assume should be the same for all the subjects faced with the same evidence.

The fact that the experimenter assumes he is able to calculate what the subjective probabilities *should* be for all of the subjects was absolutely necessary if one was to be able to judge judgement. However, it is also an indication of the artificiality of the task, and is at the root of the difficulties that were to emerge with interpreting the subjects' behaviour. The experiments conducted with this procedure produced a good deal of evidence that human judgement under these conditions is not well described by Bayes' theorem. Although subjects' opinion revisions were proportional to the values calculated from Bayes' rule, they did not revise their opinions sufficiently in the light of the evidence, a phenomenon that was labelled *conservatism*. The clear suggestion was that human judgement was to this extent poor, although there was some debate as to the precise reason for this. It might be due to a failure to understand the impact of the evidence or to an inability to aggregate the assessments according to Bayes' theorem. Aside from any theoretical interest in these possibilities there were practical implications of this debate. If people are good at assessing probabilities but poor at combining them (as Edwards, 1968, suggested) then perhaps they could be helped; a relatively simple remedy would be to design a support system that took the human assessments and combined them using Bayes' theorem. However, if they were poor at assessing the component probabilities then there wouldn't be much point in devising systems to help them aggregate these. "Garbage in garbage out" used to be a popular aphorism for summarizing this sort of predicament.

8.3 CONSERVATISM DISAPPEARS

Before any firm conclusions were reached as to the cause of conservatism, however, the research exploring the phenomenon rather fizzled out. The reasons for this seem to be twofold. One cause, which we consider in the next section, was the emergence of the heuristics and biases research and, in particular, the discovery of what Kahneman & Tversky (1973) called base-rate neglect. Before this development occurred, however, there was growing disquiet as to the validity of this sort of study as a model for judgement in the real world.

A number of studies had shown that there was considerable variability in the amount of conservatism manifested according to various quite subtle differences in the task set to subjects. For example, the diagnosticity of the data seemed an important variable. Imagine, instead of our two bags with a 70/30 split in the proportions of blue and red poker-chips, the bags contained 49 red and 51 blue or 49 blue and 51 red chips. Clearly, two consecutive draws of a blue chip would not be very diagnostic as to which of the two bags we were sampling from. Experiments have shown that the more diagnostic the information the less optimal is the subject. When the information is very weakly diagnostic, as in our example, human probability revision can be too extreme (Phillips and Edwards, 1966).

Another factor is the way in which the information is presented. Presenting the information about draws sequentially or all at once is irrelevant according to Bayes' theorem but Peterson, Schneider & Miller (1965) found that presenting the information one item at a time, with revisions after each item, were less conservative than those subjects who were given all the information in one go. Pitz, Downing & Rheinold (1967) described an "inertia effect", where subjects tended not to revise their probabilities downward once the initial sequence of information had favoured one of the hypotheses under evaluation.

DuCharme & Peterson (1968) attempted to investigate probability revisions in a situation they considered nearer to real life than the standard paradigm. They argued that the fact that the information was restricted to one of two different possibilities (red chip or blue chip) meant that there were very few possible revisions that could be made. In the real world, information leading to revision of opinion doesn't have discrete values but may more fairly be described as varying along a continuum. In an experimental study, DuCharme and Peterson used a hypothesis test consisting of the population of male heights and the population of female heights. The subjects' task was to decide which population was being sampled from on the basis of the information obtained by randomly sampling heights from one of the populations. Using this task, DuCharme and Peterson found conservatism greatly reduced to half the level found in the more artificial tasks. They concluded that this was due to their subjects greater familiarity with the data generating process underlying their task.

The argument concerning the validity of the conclusions from the book-bag and poker-chip paradigm was taken further by Winkler & Murphy (1973). Their paper, entitled "Experiments in the laboratory and the real world", argued that the standard task differed in several crucial aspects from the real world. Firstly the bits of evidence that are presented to the subjects are conditionally independent. That is, two or more pieces of information have an identical implication for the posterior probability to be credited to the hypotheses regardless of the order in which they are produced. Knowing one piece of information does not change the likelihood of the other; producing one red chip from the urn and then replacing it does not affect the likelihood of drawing another red chip. However, in real world probability revision this assumption often does not make sense. Suppose I see a football supporter wandering along the street wearing a blue scarf. This may well cause me to revise my opinions about which team may be visiting my home team; I would now be more confident that it would be a team that wears blue. However, the sight of another supporter also wearing a blue scarf will hardly change my views very much more—having seen the first blue scarf, the sight of another blue scarf is very much more likely.

For another example consider a problem posed by medical diagnosis. Loss of appetite is a symptom which, used in conjunction with other symptoms, can be useful for identifying the cause of certain illnesses. However, if I know that a patient is nauseous I know that the patient is more likely (than in the absence of nausea) to experience loss of appetite. These two pieces of information therefore are not conditionally independent and so, when making my diagnosis, I should not rely on the loss of appetite symptom as much as I might, in the absence of nausea, to diagnose diseases indicated by loss of appetite. Winkler and Murphy argued that in many real-world situations lack of conditional independence of the information would render much of it redundant. In the standard tasks subjects may have been treating the data as if it was conditionally dependent and so one possible explanation for conservatism is that the subjects are behaving much as they do in more familiar situations involving redundant information sources.

A second difference between experimental environments and reality was that in most experiments the data generators (the book-bags) are "stationary". The contents of the bags are fixed but in reality our hypotheses are not always constant; indeed the evidence may cause us to change the set of hypotheses under consideration. A third difference is that in reality the information may be somewhat unreliable and therefore less diagnostic than the perfectly reliable colours of the poker-chips. In support of this argument Yousseff & Peterson (1973) found that, when laboratory tasks included unreliable data, probability revision was less conservative. A fourth difference is that the typical experiments have offered very diagnostic evidence—clearly favouring one hypothesis—whereas in reality the evidence may very often be weakly diagnostic. Again the result of generalizing from experience may be the appearance of conservatism. Recall that Phillips & Edwards (1966) found that probability revision can be too extreme with very weakly diagnostic evidence. In summary, the arguments considered by Winkler and Murphy led them to conclude that "conservatism may be an artifact caused by dissimilarities between the laboratory and the real world."[3]

An early supposition from this line of experimentation was that subjective probabilities were inappropriate by virtue of the observed discrepancies with the probabilities derived from the Bayesian normative standard. However, over the decade of research that we have just described, a curious reversal of this conclusion was arrived at; now the normative standard is considered inappropriate and thereby subjective probabilities may be appropriate. As we shall see, the nature of contemporary criticisms of more recent research, which documents suboptimality in other aspects of probabilistic judgement, suggests that this cycle is about to repeat itself: Once again it is being claimed that inappropriate presumption lies behind the conclusion that poor judgement is responsible for the experimentally observed disparities between judgement and the normative standard (e.g. Beach & Braun, Chapter 6, and Gigerenzer, Chapter 7). The notion that judgement is valid and that poorly chosen experimental tasks have led to a misconceived view of poor human capability has strong advocates.

8.4 THE APPEARANCE OF BASE-RATE NEGLECT

In reporting studies of peoples' intuitions of random sampling Tversky & Kahneman (1971) commented, in a footnote, that their respondents "...can hardly be described as conservative. Rather in accord with the representation hypothesis they tend to extract more certainty from the data than the data, in fact, contain." Their discovery of base-rate neglect—the antithesis of conservatism—seems to have been the final nail in the coffin for the hypothesis that we are all conservative Bayesians. In Kahneman & Tversky's (1973) experiments demonstrating neglect of base-rates, subjects were found to ignore information concerning the prior probabilities of the hypotheses. For example, in one study subjects were presented with this brief personal description of an individual called Jack and told that the description was drawn at random from those of seventy engineers and thirty social scientists.

> Jack is a 45-year old man. He is married and has four children. He is generally conservative, careful and ambitious. He shows no interest in political and social issues and spends most of his free time on his many hobbies which include home carpentry, sailing and mathematical puzzles.

Half the subjects were told that the description had been drawn from a sample of 70 engineers and 30 lawyers while the other half were told that the description was drawn from a sample of 30 engineers and 70 lawyers. Both groups were asked to estimate the probability that Jack was an engineer (or a lawyer). The mean estimates of the two groups of subjects were only very slightly different (50% vs. 55%). On the basis of this result and others

Kahneman and Tversky concluded that prior probabilities are largely ignored when individuating information was made available.

Although subjects used the base-rates when told to suppose that they had no information whatsoever about the individual (a "null description"), a description designed to be totally uninformative with regard to the profession of an individual called Dick produced complete neglect of the base-rates.

> Dick is a 30-year-old man. He is married with no children. A man of high ability and high motivation, he promises to be quite successful in his field. He is well liked by his colleagues.

When confronted with this description, subjects in both base rate groups gave median estimates of 50%. Kahneman and Tversky concluded that when no specific evidence was given the base-rates were properly utilized; but when worthless information was given base-rates were neglected.

This phenomenon was attributed to the operation of the representativeness heuristic. Subjects were not judging by any kind of statistical reasoning but were judging the probability of the professions by the extent to which the descriptions were similar to the stereotype of the profession. Kahneman and Tversky argued that people did not engage in statistical reasoning as such but instead invoked heuristics such as representativeness to judge uncertainties. Their 1972 paper on representativeness argued that the Bayesian approach to the analysis and modelling of subjective probability did not capture the essential determinants of the judgement process. This was because "in his evaluation of evidence man is apparently not a conservative Bayesian: he is not Bayesian at all."

8.5 THE DISAPPEARANCE OF BASE-RATE NEGLECT

Later research established that base rates might be attended to more (though usually not sufficiently) if they were perceived as relevant (Bar-Hillel 1980) had a causal role (Kahneman & Tversky 1982b) or were "vivid" rather than "pallid" (Nisbett & Ross, 1980). However, Gigerenzer, Hell & Blank (1988) have argued that the real reason for variations in base-rate neglect is nothing to do with any of these factors *per se*, but because the different tasks may to varying degrees encourage the subject to represent the problem as a Bayesian revision problem. They claimed that there are few inferences in real life that correspond to Bayesian revision where a known base-rate is revised on the basis of new information. Just because the experimenter assumes that he has defined a Bayesian revision problem does not imply that the subject will see it the same way. In particular, the subjects may not take the base-rate asserted by the experimenter as their subjective prior probability. In Kahneman and Tversky's original experiments the descriptions were not of course actually randomly sampled (as the subjects were told) but especially selected to be "representative" of the professions. To the extent that the subjects suspected that this was the case then they would be entitled to ignore the offered base-rates.

In an experiment Gigerenzer, Hell & Blank (1988) found that when they let the subjects experience the sampling themselves base-rate neglect disappeared. In the experiment their subjects could examine ten pieces of paper each marked lawyer or engineer (according to the base-rates). Subjects then drew one of the pieces of paper from an urn and it was unfolded so that they could read a description of an individual without being able to see the mark defining it as being that of a lawyer or engineer. In these circumstances subjects clearly used the base-rates in a proper fashion—including for the "uninformative" description Dick.[4] However, in a replication of the verbal presentation where base-rates were asserted, rather than sampled, Kahneman and Tversky's base-rate neglect was replicated.

8.6 THE GAMBLER'S FALLACY CAN DISAPPEAR TOO

It is perhaps worth noting that the phenomenon in subjective probability judgement known as negative recency also "disappears" in similar contexts. Recently we (Ayton, Hunt & Wright, 1989; 1991) reviewed the evidence that people had difficulty in recognizing and generating random sequences. The typical finding reported in the psychological literature over the past thirty years has been that subjects show a degree of negative recency. That is they appear to believe that alternations of the elements of a random sequence are more likely than they really are; by the same token subjects under-estimate the degree of repetition that there is. This is also known as the gambler's fallacy after Dostoyevsky's (1866/1966) observation that players of roulette falsely assume that after a given number has come up it is much less likely to occur next time. In our review we discussed the possible reasons for this observation. We noted that there were very few occasions when people would encounter an assuredly random process and that consequently the apparent bias might actually be a function of generalization from experience of encounters with non-random sequences. We also considered evidence that the effect was extremely sensitive to subtle variations in the instructions given to subjects. We mentioned a study by Winefield (1966) which showed that if subjects had to guess the suit of a card drawn from the deck the usual measure of negative recency disappeared *if they could see the card being placed back in the deck and the deck being well shuffled*. If this was not the case then they continued to display negative recency. The experiment seems very similar in rationale and findings to that of Gigerenzer, Hell & Blank's (1988) test of the robustness of base-rate neglect.

Recently a number of authors have reported that negative recency varies quite considerably with the task and that tasks can be devised where the subjects perform quite well. A number of authors have found that using an instructional set that avoids reference to probabilistic concepts of chance or randomness will improve performance. For example, Finke (1984) found that subjects asked to produce responses that are as unpredictable as possible produce responses that more closely approached the frequency of repetitions expected by chance than do those subjects asked to produce responses as randomly as possible. If, in a competitive game, subjects are motivated to produce unpredictable sequences of binary responses then they can do so in a way that satisfies standard tests of randomness more successfully (Rapoport & Budescu, 1992). Kareev (1992) reports data that also shows variability of performance according to task and concludes that people have a basically correct notion of randomness, with apparently non-random behaviour being the result of attempts by a capacity-limited information-processing system to optimize performance given its interpretation of the standard tasks. In the words of the title of Kareev's paper, maybe human randomness is not so bad after all.

In our comments on the research into randomness (Ayton, Hunt & Wright, 1989, 1991) we also noted that there is evidence of some confusion about the normative standards used to judge human conceptions of randomness. Curiously the same basic concept (representativeness) is often used both to define the basis for objective "tests for randomness" and to explain why it is that subjects deviate from this standard. The statistical tests commonly assume that a random sequence should contain a *representative* sample of all the possible configurations (e.g. all the possible pairs of outcomes resulting from tossing a fair coin: HT; HH; TH; TT). However, subjects are typically berated when they can be assumed to be applying the same heuristic. Indeed, negative recency has often been assumed to result from the application of representativeness. Subjects have commonly been judged to be suffering from the misconception that even small samples of random output should contain a (representative) number of the basic elements and should also show disorder and absence of any obvious "patterns". In the light of this contradiction it is perhaps not altogether surprising that the instructions given to experimental subjects by experimental psychologists therefore have sometimes appeared a little confusing, and possibly, in terms of the negative recency hypothesis, rather self-fulfilling. Subjects have sometimes been explicitly told that they should produce responses which appear "jumbled" or that don't contain any patterns. Aside from being a potential source of the inappropriate "gambler's fallacy", such instructions suggest that the psychologists have been a little uncertain as to quite what the objectives are for their subjects with this task.

8.7 IS HUMAN JUDGEMENT UNDER UNCERTAINTY BAYESIAN, HEURISTIC OR FREQUENTIST?

Gigerenzer (Chapter 7) discusses several cases where, with frequentist representations, putative fallacies of probabilistic judgement disappear, and he attributes this to the idea that subjects are better equipped to process information concerning frequencies of events than they are single event probabilities. An evolutionary speculation supports the argument: in the course of their evolutionary development humans (and perhaps other creatures) have acquired the means to effectively represent and manipulate information about frequencies of events—but not their probabilities. After all, probability theory has, relative to the evolutionary scheme, only very recently emerged as a means by which to represent and communicate information. According to Hacking (1975) what we now recognize as the mathematical calculus of probability theory was only formalized in the seventeenth century. It would not be altogether surprising then if human cognition did not naturally compute probabilities, but instead used stored frequencies of events. So, requesting subjects to give their responses to problems in the form of probabilities may be asking them to understand and speak a rather foreign language. Just as it would be unrealistic to expect one's pocket calculator to accurately compute the answer to arithmetic problems entered with Roman numerals, it may be unreasonable to judge the general competence of human judgement under uncertainty on the performance of problems requiring the assessment of subjective probabilities rather than frequencies.

We have seen that Kahneman and Tversky suggested that people are not conservative Bayesians but judge and reason with probabilities using mental heuristics. Gigerenzer argues that people do not use heuristics when experimental problems are re-cast into a relative frequency paradigm and, indeed, are not equipped to reason about uncertainty using single-event probabilities at all—but they can reason successfully about uncertainty with frequencies. These frequency estimates can, under appropriate task conditions, be translated into a valid probability metric. Note though one emergent point of consensus in this dispute about human inference under uncertainty. Gigerenzer's conclusion from experiments that test subjects with frequentist versions of Kahneman and Tversky's problems is, in one sense at least, strikingly similar to that of Kahneman and Tversky's. All parties would appear to agree that human reasoning under uncertainty is "... not Bayesian at all".

Kahneman and Tversky, and others, (see Kahneman, Slovic & Tversky 1982) assumed that mental heuristics were utilised in order to reduce the complex tasks of assessing probabilities to simpler judgemental operations. Such an assumption followed very naturally from Simon's (1957) proposal of bounded rationality—the notion that, because of their limited information processing capacities, people don't use optimal methods for reasoning but instead take short cuts, or "satisfice" in order to produce judgements and decisions accurately and efficiently. Simon's work was concerned with the simplifying mental strategies that reduced the complexity of tasks to make them manageable by the kinds of minds that people actually have. However, Gigerenzer's view of reasoning under uncertainty is based on the premise that the human mind has no need to make use of such heuristic judgements *for assessing probabilities*. The claim is that people's memory for frequency is reliable enough for them to utilise past records of events as statistics for their judgements. There is in fact considerable evidence that memory for frequency information is automatically encoded and extremely good. The evidence has been well summarized by Hasher & Zacks (1979, 1984) who conclude that experiments "reliably and uneqivocably demonstrate remarkable knowledge of the frequency of occurrence of all events so far tested." (1984, page 1373).

Hasher and Zacks (1984) were aware that their views about the veracity of the storage of frequency information might be seen as in conflict with the evidence from Tversky & Kahneman (1973) that people might use an availability heuristic for judging frequency. Tversky & Kahneman (1973) reported a number of experiments which suggested that people judged likelihood or frequency by the ease with which instances of the event could be brought to mind. Instances of frequent events are typically easier to bring to mind than instances of less frequent events so the availability heuristic would often be valid. However, availability and frequency are not always perfectly correlated. For example, an experiment showed that subjects were more likely to judge that the letter R was more likely to appear in the first position of a word than in the third position. In fact the letter R occurs more often in the third position of a word than the first and Tversky and Kahneman attributed the error to the fact that words beginning with R are easier to recall.

Hasher & Zacks (1984) comment that:

> ...the conflict between our view and that of Tversky and Kahneman is more apparent than real. First of all, in most instances frequency and availability (like frequency and probability) are highly correlated: More frequent events are, other things being equal, more recallable or "available" than less frequent events. In such situations, any biasing effects of the availability heuristic will not be seen. Use of availability will bias frequency estimates most clearly when the retrieval cue (i.e. the event to be judged) is a weak one (as in the example of words having a particular letter in the middle position). It is worth noting, in addition, that in other illustrations of the availability heuristic the actual frequency differentials

> are small (e.g. 19 versus 20) and the countervailing effects of other variables (e.g. stimulus familiarity) are strong. Such is the case in Tversky and Kahneman's experiment in which subjects misjudged as being more frequent 19 famous names scattered in a list that also included 20 non-famous names. (pages 1383–4.)

Hasher and Zacks' comment that there is no real conflict in their frequency proposal and Kahneman and Tversky's heuristic model of judgement cannot be assumed to apply to those who argue that judgements of likelihood are made by reference to memorized frequencies. Either a particular judgement is based on stored frequencies or it is heuristic.[5] There are some experiments reported in the literature which attempt to unconfound the correlation between frequency and probability to determine if judgement relies on heuristics or stored frequencies. For example, Estes (1976) reported a study which led him to conclude that the basis for predictive behaviour is not a probability estimate but rather a record in memory of the past frequencies of events. In a simulated opinion poll, on each of a series of trials, subjects observed the outcome of a mini-poll contrasting pairs of alternatives (e.g. two political candidates). The exposure sequence was designed such that in some cases one candidate had a lower probability of winning the paired comparisons than another but, because the candidate appeared more often, a higher overall frequency of winning. When two such candidates were paired directly, subjects picked as the likely winner the candidate whose absolute frequency of winning was higher.

In summarising the probability learning literature Estes commented:

> ...the results suggest that the term "probability learning" is in a sense a misnomer. I have found nothing to encourage the tendency to think of probability learning as a basic or unitary process or as a direct manifestation of a capacity for perceiving the statistical structure of sequences of events. The subjects clearly are extremely efficient at acquiring information concerning relative frequencies of events. (page 51).

Other research is less supportive of the notion that judgements under uncertainty rely on stored frequencies and not heuristics. A number of studies have shown that mere repetition of the presentation of large sets of statements causes the subjective degree of belief in their validity to increase compared to non-repeated control statements (e.g. Hasher, Goldstein & Toppino, 1977; Gigerenzer, 1984). This phenomenon has been attributed to the automatic and accurate encoding of frequency information (e.g. Gigerenzer, Hoffrage & Kleinbölting. 1991). However, two studies suggest a heuristic account for the effect. Bacon (1979) demonstrated that higher levels of rated validity occurred for statements that subjects judged to be repeated—whether they had been or not. Arkes, Hackett & Boehm (1989) found that the effect did not generalise to all repeated statements. Specifically they found that statements concerning topics with which their subjects were unfamiliar did not increase in perceived validity with repetition. Arkes, Hackett and Boehm concluded that the repetition–validity effect was attributable to a familiarity heuristic like availability. Note, however, that it is possible to argue that the results of Bacon and those of Arkes, Hackett and Boehm are due to effects on the encoding of memory for frequency (cf. Estes, 1976), and that therefore an availability heuristic explanation may not be required.

With regard to the third general heuristic discussed by Tversky and Kahneman—the anchor and adjust heuristic—we know of no evidence or argument that specifically militates against its existence. There is of course a large body of research that has found evidence for its operation. A recent study of judgemental forecasting by Bolger and Harvey (1994) finds that the general anchor and adjust heuristic, operating in different forms depending on context, was very useful for explaining their subjects' responses. Their subjects were required to make forecasts of future data points given previous ones in the same series and appeared to alter their forecasting strategy depending on the presence and absence of trends and serial dependency. Bolger and Harvey questioned whether such an account might conflict with Gigerenzer's more "ecological" account of statistical reasoning. They suggest that, although we might not have evolved to perform the type of judgemental extrapolations required of their subjects, their tasks are now ecologically valid ones. Among business people judgemental extrapolation is the most popular method of forecasting.

The frequentist approach adopted by Gigerenzer argues strongly against the operation of mental heuristics and biases but there is, as yet, for those choosing to adopt a heuristic approach to judgement, plainly still scope for invoking general heuristics to account for judgemental behaviour in situations where a frequentist representation is inappropriate, i.e. the assessment of subjective probabilities for *unique* one-off events. However, the experimental evidence of fallacies of subjective probability, usually attributed to the operation of the representativeness heuristic, is obviously compromised by the disappearance of these effects when subjects are able to contemplate uncertainty from a frequentist perspective. Perhaps, as argued by Teigen (Chapter 10), alternative mental processes for reasoning under uncertainty may be invoked depending on the circumstances. Under one set of circumstances, defined by the problem structure and the format of information, people may use a frequentist mode of thought to generate a probabilistic response while under other circumstances mental heuristics may be used to generate a response.

8.8 LIKELIHOODS OF SINGLE EVENTS

The evidence for the use of heuristics in judgements of likelihood is based largely on tasks where subjects were required to estimate the likelihood of single events. How do they do this? For the frequentist statistician the task is nonsense. Probability only applies to the relative frequency of events and it makes no sense to consider the probability attached to the truth of a single statement. Plainly though people do feel different degrees of confidence in the truth of individual propositions and, on the face of it, don't object to providing descriptions of their confidence in terms of probability. From the standpoint of the Bayesian statistician there is no reason to discourage this practice. The difficulty according to Gigerenzer is that human information processing simply isn't suited to the task. Nonetheless, it follows that difficulties may arise if subjects are asked to assess veridical probabilities for single events, or assume that they can do so.

We could ask ourselves why it is that subjects produce responses that are so well predicted by the heuristics approach. For example, in one of their studies Kahneman & Tversky (1982b) asked one group of subjects to judge the representativeness of personality descriptions with respect to a whole series of different professions. A separate group of subjects rated the likelihood that each of the described individuals really was a member of each of the listed professions. The correlation between the two was 0.96; plainly, the judgements of likelihood were quite indistinguishable from those of representativeness. It would seem that there is a danger that when asked to assess subjective probabilities for single events subjects will report a measure of representativeness or availability.

Kahneman & Tversky (1979) explained that one way to avoid the biases of subjective probability implied by the heuristic account was to take an external rather than an internal view, by contemplating the target event in relation to a reference class of similar events and considering the distribution of likelihoods for the whole class of events. The strategy looks very much like a way of attempting to invoke a frequentist set for judging likelihood. (Indeed, Tversky & Kahneman (1983) themselves found evidence that the conjunction fallacy was largely eliminated when subjects were presented with the problem expressed with frequencies rather than percentages.) This analysis is amplified and extended by Kahneman & Lovallo (1993), who argue that we have a strong tendency to see problems as unique when they would be more advantageously viewed as instances of a broader class. They claim that the natural tendency in thinking about a particular problem, such as the likelihood of success of a business venture, is to take the "inside" rather than the "outside" view. People will pay particular attention to the distinguishing features of a particular case and reject analogies to other instances of the same general type as crudely superficial and unappealing. Consequently they will fall prey to fallacies of planning—anchoring their estimates on present values or extrapolations of current trends.

Once a forecaster takes the inside view they will not seek out relevant statistical knowledge, will be less likely to formulate a realistic estimate and will be overconfident about their forecasts. (For a discussion of the evidence for the overconfidence phenomenon and an evaluation of the frequentist "ecological" perspective we refer readers to Chapter 18.) It would seem then that proponents of the heuristic view are persisting with a pessimistic view about human judgement of likelihood. However, in advising that anyone attempting to assess probabilities should take an "outside view", it also seems that there is very little in practical terms that separates the advocates of the heuristic and the ecological frequentist approaches in terms of their attitudes to the quality of subjective probabilities. For example, Kahneman and Lovallo review evidence which suggests that, because they take an inside view, people can be unrealistically optimistic or, if failure is easier to imagine, pessimistic). They cite a study by Cooper, Woo & Dunkelberger (1988) which showed that entrepreneurs interviewed about their chances of business success produced assessments that were unrelated to objective predictors such as college education, prior supervisory experience and initial capital. Moreover, more than 80% of them described their chances as 70% or better whilst the survival rate for new businesses is as low as 33%. Such findings might be taken as evidence for poor judgement under uncertainty—or alternatively as evidence that people are better off not attempting to assess probabilities for single events. It seems possible that the entrepreneurs in Cooper et al's study were giving judgements of *plausibility* of success rather than any sort of statistical probability (cf. Teigen, Chapter 10). If, understandably, they do not naturally think about their business as one of a set of elements in a statistical sample but as a unique (to them) system that they work hard to control, it is difficult to see how else they might generate their estimates.

If we compare studies of the calibration of probability assessments concerning individual unique events (e.g. Wright & Ayton, 1992) with those where assessments have been made for repetitive predictions of weather events, e.g. rain (see Murphy & Brown, 1985), we can observe that relatively poor calibration has been observed in the former whereas relatively good calibration has been observed in the latter. Bolger and Wright (1994) argue that this differential forecasting performance is due, in part, to the existence of rapid and meaningful feedback to the weather forecasters in terms of both the relative frequency of probability predictions *and* the predicted event's occurrence. Such prediction-feedback frequency information may well be ideal for the achievement of frequentistic-based accuracy. An empirical study by Benson and Önkal (1992) found that simple outcome feedback had very little impact on the performance of forecasters' probabilistic judgements; however, performance feedback, i.e. information about the accuracy of the forecasters' judgements *and* the outcomes that occur, did improve forecasting. While, doubtless, people register outcome feedback about the stream of idiosyncratic "one off" events in their lives it is harder to believe that they will ordinarily be in receipt of performance feedback for their informal forecasts of these unrelated events.

We conclude our perusal of the issues surrounding the evaluation of human judgements of probability with a final point concerning a possible special case for subjective probabilities. On occasion there will be single events for which no obvious reference class exists and then one will plainly be unable to assess likelihood according to an outside view, or by taking the frequentist approach. Consider, for example, the possibility, in the mid 1980s, that Saddam Hussein would attack Kuwait. How could President Bush's administration have gone about assessing a subjective probability for this unique proposition? As van der Heijden (Chapter 22) discusses, such an assessment task may place unrealistic demands on the forecaster. He argues that in planning for such plausible, high-consequence, unique events, application of scenario planning techniques aid the creation of a robust strategy that works well under a *range* of *plausible* futures. As a methodology for dealing with uncertainty, scenario planning accepts and downplays the decision-maker's poor ability to make realistic probability assessments for single events. Ecologically, the acceptability of scenario-planning techniques to senior managers, and the relative disdain with which decision analysis is viewed, may reflect an intuitive appreciation of the poor quality of probabilistic judgements for the occurrence of unique events. By contrast, in the psychological laboratory, subjects will, helpfully, produce probabilities for unique future events as required by the experimenter. Since most of our knowledge about probabilistic judgement has been derived from laboratory studies, the documentation of the heuristics and biases implicated in the assessment of probability may be valid *but* unrelated to the way in which decision-makers choose to deal with uncertainty given a free choice.

How *should* a person go about assessing numerical subjective probabilities for such unique events? By definition, it is difficult to see how any reference class of similar events could be selected for such events. However, one might perfectly be able to account for the (no doubt varying) subjective probabilities offered by a sample of people by referring to various judgemental heuristics. But, note that, without any reference class, we have no means of evaluating the validity of any judgements that might be offered. A single probability that is unconstrained by reference to any parent distribution admits no standard for evaluation. The probability of unique events remains something of a mystery.

NOTES

(1) Savage (1954), introducing Bayesian ideas to the newly emerging field of decision analysis, proposed that these probabilities should be understood as a property of the person and for this reason suggested that they be known as personal probabilities. The expression has never really caught on in general usage which is perhaps a pity; for some, the term "subjective" automatically evokes negative connotations that the word "personal" would avoid.

(2) For the sake of simplicity our example neglects those outcomes such as dead heats or races where no horses finished which could nonetheless, in principle, be included.

(3) Edwards and his colleagues actually conducted many of their experiments not using book-bags at all but a display consisting of 48 numbered locations each containing a push button and a red and green light. On pushing the button, one of the lights was illuminated. Subjects are told that this is equivalent to sampling a chip of the corresponding colour from a book-bag. The program was carefully prepared so that the sample would be representative of the book-bag being sampled. It is possible that this "artificial" book-bag might induce subjects to respond in ways that they might not if they could see a real book-bag being sampled (cf. Winefield, 1966; Gigerenzer, Hell & Blank, 1988).

(4) In their experiment two, Gigerenzer, Hell & Blank (1988) found Bayesian conservatism in their subjects. This apparent reappearance of conservatism occurred in subjects asked to predict the final scores of football matches given the half-time scores. Subjects were unable to do so optimally; they underestimated the extent to which the half-time score predicted the result and tended to give too much emphasis to their estimate of the prior strength of the teams. By their own argument this is a task with which the subjects would be very familiar and yet they did not indulge in the correct amount of opinion revision. It is not clear whether such a result constitutes evidence for cognitive bias.

(5) However, it may be that there are different mechanisms which underlie different judgements. Teigen (Chapter 10) argues that subjective judgements of probability can be arrived at by a number of different processes. These processes may produce responses that are at variance with one another or some standard. As we mention later, postulated heuristics are very successful at accounting for the responses generated by some tasks, while the notion that people can utilise encoded frequencies is very helpful for explaining the responses from other experimental tasks.

REFERENCES

Arkes, H.R., Hackett, C. & Boehm, L. (1989) The generality of the relation between familiarity and validity. *Journal of Behavioral Decision Making*, **2**, 81–94.

Ayton, P., Hunt, A. & Wright, G. (1989) Psychological conceptions of randomness. *Journal of Behavioral Decision Making*, **2**, 221–38.

Ayton, P., Hunt, A. & Wright, G. (1991) Randomness and reality *Journal of Behavioral Decision Making*, **4**, 222–6.

Ayton, P. & Wright, G. (1987) Assessing and improving judgmental probability forecasts. *OMEGA: International Journal of Management Science*, **15**, 191–6.

Bacon, F.T. (1979) Credibility of repeated statements: memory for trivia. *Journal of Experimental Psychology: Human Learning and Memory*, **5**, 241–52.

Bar-Hillel, M. (1980) The base-rate fallacy in probability judgements. *Acta Psychologica*, **44**, 211–33.

Benson, P.G. & Önkal, D. (1992) The effects of feedback and training on the performance of probability forecasters. *International Journal of Forecasting*, **8**, 559–73.

Bolger, F. & Harvey, N. (1993) Context-sensitive heuristics in statistical reasoning. *Quarterly Journal of Experimental Psychology*, **46**A, 779–811.

Bolger, F. & Wright, G. (1994) Assessing the quality of expert judgment: Issues and analysis. *Decision Support Systems*, **11**, 1–24.
Cohen, L.J. (1989) Can human irrationality be experimentally demonstrated? *The Behavioral and Brain Sciences*, **4**, 317–70.
Cooper, A. Woo, C. & Dukelberger, W. (1988) "Entrepreneurs" perceived chances for success. *Journal of Business Venturing*, **3**, 97–108.
Dalkey, N. (1972) An elementary cross impact model. *Technological Forecasting and Social Change*, **3**, 341–51.
Dostoevsky, F. (1966) *The Gambler*. Penguin, Harmondsworth.
DuCharme, W.M. & Peterson, C.R. (1968) Intuitive inference about normally distributed populations. *Journal of Experimental Psychology*, **78**, 269–75.
Edwards, W. (1968) Conservatism in human information processing. In B. Kleinmuntz (ed.) *Formal Representation of Human Judgment*. Wiley, New York.
Estes, W. (1976) The cognitive side of probability learning. *Psychological Review*, **83**, 37–64.
Finke, R.A. (1984) Strategies for being random. *Bulletin of the Psychonomic Society*, **22**, 40–1.
Fischhoff, B., Slovic, P. & Lichtenstein, S. (1978) Fault trees: Sensitivity of estimated failure probabilities to problem representation. *Journal of Experimental Psychology: Human Perception and Performance*, **4**, 330–44.
Gigerenzer, G. (1984) External validity of laboratory experiments: The frequency–validity relationship. *American Journal of Psychology*, **97**, 185–95.
Gigerenzer, G., Hell, W. & Blank, H. (1988) Presentation and content: The use of base rates as a continuous variable. *Journal of Experimental Psychology: Human Perception and Performance*, **14**, 513–25.
Gigerenzer, G., Hoffrage, U. & Kleinbölting, H. (1991) Probabilistic mental models: A Brunswikian theory of confidence. *Psychological Review*, **98**, 506–28.
Goodwin, P. and Wright, G. (1991) *Decision analysis for Management Judgment*. Wiley, Chichester.
Hacking (1975) *The Emergence of Probability*. Cambridge University Press, Cambridge.
Hasher, L., Goldstein, D. & Toppino, T. (1977) Frequency and the conference of referential validity. *Journal of Verbal Learning and Verbal Behavior*, **21**, 127–41.
Hasher, L. & Zacks, R.T. (1979) Automatic and effortful processes in memory. *Journal of Experimental Psychology: General*. **108**, 356–88.
Hasher, L. & Zacks, R.T. (1984) Automatic processing of fundamental information. The case of frequency of occurrence. *American Psychologist*, **39**, 1327–88.
Kahneman, D. & Lovallo, D. (1993) Timid choices and bold forecasts. A cognitive perspective on risk taking. *Management Science*, **39**, 17–31.
Kahneman, D., Slovic, P. & Tversky, A. (eds.) (1982) *Judgement under Uncertainty: Heuristics and Biases*. Cambridge University Press, Cambridge.
Kahneman, D. & Tversky, A. (1972) Subjective probability: A judgement of representativeness. *Cognitive Psychology*, **3**, 430–54.
Kahneman, D. & Tversky, A. (1973) On the psychology of prediction. *Psychological Review*, **80**, 237–51.
Kahneman, D. & Tversky, A. (1979) Intuitive prediction: Biases and corrective procedures. *Management Science*, **12**, 313–27.
Kahneman, D. & Tversky, A. (1982a) On the study of statistical intuitions. In D. Kahneman, P. Slovic & A. Tversky (eds.) (1982) *Judgement under Uncertainty: Heuristics and Biases*. Cambridge University Press, Cambridge.

Kahneman D. & Tversky, A. (1982b) Judgements of and by representativeness. In. D. Kahneman, P. Slovic & A. Tversky (eds.). (1982) *Judgement under Uncertainty: Heuristics and Biases*. Cambridge University Press, Cambridge.

Kareev, Y. (1992) Not that bad after all: Generation of random sequences. *Journal of Experimental Psychology: Human Perception and Performance*, **18**, 1189–94.

Murphy, A.H. & Brown, B.G. (1985) A comparative evaluation of objective and subjective weather forecasters in the U.S. In G. Wright (ed.) *Behavioral Decision Making*. Plenum, New York.

Nisbett, R. & Ross, L. (1980) *Human Inference: Strategies and shortcomings*. Prentice Hall, Englewood Cliffs. N.J.

Peterson, C.R., Schneider, R.J. & Miller, A.J. (1965) Sample size and the revision of subjective probability. *Journal of Experimental Psychology*, **69**, 522–7.

Phillips, L.D. & Edwards, W. (1966) Conservatism in simple probability inference tasks. *Journal of Experimental Psychology*, **72**, 346–57.

Pitz, G.F., Downing, L. & Rheinold, H. (1967) Sequential effects in the revision of subjective probabilities. *Canadian Journal of Psychology*, **21**, 381–93.

Raiffa, H. (1968) *Decision Analysis*. Addison-Wesley, Reading, MA.

Rapoport, A. & Budescu, D. (1992) Generation of random series in two-person strictly competitive games. *Journal of Experimental Psychology: General*, **121**, 352–63.

Savage, L.J. (1954) *The Foundations of Statistics*. Wiley, New York.

Simon, H. (1957) *Models of Man: Social and Rational*. Wiley, New York.

Tversky, A. & Kahneman, D. (1971) Belief in the law of small numbers. *Psychological Bulletin*, **76**, 105–10.

Tversky, A. & Kahneman, D. (1973) Availability: A heuristic for judging frequency and probability. *Cognitive Psychology*, **4**, 207–232.

Tversky, A. & Kahneman, D. (1983) Extensional versus intuitive reasoning: The conjunction fallacy in probability judgment. *Psychological Review*, **90**, 293–315.

von Winterfeldt, D. & Edwards, W. (1986) *Decision Analysis and Behavioral Research*. Cambridge University Press, Cambridge.

Winefield, A.H. (1966) Negative recency and event dependence. *Quarterly Journal of Experimental Psychology*, **18**, 47–54.

Winkler, R.L. & Murphy, A.M. (1973) Experiments in the laboratory and the real world. *Organizational Behavior and Human Performance*, **20**, 252–70.

Wright, G. & Ayton, P. (1992) Judgmental probability forecasting in the immediate and medium term. *Organizational Behavior and Human Decision Processes*, **51**, 344–63.

Yates, J.F. (1982) External correspondence: decompositions of the mean probability score. *Organizational Behavior and Human Performance*. **30**, 132–56.

Yousseff, Z.I. & Peterson, C.R. (1973) Intuitive cascaded inferences. *Organizational Behavior and Human Performance*, **10**, 349–58.

Chapter 9

Applying a Cognitive Perspective to Probability Construction

Shawn P. Curley
University of Minnesota

and

P. George Benson
Rutgers University

9.1 INTRODUCTION

All processes display variation. As a consequence, decisions entail uncertainty. A standard procedure in decision analysis—the practice of and techniques for aiding decision-making—is to attempt to formally incorporate the uncertainty through a metric, typically probability. Even in more informal decision-making, we deliberate risks and uncertainties in arriving at our decisions. Thus, probability assessment has arisen as a major topic in the study of decision-making.

Since the product of probability assessment (i.e. a subjective probability) is clearly a judgment, it is not very surprising that assessment usually has been described as a judgmental activity or process. Judgment, as a process, involves a weighing or scaling activity by which a stimulus is evaluated against some criterion or, in this case, along some dimension (the 0–1 probability scale). However, the process by which the output judgment is constructed is not purely or even primarily judgmental. Instead, the process is dominated by the construction of reasoned arguments.

Subjective Probability. Edited by G. Wright and P. Ayton.
© 1994 John Wiley & Sons Ltd.

It is this general statement that motivates and summarizes this chapter. In Section 9.2 we characterize current accounts of probability assessment as being primarily judgmental in nature. The section ends by noting how researchers have begun to implicitly acknowledge that such an account has important limitations. Section 9.3 takes a step beyond the boundaries of a judgmental account of probability assessment; it derives a theoretical framework that encompasses reasoning and other nonjudgmental processes in a cognitive theory of probability construction. This account is applied in Section 9.4 in looking back at and reinterpreting "judgmental" phenomena that have been documented in the behavioral decision literature. Finally, Section 9.5 notes that progress consistent with the theory of this chapter is being made by researchers, and expresses the hope that this progress will continue.

9.2 JUDGMENTAL ACCOUNTS OF PROBABILITY

Most studies have focused on probabilities as measures of likelihood or covariation. As likelihood measures, probabilities have been evaluated in decision and forecasting tasks. Examples of such measures include the probability of precipitation as assessed by a weather forecaster, and the probability that a stock will increase in price as assessed by a securities analyst. A variant of the likelihood assessment is the use of probability as a confidence measure. A typical task has subjects responding to general-knowledge questions, such as: Which state has the larger population, Minnesota or Wisconsin? The subject answers the question and then assesses a probability that the answer is correct. The probability is interpreted as a measure of confidence in being correct.

As measures of covariation, probabilities are communicated in pairs. For example, one can express a covariant relationship between X and Y by stating two conditional probabilities: $P(X|\,Y)$ and $P(X|\,\text{not } Y)$. To the extent that these two quantities differ, there is a relationship between the variables. An example of such measures are the sensitivity, $P(\text{positive test} \mid \text{disease } X)$, and specificity, $P(\text{negative test} \mid \text{without disease } X)$, of a diagnostic test as used in medical diagnosis. A test is useful for diagnosing disease X to the extent that its sensitivity differs from $[1 - (\text{specificity})]$.

In either case, the predominant conceptualization of probability construction has been judgmental. The lens model of social judgment theory (see Hammond *et al.*, 1975) provides a concise representation of the judgmental account. The goal is to assess the probability of some target event: Will a company's stock price be higher six months from today? Figure 9.1 illustrates this as being accomplished by judgmentally weighing environmental cues to arrive at a summary response, e.g. a numerical probability. As illustrated by the left side of the figure, the environment supplies cues ($X_1, X_2, \ldots, X_n$) that may relate to the target event's actual occurrence in the environment (Y_e). For example, the target event may depend upon the company's current earnings,

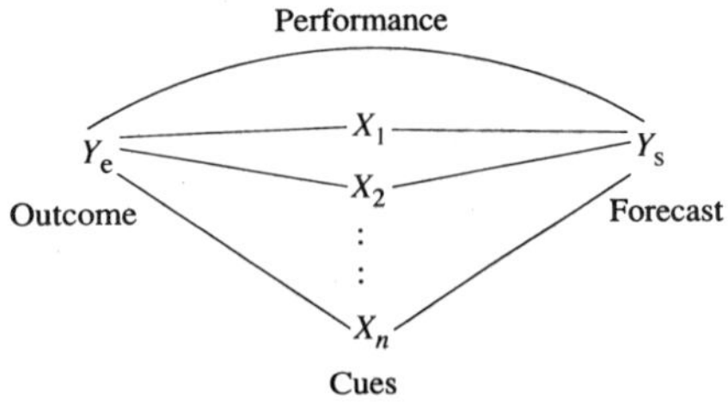

Figure 9.1 Representation of the lens model

the current stock price, projected earnings, etc. In turn, as shown at right, the forecaster uses cues to make a subjective assessment (Y_s). The forecaster's performance depends on the degree of correspondence between Y_e and Y_s, e.g. as measured by a scoring rule (e.g. Yates, 1982).

Thus, the model represents forecasting as a judgmental weighing in which evidential cues are combined through a covert process to arrive at some scale value forecast. Intended as only a paramorphic model, the lens model can afford to conceptualize the process of assessment as essentially judgmental. However, it is our contention that this conceptualization is inadequate. Both prescriptive and descriptive decision research point to this conclusion. We discuss each in turn.

Prescriptively, within the decision analysis paradigm, probabilities are elicited by a decision analyst from a domain expert through a structured interview process. The interview process described by Spetzler and Stael von Holstein (1975) and revised and amplified by Stael von Holstein and Matheson (1979) remains the standard of current practice (Kirkwood, 1990; Merkhofer, 1987). Although not originally conceptualized in this manner, the interview process consists of two distinct phases: *belief assessment*, the evocation and application of relevant knowledge to form a belief, and *response assessment*, the attachment of a numerical qualifier to the belief (Benson, Curley & Smith, 1994). Accordingly, probability assessment can be improved by supporting either belief or response assessment activity.

To date, research has focused almost exclusively on response assessment, which is essentially a judgmental scaling process involving the matching of numbers to beliefs. Unfortunately, improvements in response methods have provided only marginal improvements in assessed probabilities. In fact, analysts have argued that there is little to distinguish the different response mechanisms (von Winterfeldt & Edwards, 1986; Wallsten & Budescu, 1983). Their practical advice is to employ multiple encoding techniques as a means of identifying inconsistencies in the expert's assessments. As such, the use of multiple methods serves as an *ad hoc* stopping mechanism: When the response modes sufficiently correspond, the interview stops. If the different response modes lead to different values, then the interview returns to belief assessment. It is belief assessment that ultimately bears the burden of the quality of probability assessment. Yet, a cognitive account of belief assessment is lacking. It remains cloaked in a black box labelled "judgment."

More recently, the prescriptive literature has begun to recognize that assessment is more than a hidden scaling activity. Decision analysts have begun to pay more attention to belief assessment activity. Knowledge maps and related graphical techniques serve to aid the evocation of evidence and to organize it once identified (Benson, Curley & Smith, 1994). The main idea driving these interventions is that the probability of an event can be assessed by decomposing the target event into lower-level events. These lower-level events are then assessed through judgment and recombined through calculation. In fact, knowledge maps and related graphical techniques were developed to exploit the recombination alternatives offered by the standard probability calculus. Thus, while offering some support for belief assessment, they are still more concerned with scaling and calculation than with developing belief assessment. However, by beginning to focus attention on belief assessment, these new structuring techniques do argue the value of going beyond judgment to the study of the processes involved in belief construction.

From a descriptive perspective, the "heuristics and biases" research spearheaded by Kahneman and Tversky has served a similar function (see Kahneman, Slovic & Tversky, 1982). Heuristics are informal strategies for drawing conclusions from evidence whose applications may lead to various biases. Although still tied to a conceptualization of probability assessment as judgmental, these studies begin to address the cognitive underpinnings of assessment. They clearly indicate that there is more to assessment than weighing evidence and scaling responses.

One consequence of this realization has been a movement away from a purely subjective, judgment-oriented interpretation of probability. The subjective view is often contrasted with earlier views of probability, like the frequentist or logical views (Fishburn, 1964; Good, 1959; Weatherford, 1982). The frequentist view, by defining probability in terms of long-run frequencies, and the logical view, by defining probability as a consequence of logical relationships between propositions, both place the locus of probability external to the assessor. The locus of assessment for the subjectivist is internal; a probability is an individual's likelihood judgment and is not externally constrained. This interpretation allowed subjectivism a generality that led to its favored status in behavioral decision research. However, this pure subjectivism has recently been challenged by an emerging *constructive* interpretation of probability (Payne, Bettman & Johnson, 1992; Shafer, 1981). In the constructive view, the locus of assessment is in the relationship of the internal assessor to the external environment. An assessor constructs the probability from knowledge of the real world that relates to the target event. Construction implies a more active, conscious process, unlike the more covert, unconscious, judgmental processing of the subjective view. Again, we see an indication that the judgmental account of probability assessment is inadequate.

To address the inadequacy we must consider what besides judgment is involved in the construction of probabilities. The next section confronts this issue.

9.3 BEYOND JUDGMENT

Why is a probability assessment referred to as a "degree of belief?" How is it constructed? What does a probability capture? Most proximally, it is produced by response assessment, but there is an integral belief assessment as well. What is the connection between the two activities? How do they operate in conjunction?

The answers to these questions begin with an understanding of belief (Smith, Benson & Curley, 1991). A *belief* ("I believe that it will rain tomorrow") is a connection between a proposition ("it will rain tomorrow") and a person ("I"). Unlike wishes, desires and such, a belief attributes a correspondence between an internal state and an external reality. In language, we use the term "belief" in two senses. In a narrow sense, it communicates the adoption of a proposition as knowledge or near-knowledge. That is, the belief has a certainty about it. In a broader sense, it communicates an entertaining of a proposition. Thus, I can entertain and assess probabilities for both the proposition that it will rain tomorrow and the proposition that it will not rain tomorrow. However, I can only adopt at most one of these as a belief in the narrow sense. It is in the broader sense of entertaining a proposition that researchers use the phrase "degree of belief," as we do in this chapter.

Most beliefs that we entertain do not attain the status of knowledge. By *knowledge*, we refer to that subset of our beliefs that is justified and true (Shope, 1983). To the extent that the belief does not attain knowledge, we ascribe uncertainty to the proposition. Thus, *uncertainty* is a secondary construct indicating a gap between belief and knowledge in our evaluation of a proposition. A key lesson of all this is the relational nature of beliefs and the engendered uncertainty. A belief expresses a relation between the believer and the believed, between the internal and the external, between the subjective and the objective.

Another lesson is that beliefs are constructed. We bring our knowledge and other beliefs to bear in an attempt to justify the target proposition as true. That is, we consciously deliberate in an attempt to move from entertaining to adopting a proposition, with the goal of attaining knowledge. To the extent that this goal is not or cannot be reached, we say we are uncertain and we attempt to communicate a degree of belief.

The general means by which we bring evidence to bear in an attempt to establish a belief is called *belief processing*. Belief processing also can lead to a judgmental qualification of the belief when certainty is not attained. Framed in this way, it is clear that probability assessment is not purely judgmental, but involves other cognitive activities through which we draw conclusions as embodied in target propositions.

We turn now to a description of the different cognitive capacities that are brought to bear in establishing beliefs. Then, in the following section, we build these into a cognitive model of the belief processing from which probabilities arise.

9.3.1 Cognitive Abilities in Belief Processing

Humans draw conclusions by at least four general means: calculation, reasoning, judgment, and recall. *Calculation* is purely symbolic. The summation $2 + 2 = 4$ holds whether we mean 4 apples, 4 trolls, or 4 psychologists. The content is nonessential, the symbols themselves drive the conclusion. Logical deduction also exemplifies calculation, for example:

Premises: If A then B;
A;
Conclusion: B.

The content of A and B are irrelevant to the movement from evidence to claim.

In contrast, practical (everyday, informal) reasoning is symbolic, but the meaning or content of the symbols is relevant. In *reasoning*, the meanings of the statements, not simply their arrangement in the formal structure, drive the derivation of the conclusion. We arrive at conclusions through structural assumptions about the way the world works. For example, that events have causes forms the basis of causal reasoning. Or, if A and B are similar in some respects, they may be similar in other respects: This is the core of similarity-based reasoning by parallel case. In reasoning, we use our knowledge and beliefs to draw conclusions and form other beliefs. We argue from one proposition to another by applying our world knowledge concerning relationships among the propositions.

Evidence: Bill Clinton will be an incumbent candidate in 1996.
Warrant: An incumbent lost the last election.
Claim: Bill Clinton will not be re-elected president in 1996.

The evidence in this argument implicates the claim through a warrant based on experiential knowledge about the outcome of the last election. In general, practical reasoning is a movement from evidence to claim through some warrant involving our world knowledge (Toulmin, 1958). Although it is possible to rewrite this argument in the form of a logical deduction, the resulting structure is less important in deriving the conclusion than the content of the propositions. Further, such structures do not need to be involved in forming conclusions through practical reasoning. Because of its use of evidence and beliefs to derive other beliefs, reasoning plays an integral role in belief processing.

Unlike the other means of drawing conclusions, *judgment* is not symbolic. It is a scaling activity used to characterize stimuli along some scale or dimension. It may involve a combination of evidence through a subjective weighing to arrive at an aggregate judgment, or it may involve a matching or comparison process along a dimension or against some criterion. Examples are the grading of students, or Olympic scoring in diving or gymnastics. The process is relatively covert and indescribable. It is a "black box" process, the kind that has dominated conceptualizations of probability assessment.

Finally, *recall* may or may not be symbolic, and encompasses the interaction of long-term and working memories. Although memory research clearly suggests that recall is a constructive process, individuals do not generally perceive it as such (e.g. Mayer, 1992, Chapters 8 and 9). We can recall conclusions without a consciousness of any transformation from data to claim or of any evidence underlying the claim.

These distinctions among cognitive processes are believed to be useful, though there is clearly interaction among them. For example, calculation involves the abstraction of reality into some model of reality. While the model is a strictly symbolic representation, it gains power from a mapping to contentful propositions that describe reality (Chervany, Benson & Iyer, 1980; Coombs, Raiffa & Thrall, 1954). Simply put, there is a reasoning counterpart to a calculation. One can conclude from one proposition to another directly through reasoning; or, one can strip away the content and employ a parallel calculation. Calculation formalizes reasoning; it operates without content while reasoning operates with content (Curley *et al.*, 1994a).

In probability assessment, all of these means of drawing conclusions may be brought to bear in forming a belief and in constructing the probability response. Thus, to conceptualize assessment solely in terms of judgment is too limiting. A cognitive analysis can usefully supplement the judgmental account. The judgmental account connects judgments of, for example, causality, covariation, and similarity at one end to judgments of probabilities at the other end. But, between these judgments is a belief formation process dominated by reasoning.

The importance of recognizing and developing the role of reasoning is that, unlike judgment, reasoning is fairly overt. The reasons that conclusions hold can be made explicit, and so there is a better opportunity for improving reasoning than judgment. With this realization, there is motivation to open the black box and augment the judgmental account of assessment, thereby expanding opportunities for designing interventions to aid knowledge evocation and application.

A parallel can be drawn with studies of the quality of assessed probabilities, particularly their external correspondence as measured via scoring rules (Yates, 1982). The focus in this research has been on calibration, which is a scaling phenomenon. Less attention has been paid to discrimination, as measured for example by resolution, which is a content phenomenon. Training has been shown to improve calibration but not discrimination (Benson & Önkal, 1992). How can evidence better support the task of discriminating when events occur and when they do not? Our claim is that improvements in discrimination, arguably a more critical aspect of performance for probability assessors than calibration, will arise from improved understanding of the reasoning that underlies assessment. Reasoning is the critical activity.

As a first step in understanding the role of reasoning in probability assessment, we develop a model of the belief processing that supports probability construction. This model is briefly described in the next subsection.

9.3.2 Model of Belief Processing

Figure 9.2 explicitly highlights the role of reasoning in probability assessment. The figure and the following description are adapted from Smith, Benson and Curley (1991) and Benson, Curley and Smith (1994). The figure presents a competence model, depicting human capacities that can be performed in a constructive probability assessment. It describes what people are capable of doing in a deliberative assessment, not what necessarily occurs each time a probability is assessed.

The figure is divided into three modules of activity: data generation, argument construction, and qualifier construction. These modular activities are interactive, rather than sequential. The rectangles within the modules denote argument construction components, the production of which is driven by reasoning. Ovals denote judgmental assessments. Diamonds refer to the external inputs and outputs of probability assessment.

Data generation module

Probability assessment is initiated by some stimulus or expressed need for an assessment. For example, a securities analyst may need to produce an earnings-per-share (EPS) forecast for a company she follows. This stimulus triggers the belief formation process. If a belief has previously been formed, it may simply be recalled. Memory-retrieval research demonstrates that humans can recall significant conclusions and probabilities previously generated, but that the underlying details of the evidence and arguments leading to the conclusions are not fully retrievable (Wyer & Hartwick, 1980).

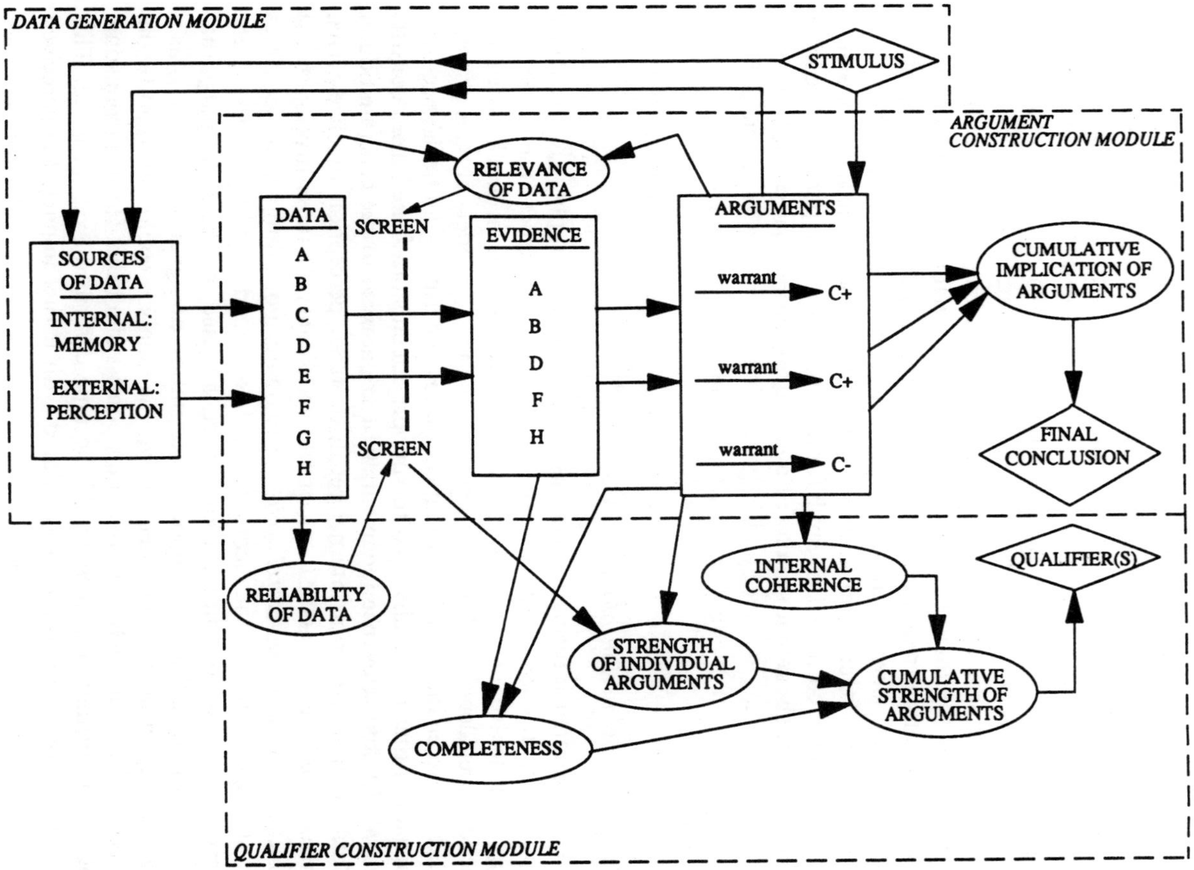

Figure 9.2 Model of belief processing (adapted from Benson, Curley and Smith, 1992)

Such lack of detail may partly result from the forecaster opting for a sparer representation to avoid the cognitive costs of increasing the precision of recalled data (Brainerd & Rayna, 1990).

If the needed belief has not already been formed, various sources are tapped for data relevant to the proposition or issue in question. The data sources are external (e.g. obtained from the company in question or from historical records) and internal (e.g. obtained through introspection: How has this company dealt with me—the analyst—in the past? What do I know about this company and its industry?). Some or all of these data ultimately support the construction of arguments. Data generation diminishes over time as sources are exhausted. Alternatively, it stops abruptly if a definitive argument arises.

Data generation may also be stimulated by particular arguments or argument types. There is an interdependence between arguments and data: arguments are formed from data, but also serve to evoke data. For example, the analyst might consciously search for data that support a particular possible conclusion; or, she might search for a similar situation that could be used in identifying relevant evidence and in generating convincing arguments. Thus, a consideration of possible arguments could itself prompt and guide data evocation.

Argument construction module

Having generated information, the assessor must now apply it to the target propositions. The major function of the argument construction module is to bring evidence to bear in forming arguments. An initial need is for the selection of evidence for further processing. Thus, the model contains a screen or filter between data—the information that is generated—and evidence—the information that is actually used to support conclusions. The screen consists of reliability and relevance judgments that are applied to the generated data.

Data might arise from internal recall or from perception of an external source. A judgmental assessment of the veridicality of these processes and sources induces a perceived *reliability* of the data. For example, the securities analyst might think that the memory trace had eroded over time, or she might believe that an external information source is biased, or she could judge that the perceptual conditions are poor (a recognition of signal-to-noise ratio as a perceptual determinant). Consequently, the analyst could downgrade the reliability of the data. Also, the data may consist of a belief formed from some prior belief formation processing of the type described by Figure 9.2. This belief may have a qualification associated with it that impacts the perceived reliability of the belief as data.

Relevance is a relational concept between a datum and a propositional conclusion, and is evaluated through the argument in which the datum and proposition are embedded. It concerns the applicability of the data for argument construction; a datum is relevant if it bears on the truth of the proposition. Thus, the securities analyst may recall that her uncle owns stock in the company being analyzed, but this datum is unlikely to be perceived as relevant for use in an argument concerning the company's expected EPS. The relevance mechanism determines which arguments are developed and analyzed in the belief formation process. Accordingly, it bears major responsibility for beliefs that are formed and probabilities that are assessed.

If the data are unreliable or irrelevant (below some threshold level), they may be screened out. Alternatively, low reliability and low relevance may only serve to weaken the argument for which the data are used as evidence. Such an effect is captured in the qualifier construction module by the arrow from the data screen to argument strength. In this way, low reliability and relevance can impact the assessed qualifier, e.g. the probability assessment.

Next, the module's argument component shows the movements from evidence to conclusions through warrants. The resulting arguments, constructed through reasoning, bring evidence to bear on the target proposition. Some arguments may conclude one way, for example, in favor of increasing EPS (C+), whereas others may imply the contrary claim, that EPS will decrease (C−).

Having constructed arguments, an overall conclusion must still be reached: Will the EPS rise or decline? Weighing arguments toward this final conclusion is judgmental. The final conclusion expresses one's belief regarding the originating stimulus: "I believe that EPS will rise."

Qualifier construction module

Paralleling the argument construction process are cognitive activities within the qualifier construction module that assess aspects of the data generation and argument construction activities. These generate one or more responses or qualifications. For example, the final conclusion or one or more of the intermediary conclusions can be qualified with a probability judgment.

The relative strengths of the individual arguments are the major source of such qualifications. An argument's relative strength is affected by the assessed reliability and relevance of the evidence within the argument and by factors intrinsic to the argument itself, such as the type of warrant employed. Thus, an argument from analogy might be regarded as having less strength than a causal argument.

Factors arising from a consideration of sets of arguments also can serve to qualify beliefs. Completeness is one such factor. Not all recognized needs for data will be satisfied; the securities analyst may wish to know what new products are in development, but this evidence may be unavailable. The final conclusion will be qualified accordingly. In addition, completeness can be assessed by considering the time and effort devoted to argument construction.

Even if all arguments favor the conclusion that EPS will rise, the analyst may acknowledge that she had insufficient time to conduct a full analysis and qualify the conclusion. Another factor is the internal coherence of the generated arguments. If a set of arguments does not form a cohesive whole, the strength of the conclusion that they support is lessened.

Eventually, all the various assessments converge to form a judgment of the cumulative strength of the arguments as a whole. This, in turn, is translated into an appropriate qualifier of the final conclusion, e.g. a probability number expressing judged likelihood or a rating of confidence.

9.4 LOOKING BACK: EXPLAINING "JUDGMENTAL" PHENOMENA

As Figure 9.2 indicates, probability assessment, which traditionally has been described largely in judgmental terms, involves considerably more than simply judgment. The goal of this section is to use examples of "judgmental" behavioral decision phenomena to indicate how the account illustrated by Figure 9.2 helps to usefully reconceptualize these phenomena. In so doing we are opening the black box labelled "judgment" and expanding our ability to study the underlying cognitive processing.

Opening the black box also leads to a cohesion of otherwise distinct phenomena. The heuristics-and-biases research of the past twenty years has identified numerous behavioral biases that can result from individuals' use of heuristics to process information for decision making (Kahneman, Slovic & Tversky, 1982; Tversky & Kahneman, 1974). One shortcoming of this important body of work is the lack of an integrating framework (Einhorn & Hogarth, 1981). Without integration, we are left with a disparate collection of heuristics and limited understanding. A theory like that in Figure 9.2, identifying the cognitive activity underlying decision tasks, moves us in the direction of such a framework. We begin by applying the theory to some of the more prominent heuristics to demonstrate this capability.

9.4.1 "Judgmental" Heuristics

In a recent review of introductory psychology texts, Van Wallendael (1992) noted that students' principal exposure to behavioral decision theory research is through the heuristics literature, and more specifically through the representativeness and availability heuristics. We begin by looking at these two judgmental heuristics.

Representativeness

The representativeness heuristic was first identified by Kahneman and Tversky (1972) and continues to be of interest to researchers (e.g. Bar-Hillel 1984; Nisbett et al., 1983; Saks & Kidd, 1986; Well, Pollatsek & Boyce, 1990). As applied to judgments of probability, the heuristic essentially states that the probability of an event reflects a similarity relationship between a sample and its parent population or between an event and its generating process. For example, after being told that a sample of five will be drawn from a population of 70 lawyers and 30 engineers, subjects responded to descriptions of members of the sample, such as the following:

> Jack is a 45-year old man. He is married and has four children. He is generally conservative, careful, and ambitious. He shows no interest in political and social issues and spends most of his free time on his many hobbies which include home carpentry, sailing, and mathematical puzzles. The probability that Jack is one of the 30 engineers in the sample of 100 is ____%. (Kahneman & Tversky, 1973, p. 241)

Other subjects responded to the same descriptions after being told that the population consisted of 70 engineers and 30 lawyers. The modal response in the experiment, with either base rate, was to respond in a manner consistent with ignoring the base-rate evidence and judging the similarity of the description of Jack to the target events, lawyer and engineer.

While this explanation is indeed valid, having been reliably established experimentally, we believe that it can be amplified with the belief-processing model described in the previous section. In assessing the probability of Jack being an engineer, we attempt to form a belief as to Jack's profession. We do so by constructing arguments, that is, by reasoning. We establish a belief regarding Jack's profession; and then, to assess the strength of our belief, we apply judgment to evaluate aspects of our belief formation process. Do we have justification for believing Jack is an engineer? How strong is the justification, i.e. to what extent do we believe it? Finally, the strength assessment is scaled to a probability response.

Framed in this fashion, representativeness is a means by which we reason to form beliefs, by which we move from one proposition (evidence) to another (conclusion). In rhetoric, Brockriede and Ehninger (1960) observed that arguments can be classified in terms of the means by which they accomplish this movement among propositions, and they developed a preliminary classification scheme based on this insight. If valid, such a classification system would allow us to more specifically identify representativeness with the type of reasoning employed.

Curley et al. (1994a) provided this validation while extending Brockriede and Ehninger's analysis in three important ways. First, they identified that arguments arise from the application of our world knowledge about the relationships among propositions. This reframing allowed an expansion of Brockriede and Ehninger's typology, and it allowed the typology to be grounded in the psychology of a cognitive representation of knowledge. Table 9.1 summarizes the argument typology and provides examples of each argument type. These types are organized by the form of the underlying knowledge relationship from which the arguments arise (Curley et al., 1994b, has more detailed descriptions and other examples of the arguments). Second, they established that these argument types could be reliably coded from verbal protocols of subjects' reasoning. This provided support for the framework's validity. Third, they observed that reasoning involving these argument types actually occurred in probability assessment; and, indeed, that reasoning was the central activity in assessment, supporting the validity of the general account illustrated by Figure 9.2.

The representativeness heuristic stems from the use of knowledge connections that are hierarchical. Hierarchical relationships represent instances, categories, and subcategories with associated features (Collins & Michalski, 1989). Thus, in the example above, we reason as to whether Jack is an instance of the category lawyer or of the category engineer. This reasoning is based on the content of the description as it relates to our world knowledge about lawyers and engineers as stored in a hierarchical knowledge base.

A key factor in such arguments is the homogeneity among instances grouped under a category. *Homogeneity* refers to the uniformity of instances within a category; heterogeneity, its complement, describes perceived variation in the instances (Nisbett et al., 1983). Our ability to reason from instances to categories, and vice versa, depends on a perception of homogeneity among instances (Curley et al., 1994a). We judge category membership by the presence of commonalities among instances; these commonalities are then used in reasoning.

The bias is not in applying this reasoning; the reasoning is appropriate and has a sound basis as a means of establishing a belief in Jack's occupation. The bias results from ignoring useful base-rate, relative frequency evidence, and using only reasoning from hierarchical world knowledge. Subjects show a preference in this task for reasoning from hierarchical world knowledge rather than from recalled or observed relative frequency data. Knowledge of such tendencies is important for understanding probability assessment, but cannot be fully exploited until we frame and understand assessment in terms of the reasoning being employed.

Let us be clear about how we perceive the theoretical account of the representativeness phenomenon that we have just put forth. The contribution is not in providing an alternative explanation. Our account is completely

Table 9.1 A taxonomy of argument types organized by their underlying relationships

Relationship	Argument type	Description	Example
Causal	Causal	Non-intentional causal link between data and claim	The product will be profitable because those responsible are very competent.
	Motivational in cause	Intentions of human agents used as a cause of some result	The product will be successful because the CEO is determined for it to succeed.
	Motivational in effect	Concludes that some human intention exists	The CEO is determined to succeed because his job is dependent upon success.
Covariational	Sign	Interprets data as symptoms or clues and establishes a covariant relationship	Japan has a high GNP because they have a developed high-tech industry.
Hierarchical	Generalization	Induction from a sample to its population	Roughly 10% of the parts in the warehouse are defective because 10% of the sampled parts were defective.
	Individuation	Concluding from the general population to a specific individual	All employees want job security. Thus, Chuck, being an employee, wants job security.
	Categorization	Establishes membership in a category from the presence of certain features	Chuck is an employee because he has a 75% time appointment with our firm.
	Hierarchical exclusion	A necessity appeal among instances that are mutually exclusive	The product will not have a demand in department or hardware stores, so its demand must come from service stations and auto parts stores.
	Hierarchical combination	The conjunction of an exhaustive set of instances implies their superset	The market is covered because we are selling in grocery and convenience stores, discount chains, and drug stores.
Resemblance	Parallel case	Concludes based on a similarity between instances	It will take 25 minutes to drive to the airport because that is what it took last time.
	Analogy	Concludes based on a similarity of relationship with something outside of the current domain of interest	Our corporate strategy is like a stool; therefore, we need all three components to be successful.
External Domain	Authority	Appeal to an external source with potentially relevant knowledge	The economy will rebound because the leading economists have so predicted.

consistent with the representativeness explanation. It is also not merely a rewording of an existing phenomenon. Our explanation moves the focus of attention from the black box of judgment to existing work in cognitive psychology. By so doing, representativeness is recognized as a label for one specific form of reasoning. Rather than trying to obtain or study judgments of representativeness, this account encourages a movement of researchers' attention to the study of the underlying cognitive processes. The explanation is thus intended to broaden the understanding of the representativeness process by grounding it in the cognitive roots from which it arises. Representativeness and other heuristics need not be treated in isolation. They reflect the manifestation of basic cognitive activities that have been and are the study of cognitive psychologists. Let us demonstrate this perspective with another heuristic: availability.

Availability

The availability heuristic was introduced by Tversky and Kahneman (1973) and, like representativeness, has received continuing attention (e.g. Brown & Siegler, 1992; Nisbett et al., 1983; Poses & Anthony, 1991). The basic idea is that an individual will use ease of recall of instances to judge the frequencies and likelihoods of target events. The heuristic and associated bias were described by Tversky and Kahneman (1973) through the following example:

> Suppose you sample a word at random from an English text. Is it more likely that the word starts with a *K*, or that *K* is its third letter? According to our thesis, people answer such a question by comparing the availability of the two categories, i.e., by assessing the ease with which instances of the two categories come to mind. It is certainly easier to think of words that start with a *K* than of words where *K* is in the third position. If the judgment of frequency is mediated by assessed availability, then words that start with *K* should be judged more frequent. In fact, a typical text contains twice as many words in which *K* is in the third position than words that start with *K*. (page 211)

As predicted, their empirical study demonstrated that subjects' modal response was to judge the first position as more frequent, supporting use of the availability heuristic.

Tversky and Kahneman were concerned with why such responses were biased. The availability heuristic is one explanation. We believe that the framework presented in this chapter enables us to go a step beyond this explanation. By viewing the phenomenon through a broader cognitive framework, we can begin to address how, why, and when people use the availability heuristic.

The key task characteristic in the above example is that the question is perceived as a general knowledge question. Such questions also have been termed "almanac questions" since the correct responses can be obtained by consulting an almanac or similar source. (In the word-count example, the authors' source of the definitive answer was a word-count study by Mayzner & Tresselt, 1965.) Tversky and Kahneman described these as retrieval tasks. Thus, the respondent perceives that a correct answer exists, at least in theory, which prompts the use of recall as an appropriate means of drawing conclusions for the task.

One can then qualify this recall process. The availability heuristic proposes that the qualification reflects an ease of recall judgment. Thus, the term "availability heuristic" is a label for the application of a recall process and a judgment process to a general knowledge task.

But, there are other conditions under which the availability heuristic has been applied. Consider the following word-construction task from Tversky and Kahneman (1973):

> Each problem consisted of a 3 × 3 matrix containing nine letters from which words of three letters or more were to be constructed. . . . For each problem, they were given 7 sec to estimate the number of words which they believed they could produce in 2 min. Following each estimate, they were given two minutes to write down (on numbered lines) as many words as they could construct from the letters in the matrix. (page 209)

The consistent finding in construction tasks of this form was that subjects' estimates correlated highly with the frequencies of constructed instances. Under the label of availability, Tversky and Kahneman hypothesized that subjects could use the number of instances constructed in 7 seconds to estimate the number of items that could be identified in 2 minutes.

Of note in this example is that there is a connection being made between past time and future time, i.e. from recall to prediction. Such a connection is legitimized when there is a perception of homogeneity in recall performance over time. This allows performances at different times to be grouped as instances of a more general category that includes both past and future performances. Given this, one can use an historical frequency to conclude future frequencies. Thus, judgments of homogeneity, grounded in hierarchical knowledge bases, should impact the applicability of the availability heuristic in construction tasks.

The application to likelihood assessment of this connection across time is demonstrated by another example:

> What is the likelihood that the Minnesota Timberwolves basketball team will win its next game?

In assessing the likelihood, it is reasonable to expect an assessor to use the question:

> How frequently do the Minnesota Timberwolves win?

This second question captures a conversion from future time to past time, i.e. a conversion from prediction to recall. The conversion is allowed by a judgment of homogeneity among games, past and future. This allows reasoning by generalization (Table 9.1) from past games to other games this season, followed by reasoning by individuation from games in general to a particular future game (Curley et al., 1994b, has other examples of such reasoning). In this way, the assessor can apply the availability heuristic to the likelihood question. In other words, the recall of wins and losses and the judgment of their frequencies can be used as evidence to assess likelihood in the same way that recall is used to judge frequency in the word construction example.

In the Timberwolves example, the desired response can be constructed through deliberative reasoning, with the frequency question producing the generalization and individuation arguments outlined above. Other arguments might be constructed, as well, using other evidence as available, e.g. the team's recent performance, the team's opponent, or the site of the game. Alternatively, the reasoning may be automated through experience, as might occur with the word-construction task. Adults have considerable experience with words, even to the point where the construction of actual words in 7 seconds may not be needed to accomplish the task. Recognition processes, along with judgment, may suffice (see Hart, 1967 and Beyth-Marom & Fischhoff, 1977, regarding the separation of recognition and recall processes in memory and in availability, respectively).

This analysis broadens the generality of the availability heuristic. It does so by revealing that the heuristic captures different patterns of processing. In the word-count example, a recall process is followed by a judgment process. In the word-construction example, reasoning from homogeneous instances in a hierarchical knowledge base (possibly automatized to a judgment process) is applied. Thus, the availability heuristic can operate in at least two distinct situations: (a) in general knowledge tasks, and (b) in situations when inferences from instances available at one point in time are made to instances at a later time. Future research should acknowledge such differences. For the purposes of this chapter, it is enough to identify that such insights arise from the broader, cognitive analysis of assessment that we advocate.

Again, the benefit of the present account is not in supplanting the representativeness or availability explanations, but in complementing them. Heuristics are the manifestations of basic underlying cognitive processes. Attention should be directed at these processes. How do judgment, reasoning, recall, and calculation operate to produce a probability response? When is each triggered? What role does each play? What situational features influence their operation? The use of a cognitive account grounded in more basic cognitive processes is a subtle but important shift in the means used to understand probability assessment. Figure 9.2 represents a preliminary step in this direction.

In the next two sections, we show how the attention to reasoning and the cognitive underpinnings of assessment helps to inform our understanding of other behavioral decision phenomena. First, we consider ambiguity; then, we look at the use of probabilities as measures of likelihood and support.

9.4.2 Ambiguity

Ellsberg (1961) conjectured that individuals would prefer to wager on an urn of known composition (e.g., 50 red balls and 50 black balls) than on an urn of uncertain composition (e.g., 100 red and/or black balls in some unstated proportion). He showed that this behavior was inconsistent with even a qualitative measure of probability. Ellsberg attributed such behavior to an avoidance of ambiguity, the additional uncertainty present in the second urn. More specifically, Ellsberg defined *ambiguity* as "a quality depending on the amount, type, reliability and 'unanimity' of information, and giving rise to one's degree of 'confidence' in an estimate of relative likelihoods" (page 657). Thus, in the case of the second urn, there is greater ambiguity since there is a lesser amount of information available. The significance of ambiguity lies in the empirical regularity with which individuals' choices are influenced by ambiguity in a variety of contexts (Curley, Eraker & Yates, 1984; Curley & Yates, 1989; Einhorn & Hogarth, 1986; Hogarth & Kunreuther, 1989).

Ambiguity results from the different judgments that accompany belief processing. Figure 9.2 emphasizes that in forming a belief, its qualification can arise from several sources. For example, the assessor may wish to qualify a belief based on evidential unreliability, perceived weaknesses in the arguments, and/or incompleteness of the evidence. To form a probability response, these judgments must be merged and scaled to a single response, e.g. a number from 0 to 1. Ultimately, such a response communicates one or more aspects of the belief and the qualification process.

For example, in the two-urn situation used by Ellsberg, when asked to judge the likelihood of selecting a red ball, subjects' modal response is 0.50 with either urn. This response reflects a subject's use of the probability scale to communicate relative frequencies, along with assumptions of equal likelihood: each ball is equally likely to be drawn; each possible distribution is equally likely for the second urn.

When asked to choose an urn for wagering, another consideration merges into the communicated response: the incompleteness of evidence. Note that this aspect of the situation does not impact the probability response; subjects simply communicate the perceived relative frequencies. However, the choice response communicates different aspects of the situation; both frequency and completeness judgments impact the choice response.

In general, response measures can be conceived as languages with which decision-makers qualify their beliefs (Shafer & Tversky, 1985). Different languages may capture different aspects of belief scaling (i.e. different aspects of the means by which beliefs are qualified), and be more or less appropriate in different situations. Operationally, the different languages for communicating one's degrees of belief are presumed to arise from the various judgments surrounding belief assessment. This linking of response languages to the underlying belief assessment process provides a means of grounding these assessments cognitively in a way that is not currently done.

Additionally, it is possible to generalize the account of reactions to ambiguity by recognizing the role that reasoning plays in assessment. One of the motives that has been suggested as impacting choice is the desire to justify one's choices. Curley, Yates & Abrams (1986) observed that this motivation was at least partially responsible for subjects' reactions to ambiguity. Subjects incorporate considerations of the amount of evidence into their choice because they perceive that it will impact their ability to justify their choice to others. Hogarth (1992) also noted the role of justification and argument in reactions to ambiguity. Again, an analysis of the cognitive underpinnings of assessment promises to further our understanding of isolated phenomena in a unified way.

9.4.3 Likelihood and Support

Finally, we address an issue that traces its roots to the early history of probability: the distinction between likelihood and support. As noted earlier, in forming beliefs we aspire to knowledge; we communicate degrees of belief to the extent that this aspiration is not attained. As the goal state of belief formation, knowledge represents those beliefs that are justified and true (Shope, 1983). Thus, in defining a degree of belief, we can do so against either of two criteria: truth or justification. Historically, the conceptions of probability arising from these different criteria appear in the distinction between Pascalian probability (truth) and Baconian probability (justification) (Cohen, 1977; Shafer, 1978). We follow Smith, Benson, and Curley (1991) in referring to the distinction as likelihood (truth) versus support (justification).

Probability theory has traditionally interpreted degrees of belief as likelihoods:

> A probability, according to Bayesians like ourselves, is simply a number between zero and one that represents the extent to which a somewhat idealized person believes a statement to be true. (Edwards, 1982, p. 359)

In contrast, measures can be developed that express support, for use in situations emphasizing justification. Possible examples are legal settings (where the goal is to remove doubt, not to determine truth), and security analyses (where the emphasis is on justifying a recommendation to a client). One major structural difference between support and likelihood is that increased support for a hypothesis does not necessarily affect the support for its complement, whereas an increase in the likelihood of a hypothesis necessitates a reduction in its complement's likelihood. Shafer (1976), building on the work of Dempster (1968), proposed a set of belief function measures that accomplish this separation between an hypothesis and its complement, and so lend themselves to measuring support. Subjects have been shown to be able to assess these measures reliably (Curley & Golden, 1994).

In terms of Figure 9.2, support measures presumably tap into the judgment activity accompanying assessment differently than do probability responses. An indication of this difference is foreshadowed by Keynes (1921):

> But it seems that there may be another respect in which some kind of quantitative comparison between arguments is possible. This comparison turns upon a balance, not between the favourable and the unfavourable evidence, but between the *absolute* amounts of relevant knowledge and of relevant ignorance respectively. (page 71)

Keynes' description acknowledges the importance of support, and suggests that support is tied more closely to underlying completeness judgments than is likelihood. This conjecture is consistent with observed reactions to ambiguity.

In sum, a cognitive analysis of assessment indicates that there are clearly multiple sources of qualification arising from belief formation activity. These multiple sources of qualification are somehow reflected in whatever response is elicited. However, different responses need not tap equally into these intermediary judgments. For example, completeness assessments might affect support responses but not likelihood responses. Different measures will communicate different things and serve different purposes. To say that any one measure, such as probability, can adequately describe all that we might consider important in a degree of belief seems unwarranted. In fact, when trying to do so, certain behavioural "anomalies", e.g. reactions to ambiguity, are observed.

9.5 WHERE DO WE GO FROM HERE?

In the last fifty years, behavioral researchers have made significant advances in our understanding of probability. This book is testament to the progress made. Our purpose in writing this chapter has been to encourage a new research direction that has the potential to substantially increase our current understanding. We have briefly laid out a pathway. More details are contained in Benson, Curley & Smith (1994); Curley et al. (1994a; 1994b); and Smith, Benson & Curley (1991). This chapter provides a framework that reveals a number of issues and questions that have not been addressed due to the lack of such a framework. We have noted just a few of these in the preceding sections.

We are encouraged that research is beginning to move in this direction. Attention to the cognitive processes underlying decision making, attention to the role of argumentation and justification in decisions, and attention to constructive processes in probability and preference assessment are all current tendencies in behavioral decision research. A few specific examples will demonstrate the point.

Perkins, Allen and Hafner (1983) observed strategies that subjects used in countering their own reasoning about common problems. Their analysis offers a fuller basis for counterfactual reasoning and broadens the application of the prescription to elicit counterarguments—one of the few strategies that has been successful in improving subjects' judgments (Fischhoff, 1982). Brained and Reyna (1990, 1992; Reyna & Brainerd, 1992) have tied the role and nature of recall as a cognitive process to various "judgmental" biases. The interrelatedness between reasoning and recall and the importance of each for understanding assessment are central themes in their research. Bostrom, Fischhoff and Morgan (1992) analyzed the content of subject's reasoning about the risks of radon. They characterized the cognitive representations that the subjects used to organize their knowledge about the hazard, and were able to identify representational sources of subjects' misrepresentations of radon's risks. Fletcher and Huff (1990) analyzed arguments made by AT&T in its 1973–84 annual reports. The analysis offered a unique perspective on the company's strategic decision-making during that period, and allowed the researchers to map interdependent changes among four strategic thrusts in the organization.

It is our hope and belief that these studies represent only the beginning of what can be achieved. We look forward to a future in which the current emphasis on judgment processes in decision-making research is broadened to encompass reasoning. It is only by opening the black box of judgment that we will be able to understand and exploit the processes by which people deal with the uncertainty that is omnipresent in real-world decisions.

REFERENCES

Bar-Hillel, M. (1984) Representativeness and fallacies of probability judgment. *Acta Psychologica*, **55**, 91–107.

Benson, P.G., Curley, S.P. & Smith, G.F. (1994, April) *Belief assessment: an underdeveloped phase of probability elicitation*. Working paper, Rutgers University, Newark, NJ.
Benson, P.G. & Önkal, D. (1992) The effects of feedback and training on the performance of probability forecasters. *International Journal of Forecasting*, **8**, 559–73.
Beyth-Marom, R. & Fischhoff, B. (1977) Direct measures of availability and judgments of category frequency. *Bulletin of the Psychonomic Society*, **9**, 236–8.
Bostrom, A., Fischhoff, B. & Morgan, M.G. (1992) Characterizing mental models of hazardous processes: A methodology and an application to radon. *Journal of Social Issues*, **48**, 85–100.
Brainerd, C.J. & Reyna, V.F. (1990) Gist is the grist: Fuzzy-trace theory and the new intuitionism. *Developmental Review*, **10**, 3–47.
Brainerd, C.J. & Reyna, V.F. (1992) Explaining "memory free" reasoning. *Psychological Science*, **3**, 332–9.
Brockriede, W. & Ehninger, D. (1960) Toulmin on argument: An interpretation and application. *Quarterly Journal of Speech*, **46**, 44–53.
Brown, N.R. & Siegler, R.S. (1992) The role of availability in the estimation of national populations. *Memory & Cognition*, **20**, 406–12.
Chervany, N.L., Benson, P.G. & Iyer, R.K. (1980) The planning stage in statistical reasoning. *The American Statistician*, **34**, 222–6.
Cohen, L.J. (1977). *The Probable and the Provable*. Clarendon, Oxford.
Collins, A. & Michalski, R. (1989) The logic of plausible reasoning: A core theory. *Cognitive Science*, **13**, 1–49.
Coombs, C.H., Raiffa, H. & Thrall, R.M. (1954) Some views on mathematical models and measurement theory. *Psychological Review*, **61**, 132–44.
Curley, S.P., Browne, G.J., Smith, G.F. & Benson, P.G. (1994a, March) *Arguments in the practical reasoning underlying constructive probability responses*. Working paper, University of Minnesota, Minneapolis, MN.
Curley, S.P., Browne, G.J., Smith, G.F. & Benson, P.G. (1994b, January) *Practical reasoning in the construction of probability responses*. Working paper, University of Minnesota, Minneapolis, MN.
Curley, S.P., Eraker, S.A. & Yates, J.F. (1984) An investigation of patients' reactions to therapeutic uncertainty. *Medical Decision Making*, **4**, 501–11.
Curley, S.P. & Golden, J.I. (1994). Using belief functions to represent degrees of belief. *Organizational Behavior and Human Decision Processes* **58**, 271–303.
Curley, S.P. & Yates, J.F. (1989) An empirical evaluation of descriptive models of ambiguity reactions in choice situations. *Journal of Mathematical Psychology*, **33**, 397–427.
Curley, S.P., Yates, J.F. & Abrams, R.A. (1986) Psychological sources of ambiguity avoidance. *Organizational Behavior and Human Decision Processes*, **38**, 230–56.
Dempster, A.P. (1968) A generalization of Bayesian inference. *Journal of the Royal Statistical Society* (Series B), **30**, 205–47.
Edwards, W. (1982) Conservatism in human information processing. In D. Kahneman, P. Slovic & A. Tversky (eds.), *Judgment under Uncertainty: Heuristics and Biases*. Cambridge University Press, Cambridge, pages 359–69.
Einhorn, H.J. & Hogarth, R.M. (1981) Behavioral decision theory: Processes of judgment and choice. *Annual Review of Psychology*, **32**, 53–88.
Einhorn, H.J. & Hogarth, R.M. (1986) Judging probable cause. *Psychological Bulletin*, **99**, 3–19.

Ellsberg, D. (1961) Risk, ambiguity, and the Savage axioms. *Quarterly Journal of Economics*, **75**, 643–69.
Fischhoff, B. (1982) Debiasing. In D. Kahneman, P. Slovic & A. Tversky (eds.), *Judgment under Uncertainty: Heuristics and Biases*. Cambridge University Press, Cambridge, pages 422–44.
Fishburn, P.C. (1964) *Decision and Value Theory*. Wiley, New York.
Fletcher, K.E. & Huff, A.S. (1990) Strategic argument mapping: A study of strategy reformulation at AT&T. In A.S. Huff (ed.), *Mapping Strategic Thought*. Wiley, New York, pages 165–93.
Good, I.J. (1959) Kinds of probability. *Science*, **129**, 443–7.
Hammond, K.R., Stewart, T.R., Brehmer, B. & Steinmann, D.O. (1975) Social judgment theory. In M.F. Kaplan & S. Schwartz (eds.), *Human Judgment and Decision Processes*. Academic Press, New York, pages 272–312.
Hart, J.T. (1967) Memory and the memory-monitoring process. *Journal of Verbal Learning and Verbal Behavior*, **6**, 689–91.
Hogarth, R.M. (1992, November) Ambiguity and arguments. Presidential address at the meeting of the Society for Judgment and Decision Making, St. Louis, MO.
Hogarth, R.M. & Kunreuther, H. (1989) Risk, ambiguity, and insurance. *Journal of Risk and Uncertainty*, **2**, 5–35.
Kahneman D., Slovic P. & Tversky, A. (eds.) (1982). *Judgment under Uncertainty: Heuristics and Biases*. Cambridge University Press, Cambridge.
Kahneman, D. & Tversky, A. (1972) Subjective probability: A judgment of representativeness. *Cognitive Psychology*, **3**, 430–54.
Kahneman, D. & Tversky, A. (1973) On the psychology of prediction. *Psychological Review*, **80**, 237–51.
Keynes, J.M. (1921) *A Treatise on Probability*. Macmillan, London.
Kirkwood, C. (1990, May). *A tutorial summary of methods for applied decision analysis*. Technical Report DIS 89/90-7, Arizona State University, Tempe, AZ 85287.
Mayer, R.E. (1992) *Thinking, Problem Solving, Cognition* (2nd edn.). Freeman, New York.
Mayzner, M.S. & Tresselt, M.E. (1965) Tables of single-letter and bigram frequency counts for various word-length and letter-position combinations. *Psychonomic Monograph Supplements*, **1**(2), 13–32.
Merkhofer, M. W. (1987) Quantifying judgmental uncertainty: Methodology, experiences, and insights. *IEEE Transactions on Systems, Man, and Cybernetics*, **SMC-17**, 741–52.
Nisbett, R.E., Krantz, D.H., Jepson, C. & Kunda, Z. (1983) The use of statistical heuristics in everyday inductive reasoning. *Psychological Review*, **90**, 339–63.
Payne, J.W., Bettman, J.R. & Johnson, E.J. (1992) Behavioral decision research: A constructive processing perspective. *Annual Review of Psychology*, **43**, 87–131.
Perkins, D.N., Allen, R. & Hafner, J. (1983) Difficulties in everyday reasoning. In W. Maxwell (ed.), *Thinking: The Expanding Frontier*. Franklin Institute Press, Philadelphia, pages 177–89.
Poses, R.H. & Anthony, M. (1991) Availability, wishful thinking, and physicians' diagnostic judgments for patients with suspected bacteremia. *Medical Decision Making*, **11**, 159–68.
Reyna, V.F. & Brainerd, C.J. (1992) A fuzzy-trace theory of reasoning and remembering: Patterns, paradoxes, and parallelism. In A.F. Healy, R.M. Shiffrin & S.M. Kosslyn (eds.), *From Learning Processes to Cognitive Processes: Essays in Honor of William K. Estes* (Vol. 2), Erlbaum, Hillsdale, NJ, pages 235–60.
Saks, M.J. & Kidd, R.F. (1986) Human information processing and adjudication: Trial

by heuristics. In H.R. Arkes & K.R. Hammond (eds.), *Judgment and Decision Making: An Interdisciplinary Reader*, Cambridge University Press, Cambridge, pages 213–42.
Shafer, G. (1976) *A Mathematical Theory of Evidence*. Princeton University Press, Princeton, NJ.
Shafer, G. (1978) Non-additive probabilities in the work of Bernoulli and Lambert. *Archive for History of Exact Sciences*, **19**, 309–70.
Shafer, G. (1981) Constructive probability. *Synthese*, **48**, 1–60.
Shafer, G. & Tversky, A. (1985) Languages and designs for probability judgment. *Cognitive Science*, **9**, 309–39.
Shope, R.K. (1983) *The Analysis of Knowing*. Princeton University Press, Princeton, NJ.
Smith, G.F., Benson, P.G. & Curley, S.P. (1991) Belief, knowledge, and uncertainty: A cognitive perspective on subjective probability. *Organizational Behavior and Human Decision Processes*, **48**, 291–321.
Spetzler, C.S. & Stael von Holstein, C.S. (1975), Probability encoding in decision analysis. *Management Science*, **22**, 340–58.
Stael von Holstein, C.S. & Matheson, J.E. (1979) *A manual for encoding probability distributions*. SRI International, Menlo Park, CA.
Toulmin, S.E. (1958) *The Uses of Argument*. Cambridge University Press, Cambridge.
Tversky, A. & Kahneman, D. (1973) Availability: A heuristic for judging frequency and probability. *Cognitive Psychology*, **5**, 207–32.
Tversky, A. & Kahneman, D. (1974) Judgment under uncertainty: Heuristics and biases. *Science*, **185**, 1124–31.
Van Wallendael, L.R. (1992, November) J/DM representation in the introductory psychology course. Working paper, University of North Carolina at Charlotte.
von Winterfeldt, D. & Edwards, W. (1986) *Decision Analysis and Behavioral Research*. Cambridge University Press, Cambridge.
Wallsten, T.S. & Budescu, D.V. (1983) Encoding subjective probabilities: A psychological and psychometric review. *Management Science*, **29**, 151–73.
Weatherford, R. (1982) *Philosophical Foundations of Probability Theory*. Routledge & Kegan Paul, London.
Well, A.D., Pollatsek, A. & Boyce, S.J. (1990) Understanding the effects of sample size on the variability of the mean. *Organizational Behavior and Human Decision Processes*, **47**, 289–312.
Wyer, R.S. Jr. & Hartwick, J. (1980) The role of information retrieval and conditional inference processes in belief formation and change. In L. Berkowitz (ed.), *Advances in Experimental Social Psychology*, Vol. 13. Academic Press, New York, pages 241–84.
Yates, J.F. (1982) External correspondence: Decompositions of the mean probability score. *Organizational Behavior and Human Performance*, **30**, 132–56.

Chapter 10

Variants of Subjective Probabilities: Concepts, Norms, and Biases

Karl Halvor Teigen
University of Tromsø

Sometimes it is difficult to know what to believe, what to expect, or what is the case. It can be even harder to tell how strongly you believe it, how much you expect it, or how close it is to the truth. Yet, that is what probability judgments are about.

To make things even more complicated, most people are aware that, in addition to their intuitive probabilities and uncertainties, there are certain rules of the game. And they are right. For many investigators, these rules and their violations) have top priority, even if they are telling their subjects to go ahead and give probability estimates solely according to what they "feel", rather than to engage themselves in mental calculations (the calculations already having been performed by the experimenter, to his or her own satisfaction).

The correspondences and, particularly, the noncorrespondences, between subjects' judgments and the experimenter's calculations have become the subject of a rich literature on probability inaccuracies, shortcomings and biases, as can be testified to by several contributions to this volume. It has also become a central issue in the debate about human rationality (Cohen, 1981; Jungermann, 1983).

Subjective Probability. Edited by G. Wright and P. Ayton.
© 1994 John Wiley & Sons Ltd.

The position taken in the present chapter is that subjective judgments of probability can be, and are, arrived at by a number of different processes, which may or may not cause them to deviate from the experimenter's norms. This is not the place to enter into a debate about when they should or should not conform to those norms, i.e. the appropriateness and normativity of the rules themselves. A more descriptive approach would be to ask under which conditions are judgments and norms most often in agreement, and when are they not?

To do so, we will first discuss two different avenues, the rule-dictated (quasi-mathematical) versus the intuitive approach, which may be taken to arrive at probability judgments. Next, we will present a "family" of subjective probability concepts, which can be activated under different circumstances and give rise to different judgmental biases. Thirdly, we will give a brief overview of circumstances that appear to affect the "normativity" of subjective probability estimates.

10.1 THE RULE-DICTATED APPROACH

Imagine that you are asked to draw a floor plan of your apartment. There are basically two ways of going about the task.

(1) You may start with the entrance, or any other room of your choice, reproducing it shape and proportions as well as you remember, adding windows and doors, then turn to the next room and do the same, until you discover that the totality of the plan is becoming impossible, rooms sticking out in all directions like the petals of a flower, unrestricted by any architecturally admissible exterior wall.
(2) The other, more professional (although less intuitively appealing) way, is to work according to a rule that is *known* to be correct, for instance that the house as a whole has a rectangular ground plan, that all rooms have adjacent walls and must fit within the same general framework with all space accounted for.

Similarly, probability judgments can be made either *intuitively* from the weight of arguments, feelings of conviction, degree of knowledge, or from whatever judgmental heuristic that seems to be most appropriate for evaluating the phenomenon under consideration; or more *indirectly*, via rules that are not themselves seen as integral parts of the object of evaluation, but rather belonging to the domain of probability calculus, as far as this is understood by the subject. We may call the latter "rule-dictated" probability assessments, or "quasi-mathematical", to emphasize that the subject is using something like a mathematical rule, but not necessarily the appropriate one.

The vast literature on fallacies and biases of probabilistic thinking seems to indicate that probability judgments are less rule-dictated than they "ought" to be. That is not surprising, if one considers that the rules are, in some sense, external to the objects of judgment. For instance it is natural to think that the probability of a horse winning a particular race would be estimated better by someone with a knowledge about racehorses than by someone with a knowledge about probabilities. But sometimes, the latter knowledge is no less essential than the former.

At the same time, education and daily life experiences have taught most people a few elementary probability rules. Indeed, the traditional claim is that the theory of probability itself can be regarded as a formalization and explication of "bon sense" (Laplace, 1816). Also, probability judgments are typically made on 0–100% or 0–1.00 probability scales, making people aware that there are some similarities between probability estimates, percentages and/or proportions. At the very least, it suggests a scale with a lower and an upper limit—for instance, one is not allowed to be 300% sure.

Perhaps because most attention has been given to deviations from the rule, few investigators seem to have focused upon which rules people generally seem to accept as valid (even if they do not always abide by them).

Provisionally, I will suggest the following ones:

The 50/50 rule. When two outcomes, or alternatives seem equally possible, the probability of either is $\frac{1}{2}$, or 50%.

The proportion rule. When one result, or alternative, can be obtained in a number of equivalent ways, its probability can be estimated as the proportion a/n, where a is the number of outcomes yielding the result in question, and n is the total number of equivalent outcomes.

This is the classical "urn" case, where e.g. the probability of drawing a red bead from an urn with 30 red beads out of a total of 100 is estimated to be 0.30.

Proportions can also be represented spatially. For instance the probability of getting "red" with a fortune wheel with 60% red sectors (Figure 10.1) will, by most people, be estimated to be 0.60.

The 1/n rule. When several outcomes, or alternatives, seem equally possible (e.g. the six sides of a die), the probability of one of them is acknowledged to be $1/n$.

These three rules are, of course, all reducible to the same general principle of classical probability calculus, namely that p values should reflect the proportion of "favourable" outcomes. But in practice, they may be not be understood as such. At least it seems that the first principle is easier to grasp and to apply than the second and especially the third one.

The relative frequency rule. When something has happened in the past with a relative frequency of a/n, its probability of happening in the future can also be estimated as a/n.

This is, of course, the basic principle of the frequentistic conception of probabilities, often assumed to establish the classical probability doctrine on a sounder and more empirical basis than the proportion rule. Our experience is, however, that people more often fail to see the relevance of relative frequencies, over time, than of relative proportions. In other words, the probability of throwing a six with a regular die is assumed to be 1/6, not because the six comes up 1/6 of the time, but because the die has 6 sides. Relative frequencies are seen to *reflect* probabilities (and can hence be taken as indications or approximations), but do not *create* them. If there is any cause–effect relationship at work here, it goes from probabilities to frequencies and not the other way around.

The complementarity rule. If the probability of something happening is calculated or estimated to be p, then the probability for something else happening is, by implication, $1 - p$. This assumes additivity, or a "distributive" probability concept, which seems to be attended to only under favourable circumstances, even if it also follows from the proportion and the relative frequency rules (for nonadditivity see e.g. Teigen, 1974a, 1974b, 1983c; Robinson & Hastie, 1985). But at least occasionally, subjects will calculate a probability by this rule instead of intuiting it. For instance, somebody would say: "Since I am 99% sure, I assume I have a 1% chance of being wrong" (or the other way around, if the 99% figure was due to the 1% estimate). This rule is also in many cases forced upon subjects when they are told to make sure that their p estimates add up to 100%, or are required to make graphical estimates on a 100% line, or on a pie chart which does not permit violations of the complementarity principle.

Students with a basis in statistics may use more advanced probability rules, as for instance the product rule for arriving at the p value of a conjunction, but such applications seem to be extremely rare (cf. Tversky & Kahneman, 1983; Nahinsky, Ash & Cohen, 1986). And even when computations are used, the resulting estimate may be incorrect, as when the 50/50 rule is misapplied to situations involving uneven proportions, or more than two alternatives.

In Kahneman and Tversky's well-known study of base-rate negligence (1973), Subjects were confronted with a hypothetical individual drawn from a sample of 30 engineers and 70 lawyers. Only when given no information whatsoever about this individual did the majority of subjects correctly apply the proportion rule, estimating his probability of being an engineer as 30%. When given an uninformative description (that could equally well characterize engineers and lawyers), most subjects switched to a 50/50 rule, regardless of base rates. With somewhat more diagnostic descriptions, no calculations seemed to be performed, probability judgments apparently following an intuitive "representativeness" heuristic.

In other cases one can observe how a calculated estimate is felt to be in conflict with one's less educated intuitions. For instance, if Tom is competing for a job with four other, equally qualified applicants, his chances will be estimated—according to the *1/n rule*—to be about 20%. But when asked to select the most appropriate verbal phrase characterizing his candidacy, a majority will prefer high-probability expressions (e.g. "he has a good chance") to low-probability phrases (e.g. "it is doubtful"). Yet few people will consider 20% as a "good chance" (Teigen, 1988a).

10.2 VARIANTS OF INTUITIVE PROBABILITIES

When people leave the tidy world of rules (because they don't know how to calculate *p* values, or the rules don't seem to apply), and start judging probabilities on an intuitive basis, it turns out that they have several intuitions to choose from. We may speak of them as a family of subjective probability concepts, or, following Kahneman and Tversky (1982), "variants of uncertainty". In their article, Kahneman and Tversky distinguish between external and internal attributions of uncertainty, and go on to subdivide the former into frequency-based and propensity-based probabilities; the latter into probabilities based on arguments vs probabilities based on introspective convincingness.

The variants of uncertainty discussed in the present chapter follow a related scheme, partly based on the external/internal distinction. However, instead of the further subdivisions described by Kahneman and Tversky, which seem to refer primarily to which kind of *evidence* that can be marshalled in favor of a particular probability judgment, we will suggest an indetermination/determination distinction cutting across the external/internal dimension, creating altogether four conceptually different uncertainties. In addition, we will discuss uncertainty in the world of action, in terms of control and lack of control; and in the world of fiction, in terms of plausibility and coherence (or lack thereof). An overview of these six species of probability concepts is given in Table 10.1.

External vs internal uncertainties. This distinction refers to the fact that we sometimes think of tendencies and lack of determination as being a part of the external world, whereas at other times, it is more due to our "internal" state of knowledge and belief. For instance, many people would say that the outcome of a lottery is undecided because of the random character of the drawing process. Similarly, downhill skiing may be considered to be inherently risky because of uncontrollable external factors that can easily turn an enjoyable activity into its opposite. On the other hand, if I say that New York is "probably" north of Paris, it is not because it is hard to tell New York's whereabouts these days, but because my geographical knowledge is not precise enough for me to offer a decisive opinion.

Table 10.1 Intuitive probability concepts

	External	Internal
Emphasis on indetermination	I Chance	IV Ignorance, uncertainty
Emphasis on tendency	II Disposition, propensity	III Confidence, belief
Pragmatic	V Controllability	
Fictional	VI Plausibility	

Determinists may of course claim that even a throw of dice is only uncertain because we do not *know* all factors involved (all probabilities being internal), whereas frequentists will have a hard time accepting "my" probability of the location of New York, unless it is taken to mean my estimate of how often I will be right when answering this type of question, as all probabilities, according to this view, need an objective relative frequency as their external reference (this is in fact the logic behind all so-called calibration curves). In the history of probability calculus, we find a similar distinction between "epistemic" and "aleatory" probabilities (Hacking, 1975). For our purpose, however, the point is that, under different circumstances, people seem to acknowledge and make use of both kinds of uncertainties, although the distinction may at times be blurred. Consider the following example.

After a storm, a ship is reported missing. The newsreporter writes: "For every hour that goes by, the chances of finding the crew alive are decreasing." The statement is ambiguous. The decrease in chances can refer to external probabilities, if we think of the crew as drowning or freezing to death if not found within the next few hours. But it may also refer to internal probabilities: a disaster may have happened, but we do not know for sure. For every hour that goes by without a sign of life, *our* probability of finding survivors decreases because we will have stronger reasons to believe the worst.

Tendencies vs. indetermination. Whether one has to do with judgments of external or internal probabilities, the task will often be one of focusing upon a particular outcome or possibility, trying to determine the extent to which this particular alternative can be counted on as being (or becoming) the case. Do we have strong or weak reasons to believe it? Is it close to or far from actually taking place? In these cases, probability judgments can be regarded as estimations of the strength of a tendency: for external probabilities, this will be the disposition or propensity for something to occur, as when we speak of a disease as "probably fatal", meaning that it is within its power to kill those afflicted. For internal probabilities, we can similarly speak of a tendency *in my mind* to believe or feel confident about something.

Indetermination refers to the other side of the coin, or, to use a slightly different metaphor, it refers to the hole rather than to the doughnut. There may be reasons to believe, or to think in terms of tendencies, but there are also reasons for agnosticism and ambiguity. From an external perspective, we may claim that many phenomena in themselves are indeterminate, with more or less randomness involved. From an internal perspective, uncertainty arises due to lack of essential information or because of conflicting knowledge. In both cases, we may be concerned with the existence of *several* possibilities rather than just one.

The feeling that uncertainty and probability do not refer to exactly the same thing has struck several authors who have been grappling with the problem of finding a general term to cover all phenomena of subjective probabilities and uncertainties (Smithson, 1988; Howell & Burnett, 1978; Peterson & Pitz, 1988). Hacking (1975) suggested that the terms "probability" and "uncertainty" should be used to refer to external and internal probabilities, respectively. I think it is more in line with common usage to identify these words with the tendency/indeterminacy distinction. When we speak of probabilities, we usually think of the tendency of an outcome to occur, whereas uncertainty more often refers to the existence of alternatives, and the lack of (external or internal) determination. Peterson and Pitz (1988) draw a similar distinction between "confidence" and "uncertainty", the latter referring to "a set of outcomes that has a fixed probability" (page 85). This is a question of the range of possibilities, rather than degree of expectancy or belief associated with one in particular.

10.2.1 Chance Probabilities (Type I)

As is well known, probability calculus evolved from studies of games of chance. These are basically situations of external indetermination, characterized by a number possible outcome, the two sides of a coin, the six sides of a die, the 37 sectors of a roulette wheel, or the 52 cards in a deck, there being no obvious way of deciding in favour of one rather than another alternative. Under such circumstances, people seem relatively willing to follow elementary laws of probability, and to calculate *p* values in accordance with the "rules". These are of course the situations Laplace had in mind when he referred to probability calculus as an extension of common sense, yet it is well worth noticing that he did not refer to the intuitions of just anybody, but to those of "l'esprits justes" (the just spirits). Moreover, Laplace included a chapter "Concerning illusions in the estimation of probabilities", discussing typical errors of chance like the gamblers' fallacy. Still, it seems that probabilistic reasoning is at its best (or rather at its most normative), when the situation is conceived as one in which chance plays a role in producing the results (Nisbett et al., 1983; Hoch, 1985; Ginossar & Trope, 1987; Wasserman, Lampert & Hastie, 1991).

The concept of chance probability has, however, two serious limitations. One is the expectation that a chance event should "look" random, which for most people means an absence of any apparent biases and regularities. That means for instance that a random number between 1 and 20 will not be expected to be close to the extremes, like 1, 2, 19, or 20, nor will it be in the exact middle, like 10, or too regular, like 5 and 15 (Teigen, 1983a). In a random series, repetitions, symmetries, orders, and other apparent regularities are not expected to occur. This means that even when equiprobability of outcomes is admitted on principle, some outcomes are, to paraphrase Orwell, more equal than others. These are the outcomes that, in Kahneman & Tversky's terms seem to be "representative" of the set we are sampling from, as well as of the sampling process.

Problems may also be due to a process we could call "subjective significance testing". People, and researchers, seem to share the principle that "If something has a low probability of occurring by chance, it is probably due to something *other* than chance." This generally sound heuristic is limited by the fact that even low probability events *can* happen by chance. In fact it is in the very nature of chance to make improbable things happen, at least from time to time. For instance every single hand of cards dealt in a game of bridge has a vanishingly low probability of occurring. Yet it takes some reflection to accept Agathon's warning (as quoted by Aristotle):

> One might perchance say this was probable:
> That things improbable oft will happen to men.
> (Rethorics, 1402b).

The "significance heuristic" also fails to take into account that in many cases, there is no plausible alternative hypothesis. Five heads in a row by chance may be quite unlikely (although not more than any other sequence of coin tosses). It may be more compatible with an alternative hypothesis about *luck*, but then we have to forget that the luck hypothesis itself is not very plausible, being rather incompatible with our other, more scientific beliefs.

These two heuristics, "representativeness" and "subjective significance testing", can together create a climate in which the chance concept of probabilities is too easily discarded. According to the former, a number of chance outcomes will seem improbable, and according to the latter, this is taken to indicate that they are not due to chance after all. In areas in which no other explanations are readily available, the concept of good and bad luck fills the gap. According to Proctor (1887), gamblers' theory of luck contain two predictions: (a) it follows the player who is "in vein", making it likely that after winning, he will go on winning, but (b) it may be used up, making it more likely that after some winning, he will suddenly start losing again. The unpredictable part is that one never knows when it will change from (a) to (b), making the theory practically useless and theoretically unfalsifiable. But similar theories of luck seem still to be popularly accepted (Wagenaar, 1988; see also Chapter 19).

10.2.2 Dispositional Probabilities (Type II)

Researchers of subjective probabilities have often been struck by the fact that people seem to treat probabilities as attributes of particular outcomes rather than relative to the set of outcomes. Evidence of this comes from studies showing subjects' neglect of base rates as well as violations of the complementarity rule (nonadditive probabilities).

It can, for instance be shown that estimated probabilities of a particular outcome and a group of outcomes (containing that particular value) will be nearly the same (Alberoni, 1962; Teigen, 1974b). Also a person's probability to choose a particular occupation, or to win a particular contest, will not change much by adding or subtracting occupations, or contestants, from the list of alternatives (Teigen, 1983c). Similarly, the probability of a particular hypothesis may not change when evidence is supplied that changes the likelihood of rival hypotheses (Robinson & Hastie, 1985). This phenomenon may also throw some light on the neglect of the category of "other" (unspecified) outcomes or alternatives. In studies of "pruned" decision trees it has been shown that the subjective probabilities of unspecified outcomes are easily underestimated, and this is all the more so the more they cover (Fischhoff, Slovic & Lichtenstein, 1978). While it is easy to see that there is a substantial probability that your car will not start for this or that particular reason, which had an obvious potential for interfering with its normal functioning, it is harder to see how "other" (unspecified) factors carry the same (or an even greater) potential.

Dispositional probability thinking may come close to a reification of probabilities. Some behaviors or substances may for instance be thought of as *inherently* dangerous, regardless of how they are handled. Some people are "born winners", regardless of the number and quality of their competitors. Even a chance probability, originally calculated by a proportion or relative frequency rule, may later on be regarded as dispositional. The gamblers' fallacy can perhaps be interpreted along these lines, as a particular *disposition* on the part of the coin to keep the pattern of outcomes in a kind of homeostatic balance.

How do people estimate dispositional probabilities?

Since these probabilities are conceived as tendencies to actual occurrence, their magnitude should be most directly estimated by some measure of how *easily* the outcome in question may occur, or how *close* it is to becoming realized. For instance, a cup balanced on the edge of a table will seem to have a high probability of falling down because it is so close to toppling over, and a fall can so easily happen (a light push or shake will be enough). Thus we may overestimate the probability of a particular outcome, e.g. an airplane crash, if we think of all the things that can go wrong and how small the safety margins seem to be.

Descriptions of accidents and near misses offer rich material for studying the easiness and closeness mechanism. For instance in a Norwegian student population, 50–75% claim to have experienced (and survived) a life-threatening situation (Teigen & Brun, unpublished data). When asked to estimate their probabilities of actually being killed, nearly all gave *p* values in the range of 0.50–0.99, indicating, if taken literally, that a substantial percentage (at least one third) of the Norwegian population should not be expected to achieve adulthood. From the descriptions, probabilities seem to reflect judgments of closeness (often in terms of seconds and centimeters) to catastrophe.

Even if not imminent, an outcome may be considered highly probable if there is a tendency clearly pointing in its direction. To stand one foot away from the cliff may be considered dangerous, to approach it (even several feet away) may appear even more so. Near accident stories are replete with such descriptions, the situations being of a kind that seems *headed* toward disaster. Again, this is a kind of scenario in which the probability of disaster tends to become overestimated, compared to actual accident frequencies.

To support the claim that a particular outcome can be expected or not expected to happen, people will draw upon their knowledge of how *common* the phenomenon seems to be, and its *typicality*, or "representativeness" (Kahneman & Tversky, 1972). Both principles have been endorsed as legitimate ways of establishing probabilities by classical writers like Hume (1748/1976) and Laplace (1816). But these principles have also a psychological counterpart, in the two basic laws of association: contiguity and similarity, making our subjective judgments of probabilities related to, but not identical to what they "ought" to be (the strength of an association also obeying secondary laws, like recency and vividness). There is a clear parallel between these principles and the judgmental heuristics of availability and representativeness introduced by Kahneman and Tversky (1972; Tversky & Kahneman, 1973; cf. Teigen, 1989).

10.2.3. Confidence (Type III)

Confidence, or *degree of belief*, can be regarded as an internal counterpart of dispositional probabilities. In both cases, we are trying to quantify a particular tendency, but in one case it is the tendency for something to take place outside of the observer, in the other it is about its existence in his mind, as a more or less strong conviction.

It is usually assumed that these two tendencies should mirror each other, or specifically, that the internal should mirror the external one. "The wise man," wrote Hume, "proportions his belief to the evidence. . . . He weighs the opposite experiments: He considers which side is supported by the greater number of experiments: to that side he inclines, with doubt and hesitation" (1748/1957, pages 110–111). But even the wise man may run into some problems, particularly when he does not know the evidence. For instance, after election day, most external uncertainty about who is going to be the next president is gone, the candidates' probabilities being either zero or close to one. The wisest of hermits, without an access to newspapers and broadcasts, must however stick to his internal probabilities, perhaps with an acute feeling that they now are becoming a very private affair with no external counterparts.

The difference between estimated external probabilities and estimated confidence refers to the difference between the two questions: "How certain is it?" and "How certain are you?", two questions that seem to invite one to an objective vs. a subjective (introspective) orientation, reminiscent of the Gibsonian distinction between perception and sensation (Gibson, 1966, referring back to Reid, 1785). The two may match when subjects have access to all relevant information and are trying to base their judgments on "the whole truth and nothing but the truth" (cf. Beach & Wise, 1969). They will be out of step when relevant knowledge exists that is for the moment unavailable for the person. Such situations have been referred to as "ambiguous" (Frisch & Baron, 1988), and are often described as more uncomfortable than other equally uncertain situations, but where the internal and external probabilities are more comparable. For instance, people have a clear preference for guessing the outcome of an event before rather than after it has taken place (Brun & Teigen, 1990).

There are also occasions in which subjective confidence *exceeds* the estimated external probabilities. Such cases have not received much attention in the probability judgment literature, but is well known from other fields, including clinical psychology and the study of religious beliefs. In fact, faith is often defined as "belief despite the absence of external evidence", and has often been defended precisely on the grounds of its irrationality (cf. Tertullian's "credo quia absurdum"). In line with the above reasoning, it is natural to guess that in these cases, the person may feel that he has a privileged access to more information (perhaps of a highly private nature) than is generally available.

Thus the difference between confidence and probability becomes especially clear when there is a discrepancy between what we think *we* know and what *is* to be known, either because *our* knowledge is scarce, or because the facts themselves are unknowable, making everything "a question of belief". As is well known from the history of religious ideas, agnosticism is a hard position to hold, although easy to justify.

Confidence estimations are often made as a kind of secondary, or second-order judgments, after subjects have already made their choice as to what they think is, or will be the case. Keren (1991), in his conceptual and methodological review of the calibration literature, describes confidence judgments as a two-step affair (selection preceding estimation), which may or may not be the case with judgments of external probabilities. It follows that it is unproblematic to ask for (external) probability estimates of alternatives one does not really believe in, whereas a question about *confidence* presupposes an already favored candidate. This may even be the case with *doubt*: There is no reason for having strong doubts about a particular hypothesis or outcome unless it has already been singled out as an object for belief. Hence, beliefs and doubts are not necessarily of complementary magnitudes. For instance the same specific (a opposed to a more general) statement was selected by a majority of respondents as the one they were most confident as well as most skeptical about (Teigen, 1990).

Confidence judgments will thus depend upon a number of factors, in addition to estimated probabilities. In one investigation (Teigen, 1983d) they were found to be inversely related to the amount of chance assumed to be involved. They have often been found to increase with subjects' degree of expertise (Bradley, 1981), and with the amount of information available, even without a parallel increase in judgment accuracy (e.g. Oskamp, 1965). This phenomenon has been explained as partly based on an expected link between knowledge and accuracy, carried over to areas which cannot be safely predicted, not even by experts. In addition, the more information, the easier it is to find *some* good reason for the phenomenon to occur exactly as predicted. It should of course also be easier to see that something *else* may be the case. But in confidence judgments, this is a less pressing question, unless specifically asked for (Hoch, 1985). The most "focalized" phenomena (to use a term borrowed from Klar, 1990, 1991) are usually the *belief* and its *positive* reasons.

The subjective character of beliefs make them readily influenced by our imagination. Kahneman and Tversky's "simulation" heuristic (1982), and similar mechanisms, such as availability and vividness, are all examples of attempts to go from "what is easy to think" to "what is easy to believe". The common denominator of these mechanisms was clearly expressed by John Stuart Mill as the "a priori fallacy" to believe that our imagination mirrors the world, so what is most natural for us to think must also exist, and what we cannot conceive must be non-existent. ". . .even of things not altogether inconceivable, what we can conceive with the greatest ease is likeliest to be true" (Mill, 1856, page 312).

Perhaps the most distinguishing and interesting aspect of belief (setting it clearly apart from external probabilities) is its relationship to personal *decisions*, which may affect it more strongly than external facts. We speak of *faith* in the absence of evidence, and *trust* in the face of interpersonal risks, and seem in both cases to assume that there are certainties that can (and should) be chosen as a matter of principle, not because they are intellectually convincing, but because they are morally better and/or pragmatically superior than their opposites (Gambetta, 1988).

10.2.4 Uncertainty by Ignorance (Type IV)

Apart from having greater and less confidence in our chosen hypothesis, or prediction, we can feel more or less certain about which hypotheses or predictions to choose. For all we know, there may be several different possibilities. Peterson and Pitz (1986, 1988) have introduced an interesting distinction between "confidence" and "uncertainty" of a prediction. While the first is commonly measured by asking for a likelihood or probability that a prediction is correct, the second refers to a person's belief about the *variability* of possible outcomes. What is the range of possibilities, and how much should one of them be preferred to the others? Peterson and Pitz go on to compare this distinction to the difference between hypothesis generation and hypothesis testing. In a prediction task, people may either generate possible outcomes, which can be said to reflect their uncertainty, or they may evaluate an already selected outcome, yielding their level of confidence (Peterson & Pitz, 1988).

Although confidence and uncertainty can be thought of as logical complements, it seems that they can, psychologically, vary to some extent independently of each other. There may be cases where several alternatives can be ruled out, and there is only one possibility left. We may in such cases say that "the only possibility", or "the only explanation" must be so and so, but without any positive evidence our confidence may still be quite low even if the variability of options has narrowed down considerably (cf. Chapter 15). In their investigations, Peterson and Pitz (1986, 1988) show that added information can increase *both* confidence and uncertainty, in the sense that a person attaches a stronger belief in his chosen prediction, but is also more willing to admit that several other outcomes are possible. This can be shown to apply also to one's own performances. For instance, members of a running club became more confident about their "best guess" of their own race time, the closer they came to (and presumably the more they knew about) the actual race. But at the same time their confidence intervals concerning potential race times increased significantly (Pitz & Peterson, 1987). Their explanation is that the information acquired in preparation for the race suggested more outcomes to be possible than were previously considered.

Peterson and Pitz's "uncertainty" concept obviously refers to experiences of being "open" or "undecided" among alternatives. To the extent that this is attributed to lack of diagnostic information, we can term this particular variant of uncertainty "uncertainty by ignorance". The finding that this uncertainty can increase with general knowledge can perhaps be regarded as a demonstration of the proverbial awareness of ignorance acquired by education: "the more we know the more we know we don't know".

With quantitative judgments, degree of uncertainty/ignorance can often be expressed as interval rather than point predictions. The size of the chosen interval, or the prediction's coarseness, indicates the person's range of possibilities. Yaniv and Foster (1992) refer to this aspect of uncertainty as "graininess of judgment", and report a close relationship between subjects' choice of judgment scale (with broad rather than fine intervals), width of confidence intervals, and lack of expertise. But grain is not chosen just to make sure there is no error. In their example, somebody who is asked to predict next year's inflation rate would not say "between 1 and 150%" just to be on the safe side. Since people also want to be as informative as possible, they will probably choose an interval (grain) that is precise enough to be regarded as informative, yet wide enough to be approximately right. This "tradeoff" between accuracy and informativeness will often lead to hyperprecision, particularly since confidence intervals are usually underestimated. Somewhat paradoxically, a precise statement is also more readily believed than a more vague or unspecified prediction, even if the latter include the former, and thus has a better chance of actually being correct (Teigen, 1990). This means that the person's choice of "grain" in prediction and judgment will not only reflect his relative ignorance and subjective chances of getting it right, but also his wish to be informative and to be believed. Of experts, we don't just expect truth, but confidence and precision (Shanteau, 1987).

10.2.5 Controllability

In theories of attribution and locus of control, the internal/external distinction is used in a different sense than the one introduced above. Instead of referring to the causes of our uncertainty, attribution theorists refer to the causes of the event itself, which can be regarded as external, if independent of the perceiver, and internal, if his qualities or actions are seen to be instrumental in bringing it about. If a person believes that what is happening in his life is largely dependent upon his own behavior, and his own decisions, he is said to have an internal locus of control (Rotter, 1966; Lefcourt, 1982).

Perceptions of degree of control can be regarded as an aspect of uncertainty not readily explained by the variants of uncertainty discussed till now. In their "cognitive taxonomy" of uncertainty measurement, Howell and Burnett (1978) refer to degree of control as an internal/external dimension, whereas in our terminology, controllability refers chiefly to external factors, in the sense that the probabilities of controllable as well as uncontrollable outcomes are *not* dependent upon what the person knows or thinks, but what he (or somebody else) more or less successfully *does*. At the same time, personal control gives a sense of certainty that is phenomenologically different from knowing for sure what is going to happen as a result of outside forces. At the positive end of the scale, this is reflected in the distinction between "self-confidence" and "trust", or more abstractly, belief in free will vs. determinism (or fatalism). At the negative end we can distinguish between uncertainty due to an indecisive or incompetent person or to an unpredictable environment.

On this dimension, uncertainty refers to the status and the efficacy of one's own decisions. Uncertainty is felt to be high when I don't know "what to do". On closer examination, this may mean two slightly different things. (a) That I don't fully know *what* I intend to bring about, I have not yet made up my mind. For example, I am still debating on whether I shall say yes or no to write a book chapter. This is a question of the *status* of my decision. Is it made or am I still hovering around in a predecisional state? (b) That I don't know fully *how* to bring about something I have decided to do. For example, I have agreed to write the chapter but for various reasons of an external or internal nature, I am not sure I can make it before the deadline. This is a question of the *efficacy* of my decisions, whether I really can master myself and my environment according to my choice.

The celebrated and much envied "decisive" person is usually believed to be privileged in both respects, certain about *what* to do as well as *how* to do it. The distinction may be blurred because of the belief most people have (rightly or wrongly) about the dynamic character, or inherent efficacy of decisions. Unlike decision theorists, who usually regard a decision as the end point of a complicated cognitive process, ordinary people seem more inclined to think of a decision as a first step towards action, the decision itself taking an active part in producing the result (Teigen, unpublished data).

Subjective probabilities of preferred events have regularly been found to be overrated (compared to the probabilities of neutral or negative events), a phenomenon often referred to as "wishful thinking", "preference–expectancy link", or "unrealistic optimism" (Weinstein, 1980), and has been given a variety of possible explanations. In this connection, it is noteworthy that such overoptimism is particularly likely to occur with events believed to be to some degree personally controllable (Weinstein, 1980; Zakay, 1984). Howell and Burnett (1978) suggest that a person will overvalue the certainty of an outcome the more it can be attributed to his own ability or effort. For instance, students regarded their own probability of becoming involved in a car accident when driving as less than that of a fellow student, whereas their chances as back seat passengers were not estimated differently (Zakay, 1984).

10.2.6 Plausibility

The variants of subjective probabilities discussed thus far all pertain to the relationship between a tentative description of reality on one hand, and what is really the case, on the other. A highly probable state of affairs is "almost" real, either because it can readily happen (external probabilities), or the person has almost enough knowledge to be sure (internal probability), or almost enough power to make it happen at will (controllability). Yet there are situations in which we speak about probabilities with no reference to outer reality. For instance it is common to describe a completely fictional character or the plot in a novel as being more or less "probable", without implying any belief in its existence outside the world of fiction. The ambiguity of the probability concept in literature can be readily observed in documentary books, for instance biographies, where we can sometimes wonder if the author has succeeded in creating a character that is "too good to be true", painting a picture that is very convincing in and of itself, but not necessarily truthful of the real person to be portrayed.

On the other hand, a description of an event known to have taken place can sound so unconvincing that we have to admit that even if true, it does not appear very probable. In this and other contexts one is sometimes asked to "make" the story more probable, *not* by changing its content but the way it is told. This implies that the probability of a particular event, prediction or explanation, is not fixed but can be considered relative to the way it is presented.

How do we assess plausibilities? If other probabilities are about closeness to reality, or closeness to truth, these fictional probabilities can be conceptualized as closeness to a fictional "truth"; how well a given description fits into a created rather than external reality. The means with which this can be done have not been systematically explored in the probability judgment literature. It seems reasonable to believe that plausibilities are affected, among other factors, by the following:

(1) Completeness of a description (it sounds convincing if enough concrete details are added to make it "come alive").

(2) Coherence. The description is perceived to be logical and does not contain internal contradictions.

(3) Causal elements. The description is more than a mere enumeration of facts, but contains also explanations (causes, motives) for the events.

(4) Suggestiveness. The description gives rise to expectations of events and explanations not explicitly stated, as when a novelist tells us that his characters are starting to have a life of their own, telling *him* what will happen instead of the other way around.

(5) Familiarity, acceptability. The description contains enough familiar elements to be recognized as something that "could" exist under the given circumstances. This means that the story can be assimilated to schemas (e.g. implicit personality theories) already existing in the listener's or reader's mind.

(6) Explicitness of premises. It follows from the previous point that plausibility requires that something sounds natural under the given circumstances. It is therefore important to distinguish between the circumstances on one side, which do not have to be natural, and their consequences, which should follow more naturally. A surrealistic story, like one of Kafka's, can have a great deal of plausibility if we are led to believe that it all follows from the "given" premises (which may be weird). We feel less comfortable with a description in which the premises are changing throughout (the author constantly pulling the strings).

One reason for mismatch between subjective and "objective" probability estimates may be due to subjects offering a plausibility rather than an external probability judgment. This may be the mechanism behind at least some instances of the so-called "conjunction fallacy" (Tversky & Kahneman, 1983). In one of their examples, describing John P., a rather suspicious businessman, subjects gave a higher probability rating to the statement "Mr P. killed one of his employees to prevent him from talking to the police", than to the simpler statement "Mr P. killed one of his employees", even if the latter includes the former. According to Macdonald (1986) this can only be regarded as a fallacy if we assume that Kahneman and Tversky and their subjects are referring to the same probability concept. From a plausibility point of view, "a story can become more believable as it develops despite the fact that there is necessarily more to believe" (Macdonald, 1981, p. 19). In line with the present analysis, this interpretation of probabilities in terms of plausibility is particularly defensible, as the subjects are probably right in assuming that they are estimating the probability of a completely fictional event.

10.3 VARIANTS OF UNCERTAINTY AND BIASES OF PROBABILITY ESTIMATION

Table 10.1 gives an overview of the family of uncertainty concepts discussed in this chapter. By "family" I mean that they are sufficiently related to affect each other and to be confused with each other, while at the same time being independent enough to follow different principles and sometimes yield divergent results.

The difference between "statistical" and "causal" thinking is a case in point, the former typically implying probabilities of Type I, and the second probabilities of Type II or V. An individual event will usually be explained and predicted from causal considerations, yielding a dispositional probability, or a judgment of controllability. If asked what "my chances" are to succeed in one particular business venture, others' failure frequencies will often be considered irrelevant, or at most as a rough indication of the level of difficulty in the area (if not as an indication of my competitors' lack of skill). Repeated events are easier to conceptualize along statistical lines, at least if there are no obvious reasons why I should succeed on occasion 2, 5 and 6, but not on 1, 3, and 4. This may be one of the reasons why subjects behave differently in situations involving unique versus repeated gambles (Keren & Wagenaar, 1987). In an attempt to "make statistical illusions disappear", Gigerenzer (1991; Gigerenzer, Hoffrage & Kleinbölting, 1991) shows how typical "errors" in probabilistic reasoning can be reduced when the problems are recast in terms of relative frequencies, rather than as single cases. For instance, if people are asked how many times out of *n* they think they are right, they will display less overconfidence than whey asked to state their degree of confidence in one particular answer (Gigerenzer, Hoffrage & Kleinbölting, 1991). This does not, however, invalidate the overconfidence phenomenon, since one may well ask (a) why the problem is changed by reformulating it in terms of frequencies, and (b) whether subjects are willing to accept their relative frequencies as probabilities. The idea of a "family" of probability concepts supports Gigerenzer's claim that the usual calibration studies, in which subjective confidence judgments are checked against hit frequencies, are not comparing apples to apples, but it goes further by encouraging the study of differences between apples and pears.

The problem of apples and pears seems also to be partly responsible for some other puzzles of probabilistic reasoning. Why do people neglect base rates in the original versions of the engineers' and lawyers' problem (Kahneman & Tversky, 1973), but pay attention to them in some other studies (Manis et al., 1980; Fischhoff & Bar-Hillel, 1984; Gigerenzer, Hell & Blank, 1988). One obvious problem in the original version is that subjects are invited to combine a probability due to random sampling (a description being drawn from a 30/70 set), with a completely different kind of probability based upon the *content* of the descriptions. So the *a priori* odds have to do with chance probabilities (Type I), whereas the more or less diagnostic likelihood ratio is dispositional (Type II), if not completely internal (Type III–IV) or fictional (Type VI). From the literature on social stereotypes it is evident that *a priori* probabilities (even fictional ones) will be taken into consideration when considered "causally relevant"—i.e. when they are seen as dispositional rather than chance. Base rates in the 30/70 type of design will also be attended to if descriptions are less diagnostic (Manis et al., 1980; Fischhoff & Bar-Hillel, 1984) and accordingly less relevant, or when the problem becomes recast as primarily a sampling task by letting subjects actually *draw* the description from an urn (Gigerenzer, Hell & Blank, 1988).

The probability of a complex event will also be judged differently depending upon the type of probability involved. Subjects in Tversky and Kahneman's original study on "conjunction fallacy" (1983) strongly believed that Björn Borg's probability of losing the first set but winning the match in the

Wimbledon finals was higher than his losing the first set, period (nothing said about winning and losing the match). Obviously they were not thinking in terms of pure chances but according to the champion's alleged propensities for winning. In the world of chance, conjunctions are viewed completely differently, namely as "coincidences", which have often been singled out as examples of extremely low-probability events. Indeed, it has been argued that the probability of coincidences is usually underestimated, especially when they are experienced personally (Falk, 1982–83; 1989).

Not only are probabilities estimated differently according to which probability concept is being activated, the meaning of the same *p* value will also differ. For instance, most people prefer to bet upon an uncertain event they know something about, or can control to some extent, compared to an equally uncertain chance event (Heath & Tversky, 1991). Heath and Tversky refer to this as a "competence" effect, and speculate that it has to do with attributions of credits and blame. In an area that can to some extent be known or controlled, successes and failures are readily attributed to the person, which would make him prefer such bets over chance bets (where nobody can be credited or blamed), as long as the probabilities of succeeding are substantial, whereas it would make him prefer chance bets if probabilities of success are small.

One bias that seems to be widespread among all variants of probability is a tendency to underestimate variability and unpredictability. Our minds are generally on the side of law and order. This takes, however, different forms with different probability concepts and should be named accordingly. With uncertainties of Type I and IV it takes the form of under-variability, or hyper-precision. Random events are not assumed to fluctuate too widely from "ideal" randomness, and confidence intervals for imperfectly known quantities are usually underestimated. When it comes to dispositional probabilities (Type II) overestimation is a common occurrence, especially in areas in which there are more than two alternative outcomes (Teigen, 1983c). For probabilities of Type III, overconfidence is more often reported than underconfidence except with very easy tasks. In the area of control (Type V), overestimation of controllability (e.g. illusion of control) is more often reported than the opposite (cf. Langer, 1975).

Overestimation of plausibility (gullibility) in the realm of fiction (VI) is harder to prove, due to the lack of norms. What is a "well-calibrated" reader of fiction? Still, the ease with which we seem to accept the contents and "internal logic" in most works of fiction, even if incomplete and unfamiliar, being the creation of a different mind than our own, attests to the fact that we are more ready to believe than to doubt. To put on the attitude of a critic, pointing out weaknesses and inconsistencies, requires a determined, conscious effort and seems less natural than going along as far as possible, trying to accept what is being offered at face value. According to some classical moral philosophers (Montaigne, 1588/1885; Holberg, 1748/1945), our minds are susceptible to a peculiar bias when being told something "incredible" or strange: People ask *why* it is the case, instead of questioning *what* is really the case. "They leave things and runne for causes. ... They commonly beginne thus: How is such a thing done? Whereas they should say: *Is* such a thing done?" (Montaigne, 1588/1885, page 526).

This general "belief bias" can be given motivational explanations in terms of "basic trust" and similar concepts. It may also be defended on the grounds that it feels better to be convinced about what one thinks and does, than to be constantly troubled and inhibited by nagging (even legitimate) doubts. From a cognitive point of view it has been explained primarily as an attentional bias: it is difficult to focus on all alternatives at the same time, and the easiest choice is usually the one that is suggested to us (by others or by ourselves), along with what speaks in its favour (Klar, 1990, 1991). Baron refers to "insufficient search, or the failure to consider alternative possibilities, goals, and additional evidence" (1988, page 217) as a central bias, explaining the typical overestimation phenomena.

Another general, perhaps related bias, has recently been described by Griffin and Tversky (1992) as a tendency to focus on the strength or extremeness of the available evidence, with insufficient regard for its weight or credence. This means in many cases overconfidence or overestimation of probabilities, but also in some cases underconfidence (when tendencies are not strong, but reliable). Griffin and Tversky show this phenomenon to occur with chance probabilities as well as confidence estimates. It seems also akin to the well-known psychometrical "fallacy" of believing in validity as independent from (occasionally even opposed to) reliability. Again it may be a question of attentional focus: Upon the magnitude of the tendency one can "see", as opposed to upon the strength of its less visible basis.

10.4 CONDITIONS OF NORMATIVITY

Our excursion into the numerous variants of intuitive uncertainties was prompted by an initial question: When are subjects' probability estimates in agreement with the experimenter's calculated p values, and when are they not?

In line with Ginossar and Trope's (1987) conception of probability estimation as a problem-solving process, it is suggested that any judgment is dependent upon (a) externally available information; (b) subjectively available concepts and strategies; (c) the kind of task to be solved, and the goal to be achieved. From the research reviewed, and the concepts discussed in this chapter, it seems reasonable to suggest the following conditions as favorable for "normative" probability estimates.

(1) Availability of relevant numerical information. For instance, when subjects are explicitly informed about the percentages covered by different sectors of a gambling wheel, the majority will use these percentages as their probabilities. For the wheel shown in Figure 10.1, the probability of sector B is (correctly) estimated to be 0.25, while the estimated probability of a *red* sector (A, C, E, G) is 0.60.

(2) Focusing on the randomness of the sampling process (perceived chance rather than causal determination). Traditional "Pascalian" probabilities came from the study of gambles, and seem easiest to understand and to apply in the context of chance probabilities (Nisbett *et al.*, 1983; Hoch, 1985; Ginossar & Trope, 1987). These are also situations in which subjects are most willing to use a rule-dictated approach, whether it be because of training, readily available numerical solutions, or because the "correct" values somehow match their intuitions. But even in a typical chance situation, like the fortune wheel, with the relevant numerical information clearly in sight, a varying number of subjects will find reasons to deviate from the norms, depending upon the circumstances, as shown below.

(3) Estimation vs. choice. When estimation of probabilities is preceded by a choice, normative probabilities seem less relevant. In the fortune wheel example, subjects asked to make a choice of sector (or color) often failed to use the corresponding percentages as their subsequent probability estimates. This was especially the case for those who chose one of the smaller sectors,

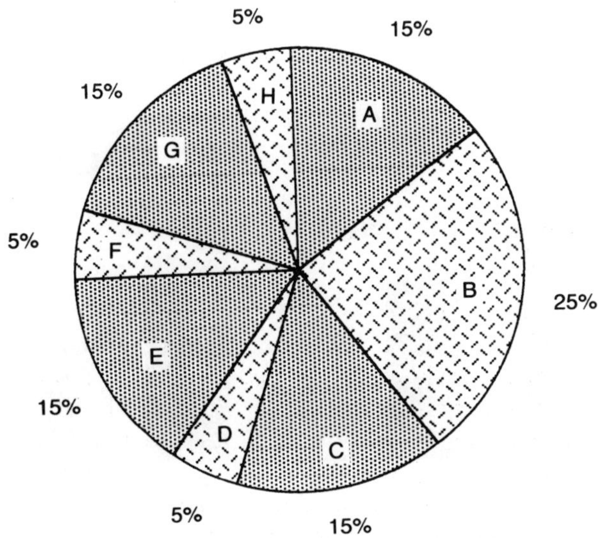

Figure 10.1 Fortune wheel with four red sectors (A, C, E, G) and four blue sectors (B, D, F, H) used in experiment on subjective probabilities

or blue color rather than red. These non-normative choices (favored by 20–50% of the subjects) were nearly always followed by a non-normative p estimate.

(4) Two vs. several alternative outcomes. When three or more alternatives are to be considered, subjects frequently fail to use the "complementarity rule", particularly with non-chance probabilities, and this is more so if more alternatives are available (Teigen, 1983c).

(5) External vs. internal probabilities. Subjects who were asked to state the "probability" of their chosen color (or sector) in the fortune wheel experiment gave more consistently estimates corresponding to its relative area (cf. 1), than did subjects who were asked to estimate how "certain" they were (both on a scale from 0–100%).

(6) Simple vs. complex situations. A complex situation can be one where several probabilities have to be combined (as in base-rate problems), or where the probability of a combined event has to be evaluated (as in conjunction problems), or where the answer is arrived at by a complex decision process. For instance, in the gambling wheel situation, subjects who were asked to predict color before choosing their favourite sector (or vice versa), often felt trapped by their first answer, "red" being taken to imply one of the less probable letters (Teigen, 1983b). In such cases, very few subjects anchored their subsequent probability (or certainty) estimates in the area percentages, even among those who finally chose the normatively correct sector B (Figure 10.2).

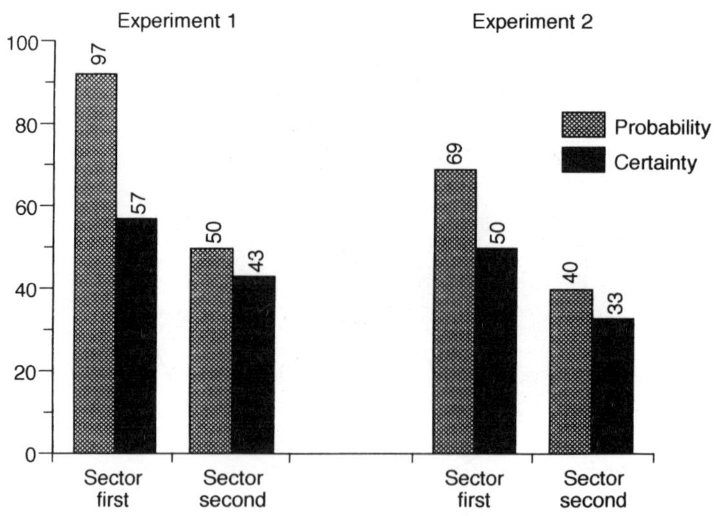

Figure 10.2 Percentages of subjects agreeing with the normative probability (0.25) as an estimate of the "probability" and degree of "certainty" of obtaining sector B. Subjects in the "sector second" condition were first asked to guess color. Data from two experiments ($n = 86$ and $n = 71$)

(7) Condition focus vs. outcome focus. In some judgment situations the emphasis will be upon what will happen under the given circumstances, or what are the explanations of a given event. In both cases, subjects are invited to consider and compare several alternatives, which may be conducive to a more realistic attitude than when a particular outcome (or a particular hypothesis) is already selected and judged in isolation. Probabilities of Type I and IV will typically be more "condition focused" than probabilities of Types II and III. This principle is obviously related to the estimation vs. choice difference described above.

(8) Numerical vs. verbal estimates. Occasionally the argument has been made that probability judgments are inaccurate or biased because subjects are required to make a numerical response even if they are not particularly familiar with thinking in terms of numbers, and they may find it more natural to convey their ideas of chances and uncertainties through verbal phrases (like "perhaps", "very likely", "somewhat doubtful", etc). But when attempts are made to translate these phrases into numbers, it is hard to find clear cut correspondences, at least at an individual level (Lichtenstein & Newman, 1967; Beyth-Marom, 1982; Brun & Teigen, 1988). One is left with an impression that linguistic probabilities are very imprecise, or very idiosyncratic, *or* that their primary purpose is to convey something different from degrees of probabilities (Fox, 1987; Teigen, 1988b). For example: "It is uncertain" may indicate unwillingness to come up with *any* precise probability estimate. Other expressions, like "there is a small probability", may indicate, literally, a small *p* value, but at the same time, its "attentional focus" (Moxey, Sanford & Barton, 1990), or "argumentative function" (Champaud & Bassano, 1987) concerns the possible *occurrence* rather than the non-occurrence of the phenomenon in question. By referring to it as "somewhat doubtful", or "not quite certain" we may have a higher probability in mind but at the same time feeling more right if the phenomenon in question does not occur (Teigen, 1988b; Teigen & Brun, in press).

The impreciseness of verbal probabilities can make them a less adequate basis for making decisions (Budescu, Weinberg & Wallsten, 1988). Also, verbal and numerical probabilities may activate different probability concepts. A point in case was given on p. 215 in relation to equiprobable outcomes, whose numerical probabilities are usually estimated according to the $1/n$ rule, whereas the corresponding verbal expressions tend to emphasize the "good" chances associated with the particular alternative (Teigen, 1988a).

A careful investigation of how people *talk* about uncertainties may be a promising way of finding out more about subjective probability concepts and their relationships. Researchers in the heuristics and biases tradition have occasionally expressed their amazement that people are so deficient in estimating numerical probabilities despite the fact that we are living in an uncertain (probabilistic) world (Kahneman, Slovic & Tversky, 1982; Nisbett & Ross, 1980). The variety of terms and concepts used to express and communicate uncertainty in language tells a somewhat different story. My guess is that we are very sophisticated probabilists in most respects except the quantitative one. (To some, this may seem a facetious statement, akin to Wilde's "I can resist everything but temptations".) But even in the sphere of action, some of the "non-normative" probability concepts may prove their practical usefulness. For instance it can be more important to know whether something is "close" to, or can "easily" happen, than to think of its relative frequencies in the long run. At the same time psychological research is needed to establish the conditions, persistence and consequences (positive *and* negative) of such beliefs.

Obviously, probability judgment studies have made a habit out of comparing apples and pears (cf. above). This is no objection, on the contrary the study should be extended to oranges and bananas. One reason is to find out how many fruits there are to be had, another is that only by comparison do we learn about their different qualities and how easily, or with how much difficulty, they mix.

ACKNOWLEDGEMENTS

Thanks are due to Sigrid Aspen, Hanne-Trine Engdal, Nina Hermann, and especially Else-Janie Karlsen and Margoth Kristiansen for valuable assistance in conducting the wheel of fortune experiments. The study was supported by grants from the Norwegian Research Council for Science and the Humanities and the University of Tromsø.

REFERENCES

Alberoni, F. (1962) Contributions to the study of subjective probability, I. *Journal of General Psychology*, **66**, 241–64.

Aristotle (1941) *Rethoric*. In R. McKeon (ed.), *The Basic Works of Aristotle*. Random House, New York.

Bar-Hillel, M. & Fischhoff, B. (1981) When do base rates affect predictions? *Journal of Personality and Social Psychology*, **41**, 671–80.

Baron, J. (1988) *Thinking and Deciding*. Cambridge University Press, Cambridge.

Beach, L.R. & Wise, J.A. (1969) Subjective probability estimates and confidence ratings. *Journal of Experiment Psychology*, **79**, 438–44.

Beyth-Marom, R. (1982). How probable is probable? Numerical translation of verbal probability expressions. *Journal of Forecasting*, **1**, 257–69.

Bradley, J.V. (1981) Overconfidence in ignorant experts. *Bulletin of the Psychonomic Society*, **17**, 438–44.

Brun, W. & Teigen, K.H. (1988) Verbal probabilities: Ambiguous, context-dependent, or both? *Organizational Behavior and Human Decision Processes*, **41**, 390–404.

Brun, W. & Teigen, K.H. (1990) Prediction and postdiction preferences in guessing. *Journal of Behavioral Decision Making*, **3**, 17–28.

Budescu, D.V., Weinberg, S. & Wallsten, T.S. (1988) Decisions based on numerically and verbally expressed uncertainty. *Journal of Experimental Psychology: Human Perception and Performance*, **4**, 281–94.

Champaud, C. & Bassano, D. (1987) Argumentative and informative functions of French intensity modifiers. *Cahiers de Psychologie Cognitive/European Bulletin of Cognitive Psychology*, **7**, 605–31.

Cohen, L.J. (1981) Can human irrationality be experimentally demonstrated? *Behavioral and Brain Sciences*, **4**, 317–70.

Erev, I. & Cohen, B.L. (1990) Verbal versus numerical probabilities: Efficiency, biases, and the preference paradox. *Organizational Behavior and Human Decision Processes*, **45**, 1–18.

Falk, R. (1981–82) On coincidences. *Skeptical Inquirer*, **6**, 18–31.

Falk, R. (1989) Judgment of coincidences: Mine versus yours. *American Journal of Psychology*, **102**, 477–93.

Fischhoff, B. & Bar-Hillel, M. (1984) Diagnosticity and the base-rate effect. *Memory and Cognition*, **12**, 402–10.

Fischhoff, B., Slovic, P. & Lichtenstein, S. (1978) Fault tree: Sensitivity of estimated failure probabilities to problem representation. *Journal of Experimental Psychology: Human Perception and Performance*, **4**, 330–4.

Fox, J. (1987) Making decisions under the influence of knowledge. In P. Morris (ed.), *Modelling Cognition*. Wiley, London.

Frisch, D. & Baron, J. (1988) Ambiguity and rationality. *Journal of Behavioral Decision Making*, **1**, 149–57.

Gambetta, D. (ed.). (1988) *Trust: Making and breaking cooperative relations*. Blackwell, London.

Gibson, J.J. (1966) *The Senses Considered as Perceptual Systems*. Houghton Mifflin, Boston.

Gigerenzer, G. (1991) How to make cognitive illusions disappear: Beyond 'heuristics and biases'. In W. Stroebe & M. Hewstone (eds.), *European Review of Social Psychology*, Vol. 2. Wiley, London.

Gigerenzer, G., Hell, W. & Blank, H. (1988) Presentation and content: The use of base rates as a continuous variable. *Journal of Experimental Psychology: Human Perception and Performance*, **14**, 513–25.

Gigerenzer, G., Hoffrage, U. & Kleinbölting, H. (1991) Probabilistic mental models: A Brunswikian theory of confidence. *Psychological Review*, **98**, 506–28.

Ginossar, Z. & Trope, Y. (1987) Problem solving in judgment under uncertainty. *Journal of Personality and Social Psychology*, **52**, 464–74.

Griffin, D.W. & Tversky, A. (1992) The weighing of evidence and the determinants of confidence. *Cognitive Psychology*, **24**, 411–35.

Hacking, I. (1975) *The Emergence of Probability*. Cambridge University Press, Cambridge.

Heath, C. & Tversky, A. (1991). Preference and belief: Ambiguity and competence in choice under uncertainty. *Journal of Risk and Uncertainty*, **4**, 5–28.

Hoch, S.R. (1985) Counterfactual reasoning and accuracy in predicting personal events. *Journal of Experimental Psychology: Learning, Memory, and Cognition*, **11**, 719–31.

Holberg, L. (1945) *Epistler*, Vol. II. Hagerup, Copenhagen (original work published in 1748).

Howell. W.C. & Burnett, S.A. (1978). Uncertainty measurement: A cognitive taxonomy. *Organizational Behavior and Human Performance*, **22**, 45–68.

Hume, D. (1976) *A Treatise of Human Nature*. Clarendon Press, Oxford, (Original work published in 1739).

Hume. D. (1957) *Enquiries Concerning Human Understanding and Concerning the Principles of Morals*. Clarendon Press, Oxford, (Original work published in 1748).

Jungermann, H. (1983) The two camps on rationality. In R.W. Scholz (ed.), *Decision Making under Uncertainty*. Elsevier, Amsterdam.

Kahneman, D., Slovic, P. & Tversky, A. (eds.) (1982) *Judgment under Uncertainty: Heuristics and Biases*. Cambridge University Press, Cambridge.

Kahneman, D. & Tversky, A. (1972) Subjective probability: A judgment of representativeness. *Cognitive Psychology*, **3**, 430–54.

Kahneman, D. & Tversky, A. (1973) On the psychology of prediction. *Psychological Review*, **80**, 237–51.

Kahneman, D. & Tversky, A. (1982) Variants of uncertainty. *Cognition*, **11**, 143–57.

Keren, G. (1991) Calibration of probability judgment: Conceptual and methodological issues. *Acta Psychologica*, **77**, 217–74.

Keren, G. & Wagenaar, W.A. (1987) Violation of utility theory in unique and repeated gambles. *Journal of Experimental Psychology: Learning, Memory, and Cognition*, **13**, 387–91.

Klar, Y. (1990) Linking structures and the sensitivity to judgment-relevant information in statistical and logical reasoning tasks. *Journal of Personality and Social Psychology*, **59**, 841–58.

Klar, Y. (1991) Logical, statistical, and causal reasoning: Compartmentalized or generic process? In W. Stroebe & M. Hewstone (eds.), *European Review of Social Psychology*, Vol. 2. Wiley, London.

Langer, E.J. (1975) The illusion of control. *Journal of Personality and Social Psychology*, **32**, 311–28.

Langer, E.J. (1978) The psychology of chance. *Journal for the Theory of Social Behaviour*, **7**, 185–207.

Laplace, P.S. (1816) *Essai philosophique sur les probabilités*. Courcier, Paris.

Lefcourt, H.M. (1982) *Locus of control: Current trends in theory and research*. Erlbaum, Hillsdale, NJ.

Lichtenstein, S. & Newman, J.R. (1967) Empirical scaling of common verbal phrases associated with numerical probabilities. *Psychonomic Science*, **9**, 563–4.

Macdonald, R.R. (1986) Credible conceptions and implausible probabilities. *British Journal of Mathematical and Statistical Psychology*, **39**, 15–27.

Manis, M., Dovalina, I., Avis, N.E . & Cardoze, S. (1980) Base rates can affect individual predictions. *Journal of Personality and Social Psychology*, **38**, 231–48.

Mill, J.S. (1856) *A System of Logic*. Parker, London.

Montaigne, M. (1885) *The Essayes of Michael Lord of Montaigne*. Routledge, London, (Original work published in 1588).

Moxey, L.M., Sanford, A.J. & Barton, S.B. (1990) Control of attentional focus by quantifiers. In K. Gilhooly, M. Keane, R. Logan and G. Erdos (eds.), *Lines of Thinking: Reflections on the Psychology of Thought*. Wiley, Chichester.

Nahinsky, I.D., Ash, D. & Cohen, B. (1986) The conjunction fallacy: Judgmental heuristic or faulty extensional reasoning? *Bulletin of the Psychonomic Society*, **24**, 186–8.

Nisbett, R. & Ross, L. (1980) *Human Inference: Strategies and Shortcomings*. Prentice Hall, Englewood Cliffs, NJ.

Nisbett, R.E., Krantz, D.H., Jepson, C. & Kunda, Z. (1983) The use of statistical heuristics in everyday inductive reasoning. *Psychological Review*, **90**, 339–63.

Oskamp, S. (1965) Overconfidence in case-study judgments. *Journal of Consulting Psychology*. **29**, 261–5.

Peterson, D.K. & Pitz, G.F. (1986) The effects of amount of information predictions of uncertain quantities. *Acta Psychologica*, **61**, 229–41.

Peterson, D.K. & Pitz, G.F. (1988) Confidence, uncertainty, and the use of information. *Journal of Experimental Psychology: Learning, Memory, and Cognition*, **14**, 85–92.

Pitz, G.F. & Peterson, D.K. (1987) Uncertainty in estimating one's own performance: The prediction of race times. Unpublished manuscript.

Proctor, R.A. (1887) *Chance and Luck: A Discussion of the Laws of Luck, Coincidences, Wagers, Lotteries, and the Fallacies of Gambling*. Longmans Green, London.

Robinson, L.B. & Hastie, R. (1985) Revision of beliefs when a hypothesis is eliminated from consideration. *Journal of Experimental Psychology: Human Perception and Performance*, **11**, 443–56.

Rotter, J.B. (1966) Generalized expectancies for internal versus external control of reinforcement. *Psychological Monographs*, **80** (No. 609).

Shanteau, J. (1987) Psychological characteristics of expert decision makers. In J.L. Mumpower, L.D. Phillips, O. Renn & V.R.R. Uppuluri (eds.), *Expert Judgments and Expert Systems*. Springer, Berlin.

Smithson, M. (1988). *Ignorance and Uncertainty*. Springer, Berlin.

Teigen, K.H. (1974a) Subjective sampling distributions and the additivity of estimates. *Scandinavian Journal of Psychology*, **15**, 56–62.

Teigen, K.H. (1974b) Overestimation of subjective probabilities. *Scandinavian Journal of Psychology*, **15**, 56–62.

Teigen, K.H. (1983a) Studies in subjective probability I: Prediction of random events. *Scandinavian Journal of Psychology*, **24**, 13–25.

Teigen, K.H. (1983b) Studies in subjective probability II: Dilemmas of chance. *Scandinavian Journal of Psychology*, **24**, 27–35.

Teigen, K.H. (1983c) Studies in subjective probability III: The unimportance of alternatives. *Scandinavian Journal of Psychology*, **24**, 97–105.

Teigen, K.H. (1983d) Studies in subjective probability IV: Probabilities, confidence, and luck. *Scandinavian Journal of Psychology*, **24**, 97–105.

Teigen, K.H. (1988a) When are low-probability events judged to be "probable"? Effects of outcome-set characteristics on verbal probability judgments. *Acta Psychologica*, **67**, 157–74.

Teigen, K.H. (1988b) The language of uncertainty. *Acta Psychologica*, **68**, 27–38.

Teigen, K.H. (1989) From fallacies and illusions to heuristic and biases. In I.A. Bjørgen (ed.), *Basic Issues in Psychology: A Scandinavian Contribution*. Sigma, Bergen.

Teigen, K.H. (1990) To be convincing or to be right: A question of preciseness. In K. Gilhooly, M. Keane, R. Logan & G. Erdos (eds.), *Lines of Thinking: Reflections on the Psychology of Thought*. Wiley, Chichester.

Teigen, K.H. & Brun, W. (in press) Yes, but it is uncertain: Direction and communicative intention of verbal probabilistic terms. *Acta Psychologica*.

Tversky, A. & Kahneman, D. (1973) Availability: A heuristic for judging frequency and probability. *Cognitive Psychology*, **5**, 207–32.

Tversky, A. & Kahneman, D. (1983) Extensional versus intuitive reasoning: The conjunction fallacy in probability judgment. *Psychological Review*, **90**, 293–315.

Wagenaar, W.A. (1988) *Paradoxes of Gambling Behavior*. Erlbaum, Hillsdale, N.J.

Wagenaar, W.A. & Keren, G.B. (1986) Does the expert know? The reliability of predictions and confidence ratings of experts. In E. Hollnagel, G. Maneini & D. Woods (eds.), *Intelligent Decision Support in Process Environments*. Springer, Berlin.

Wasserman, D., Lampert, R.O. & Hastie, R. (1991) Hindsight and causality. *Personality and Social Psychology Bulletin*, **17**, 30–5.

Weinstein, N.D. (1980) Unrealistic optimism about future life events. *Journal of Personality and Social Psychology*, **39**, 806–20.

Yaniv, I. & Foster, D.A. (1993) Graininess of judgment under uncertainty: An accuracy-informativeness tradeoff. Manuscript submitted for publication.

Zakay, D. (1984) The influence of perceived event's controllability on its subjective occurrence probability. *The Psychological Record*, **34**, 233–40.

Chapter 11

The Origins of Probability Judgment: A Review of Data and Theories

Valerie F. Reyna and Charles J. Brainerd
University of Arizona

Developmental studies of probability judgment provide a unique and important perspective on adult conceptions of probability. Charting the origins of probability concepts deepens our understanding of adult reasoning in two ways. First, it supplies an independent body of evidence that can be used to select among competing theories. Second, it generates sensible explanations of otherwise puzzling aspects of adult reasoning by rooting them in ontogenetic mechanisms. With respect to this second point, in particular, we shall see that developmental research has helped make sense of some of the most surprising results of research on adult conceptions of probability (e.g. Kahneman, Slovic, & Tversky 1982).

During the past quarter-century, research on cognitive development has passed through three stages (Reyna & Brainerd 1990): a Piagetian period of grand theory, followed by a second stage that focused on information processing, followed, most recently, by a third stage that emphasizes intuitive reasoning. These same three stages are apparent in developmental research on probability judgment, and we exploit this fact to review such research. We present work from each of the three stages seriatim, reviewing major theories, paradigms, and findings. We conclude by discussing the implications of

Subjective Probability. Edited by G. Wright and P. Ayton.
© 1994 John Wiley & Sons Ltd.

research in the third stage for broader theoretical questions of rationality and cognitive competence.

11.1 STAGE 1: THE LOGICIST APPROACH TO PROBABILITY JUDGMENT

11.1.1 Theory

The developmental analysis of probability judgment begins with the work of Piaget and his colleagues (Inhelder & Piaget, 1958; Piaget, 1950; Piaget & Inhelder, 1951). According to Piaget, conceptions of probability develop from a preoperational (prelogical) stage in which chance and non-chance events cannot be distinguished, through a subsequent concrete operational stage in which such events can be distinguished, to an ultimate stage of formal operations in which the mathematics of probability is at last understood. During the preoperational stage (roughly ages 2–7), children do not grasp the logic of cause and effect, and so the origin of non-chance events is mysterious. The concept of chance emerges from the insight that a non-chance event is causally determined whereas a chance event is not. Until that insight is achieved, chance events are viewed either as miraculous, or, at the other extreme, as inevitable, rather than random. The first achievement in the development of probability concepts, then, is the ability to discriminate random events from those that are causally determined.

Although concrete operational children understand the logic of cause and effect, and they appreciate the distinction between chance and non-chance events, their ability to think logically has limitations. These children judge probabilities based on the odds of alternative outcomes, so long as the logic does not become too complicated. Two complications include keeping track of possible outcomes and relations among sets, both essential to working with ratios or proportions. For example, given a sample consisting of three sets—blue tokens, red tokens, and yellow tokens—the probability of randomly drawing a blue token can be expressed as the ratio of blue tokens to the total number of possible outcomes. Concrete operational children have difficulty keeping track of such outcomes when the number of outcomes becomes large, and, it is claimed, they have difficulty thinking of sets separately and, at the same time, as part of the total (i.e. assigning blue tokens to both numerator and denominator simultaneously). It is only at the final stage of formal operations that these complications are assumed to be fully mastered. Correct quantification of probabilities follows directly on sorting out the logic of possible outcomes and their relations. Thus, development progresses from an initial awareness that events can be causally determined; during middle childhood, that awareness sets the stage for the analysis of events that are not determined, until the logical quantification of probabilities becomes perfected in formal operations.

11.1.2 Data

The Piagetian account of the development of probability judgment was compelling and, for many years, dominated theorizing. The evidence for this view, however, was sketchy, and not entirely consistent. Most of the supportive findings were contained in one volume (Piaget and Inhelder, 1951). The key experiments were as follows: in the first experiment, two rows of beads were lined up on opposite ends of a tray, red beads on one side and white ones on the other. The tray was then tilted so that the beads were mixed. This process was repeated until the red and white beads became thoroughly intermingled (ostensibly randomly). Children were asked to predict the outcome of the first tilting, of subsequent tiltings, and of a large number of tiltings. Preoperational children ascribed a lawfulness to the trajectories of the beads—either that they would eventually return to their original segregated state, or that they would cross over symmetrically.

In other studies, children were asked to predict the shapes of distributions of marbles poured through a funnel, of beads shaken onto a piece of paper, or of spins of a roulette wheel (which was subsequently rigged; younger children failed to differentiate the random spins from the nonrandom ones). Although older children had predicted greater *ir*regularity in the mixture of beads in the first experiment, they predicted greater regularity in the distribution experiments, especially as sample size increased (e.g. as the number of marbles increased). Predictions of greater regularity with larger sample sizes were said to reflect an appreciation of the law of large numbers (itself predicated on understanding proportionality).

The last group of experiments involved random draws from sampling spaces. In one experiment, children were shown tokens with a circle on one side and a cross on the other. The task was analogous to predicting the flip of a coin. Children were asked to predict the outcome (circle or cross) of tossing one token, and the distribution of outcomes if the entire set of tokens was tossed. However, the experimenter substituted tokens with crosses on both sides. As with the rigged roulette wheel, younger children did not suspect that outcomes were nonrandom despite consistent outcomes of only one type, and they predicted similar results for two-outcome and one-outcome (crosses on both sides) situations. In the remaining experiments, children were asked to predict the outcomes of random draws either from one or two sampling spaces. For example, colored tokens of varying frequencies (e.g. 15 blue, 10 yellow, 5 red, and 2 green) might be placed in a bag, and children would predict the colors of successive pairs drawn from the bag. In the two-sample experiments, children judged which of two containers was more likely to yield a token of a certain type, say a blue token (see Figure 11.1).

In both the one-sample and two-sample experiments, frequencies of outcomes were varied, a feature that would become a focus of later research. Piaget and his colleagues observed that older children handled frequency information better than younger ones. Younger children's difficulties were taken to be logical rather than computational, however. Specifically, it was argued that preoperational children failed to exploit frequency information because of their inability to differentiate probabilistic from nonprobabilistic causes. On the other hand, although concrete operational children used frequency information, they did not understand the complexities of relations among outcomes. For example, they failed to keep track of the effects of successive draws—that early outcomes constrained later ones when tokens were not replaced. At the formal operational stage, however, such complexities were supposedly understood, including the relation between absolute frequencies of separate outcomes and the total frequency across outcomes (i.e. proportions).

Initial studies (e.g. Pire, 1958) were globally consistent with the Piagetian account of the development of probability judgment. Younger children were confused, sometimes imputing miraculous causes for chance events; and, older children's judgments were more likely to be based on relative frequencies. However, the theoretical rationale proposed by Piaget was not specifically tested in these early studies. At about the same time, there was growing

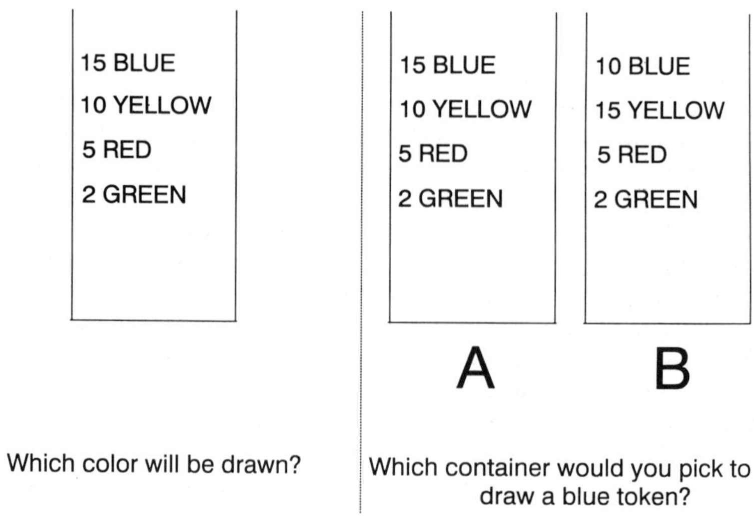

Figure 11.1 One- versus two-sample probability judgment tasks.

criticism of Piaget's methods. Braine (1959), for example, argued that requiring verbal explanations, resistance to countersuggestions, and the like, in order to be credited with a correct response, led to underestimation of children's reasoning competence. In other words, such procedures resulted in a high rate of false negative errors (see Reyna & Brainerd, 1990, for a contemporary review of these methodological arguments). A new era of assessment was ushered in, characterized by greater concern for the extraneous demands of different methods, and by more careful attention to the link between experimental manipulations and specific hypotheses.

11.2 STAGE 2: DETAILS, DETAILS

In the first stage, research on probability judgment led to important theoretical and empirical discoveries. Although research in the second stage was initially motivated by the need to fill in details in the Piagetian program, these "details" ultimately became crucial elements in theoretical debates. Researchers systematically varied the nature of the subject's response (e.g. verbal versus nonverbal), the task (e.g. one- versus two-sample), and the problem (e.g. the discrepancy between probabilities of alternative outcomes). Although some theorists might dismiss the first-stage data as indeterminate because they were gathered using Piaget's "clinical method," the more finely tuned experiments of the second stage cannot be similarly dismissed. The bulk of our knowledge about probability judgment was generated in Stage 2, and any theory purporting to explain it must account for these data.

11.2.1 Varying Spurious Performance Factors

Yost, Siegel, and Andrews (1962) assessed the contribution of extraneous performance demands to children's probability judgments by explicitly comparing Piaget's methods to more direct tests of underlying competence. There were five performance factors that Yost et al. identified as potential barriers to displaying competence, including requiring elaborate verbal explanations from young children, failing to control for color preferences (so that children made preference judgments rather than probability judgments), and failing to provide reinforcement for correct answers (so that children were insufficiently motivated to perform difficult reasoning operations). To address these concerns, Yost et al. embedded probability judgment in a choice task in which children selected one of two containers more likely to deliver a reward. After the child's selection, a token was drawn from the designated container, and the child received a toy if the payoff color was indeed drawn. In addition to using a more motivating task, Yost et al. controlled for other performance biases (e.g. color preferences). They found that children's predictions were correct 75% of the time and concluded that even "4-year-olds have some understanding of probability" (page 779).

Because Yost et al. had changed Piaget's procedures in multiple ways, researchers subsequently attempted to isolate which changes had produced the large improvements in performance. Davies (1965), for example, compared performance with verbal versus nonverbal responses, and obtained the now familiar result that performance was vastly superior with nonverbal methods: With nonverbal responses, 3-year-olds showed some evidence of understanding probability, but it was not until age 9 that children were able to consistently verbalize that understanding. Goldberg (1966) separated the motivational effects of receiving a reward from the informational effects of knowing the outcome of a draw. Performance in a Piagetian prediction task was compared to that in Yost et al.'s two-sample decision-making task—except that the child did not receive a reward. In both the prediction and the decision-making tasks, however, children were informed of the outcome of each draw. Despite the absence of a reward, 75% of the children (3 to 5 years of age) performed above chance in the decision making task. However, information about draws was not the key ingredient for success because only 35% of children in the Piagetian prediction task performed above chance. In addition to the facilitation provided by nonverbal methods, something else about the two-sample decision-making task apparently made it easier than the Piagetian prediction task.

11.2.2 Examining the Two-sample Facilitation Effect

Researchers were particularly interested in solving the mystery of the two-sample facilitation effect because many believed its solution had direct implications for the question of early competence. Unfortunately, subsequent research failed to entirely resolve this question. Explanations for the two-sample facilitation effect fell into two categories, differing according to whether competence was believed to be overestimated in such tasks. Piaget and Inhelder (1951) initially suggested that the two-sample task overestimated competence. According to this view, although the two samples each contain targets (e.g. blue tokens) and nontargets (e.g. tokens of other colors), the two-sample task does not require decomposition of possible outcomes. If the total frequency is the same for the two samples, children need only compare the number of targets (e.g. blue tokens) in one sample to the number of targets in the other; nontargets can be ignored. On the other hand, in the one-sample task, the number of targets and nontargets must be compared in order to predict which will be drawn. Thus, the one-sample task may offer a truer picture of conceptual competence because its solution involves some understanding of relations among different classes of outcomes.

Hoemann and Ross (1971) tested the hypothesis that the two-sample task elicited magnitude comparisons of targets, as Piaget and Inhelder (1951) had claimed, rather than true probability judgments. In their first experiment, virtually identical situations were presented to two groups of children, a two-spinner task in which the relative proportions of black and white areas varied (see Figure 11.2). One group received probability instructions; they were asked to point to the spinner where black would be the winner (white was also used as the target on half of the trials). The group receiving magnitude estimation instructions, however, was told to point to the circle (the spinner had been removed) which had the "most" black (white was also used on half the trials). There were no differences between instructional groups at any of the ages tested (from 4 to 10 years), prompting Hoemann and Ross to conclude that children could have relied on magnitude estimation in the probability task.

In Experiment 2, children made both probability and magnitude judgments for a single spinner. Children were asked which color the spinner would point to for probability questions and which color was "most" for magnitude questions. In contrast to the first experiment, probability instructions produced more errors than magnitude instructions at all ages for which comparisons were available (magnitude judgments were not obtained for the two oldest groups because they were assumed to be at ceiling). Four-year-olds did not perform significantly above chance on the probability questions, and errors continued to decrease through age 12. Consistent with Piaget and Inhelder, Hoemann and Ross interpreted the greater difficulty of probability versus magnitude judgments, and the improvements with age, as evidence in favor of the magnitude estimation hypothesis: Younger children did not really understand probability, and made magnitude rather than probability judgments in two-sample tasks, leading to overestimates of their true competence (see also Hoemann & Ross 1982).

Perner (1979) also systematically examined the basis of the "double spinner facilitation effect," which, he argued, was not always replicable, and must

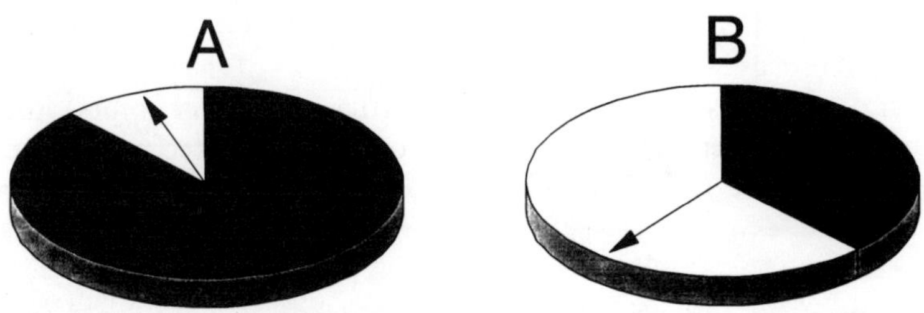

Figure 11.2 Two-sample spinner task in which (if black is designated the winning color) a 7/8 chance of winning (Panel A) is pitted against a 3/8 chance of winning (Panel B).

therefore, depend on more than magnitude estimation. He offered two alternatives to the magnitude estimation hypothesis. First he noted that those studies finding superior performance with two spinners had also induced a "set" to express a preference. Hence, choices would tend to reflect subjects' preferences for more of the winning color, facilitating performance in two-sample tasks. Second, Perner pointed out that there had been more variation in the sizes of spinner segments (or number of objects) in the two-sample than in the one-sample experiments. Such variation, he reasoned, might have focused greater attention on the critical dimension (e.g. area size) in the two-sample tasks. These hypotheses were not supported, however. Neither preference instructions nor greater variability along critical dimensions facilitated performance. Although failures to find facilitation effects remained unexplained, the magnitude estimation hypothesis seemed to account for such effects when they occurred.

Two-sample tasks can easily be modified so that comparing magnitudes no longer affords a correct solution to probability questions. For example, as Figure 11.3 shows, if blue tokens are designated as winners, a sample of 1 blue and 2 red tokens can be pitted against a sample of 2 blue and 6 red tokens (see Surber & Haines, 1987, for problem types). Although absolute frequency favors the second sample, relative frequency (proportions) favors the first sample. Clearly, if children use the magnitude estimation strategy, they will erroneously choose the second sample because of the greater magnitude of targets. Chapman (1975), Ross and Hoemann (1975), and Hoemann and Ross

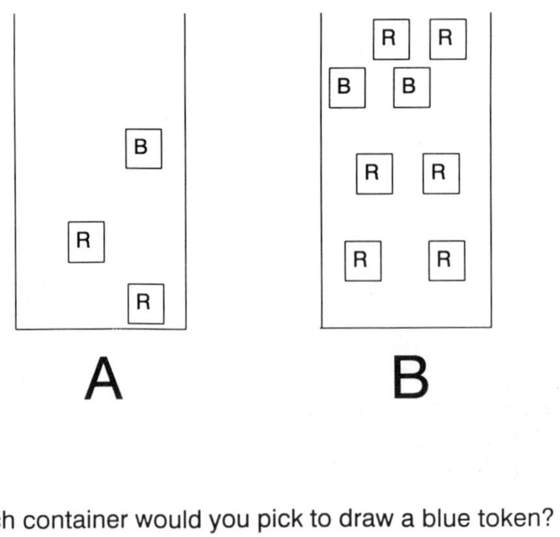

Figure 11.3 Two-sample task in which targets (i.e blue tokens) are more probable in option A but more numerous in option B.

(1971, in their third and fourth experiments) included modified problems, and found that they were more difficult than the standard type in which total frequency was kept constant across the two samples. Apparently, children relied on magnitude estimation in two-sample tasks. The implication that children were *incapable* of making true probability judgments under any circumstances, however, was still a matter of dispute.

11.2.3 Information-processing Approaches to Probability Judgment

The Stage 2 research we have reviewed thus far can be characterized as "neo-Piagetian" in the sense that it dealt primarily with phenomena and with theoretical hypotheses that had been introduced by Piaget. In these studies, age variations in probability judgment were generally attributed to logical deficits (e.g. the magnitude estimation hypothesis). Eventually, however, there was a shift towards information-processing explanations of age variability. The reasons for that shift were many, and cannot be reviewed in any depth here (but see Siegler, 1983). It is useful to mention, however, that the kind of experiment performed by Yost et al. was conducted for many of the other reasoning paradigms introduced by Piaget with similar results: removing spurious performance obstacles extended estimates of operational competence well into what should have been the preoperational stage (Siegel, 1978). Moreover, training studies of various types showed that such competence could be instilled in young "preoperational" children, with ample evidence of conceptual understanding (Brainerd, 1978). For example, such children could correctly explain the basis for their judgments, and could correctly solve new tasks that differed superficially from training tasks (Brainerd, 1982). In the face of such contradictory evidence, the notion that young children lacked certain logical structures (making it impossible for them to know, or even to learn, certain ways of thinking) was difficult to sustain.

Researchers were attracted to the information-processing approach not only because of the perceived shortcomings of Piagetian theory, but because of the richness of the "mind-as-computer" metaphor. A host of new constructs became available to explain variations in children's performance. These included perceptual processes, working-memory capacity, retrieval as opposed to storage, and so on (e.g. Bjorklund, 1989; Brainerd, 1981, 1983; Siegler, 1981, 1983). In contrast to Plaget's stage approach, limitations in information processing were generally viewed as *nonconceptual* deficits.

The processing of quantitative information was a particular focus of research. Although Perner (1979) had failed to support his main hypotheses about two-sample effects, later reviewers (e.g. Brainerd, 1981; Hoemann & Ross, 1982) made much of effects that he did find for differences in magnitudes. Perner had evaluated the performance of younger (4–5) and older (6–7) children in standard one and two-spinner tasks, as well as in a "deceptive disk" condition. In both of the standard tasks, older children performed better when odds were highly discrepant (7 : 1) than when they were close (5 : 3). Younger children also showed an effect of discrepancies in odds for the two-spinner task, but not for the one-spinner task. These results replicate Hoemann and Ross's (1971) findings that—except for the 4-year-olds in the one-spinner task—children were sensitive to the degree of disparity between outcome frequencies in one- and two-spinner tasks.

In Perner's deceptive disk condition, spinner segments were made to appear to vary in area contrary to the actual odds. In this condition, older children again performed better with discrepant odds in both the one-spinner and two-spinner tasks, and the younger group was again inconsistent. Now, however, the younger group showed the effect for the one-spinner, rather than the two-spinner, task. Although these effects of odds differences have been interpreted as supporting the magnitude estimation hypothesis in the two-sample case (e.g. Hoemann & Ross, 1982), differences of the same order of magnitude have been found in one-sample tasks. Moreover, such differences were found for children from 5 through 13 years of age (e.g. Hoemann and Ross, 1971). Thus, older children, in a task that is widely agreed to involve probability judgment, were also sensitive to disparities in odds. Of course, in these kinds of tasks, predictions should ideally be affected only by the direction of differences in odds, not by their magnitude.

In an unpublished dissertation, Callahan (1989) reported a study of quantitative processing in older Anglo and Apache children (third and sixth graders). He presented two-sample probability problems via computer using a display with two buckets, each containing targets and nontargets. Children were instructed to choose the bucket that would afford them the best chance of randomly drawing a target. The computer was programmed to measure response latencies, as well as to analyze patterns of responses and assign them to strategic categories, such as the magnitude estimation strategy ("most winners"). As the odds-disparity results in prior studies indicated, even older children tended to focus on comparing the magnitudes of winners.

Children received computer-administered tutorials that specifically targeted weaknesses in their individual strategies. For example, children using a magnitude estimation strategy would receive critical problems of the type described in Figure 11.3, namely problems in which absolute frequencies differed in a misleading direction (the two sample sizes were unequal). Such children shifted to strategies focusing on both targets and nontargets, but tended to return to the simpler unidimensional strategies after a delay. Interestingly, simpler strategies were associated with faster response latencies, even after the tutorial, suggesting that subjects "knew" the correct strategy but traded accuracy for speed. Similarly, Offenbach, Gruen, and Caskey (1984) reported that, although sixth-graders preferred a "more target" strategy, using it on a majority of trials, they occasionally used a proportional strategy. In the

Offenbach et al. study, as in the Callahan study, most problems could be correctly solved using the simpler strategy. Again, the implication is that older subjects were aware of more complex and universally applicable strategies, but chose to make simple frequency estimates.

Brainerd (1981) found that young children (4- and 5-year-olds) also processed frequency information (contrary to Piagetian theory). In a series of twelve experiments, he used Markov models to estimate the contributions of various information-processing factors to probability judgments. In these experiments, there were 12 types of tokens affixed with colored pictures of different animals. Children placed tokens of two or three types (e.g. seven monkeys and three cows) into an opaque container. After the two or three sets were placed into the container, the experimenter shook it (mixing the tokens). Children were then asked to predict a series of one-element draws from the container. Thus, this was the more stringent one-sample task; sampling was done with and without replacement.

On the first trial, younger (4–5 years) and older (7–8 years) children based predictions on frequencies. That is, choice probabilities on Trial 1 preserved the ordering of the event probabilities. Children were more likely to pick the more frequent type of animal, but even older children's responses did not conform entirely to event frequencies. After Trial 1, however, children at both ages shifted to nonfrequency-based strategies. When feedback about previous draws was unavailable, children used a response-alternation strategy roughly similar to the gambler's fallacy exhibited by adults (e.g. Baron 1988). They predicted the response that they had not predicted on the previous trial. When feedback about outcomes was available, however, children perseverated with previous outcomes. Because responses on Trials 2–5 were not frequency-based, summing across all trials (as Piaget and Inhelder did) would have given the erroneous impression that children had not used frequency information. Nevertheless, it is clear that children's performance was not optimal.

Brainerd tested several information-processing explanations for children's nonoptimal performance. First, he eliminated storage failure (failure to maintain accurate frequency information in memory across trials) as a source of errors by presenting children with an "external store." The same number of animal tokens of a given type that had been placed in the container was also placed in a row in front of the container. Response probabilities were unaffected by this manipulation for either age group suggesting, that failure to store frequency information did not explain the failure to use such information on Trials 2–5.

Since storage failure did not appear to be the culprit, other experiments were designed to determine whether failure to retrieve stored frequency information might be a problem. The retrieval-failure hypothesis was pitted against a rule-failure hypothesis. According to the rule-failure hypothesis, even if children stored and retrieved frequency information on later trials, they simply did not know how to apply it correctly. (The reader will recognize the latter as an instance of a conceptual deficit explanation in the spirit of Piagetian theory.) In order to discriminate between these hypotheses, Brainerd presented younger and older children with three event classes of unequal frequency. The two hypotheses lead to different predictions about performance after Trial 1.

Taking the first hypothesis, let us assume that children do not retrieve frequency information after Trial 1. Because of response alternation, the most frequent class would not be selected on Trial 2. However, if frequency information is not retrieved, then there is no principled basis on which to decide between the two remaining classes. Therefore, children might guess in order to decide between the two remaining classes, or engage in idiosyncratic strategies. On the other hand, if frequency information is retrieved after Trial 1, choices should reflect that information (if children know how to use it) because of the indeterminacy of the response alternation rule. In other words, with three rather than two classes, the response alternation rule does not uniquely identify which of the remaining classes should be chosen on Trial 2. If children retrieve frequency information on Trial 2 (and later), and know that it is applicable, indeterminacy should be resolved on the basis of relative frequencies.

For younger children, the probabilities of choosing the remaining two classes were equal, consistent with a strategy of guessing between the classes that had not been selected on the previous trial. For older children, however, there was evidence of retrieving and implementing frequency information on later trials; 7- and 8-year-olds tended to choose the more frequent of the two classes that had not been selected on the previous trial. Although such results can be interpreted as evidence that older children are more likely to spontaneously retrieve frequency information on later trials, it is still possible that younger children were unable to apply information appropriately on later trials (i.e. rule failure could also be a problem). In order to determine the extent to which retrieval failure as opposed to rule failure accounted for errors as well as developmental differences, Brainerd induced retrieval of frequency information prior to probability judgments.

Inducing retrieval of frequency information produced large increases in the accuracy of probability judgments for both age groups. For younger children, correct responses jumped to 94% (not significantly different from unity) on Trial 1, and to 86% on Trials 2–5. For the older children, the percentage of correct responses was 99% on Trial 1, and 96% on Trials 2–5. Conditional probabilities revealed that better retrieval in older subjects could account for their slight superiority in probability judgments. Since children were never provided with any explanation of how to use the retrieved information (retrieval amounted to a re-presentation of the original problem information), it could be inferred that rule failure (inability to apply frequency information correctly) had not been a source of prior poor performance. However, if children knew how to apply frequency information correctly once it had been retrieved into working memory—if they had the correct rule—why had they not used it in earlier experiments?

Brainerd concluded that children lacked sufficient working memory capacity to accommodate frequency information as well as the other information that they maintained about the problems (e.g. previous responses). Similarly, developmental differences were explained as differences in memory capacity. Older children had more memory capacity than younger ones, and, so could maintain retrieved frequencies as well as other ancillary information. He based these conclusions on two kinds of evidence. First, accuracy declined for probability judgments when memory load was increased by presenting (additional) irrelevant information, especially for younger children (Experiments 4 and 10). Although it is clear that younger children are more distractable than older children, this does not imply that working memory capacity is a source of errors in probability judgment under less distracting conditions. The second type of evidence implicating memory capacity seemed unassailable, however. When children were probed for their memory for frequencies, and probability judgments were conditionalized on those probe responses, judgments were found to be stochastically dependent on memory performance. If judgments were correct, the probability was high that memory probes had been answered correctly. If judgments were wrong, memory responses tended to be random.

In sum, studies by Perner (1979), Hoemann and Ross (1971), and Brainerd (1981) appeared to demonstrate that frequency information was critical for probability judgments. Callahan's (1989) and Offenbach, Gruen & Caskey's (1984) data suggested that older children compared frequencies of targets in two-sample tasks for the sake of simplicity, despite awareness of more universally accurate strategies. Brainerd's results further indicated that failure to retrieve frequency information was the proximal source of errors in probability judgments, and that younger children were less likely to retrieve such information than older ones. Although earlier workers interpreted the failure to use frequency information by younger children as a logical or conceptual deficit, Brainerd showed that children could use such information with a high degree of accuracy when its retrieval was prompted. The fact that the conditional probability of predicting the most frequent class, given success on a memory probe, was not reliably different from unity appeared to provide strong quantitative evidence of the link between memory for frequency and probability judgment.

11.3 STAGE 3: INTUITIVE PROCESSES

As we have seen, the first stage of research on probability judgment had a monolithic quality to it, having been dominated by the only available theory at the time, Piagetian logicism. The second stage was concerned primarily with filling in empirical details, and with methodological controversies. Most studies focused on quantitative processing, namely whether and how frequency information was used by young children. Two viewpoints emerged from the second stage: that probability judgments were based on simple magnitude estimation—in which case young children (and maybe older ones too) lacked logical competence; and, alternatively, that frequency processing was limited by working memory and retrieval failure—in which case conceptual competence was present in young children, and performance variations reflected nonconceptual factors.

This conflict between findings of precocity on the one hand, and late emergence on the other, has continued into the third stage. Using increasingly sensitive measures, some studies have shown that younger children are able to demonstrate competence. At the same time, older children *and adults* have exhibited a number of biases that place their competence in question. In contrast to the homogeneity of the Piagetian period, variability in results and in theoretical explanations has become the rule. The neat developmental divisions that earlier researchers envisioned, even the idea that thinking necessarily progresses in a linear fashion, are now being questioned. In order to account for this variability, several researchers have introduced a distinction between intuitive (or qualitative) processing and quantitative processing. These different ways of thinking are not strictly identified with ages or with developmental stages, however. Although some researchers have become resigned to task and domain specificity, invoking qualitative and quantitative thinking as selections from a cognitive menu, others are attempting to integrate these disparate findings into a general theory of cognitive development—a "new intuitionism."

11.3.1 Task Variability

For any regular reader of psychological research, it is by no means surprising to discover that performance differs across tasks that ostensibly tap the same competence (see Ceci 1990, for an excellent review). This is true also in children's probability judgment. Research on task variability in probability judgment has examined effects of the context of judgments (social versus object judgments), problem content (whether extraneous information is presented), and methods of assessment (information-integration techniques or rule assessment). The upshot of these studies is that young children can perform advanced processing operations (under certain conditions). The implications of task variability, however, are controversial, with some theorists claiming that variability itself proves that children (and perhaps adults) do not really understand probability.

Jacobs and Potenza (1991) compared probability judgments for scenarios involving objects as opposed to social judgments. For object judgments, children were told, for example, that a dresser drawer contained three pairs of white socks and six pairs of colored socks, and were asked to predict which kind of socks would be obtained if someone reached in and grabbed a pair without looking. An example of a social judgment involved 10 girls trying out to be cheerleaders, and 20 trying out for the band, the task being to predict which activity a given individual was likely to be trying out for. In addition, some scenarios included qualitative information designed to bias predictions (e.g. Juanita is "very popular and very pretty. . . . Do you think Juanita is trying out to be a cheerleader or for the band?"). This information was designed so that subjects who relied on the *similarity* between the qualitative description and a stereotype for a category (the "representativeness heuristic") would choose the less frequent category.

Jacobs and Potenza found that predictions were less apt to be based on frequency information for social judgments than for object judgments. In fact, frequencies were likely to be ignored for social scenarios when biasing information was included (the bias effect was smaller for object judgments). Interestingly, developmental trends differed depending on the domain, social or object, and the information provided. When biasing information was not provided, object judgments did not differ from first grade through college age; all age groups predicted the more frequent class (e.g. colored socks). However, for the social scenarios, there was an age trend. Frequency-based judgments increased with age for the social scenarios that lacked biasing information. Thus, even when biasing information was omitted, the domain of the judgment—objects or a social situation—affected whether children used frequencies to make probability judgments.

When biasing information was included, developmental differences emerged for object judgments as well. Younger children were more likely to be distracted by such information than were older children and adults. This result is consistent with Brainerd's (1981) finding that young children used presented information even when it was irrelevant for judging probability. (See Dempster, 1992 for an excellent review of interference effects in a variety of cognitive developmental tasks.) Surprisingly, the age trend reversed when biasing information was provided in the social scenarios. Older subjects, including adults, generally ignored frequencies when the qualitative information matched the stereotype of the less frequent category. Although the youngest group was prone to the same bias, its magnitude was somewhat smaller. (Jacobs and Potenza also elicited verbal explanations for predictions, obtaining the usual finding that choices consistent with the use of a given strategy—either frequencies or the representativeness heuristic—preceded the ability to articulate that strategy.) With respect to adults, this study is another replication of so-called errors and biases in probability judgment (e.g.

Kahneman, Slovic, & Tversky, 1982). Jacobs and Potenza have provided the first evidence, however, that such errors increase with age.

Jacobs and Potenza's study illustrates the effect of problem content, how the context and type of information affects probability judgments (see also, Reyna & Brainerd, 1991a, 1993). Most of the discussion about task variability in probability judgment, however, has centered around comparisons of two assessment methods: Siegler's (e.g. 1981) rule assessment and Anderson's (e.g. 1980) information integration procedures. Kerkman and Wright (1988) compared results using the two methods for probability judgment, as well as for other tasks. They concluded that different methods yielded different results, and recommended rule assessment for younger children, and information integration theory for older children because of its greater sensitivity to advanced computational strategies. Surber and Haines (1987), in their review of proportional reasoning tasks, also discussed differences between the two methods, but they reached the opposite conclusion: "The rule assessment approach emphasizes analytic or computational knowledge of proportional reasoning, whereas the information integration approach is better suited to intuitive or qualitative knowledge" (page 52).

Despite disagreements about the advantages of different measures, there is consensus that results vary across methods. For example, in Siegler's (1981) study, probability judgment lagged behind other proportional compensation tasks, and a minority of 8-year-olds had the correct quantitative rule. In Anderson's (1980) study, on the other hand, five-year-olds' responses fit a linear fan pattern characteristic of the correct rule. These differences in results may have to do with processing differences induced by the methods, *in addition* to any intrinsic properties of the measures. A clue is provided in a study by Hommers (1980). Like Anderson (1980), Hommers assessed children's knowledge of the expected values of bets involving two dimensions, probabilities and payoffs. Performance in a choice task (in which children picked the option they preferred) was compared to that in a rating task (in which they rated the "likeability" of bets). Children in the choice task who processed only one dimension were able to combine two dimensions in their ratings of bets. Similar findings have been obtained with adults (Payne, 1982). Thus, choice tasks, such as those used in Siegler's rule assessment procedure, may encourage simpler information-processing strategies, compared to rating tasks, such as those typically used in information integration experiments (Reyna & Brainerd, 1991b, 1992).

This interpretation is consistent with the results of Acredolo et al. (1989). First, third, and fifth graders were asked to estimate the probabilities of a series of events based on the number of potential successful outcomes and the total number of possible outcomes (numerator and denominator values, respectively). Three levels of numerator (1, 2, and 3) and of denominator

(6, 8, and 10) values were crossed, producing nine problems, each of which was presented three times. The task was presented as one in which a teacher wanted to draw a jelly bean of a given color from a bag without looking. Children rated probability by sliding a marker along a "happy-face" scale. A bag with none of the designated color (sad) and one with only the designated color (happy) were used to indicate the endpoints of the scale. According to information integration theory, children possess tacit knowledge of the relations between dimensions long before they are capable of applying the appropriate mathematics. This knowledge is reflected in the pattern of their estimated values. The happy-face scale provided 32 points with which children could convey such estimates of probability.

Results indicated that all three age groups took into account variations in the numerator and in the denominator, as well as the interaction of these two dimensions. Numerators influenced ratings more than denominators, but greater variance among the probabilities was in fact associated with the numerator because of the specific values that had been presented. So, for example, children assigned a higher rating to 3/10 than to 2/6 (a typical error when actual differences in probability are small). Acredolo et al. conducted several types of individual subject analyses, categorizing the ordering of subjects' ratings as consistent or inconsistent with their objectively correct ordering. As in the group analysis, category assignments were not associated with grade.

In a second experiment, Acredolo et al. changed the problem set to include matched pairs of "critical" problems of the sort noted above (3/10 vs. 2/6). Instead of judging the probability of drawing a jelly bean from a bag, children judged whether a bug who jumped into a box would land on a flower rather than a spider. Boxes contained varying frequencies of "good" (i.e. flowers) and "bad" (i.e. spiders) outcomes. Again, there was no effect of grade, and children took both dimensions and their interaction into effect in assigning probability ratings. Numerator and denominator were weighted more appropriately than they had been in the first study; not surprisingly, therefore, children seldom erred on the critical problem pairs. However, children did exhibit a consistent bias when probabilities were *equal*, as in 5/10 versus 3/6, assigning the higher rating to the more numerous stimulus (5/10 in this example). Fischbein, Pampu, and Manzat (1970) reported similar results. As Acredolo et al. and Reyna and Brainerd (1993) discuss, analogous behavior occurs in adults. Despite equal expected values, a choice between $100 for sure and a 50% chance of $200, for instance, does not elicit indifference (Tversky & Kahneman, 1981). In sum, then, in contrast to earlier studies using a binary choice paradigm (see also Wilkening & Anderson, 1982), Acredolo et al. concluded that "children do have the ability to choose on the basis of odds generated through the construction of proportions" (pages 943–4).

11.3.2 Intuitive Thought

Our review of research on task variability suggests that younger and older children process information in qualitatively similar ways, but that younger children are more sensitive to certain features of the task. Younger children generally perform better when they are not distracted by irrelevant information, and when, according to Acredolo et al., they estimate likelihoods rather than make choices. According to Acredolo et al., the high level of performance they observed is achieved by "rough estimation" (page 944) before children are capable of precise mathematical computations. The nature of "rough estimation" is not completely spelled out, however. In particular, Acredolo et al.'s results do not pinpoint the degree to which children's processing is quantitative. For example, children might process exact frequencies—they might count the number of targets and non-targets (and then combine them intuitively)—or they might simply estimate their relative magnitude.

In an ingenious follow-up study, Lovett and Singer (1991) examined whether children estimated probabilities in an essentially quantitative or non-quantitative manner. Using Acredolo et al.'s second task, subjects ranging in age from kindergarten through college estimated "how good or bad the box is" that a bug might jump into on a happy-face scale. In addition to recording probability estimates, however, Lovett and Singer also measured the time it took subjects to make their estimates. Across three experiments, they did obtain age differences in children's estimates. However, although the younger subjects (especially kindergartners) deviated somewhat from an ideal response pattern, results were similar to Acredolo et al.'s in that first graders' and older children's responses demonstrated a sophisticated knowledge of probability.

The key results involved response times. Lovett and Singer reasoned that, if children counted the items in a box, response times should increase as the number of items increased. So, they performed a regression analysis for each subject using the size of the denominator (total number of items) as the predictor variable and response time as the criterion variable. The regression analysis reached significance (indicative of counting) for only about half of the subjects at each grade level (and for even fewer kindergartners). Thus, although children performed surprisingly well at estimating probability, a sizeable number of subjects did not appear to count the number of items in a box.

In order to discover how children managed to perform so well without processing exact frequencies, Lovett and Singer decided to test the limits of children's abilities to solve the task nonquantitatively. In a second experiment, children saw one of two displays. The first display was a pond with an outcropping of rock. Children were asked to estimate how likely it was that "the bug will land on the rock and not in the water when it jumps into the pond." Both the length of rock and the water were drawn as continuous variables, and therefore their magnitudes could not be estimated by counting (children were not allowed to use tools, such as a ruler, to measure the display). The second, quantitative, display, was again a box with flowers and spiders. Here, children were explicitly instructed to count the items, and the items were scrambled to make it difficult to estimate their frequencies perceptually.

Despite using an array of performance measures, no effects of display condition were detected. Children and adults performed equally well when quantitative cues were available, and subjects were told to use them, as when such cues were unavailable. In a third experiment, Lovett and Singer presented a single display that supported either counting or perceptual estimation. In contrast to Experiments 1 and 2, in which items had been positioned randomly, flowers were grouped together and spiders were grouped together. Regression analyses on response times in this task revealed that most subjects failed to count the number of flowers and spiders, presumably estimating their relative magnitudes perceptually. Interestingly, the developmental trend was nonmonotonic. The age groups that were least likely to count were at each extreme: kindergartners and adults. Indeed, only 25% of adults processed the display quantitatively when they had the option to estimate magnitudes perceptually. This result contrasts with the increase with age in the use of counting observed in Experiment 1, a task in which scrambling virtually eliminated the option of perceptual estimation. Thus, older children and adults apparently estimate probabilities both intuitively and computationally, depending on the task, but intuition is the default option.

Surber and Haines (1987) make a similar distinction between computational strategies (e.g. counting) and intuitive, or qualitative, strategies. They assume, with Inhelder and Piaget (1958), that intuitive strategies are more primitive than computational ones. However, they acknowledge that most studies of probability judgment used problems that do not require computational solutions, making it difficult to determine whether young children are capable of such advanced processing. Moore, Dixon, and Haines (1991) compared intuitive and computational processing in a proportional reasoning task. Like Lovett and Singer, Moore et al. presented two versions of a task, one in which estimation was required (i.e. numerical information was unavailable) and one in which numbers were presented and subjects were explicitly instructed to compute their responses. Moore et al. compared performance in these two types of task for second, fifth, and eighth graders, and college students. At every age level, the intuitive version of the task was easier than the computational version.

Moore et al. also analyzed the quality of each subject's understanding of the task. Six prototype categories of levels of understanding were developed (each prototype being composed of various components of understanding).

Individual subjects were assigned to categories based on the degree to which their responses fit the prototype for that category. As expected, subjects in the intuitive condition were less likely to exhibit a computational strategy and those in the computational condition were less likely to exhibit an intuitive strategy. Moreover, lower-level strategies were more likely to be used by subjects in the computational condition. Thus, the quality of children's strategies (especially for younger children) was generally higher in the intuitive condition.

The Moore et al. data conflict with the traditional Piagetian view that intuitive thinking forms the basis for, and is supplanted by, concrete and formal operational thinking. First, significant differences remained for the older subjects (long past the preoperational stage) between the intuitive and computational conditions. This is consistent with Lovett and Singer's finding that adults, when they had both options, preferred intuitive over quantitative processing. Second, quantitative proportional reasoning apparently does not follow automatically from mature intuitive proportional reasoning. As Moore et al. indicate, "The relation between intuitive understanding and the process of arriving at a computational scheme deserves further research" (page 456). Although Moore et al. differ from the traditional view in asserting that computational reasoning does not derive directly from qualitative understanding—that, indeed, there may be "multiple developmental paths" (page 457), they adhere to the traditional assumption that computational reasoning is the apex of development. Siegler's model (e.g. 1981) also incorporates the assumption that correct reasoning (as opposed to performance) requires the application of appropriate quantitative rules. This assumption, however, is by no means universal.

Fischbein (1975) was the first to advance the idea that children could *correctly* solve probability problems intuitively and, moreover, that such thinking could reflect an initial understanding of probability. Of course in Piagetian theory, "intuitive understanding" is an oxymoron because intuitive thought presupposes a lack of appreciation for the defining features of the probability concept. The hegemony of Piagetian theory probably accounts for the fact that, as Hoemann and Ross (1982) note, Fischbein's proposal received little attention when it was first introduced. In the third stage, however, Fischbein's ideas have found increasing acceptance.

Fischbein distinguished between a natural intuition of probability and the ability to quantify that intuition. According to Fischbein, preschoolers respond to probability questions appropriately given the limits of their quantitative skills. Such children estimate relevant quantities or areas perceptually, comparing subjective odds, but they are unable to construct ratios or evaluate them computationally. For younger children, then, task difficulty is a function of perceptual complexity (see also Spinillo & Bryant 1991). Consistent with Fischbein's interpretation, Huber and Huber (1987) found that young children affirmed six formal axioms of *qualitative* (nonnumerical) probability.

11.3.3 Reconciling Earlier and Later Findings

As we have noted, research on intuitive estimation of probability indicates that children exhibit advanced knowledge of how numerators and denominators should be integrated before they are able to perform precise computations. They seem to achieve this level of performance by engaging in qualitative, often perceptual, processing in proportional reasoning tasks generally, including probability judgment. Moreover, intuitive strategies persist into adulthood. This emphasis on nonquantitative reasoning contrasts sharply with the emphasis on quantitative reasoning in the second stage. In that stage, even children who supposedly misunderstood probability were said to process frequencies (e.g. they compared the number of targets), and it was universally agreed that adults processed frequencies. The question in the third stage, then, is how to reconcile these seemingly contradictory interpretations.

If children process probability qualitatively, how can one explain the effects of such numerical nuances as, for example, discrepancies in odds? Fischbein explains variations across one-sample and two-sample tasks (e.g. in Hoemann and Ross's, 1971, study) in terms of perceptual discriminability. The same argument can be applied to differences in odds; as such differences get bigger, they are also more perceptually discriminable. A difference between 7/8 and 1/8 is more salient than a difference between 7/8 and 5/8 (see Spinillo & Bryant, 1991). Moreover, spinners at the same level of differences are not numerically identical; absolute frequencies differ, for example, 7/8 and 3/8 versus 3/4 and 1/4. [Absolute frequencies do not appear to predict probability judgments in adults either (e.g. Allen & Estes, 1972; Estes, 1976).] Thus, global differences that can be estimated perceptually, rather than exact frequencies, could explain effects of odds differences. Such a hypothesis would also explain biases favoring the more numerous (hence, more perceptually salient) stimulus for close ratios, such as 3/10 and 2/6, and for equivalent ratios, such as 3/9 and 2/6.

Further support for this proposal is provided by research on memory for frequencies. Marx (1985, 1986) varied presentation frequencies for different stimuli (e.g. words), instructing subjects to remember the exact number of times each stimulus appeared. Subjects then judged relative frequency, such as which word of a pair had been presented more often. He found that, for adults, judgments were not related to retrieval of presentation frequencies, but were associated with global, gist-like, impressions of relative frequency. Marx compared subjects' reports of having used these two kinds of representations, frequency memory versus global impressions, across age groups. Among the youngest subjects tested, fifth graders, few reported using global impressions, and most justified their judgments by referring to exact frequencies of presentation. The reverse was true for adults. Surprisingly, although fewer judgments were reportedly based on retrieval of frequencies with age, the accuracy of those judgments improved.

We might account for effects that seemed quantitative in origin, therefore, with assumptions about perceptual processing and the nature of memory for frequencies. However, under conditions in which memory was not a factor (i.e. frequencies were displayed in the external store conditions), Brainerd (1981) appeared to have proved that young children's probability judgments were dependent on frequencies (at least for Trial 1; nonnumerical strategies predominated thereafter). However, Brainerd did not actually assess memory for frequencies. For memory probes, children were simply asked which of the sets had been more numerous. These global judgments about which set had more were then related to probability judgments. The probability judgments were similarly ordinal. The Markov models that were fitted to these data only evaluated whether children's choices (of which set was more probable) went in the correct direction (i.e. whether choices corresponded to the correct *ordering* of the sets according to their probabilities). Relationships between numerical frequencies and numerical probability judgments were not, in fact, measured.

In 1985, Brainerd and Kingma directly tested the proposition that probability judgments were dependent on memory for exact frequencies, with surprising results. They found that judgments and memory for frequencies were independent. Alternative explanations for these findings, such as measurement insensitivity, were ruled out (see also Brainerd & Kingma, 1984; Brainerd & Reyna, 1992). Brainerd and Kingma (1985) suggested that reasoning performance might be based on memories for the gist of presented problem information (as opposed to details such as exact frequencies). There was no explanation, however, for why this might be so, and no comprehensive theory was offered to reconcile this anomaly with the tenets of the information-processing approach.

11.3.4 Fuzzy-trace Theory: The New Intuitionism

Attempts to reconcile the reasoning–remembering independence effect, as it came to be called, with information-processing assumptions were ultimately unsuccessful. The independence effect called into question the assumption that, in reasoning tasks such as probability judgment, problem information was processed in a working memory with limited capacity. If that were so, reasoning and memory would *necessarily* be dependent, rather than independent (as observed). For example, in probability judgment, the argument that judgments are based on global impressions rather than memory for frequencies does not explain away the independence effect in a manner that is consistent with basic information-processing assumptions. This is because those global impressions must, in turn, have been derived from problem information that was initially represented in working memory. Thus, if probability judgments were only *in*directly linked to initial representations of frequency information in working memory, dependency would still be detected—and it was not (see Brainerd & Reyna, 1992 and Reyna 1992 for reviews of the arguments and evidence on this point). These kinds of disconfirmations eventually forced a reassessment of information-processing accounts of reasoning.

In 1990, two papers appeared (Brainerd & Reyna, 1990a; Reyna & Brainerd, 1990) that introduced fuzzy-trace theory. The initial impetus for fuzzy-trace theory was the need to resolve contradictions that had arisen in the wake of information-processing theory (Reyna & Brainerd, 1992). For example, despite findings of reasoning–remembering independence, it is self-evident that children's reasoning must somehow "depend on" memories that are related to problem information. Children's responses do vary with problem information—such as frequencies in probability judgment—and their responses are often correct. How could they manage this if they failed to process problem information?

The solution to this dilemma that is advanced in fuzzy-trace theory is to assume that there is another kind of memory representation that is separate from memory for the problem *per se*. These other memories are for the gist of the problem information, the global patterns instantiated by the stimuli (e.g. "more than," "increasing") rather than the stimuli themselves (i.e two spinner segments with differing areas, a row of rods increasing in length, respectively). After reasoning paradigms of many types were surveyed, it became clear that reliance on gist, as opposed to verbatim traces, was pervasive (for a review, see Reyna & Brainerd, 1991a). This reliance was evident even when verbatim memory was found to be highly accurate (e.g., Reyna, 1991). This led to a reorientation of theory about reasoning, memory, and especially, their relationship.

Hitherto, the relationship between memory and reasoning was presumed to be hand-in-glove. One of the contributions of the information-processing approach had been to emphasize the role of memory in problem solving, a role that many theorists claimed had been neglected by Piaget. A classic demonstration was provided by Bryant and Trabasso (1971), who found that the probability of correct reasoning could be predicted by the probability of remembering presented information. Dual-task interference effects (that performing a concurrent task such as finger tapping interferes with reasoning) were also seen as favoring the importance of accurate memory for accurate reasoning (e.g. Bjorklund, 1989). As we have noted, however, direct tests of the relationship between reasoning and memory failed to uncover expected dependencies. In addition to accounting for independence results, fuzzy-trace theory offered alternative explanations for dual-task interference and for supposed memory effects in reasoning tasks, bringing seemingly contradictory findings together under the same theoretical umbrella.

Although fuzzy-trace theory has been extended in a variety of ways (see Brainerd & Reyna, 1993a; Reyna & Brainerd, 1991a, for reviews), incorporating a set of interrelated principles about encoding, retrieval, processing, and retention, we will focus here on just three principles that are especially relevant to probability judgment. These are *fuzzy-to-verbatim representational continua*, *the fuzzy-processing preference*, and *memory-to-reasoning interference*. According to fuzzy-trace theory, reasoners encode multiple representations of problem information that vary in specificity (or fuzziness). These representations are functionally dissociated early in processing. That is, gist is not derived from verbatim representations of problem information; instead, they are encoded in parallel. Given multiple representations, reasoning is governed by a fuzzy-processing preference. Reasoning gravitates to the fuzziest, least specified, level of representation that can be used to successfully solve a problem. So, for example, reasoners deciding which number is larger, 294 or 7, would tend to rely on gross magnitude estimation, as opposed to processing the exact numbers (e.g. Link, 1990). Functional dissociation and the fuzzy-processing preference, then, explain why verbatim memories and gist-based reasoning would be independent of one another in probability judgment (as well as other tasks).

Although it is a fundamental assumption of information-processing theories that memory accuracy is positively related to reasoning accuracy, according to fuzzy-trace theory, accurate memory can adversely affect reasoning. In class-inclusion reasoning, for instance, when children are asked if there are more cows or more animals, focusing on problem facts, including that there are more cows than horses, leads to systematic errors. The ability to ignore verbatim information, and operate on the gist of relationships, increases with age (Dempster, 1992). Not coincidentally, the functional dissociation between gist and verbatim memory, as well as the fuzzy-processing preference, also increase with age (Brainerd & Reyna, 1993; Reyna, 1991).

In sum, rather than explaining reasoning accuracy in terms of local breakdowns in memory (e.g. effects of memory "load"), although these can be demonstrated, reliance on gist implies that reasoning is not mainly determined by such errors (see Estes, 1980, for similar arguments). Thus, studying variations in working memory for "bits" of information is unlikely to illuminate the origins of intelligent behavior. Fuzzy-trace theory shifts the explanatory focus away from memory for information to the ability to recognize and represent the global patterns embedded in that information. The theory departs sharply from traditional approaches in assigning a central role to gist in advanced reasoning. In this view, a "fuzzy-processing preference" represents a system-wide adaptation to the limits of information processing, a means of avoiding systematic errors caused by poor verbatim memory.

In fuzzy-trace theory, therefore, thinking is seen as fundamentally intuitive. This contrasts with the Piagetian conception of thinking as the rigid application of logical rules to premise-like inputs. It also contrasts with the mind-as-computer metaphor of information-processing, which emphasizes rich elaborations ("knowledge representations") of inputs, and identifies precision and quantification with accuracy. Not only does the conception of thinking differ in fuzzy-trace theory, so do ideas about development. The fuzzy-processing preference is ascribed to mature reasoners, and represents a flexible and adaptive approach to reasoning that, overall, has the effect of reducing errors. This is the opposite of the traditional view that development progresses away from intuition towards greater logic or computation.

11.3.5 Explaining Probability Judgment Intuitively

It is clear from several studies that children as young as five or six can make accurate probability judgments, and that they can apply the logic of ratios to these judgments—despite the fact that they cannot yet divide. The data of Brainerd (1981), Acredolo et al. (1989), Lovett and Singer (1991) and Jacobs and Potenza (1991) support such a conclusion. In addition, Huber and Huber (1987) showed that young children understood qualitative axioms of probability, and Anderson (1980) and Hommers (1980) showed that they could apply their knowledge of probability in rating tasks (see also Fischbein, 1975). However, it is also clear that younger children are sometimes distracted by irrelevant information, including their own prior responses (Brainerd, 1981; Jacobs & Potenza, 1991).

If our review stopped here, we would conclude that the probability concept is present early in development, although implementing that competence steadily improves with age. The implication would be that once those same children became adults, they would perform at least as well, if not better than, young children. However, if we define "better" as processing ratios of frequencies, that implication is false. Second stage research showed that, although the magnitude estimation hypothesis was initially supposed to explain why younger children's judgments were inferior to older children's, evidence (e.g., odds disparity effects) indicated that this simple strategy was used by older subjects. Callahan's (1989) and Offenbach, Gruen & Caskey's (1984) findings further suggested that older subjects were aware of, and able to use, ratio strategies, but they preferred magnitude estimation instead.

More recent research concurred in demonstrating that older reasoners often failed to compute ratios. Older children and adults were more likely to reject frequency information in a biasing context than younger children (Jacobs & Potenza, 1991); they were more likely to make intuitive rather than frequency-based judgments when the latter were optional (Lovett & Singer, 1991); and intuitive tasks were easier (and elicited more advanced reasoning) compared to computational tasks even for adults (Surber & Haines, 1987). Finally, adults' errors in probability judgment have been demonstrated in numerous studies on "heuristics and biases" (Kahneman, Slovic & Tversky, 1982). It is not just that adult cognition is less advanced than once thought. The heuristics and biases research and the developmental research imply diametrically opposed views of adult competence in judging probability.

One might argue, based on this collection of findings, that children and adults select from a menu of rules (Siegler, 1988) depending on the task, and that the variability across tasks cannot be spanned by domain-general assumptions. Alternatively, one might argue, with Piaget, that such variability proves that formal competence has not been acquired because such competence, by definition, spans tasks that differ superficially. Of course, by this standard, one would be forced to conclude that adults lack the probability concept.

In contrast, fuzzy-trace theory would argue that adults are able to process information in a quantitative fashion, but that they tend not to engage in such processing unless the task requires it. Thus, the probability concept is present in adults (as shown in studies in which sensitive measures reveal advanced competence), and advanced understanding of that concept can be achieved intuitively (Acredolo et al. 1989; Fischbein 1975; Huber & Huber 1987). Younger children, at least as young as first graders, also have the probability concept, but they cannot inhibit output interference and interference from irrelevant problem information (e.g. Brainerd & Reyna, 1993; Reyna & Brainerd, 1989). Development, in this view, consists of increasing resistance to interference, and greater reliance on the gist of magnitude relationships—as opposed to precise computation (although computational skills clearly improve during childhood and, for some tasks, must be invoked).

Naturally, intuitive processing has predictable pitfalls such as "denominator neglect," the tendency to assume equal denominators when comparing ratios, in order to apply magnitude estimation (which if denominators were actually equal, would deliver consistently correct responses). Such denominator neglect was observed by Offenbach, Gruen & Caskey (1984) and by Callahan (1989), and similar examples from everyday cognition are discussed by Reyna & Brainerd (1993). Another bias likely to result from intuitive probability judgment is analogous to rounding error, namely the confusion of close ratios (e.g. 3/10 and 2/6). Ideally, however, probability judgments in the standard task should be entirely dissociated from numerical disparities such as odds differences or the exact level of probabilities. In fuzzy-trace theory, as Surber and Haines (1987) argue, quantitative and intuitive processing are indeed separate and independent, especially for adults.

11.3.6 Is Intuition Irrational?

To say that adults possess underlying competence, that they have the probability concept, and have had it from early childhood, is not necessarily to claim that their reasoning is rational. The definition of rationality has been debated by great minds, and we cannot begin to do the concept justice here. However, with respect to probability judgment, there are several distinct features of rationality that have been proposed, and that bear on the issue of whether intuition is rational.

The question of rationality is central to developmental research for two reasons. First, theorists have assumed that cognitive development is progress toward rationality (e.g. Piaget, 1931; Werner, 1948). Second, intuition is assumed to be irrational. As we have seen, however, this places us on the horns of a dilemma: We must accept experimental data (showing that adults reason intuitively and that young children reason logically), but we must also somehow explain the course of development. Like many situations in which theoretical conflicts occur, however, it turns out that implicit definitions of key concepts differ—in this case, critically, of rationality. By carefully exploring definitions of rationality, these conflicts can be reconciled.

Tversky and Kahneman (e.g. 1983) have used the term "intuitive" to refer to the kind of natural processing that leads to biases and errors in probability judgment (e.g. the conjunction fallacy). In this usage, they can be grouped roughly with Piaget, for whom intuition was, by its nature, the absence of rationality (i.e. the absence of logic). However, Tversky and Kahneman (and others) have discussed at least three kinds of rationality, including correspondence to extensional reality, correspondence to formal rules (e.g. principles of inferential statistics), and various forms of internal coherence among judgments (e.g. the invariance axiom).

Extensional reality refers, for example, to whether preferences between choices map onto *actual* differences in wealth (Tversky & Kahneman, 1981, 1986). To oversimplify somewhat, to pick an option that makes one poorer in reality (compared to its alternative) is irrational (assuming that one would prefer to be richer). Nonmonetary consequences, including emotional consequences such as anticipated regret should be taken into account. But there is still something disquieting about asserting that happiness predicated on an illusion of greater wealth is rational simply because the dupe thinks he is happy. If reality were no concern, then the paranoid who believes that there is a plot to get him, when there is no plot, would be rational. In this respect, reality separates the rational from the irrational.

Gist does not represent literal reality; it is an abstract representation of underlying patterns. Nevertheless, one can evaluate those representations for their *consistency* with reality. So, there is nothing intrinsic to gist-based representations that would entail any contradiction with reality. Moreover, the fuzzy-processing preference is generally constrained by the level of precision demanded in a task. For example, when alternatives begin to diverge in expected value, subjects do note the nonequivalence of options and shift responses (though not necessarily sufficiently, see Reyna & Brainerd, 1991b).

Further, processing fuzzy, or gist-level, representations tends to minimize reasoning errors (Estes, 1980; Reyna & Brainerd, 1991a). Taken together, these considerations suggest that intuitive processing allows the individual to maximize overall performance by trading off precision and simplicity.

For many judgments, it is difficult to determine what reality is. In these cases, the second criterion of rationality—correspondence to formal rules—is used. In Jacobs and Potenza's (1991) example given earlier about cheerleaders, for instance, the probability that cheerleaders are "pretty" or "popular" is unknown; in Tversky and Kahneman's (1983) well-known Linda problem, there is no way to know the probability that Linda (or someone fitting her description) is a feminist. Logic, mathematics, and other formal systems, however, allow us to place constraints on relations among judgments without knowing the exact state of reality. Tversky and Kahneman qualify their attributions of biases and errors in probability judgments *relative* to specific canons of logic, Bayesian probability theory, and other formal systems. In other words, if the validity of those systems is not granted, then subjects' responses are not necessarily errors.

Gigerenzer and colleagues (e.g. Gigerenzer & Murray, 1987) have challenged Tversky and Kahneman's formal analyses of some probability problems, and argued that subjects are not really making errors. The general thrust of these arguments is that problems are "ambiguous," and so subjects make additional assumptions that go beyond the presented information. Given these assumptions, there are alternative conceptions of probability that support subjects' responses. Similar arguments could be applied here to explain the variability in results that we have discussed. However, there is little direct evidence that subjects make these additional assumptions, or that they adhere to alternative theories of probability—except that their responses violate standard formalisms. The theoretical tack is reminiscent of Henle's (1962) arguments that subjects import additional premises from their pragmatic knowledge of the world, and once these are accepted, their logic is impeccable. The problem with such an approach, of course, is that all errors can be construed *post facto* as successes (especially if the theorist is not limited by considerations of parsimony), yet the fact remains that the reasoner has not solved the problem *as presented* correctly by some standard, often one the reasoner claims to adhere to.

As our review indicates, this criterion—adherence to logical and computational formalisms—has been a prime concern of developmental research on probability judgment. It is obvious from this research that, however subjects manage magnitude estimation and dimensional integration (e.g. Acredolo et al., 1989), they do not necessarily process information logically or quantitatively. In that sense, intuitive processing is irrational. However, although reason has historically been defined with respect to logic and mathematics, it is conceivable that a truly psychological definition of reason might someday be derived that is not drawn from these analogies. (Such a definition might also encompass those aspects of advanced reasoning, such as creativity, that are not well captured by conventional models.) Thus, although intuition does not fit the historical definition of reason, developmental research on probability judgment establishes that intuition is a feature of its psychological definition.

The most basic criterion of rationality does not require any particular definition of reason. Nor does it require exact knowledge of reality. Whatever the basis for subjects' responses, the minimal condition that those responses must satisfy, as Tversky and Kahneman (1986) have compellingly argued, is invariance. Judgments of the same information by the same individual should at least (barring statistical error, the passage of time, or changes in circumstances) be consistent with one another. If judgments are inconsistent, it is difficult to claim that there is some rational basis for those judgments. For example, we noted earlier that most people prefer a sure gain (e.g. of \$100) when it is pitted against a gamble of equal expected value (a 50% chance of \$200). However, if the same numbers are given, but the choices involve losses, most people prefer the gamble (i.e. the chance of a higher loss over a sure but smaller loss). Even when given an endowment from which losses are to be deducted (making the net gain equal to that in the positively worded problem), subjects still prefer the gamble. So, superficial changes in wording—although the same amount of money will be gained—cause subjects to shift from risk aversion to risk seeking. This shift is taken as evidence of irrationality (e.g. Tversky & Kahneman, 1986). So, violations of invariance do not apply to situations in which problems really change—for instance, if the amount to be gained actually changes; they do not have to do with adaptation to changing circumstances. Instead, they have to do with inconsistent responses to superficially different presentations of the same information.

By this most fundamental criterion, gist is the key to rationality. By operating on the underlying gist of information, rather than on verbatim details, reasoning can be invariant across superficially different problems. But, is it? Developmentally, according to fuzzy-trace theory, reasoning becomes increasingly invariant. That is, correct reasoning is applied to a wider range of tasks that abstractly instantiate the same concepts. (Two classic examples are oddity transfer and learning sets generally, Reyna & Brainerd, 1991a, 1992.) Correct reasoning, however, involves more than just recognizing the appropriate gist in problem information. It also involves inhibiting interference from irrelevant details, editing out irrelevant gists, knowing the relevant reasoning principle (e.g., proportionality), retrieving that principle in context, and correctly implementing that principle (i.e., correctly applying the principle to the gist representation of the problem). Each of these components has been shown to make independent contributions to successful reasoning (e.g. Brainerd, 1983; Brainerd & Reyna, 1990b; 1993; Reyna, 1991; Wilkinson, 1982).

With development, children become less subject to interference from both verbatim details (Brainerd & Reyna, 1993; Reyna, 1991) and from competing gists (Reyna & Brainerd, 1991a). They also more reliably retrieve and process relevant principles. All of this fosters the development of invariance. Interestingly, across the many reasoning tasks that have been investigated, recognizing the appropriate gist and knowing the reasoning principle do not seem to be the source of much developmental variation, especially after first grade. Among both children and adults, it is common for an array of gists to be activated in a specific problem-solving context, despite the fact that most are not used. In addition, adults sometimes are systematically misled by salient patterns into responding to the wrong gist (Tversky & Kahneman, 1981, 1983). Thus, it would be incorrect to conclude that reasoning ever becomes entirely invariant. There seem to be some predictable pitfalls in reasoning that are created by the need to retain flexibility. By flexibility, again, we do not mean the ability to shift reasoning in differing circumstances, but the ability to respond to the same information in different ways. Reasoners appear to trade off two competing aims, the need to use the same information in different ways and for different purposes, and the need to respond to similar situations consistently.

11.4 SUMMARY AND CONCLUSIONS

We have divided research on the origins of probability judgment into three stages. The first stage was dominated by the theory and data of Piaget and Inhelder (e.g. 1951). Their theory provided a comprehensive view of human thinking and its ontogenesis, but its empirical foundation was shaky. Stage 2 research redressed the imbalance between theory and data. Although it began as an effort to validate Piagetian claims, Stage 2 research ultimately proceeded in unexpected directions, generating findings that contradicted the Piagetian program. These findings included that spurious performance obstacles led to underestimation of children's ability to reason about probability, that younger and older children used simple magnitude estimation to judge probabilities (despite awareness of ratio strategies), and that younger children could use ratio strategies correctly in tasks that were not solvable by magnitude estimation (e.g. if retrieval of frequency information was cued).

Thus, Stage 3 began in a theoretical quandary. Younger children were more advanced and older children were less advanced (e.g. they, too, used simple magnitude estimation) than should have been the case, according to Piagetian theory. Further research merely lengthened the list of contradictions: Findings of early proficiency such as those in Acredolo et al.'s (1989) study were offset by studies such as Jacobs and Potenza's (1991) in which certain kinds of judgment errors actually increased with age. Researchers ultimately concluded that children and adults solved probability judgment problems both by intuitive estimation and by quantitative proportional reasoning—and there was little dispute that intuition was easier.

Theoretical opinion continues to diverge, however, with respect to the implications of these differing approaches to probability judgment. Some theorists believe that reasoning is determined primarily by local task demands; theories, therefore, should be task-specific. Other theorists believe that variability in performance, vulnerability to biases, and intuitive processing, especially in adults, all indicate that the probability concept is never fully acquired by most reasoners. Still others—for example, fuzzy-trace theorists—believe that intuitive processing is the key to achieving the most basic criterion of rationality, invariance. In this view, development is progress toward invariance—the ability to screen out irrelevancies and respond consistently to the core gist of problems. Hence, reasoners understand probability at an early age, but they increasingly rely on intuitive processes as they get older. Apparently, reasoning never advances to perfect invariance because the system serves competing goals: that reality should be represented veridically but simply, that processing should be accurate but easy, and, most important, that the same information should be perceived similarly and differently.

REFERENCES

Acredolo, C., O'Connor, J., Banks, L. & Horobin, K. (1989) Children's ability to make probability estimates: Skills revealed through application of Anderson's functional measurement methodology. *Child Development*, **60**, 933–45.

Allen, G.A. & Estes, W.K. (1972) Acquisition of correct choices and value judgments in binary choice learning with differential rewards. *Psychonomic Science*, **27**, 68–72.

Anderson, N.H. (1980) Information integration theory in developmental psychology. In F. Wilkening, J. Becker & T. Trabasso (eds.), *Information Integration by Children*, Erlbaum, Hillsdale, NJ, pages 1–45.

Baron, J. (1988) *Thinking and Deciding*. Cambridge University Press, New York.

Bjorklund, D.F. (1989) *Children's Thinking: Developmental and Individual Differences*. Brooks/Cole, Pacific Grove, CA.

Braine, M.D.S. (1959) The ontogeny of certain logical operations: Piaget's formulation examined by nonverbal methods. *Psychological Monographs*, **73** (Whole No. 475).

Brainerd, C.J. (1978) Learning research and Piagetian theory. In L.S. Siegel & C.J. Brainerd (eds.), *Alternatives to Piaget: Critical essays on the theory*. Academic Press, New York.

Brainerd, C.J. (1981) Working memory and the developmental analysis of probability judgment. *Psychological Review*, **88**, 463–502.

Brainerd, C.J. (1982) Children's concept learning as rule-sampling systems with Markovian properties. In C.J. Brainerd (ed.), *Children's Logical and Mathematical Cognition*. Springer, New York.

Brainerd, C.J. (1983) Young children's mental arithmetic errors: A working memory analysis. *Child Development*, **54**, 512–530.

Brainerd, C.J & Kingma, J. (1984) Do children have to remember to reason? A fuzzy-trace theory of transitivity development. *Developmental Review*, **4**, 311–77.

Brainerd, C.J. & Kingma, J. (1985) On the independence of short-term memory and working memory in cognitive development. *Cognitive Psychology*, **17**, 210–47.

Brainerd, C.J. & Reyna, V.F. (1990a) Gist is the grist: Fuzzy-trace theory and the new intuitionism. *Developmental Review*, **10**, 3–47.

Brainerd, C.J. & Reyna, V.F. (1990b) Inclusion illusions: Fuzzy-trace theory and perceptual salience effects in cognitive development. *Developmental Review*, **10**, 365–403.

Brainerd, C.J. & Reyna, V.F. (1992) The memory independence effect: What do the data show? What do the theories claim? *Developmental Review*, **12**, 164–86.

Brainerd, C.J. & Reyna, V.F. (1993a) Domains of fuzzy-trace theory. In M.L. Howe & R. Pasnak (eds.), *Emerging Themes in Cognitive Development: I. Theoretical Foundations*. Springer, New York, pages 50–93.

Brainerd, C.J. & Reyna, V.F. (1993b) Memory independence and memory interference in cognitive development. *Psychological Review*, **100**, 42–67.

Bryant, P. & Trabasso, T. (1971) Transitive inferences and memory in young children. *Nature*, **232**, 457–9.

Callahan, P. (1989) Learning and development probability concepts: Effects of computer assisted instruction and diagnosis. Unpublished doctoral dissertation, University of Arizona, College of Education, Tucson.

Ceci, S.J. (1990) *On Intelligence...More or Less: A Bio-ecological Treatise on Intellectual Development*. Prentice Hall, Englewood Cliffs, NJ.

Chapman, R.H. (1975) The development of children's understanding of proportions. *Child Development*, **46**, 141–8.

Davies, C.M. (1965) Development of the probability concept in children. *Child Development*, **36**, 779–88.

Dempster, F. (1992) The rise and fall of the inhibitory mechanism: Toward a unified theory of cognitive development and aging. *Developmental Review*, **12**, 45–75.

Estes, W.K. (1976) The cognitive side of probability learning. *Psychological Review*, **83**, 37–64.

Estes, W.K. (1980) Is human memory obsolete? *American Scientist*, **68**, 62–9.

Fischbein, E. (1975) *The Intuitive Sources of Probabilistic Thinking in Children*. Reidel, Boston.

Fischbein, E., Pampu, I. & Manzat, I. (1970) Comparison of ratios and the chance concept in children. *Child Development*, **41**, 377–89.

Gigerenzer, G. & Murray, D.J. (1987) *Cognition as Intuitive Statistics*. Erlbaum, New York.

Goldberg, S. (1966) Probability judgments by preschool children: Task conditions and performance. *Child Development*, **37**, 151–68.

Henle, M. (1962) On the relation between logic and thinking. *Psychological Review*, **69**, 366–78.

Hoemann, H.W. & Ross, B.M. (1971) Children's understanding of probability concepts. *Child Development*, **42**, 221–36.

Hoemann, H.W. & Ross, B.M. (1982) Children's concepts of chance and probability. In C.J. Brainerd (ed.), *Children's Logical and Mathematical Cognition*. Springer, New York.

Hommers, W. (1980) Information processing in children's choices among bets. In F. Wilkening, J. Becker & T. Trabasso (eds.), *Information integration by children* Erlbaum, Hillsdale, NJ. pages 99–112.

Huber, B. L. & Huber, O. (1987) Development of the concept of comparative subjective probability. *Journal of Experimental Child Psychology*, **44**, 304–16.
Inhelder, B. & Piaget, J. (1958) *The Growth of Logical Thinking from Childhood to Adolescence*. Basic Books, New York.
Jacobs, J.E. & Potenza, M. (1991) The use of judgment heuristics to make social and object decisions: A developmental perspective. *Child Development*, **62**, 166–78.
Kahneman, D., Slovic, P. & Tversky, A. (eds.) (1982) *Judgment under Uncertainty: Heuristics and Biases*. Cambridge University Press, New York.
Kerkman, D.D. & Wright, J.C. (1988) An exegesis of two theories of compensation development: Sequential decision theory and information integration theory. *Developmental Review*, **8**, 323–60.
Link, S. (1990) Modeling imageless thought: The relative judgment theory of numerical comparisons. *Journal of Mathematical Psychology*, **34**, 2–41.
Lovett, S.B. & Singer, J.A. (1991, April) The development of children's understanding of probability: Perceptual and quantitative conceptions. Paper presented at the biennial meeting of the Society for Research in Child Development, Seattle, WA.
Marx, M.H. (1985) Retrospective reports on frequency judgments. *Bulletin of the Psychonomic Society*, **23**, 309–10.
Marx, M.H. (1986) More retrospective reports on event-frequency judgments: Shift from multiple traces to strength factor with age. *Bulletin of the Psychonomic Society*, **24**, 183–5.
Moore, C.F., Dixon, J.A. & Haines, B.A. (1991) Components of understanding in proportional reasoning: A fuzzy-set representation of developmental progressions. *Child Development*, **62**, 441–59.
Offenbach, S.I., Gruen, G.E. & Caskey, B.J (1984) Development of proportional response strategies. *Child Development*, **55**, 963–72.
Payne, J.W. (1982) Contingent decision behavior. *Psychological Bulletin*, **92**, 382–402.
Perner, J. (1979) Discrepant results in experimental studies of young children's understanding of probability. *Child Development*, **50**, 1121–7.
Piaget, J. (1931) Children's philosophies. In C. Murchison (ed.), *Handbook of Child Psychology*, Clark University Press, Worcester, MA, pages 377–91.
Piaget, J. (1950) Une expérience sur la psychologie hasard chez l'enfant: Le tirage au sort du couples. *Acta Psychologica*, **7**, 323–36.
Piaget, J. & Inhelder, B. (1951) *La genèse de l'idée de hasard chez l'enfant*. Presses Universitaire de France, Paris.
Pire, G. (1958) Notion du hasard et developpement intellectuel, *Enfance*, 131–43.
Reyna, V.F. (1991). Class inclusion, the conjunction fallacy, and other cognitive illusions. *Developmental Review*, **11**, 317–36.
Reyna, V.F. (1992) Reasoning, remembering, and their relationship: Social, cognitive, and developmental issues. In M.L. Howe, C.J. Brainerd & V.F. Reyna (eds.), *Development of long-term retention*. Springer, New York, pages 103–27.
Reyna, V.F. & Brainerd, C.J. (1989) Output interference, generic resources, and cognitive development. *Journal of Experimental Child Psychology*, **47**, 42–6.
Reyna, V.F. & Brainerd, C.J. (1990) Fuzzy processing in transitivity development. *Annals of Operations Research*, **23**, 37–63.
Reyna, V.F. & Brainerd, C.J. (1991a) Fuzzy-trace theory and children's acquisition of mathematical and scientific concepts. *Learning and Individual Differences*, **3**, 27–59.
Reyna, V.F. & Brainerd, C.J. (1991b) Fuzzy-trace theory and framing effects in choice: Gist extraction, truncation, and conversion. *Journal of Behavioral Decision Making*, **4**, 249–62.

Reyna, V.F. & Brainerd, C.J. (1992) A fuzzy-trace theory of reasoning and remembering: Paradoxes, patterns, and parallelism. In A. Healy, S. Kosslyn & R. Shiffrin (eds.), *From Learning Processes in Cognitive Processes: Essays in Honor of William K. Estes*, Vol. 2. Erlbaum, Hillsdale, NJ, pages 235–59.

Reyna, V.F. & Brainerd, C.J. (1993) Fuzzy memory and mathematics in the classroom. In G. Davies & R. Logie (eds.), *Memory in Everyday Life*. North Holland, Amsterdam, pages 91–119.

Ross, B M. & Hoemann, H.W. (1975) The attainment of formal operations: A comparison of probability concepts in deaf and hearing adolescents. *Genetic Psychology Monographs*, **91**, 61–119.

Siegel, L.S. (1978) The relationship of language and thought in the preoperational child: A reconsideration of nonverbal alternatives to Piagetian tasks. In L.S. Siegel & C.J. Brainerd (eds.), *Alternatives to Piaget: Critical Essays on the Theory*. Academic Press, New York.

Siegler, R.S. (1981) Developmental sequences within and between concepts. *Monographs of the Society for Research in Child Development*, **46**. (Whole No. 189).

Siegler, R.S. (1983) Information processing approaches to development. In P.H. Mussen (ed.), *Handbook of Child Psychology, Vol. 1. History, Theory and Methods*. Wiley, New York, pages 129–211.

Siegler, R.S. (1988) Strategy choice procedures and the development of multiplication skill. *Journal of Experimental Psychology: General*, **117**, 258–75.

Spinillo, A.G. & Bryant, P. (1991) Children's proportional judgments: The importance of "half". *Child Development*, **62**, 427–40.

Surber, C.F. & Haines, B.A. (1987) The growth of proportional reasoning. *Annals of Child Development*, **4**, 35–87.

Tversky, A. & Kahneman, D. (1981) The framing of decisions and the psychology of choice. *Science*, **211**, 453–8.

Tversky, A. & Kahneman, D. (1983) Extensional versus intuitive reasoning: The conjunction fallacy in probability judgment. *Psychological Review*, **90**, 293–315.

Tversky, A. & Kahneman, D. (1986) Rational choice and the framing of decisions. *Journal of Business*, **59**, 252–78.

Werner, H. (1948) *Comparative Psychology of Human Development*. Follett, Chicago.

Wilkening, F. & Anderson, N.H. (1982) Comparison of two rule-assessment methodologies for studying cognitive development and knowledge structure. *Psychological Bulletin*, **92**, 215–37.

Wilkinson, A.C. (1982) Theoretical and methodological analysis of partial knowledge. *Developmental Review*, **2**, 274–304.

Yost, P.A., Siegel, A.E. & Andrews, L.M. (1962) Nonverbal probability judgments by young children. *Child Development*, **33**, 769–80.

Chapter 12

Ambiguous Probabilities and the Paradoxes of Expected Utility

Jonathan Baron
University of Pennsylvania

and

Deborah Frisch
University of Oregon

Recently the ambiguity effect (Ellsberg, 1961) has received a great deal of attention from psychologists and philosophers interested in decision theory (Einhorn & Hogarth, 1985; Frisch & Baron, 1988; Gärdenfors & Sahlin, 1982; Heath & Tversky, 1991). The original ambiguity effect was the finding that people often prefer to bet on gambles with a known chance of winning as opposed to those where the chance of winning is unknown. For example, consider the following two gambles:

Gamble 1: A marble will be drawn from an urn containing 50 black marbles and 50 white marbles. You win \$100 if the marble is black. (Or, you can pick a color and you win if that color is drawn.)
Gamble 2: An urn contains 100 marbles. Between 0 and 100 are black and the rest are white. A marble will be selected at random from the urn. You win \$100 if the marble is black. (Or, you can pick a color and you win if that color is drawn.)

Subjective Probability. Edited by G. Wright and P. Ayton.
© 1994 John Wiley & Sons Ltd.

From the perspective of expected-utility theory, as we shall explain, these two gambles are equivalent. There is no reason to think that black is more or less likely than white in either case, and there is no other possible outcome. It therefore makes sense to think that the probability of winning is 1/2 in either case. Nonetheless, many people prefer Gamble 1. Ellsberg used the term *ambiguity* for the kind of unknown risk in Gamble 2. A situation in which the "probability is unknown" is called *ambiguous.*

In principle, you can make money from someone who dislikes ambiguous bets (Camerer and Weber, 1992, page 359). You can remove the ambiguity from Gamble 2 by flipping a coin in order to decide which color wins (Raiffa, 1961): the chance of winning is clearly 50% in this case. An ambiguity averter holding a ticket on "black in Gamble 2" will therefore pay you to trade it for "black if heads and white if tails". Then flip the coin. If it is heads, do nothing, and you have been paid to return the person to her original state. If it is tails, get her to trade her bet on white for a bet on black. (Surely she is indifferent between these.) Again she has paid you to get her back where she started.

Although this particular con game has apparently not been tried, Tversky and Kahneman asked subjects about the following game: "Two boxes each contain red and green marbles. A marble is drawn from each box; if their colors match, you win $60. In game *A*, both boxes have 50% red marbles and 50% green marbles. . . . In game *C*, both boxes have the same composition of red and green marbles, but the composition is unknown" (cited by Camerer & Weber, 1992, page 359). Most subjects preferred to play game *A*, but the chance of winning is higher in *C*. The decision rules that people follow thus fail to maximize their winnings in the long run. This fact suggests that aversion to ambiguity is an error. We shall examine this suggestion.

In a three-color version of the Ellsberg paradox, an urn contains 90 balls. Thirty of them are red, and 60 of them are either black or yellow—we do not know which. A ball is to be drawn from the urn, and we can win some money, depending on which ball is drawn and which option we take, as shown in Table 12.1.

Most subjects lean toward option *X*. They "know" that they have a $\frac{1}{3}$ chance of winning in this case (30 out of 90 balls). They do not like option *Y* because they feel that they do not even know what the "real probability" of winning is. It appears to them that it could be as high as $\frac{2}{3}$ or as low as 0.

Table 12.1

	30 red balls	60 balls black	yellow
Option *X*	$100	$0	$0
Option *Y*	$0	$100	$0

Now consider the following pair of options given in Table 12.2. In this example, most subjects prefer option *W*, because they "know" that their chance of winning is $\frac{2}{3}$, whereas their chance of winning with option V could be as low as $\frac{1}{3}$ or as high as 1.

Note that subjects reversed their choice merely because the "yellow" column was changed. According to the *independence principle*, you should ignore any column that has the same entries for both options. So your choice should not be affected by whether the "yellow"-column contains $100 for both options or $0. Hence, this pattern violates the independence principle.

Many people, nonetheless, feel a strong temptation to make the choices as Ellsberg's subjects (mostly economists) did, choosing *X* and *W*. Becker and Brownson (1964) have even found that subjects will pay money to avoid making choices in which the probabilities seem to be "unknown".

Kashima & Maher (1994; see also Maher, 1993) examined a modification of the Ellsberg paradox in which you are first told whether or not the ball is yellow. Then, if the ball is not yellow, you have a choice between *X* and *Y* or between *V* and *W*. Ellsberg-type subjects asked what they would choose if the ball were not yellow favored *X* and *V* (thus not violating independence) rather than *X* and *W*. (The switch cannot be explained in terms of information supplied by the fact that the ball is not yellow. If anything, that should raise the probability of a black ball and incline the subject toward *W*.) All that changed was the order in which information was revealed before the choice took effect, yet the subjects could have anticipated such revelations in the original version. Again, this fact suggests that ambiguity aversion is an error.

The Ellsberg example is a particularly clear case, but it is not isolated. Ambiguity enters many of our real decisions and opinions, such as those concerning the safety of nuclear power plants or of genetically engineered food. The ambiguity effect pits strong intuitions about an example against a powerful normative theory—that of expected-utility (EU) maximization. Many theorists (Shafer, 1976, 1981; Gärdenfors & Sahlin, 1982) have taken it, or closely related phenomena, as a starting point for the development of what they take to be alternatives to EU theory and the Bayesian probability theory that it implies. Rawls (1971) argued for the worst-case decision rule in cases of ambiguity in the "original position", and the use of this rule provided a

Table 12.2

	30 red balls	60 balls black	yellow
Option *V*	$100	$0	$100
Option *W*	$0	$100	$100

THE LIBRARY
BOLTON INSTITUTE OF HIGHER
EDUCATION-DEANE CAMPUS

major argument for the difference principle, in which primary goods are distributed so as to benefit the least advantaged group.

This phenomenon and related examples demonstrating subjects' aversion to ambiguity has led to empirical research examining the causes and effects of ambiguity (Curley & Yates, 1985, 1989; Curley, Yates, & Abrams, 1986; Einhorn & Hogarth, 1985; Heath & Tversky, 1991), as well as theoretical work attempting to specify the relevance of ambiguity for normative models of decision-making (reviewed by Camerer & Weber, 1992).

More generally, Ellsberg's (1961) seminal finding has been important because it calls into question three fundamental claims in utility theory, as presented by Savage (1954). Savage (1954) showed that the principle of maximizing EU could be derived from a set of seemingly uncontroversial axioms. Utility theory, as presented by Savage, consists of three related claims:

Measurement claim: Subjective probabilities can be defined in terms of preferences among gambles.
Descriptive claim: Utility theory describes people's behavior.
Normative claim: The rule of maximizing EU is a normative rule of decision-making.

First, Savage showed that subjective probabilities could be defined in terms of preferences (if certain "axioms" were true of sets of preferences). By defining probabilities in terms of preferences, Savage was able to develop the concept of subjective probability in a way that was acceptable to behaviorally oriented theoreticians.

Second, Savage proposed his theory as a descriptive model of choice under uncertainty. Utility theory was assumed to be a reasonably accurate model of people's choices under uncertainty. A crucial implication of this theory is that there is no meaningful distinction between "known risk" and "uncertainty".

Finally, Savage showed that the principle of maximizing EU followed from a set of intuitively compelling axioms. Thus, Savage presented a strong justification of utility theory as a normative model. In Savage's (1954) theory, choice is a function of utilities and probabilities, where probabilities are one's subjective estimates of the likelihood of states of the world.

The ambiguity effect provides first-blush evidence against all three of Savage's claims. While previous discussions of ambiguity have noted the relevance of ambiguity for each of these claims, the implications have not been distinguished very clearly. In this paper, we shall discuss the implications of ambiguity for each of Savage's claims and show how the ambiguity effect leads to new insights into the uses and limits of utility theory. We conclude with a discussion of prescriptive implications, that is, implications for practice.

12.1 THE MEASUREMENT CLAIM

There are two distinct ways of operationalizing the notion of "degree of belief" or subjective probability (Ramsey, 1931). First, one can define a subjective probability as an intuitive judgment of probability. On this view, the way to measure subjective probabilities is to ask people. We can call this tradition the *introspective interpretation* of subjective probabilities. Alternatively, one can define subjective probability as a theoretical entity that is inferred from a person's choices. We can call this tradition the *behaviorist interpretation* of subjective probabilities, since probabilities are defined in terms of choices and are inferred from choices. One of Savage's contributions was to show that if certain constraints were true of people's preferences, then probabilities could be inferred from choices. Savage's theory was very useful to researchers wanting to apply the behaviorist interpretation.

When Ellsberg (1961) first discussed the issue of ambiguity, most researchers were committed to the behaviorist interpretation of subjective probabilities. Ramsey (1931) and others (Marschak, 1975) argued that one cannot just ask people for probability judgments. They claimed that if you ask people for probabilities, the answers you get are not necessarily meaningful. Both Ramsey (1931) and Savage (1954) suggested that people may not have access to intuitions about "How likely is *X*?" They also suggested that such intuitions may have nothing to do with behavior. Ramsey argues: ". . .the quantitative aspects of beliefs as the basis of action are evidently more important than the intensities of belief-feelings." (page 171). Savage puts it quite clearly: "Even if the concept were so completely intuitive, which might justify direct interrogation as a subject worthy of some psychological study, what could such interrogation have to do with the behavior of a person in the face of uncertainty, except of course for his verbal behavior under interrogation?" (page 27).

Thus, when Ellsberg wrote his paper, many researchers believed that the only sensible way to define subjective probabilities was in terms of behavior. The ambiguity effect demonstrated that the probabilities inferred from choices are not coherent. That is, if a person states a pattern of preferences in which ambiguity is avoided (or preferred), then it is impossible to assign coherent probabilities to that person. For example, if the person prefers Gamble 1 whether black or white is associated with the payoff, then the probability of white in Gamble 2 must be less than the probability of white in Gamble 1, and the probability of black in Gamble 2 must be less than the probability of black in Gamble 1. If the probability of white and black in Gamble 1 is 0.5, then the probabilities of the two outcomes of Gamble 2 must add to less than one. Thus, Ellsberg's finding called into question the validity of the concept of subjective probability. This finding was obviously troubling to researchers in decision-making. If observable behavior (choices) is the only type of admissible data and the probabilities inferred from this observable behavior are incoherent, then one could not develop a theory of decision-making in terms of probabilities and utilities.

In the past two decades, researchers have become increasingly comfortable with the practice of asking people for probability judgments directly (e.g. Kahneman, Slovic & Tversky, 1982). Although such data were inadmissible 25 years ago (see e.g. Marschak, 1975), today the practice of asking subjects to give probability assessments is very common. For example, Kahneman and Tversky's work on judgment under uncertainty largely involves experiments in which subjects are explicitly asked for probability ratings. Perhaps this increased willingness to ask for numerical judgments is a result of the increased use of scoring rules both in theory (see Chapter 1) and practice (Murphy & Winkler, 1977): scoring rules provide a behaviorist constraint on numerical judgments.

Nonetheless, we cannot assume that all problems with probability elicitation have been solved. Until they have been solved, and so long as hypothetical decisions are used to elicit probabilities, the ambiguity effect is relevant to probability measurement as well as to decision-making.

12.2 THE DESCRIPTIVE CLAIM

Given that psychologists have become increasingly comfortable assessing probabilities directly as opposed to inferring them from preferences, it now makes sense to ask whether ambiguity affects preferences directly or through an effect on probabilities. Although this question would not have made sense to Savage or Ellsberg, it is a reasonable question today, since we accept as data both probability judgments and preference judgments.

An important goal of research on ambiguity is to explain why ambiguity influences choices in the ways it does. There have been two basic approaches to explaining ambiguity, one in which ambiguity affects beliefs (probabilities) and one in which ambiguity affects preferences directly.

12.2.1 Effects on Belief

Some authors have attempted to explain ambiguity as an effect on belief. Einhorn and Hogarth (1985) account for ambiguity effects in terms of distortion of beliefs. In particular, when subjects are given an ambiguous probability—or some results that imply one—they use that probability as an anchor and they adjust it, as if they were adjusting toward some central point by regression. Adjustment is less when the anchor is near 0 or 1 than when it is more central. The central point itself depends on the subject and the situation; it may be taken as an index of optimism or pessimism when the outcomes differ in utility. The mechanism for this adjustment is the imagination of values both higher and lower than the anchor, and the averaging of these imagined values. The adjusted probability is then entered into the decision as if it were a stated probability.

In order to obtain results that support this view, subjects must not be told the overall "marginal" probability, or else they must be discouraged from taking it too seriously. When this is done, most results support the theory. (Camerer and Weber, 1992, provide a thorough review.) The most important result is that the stated probabilities assigned to complementary events can systematically sum to less than one (Einhorn & Hogarth, 1985). For example, one subject, told that four witnesses had identified a car as blue and one had identified it as green gave 0.77 as the probability that it was blue. On another trial of the experiment, the subject gave 0.18 as the probability that the car was green, based on the same data (Table 4 in Einhorn & Hogarth, 1985).

The Einhorn/Hogarth model predicts that adjustments resulting from ambiguity will be greater for more extreme probabilities. It therefore accounts for the fact that some people prefer to bet on an ambiguous urn when the probability of winning is very low: each urn contains 1000 balls; you win if #683 is drawn; in the unambiguous urn, the balls are numbered 1 to 1000; and in the ambiguous urn, each ball can have any number in that range. (According to Becker & Brownson, 1964, such preference for ambiguity was observed by Ellsberg; Einhorn & Hogarth, 1986, present additional supporting data.) According to the model, subjects assume that the probability of drawing #683 is greater in the ambiguous urn. Although this prediction has not been directly tested by asking subjects to compare the probabilities, no other model has been proposed to account for such findings.

The regression of belief strength toward some reference level is a reasonable strategy when evidence is poor. For example, if you are told that the probability of streptococcus infection is 10% in people with fever, sore throat, and swollen glands, but a recent study of 10 patients with sore eyes in addition found that 9 of them had this infection, a reasonable estimate of the true probability for a patient with all four symptoms would be closer to 0.40 than to 0.90. This regression heuristic can be overgeneralized to cases in which it is inappropriate, however. In the context of an experiment, adjusting beliefs amounts to perversity when an experimenter specifies that they should not be adjusted, For example, when subjects are told that the probability of a disease depends on membership in a risk group, but membership in the risk group is unknowable and the overall probability of the disease is X (taking into account both members and nonmembers), the subject would be perverse not to accept the value of X as the probability. Of course, a subject who accepted the value of X might still *prefer* to bet on some other event with the same probability. That is a different issue.

In real life, however, probabilities are not so constrained. When the Food and Drug Administration tells us that the increased lifetime risk of cancer from some birth-control method is 0.003, we are *not* necessarily irrational to adjust that figure upward. In particular, we might have good reason to believe that such figures are generally underestimates and that the tendency to underestimate risk is greater when less information is available. Studies of changes in expert probability estimates as a function of increased data are needed. (Loewenstein & Mather, 1990, report such data for public perceptions.) It may well turn out that risk estimates first increase as a risk first enters our consciousness on the basis of preliminary findings and then decrease, in part because of regression to the mean (since studies would not be done if no risk is perceived) and in part because initial estimates are often based on "worst case" assumptions. These assumptions, of course, are the result of psychological processes like those described by Einhorn and Hogarth. Thus, nonexperts may be unwise to adjust reported probability estimates upward, if the estimates have already been revised upward once by the experts who produced them.

More generally, whether probability estimates should be adjusted depends on the social context in which they were generated. The rationality of adjustment depends on the facts of the matter in the social context. It could go either way. In sum, the adjustment of beliefs because information is ambiguous is a useful heuristic that may sometimes be overused.

12.2.2 Effects on Preference

Although Einhorn and Hogarth (1985) present evidence demonstrating that ambiguity can influence choice through an effect on beliefs, this is not a sufficient explanation for all ambiguity effects. In Ellsberg-type experiments, at least the more sophisticated subjects can figure out the marginal probability for themselves, so accounts in terms of belief are unlikely to account for these results. Frisch (1988), Ritov and Baron (1990), and Heath and Tversky (1991) gave subjects the marginal probabilities, or asked subjects to provide them, so their results clearly demand an explanation in terms of preference rather than belief.

Frisch and Baron (1988) provide an explanation for why ambiguity influences preferences, independent of beliefs. Ambiguity effects may be a result of our perception that important information is missing from the description of the decision (Frisch & Baron, 1988). Perhaps, then, we avoid ambiguous options because we really want to exercise another option: that of obtaining more information. (Roberts, 1963, p. 335, attributes this idea to Ward Edwards.) When this other option *is* available—as it often is—it is perfectly rational to choose it, providing that the information is worth obtaining. When the information is not available, however, or not worth the cost, we would do better to put aside our desire to obtain it and go ahead on the best evidence we have, even if it is "ambiguous". More generally, we can think of our tendency to avoid ambiguous decisions as a useful heuristic that points us toward the option of obtaining more information. From a *prescriptive* point of view, we probably do well to follow a rule of thumb that tells us to avoid irreversible commitments when information is missing. If we can learn to put this rule aside when the missing information is too costly or truly unavailable, however, we shall achieve our goals more fully in the long run.

Note that the effect of missing information is a matter of perception. In principle, an apparently unambiguous option could become ambiguous by calling attention to missing information. For example, in an urn with 50 red balls and 50 white ones, the probability of a red ball seems to be 0.5, without ambiguity. But think about the top layer of balls, from which the ball will actually be drawn. We have no idea what the proportion of red balls is in that layer; it could be anywhere from 100% to 0%, just like the proportion of black to yellow balls in the Ellsberg paradox. By thinking about the situation in this way, we have turned an unambiguous situation into an ambiguous one. The idea that some probabilities are "objective" is simply a consequence of our not paying attention to unknown determinants of each event.

Support for our proposal comes from a study of hypothetical vaccination decisions (Ritov and Baron, 1990). In one experiment, subjects were told to imagine that their child had a 10 out of 10 000 chance of death from a flu epidemic, a vaccine could prevent the flu, but the vaccine itself could kill some number of children. Subjects were asked to indicate the maximum overall death rate for vaccinated children for which they would be willing to vaccinate their child. Most subjects answered well below 9 per 10 000. Of the subjects who showed this kind of reluctance, the mean tolerable risk was about 5 out of 10 000, that is, half the risk of the illness itself. The results are also found when the subject is asked to take the position of a policy-maker deciding for large numbers of children. This result was interpreted as biased toward omission, toward the default option of not vaccinating.

Of interest here is what happened when this manipulation was combined with ambiguity. In two experiments, subjects were told that the effect of vaccination, or of the flu, depended heavily on whether the child was in a "risk group". Children not in the risk group were safe, but those in the risk group were subject to a considerable risk. The test for the risk group was not available. Thus, all that could be known was the overall probability of death in each case. The risk group was a form of salient missing information, which should, according to Frisch and Baron, induce a reluctance to choose the option in question. Subjects were in fact less willing to vaccinate when the result of vaccination was affected by membership in the risk group, thus supporting our hypothesis. Interestingly, the risk group did not affect preference when it applied to the effect of the flu. It seems that the effect of missing information reduces the tendency to act but has no effect on the tendency to omit action. This asymmetry deserves further investigation.

Heath and Tversky (1991) provide an account of ambiguity effects that is similar to that of Frisch and Baron (1988). They argue that people prefer to bet when their perceived competence is high. In several experiments, subjects were asked to give probabilities of answers to various questions, such as questions about general knowledge, football predictions, or political predictions. Subjects were then asked whether they would prefer to bet on their answers or on chance lotteries (based on coloured poker chips) with the same probabilities. Subjects chose their answers when the probability they had assigned was high (indicating competence) or when they knew a lot about the subject. Subjects chose the lotteries when their probabilities were low or when they knew little. Heath and Tversky interpreted these results as follows:

> . . . holding judged probability constant—people prefer to bet in a context where they consider themselves knowledgeable or competent than in a context where they feel ignorant or uninformed. We assume that our feeling of competence in a given context is determined by what we know relative to what can be known. Thus, it is enhanced by general knowledge, familiarity, and experience, and is diminished, for example, by calling attention to relevant information that is not available to the decision maker, especially if it is available to others (page 7).

They suggested that this competence effect has both cognitive and motivational determinants. Cognitively, the effect results from an overgeneralization of a rule that people do better in situations about which they have more information. Motivationally, Heath and Tversky suggest that the effect can result from anticipations of credit and blame: a subject would expect more blame for a wrong guess on a lottery than for a wrong guess on an equally probable item in which the subject was expert. Such an expectation, however, requires the subject to assume that *others* are committing a cognitive error. Either this is true, in which case a cognitive error is being made somewhere, or not, in which case the subject is making an error in predicting the reactions of others. We also have no reason to think that subjects would expect others to make such an error, unless the error were often made. Thus, we regard the motivational account as secondary to some sort of cognitive account, if it is true.

Regardless of the source of this competence effect, its similarity to our earlier hypothesis is striking. In both accounts, the appearance of missing information leads to an unwillingness to bet, and Heath and Tversky's cognitive account is similar to our account in terms of overgeneralization of the reluctance to act when missing information might be available.

Some effects attributed to effects on beliefs (as postulated by Einhorn & Hogarth, 1985) might be at least partially the result of direct effects of perceived missing information on choice. For example, Kunreuther and Hogarth (1989) describe the effects of ambiguity on decisions about buying and selling (hypothetical) insurance contracts. In their experiments—done with actuaries as well as business students—subjects set higher prices for insurance when risks were ambiguous. Ambiguity was manipulated by telling subjects that experts disagreed about the probability of the adverse event in question. Subjects were also told the mean of the experts' judgments, however, and this was held constant between ambiguous and unambiguous conditions. It therefore seems likely that subjects regarded the question of why the experts disagreed as missing information, so they were more reluctant to accept the risk.

In sum, two different mechanisms seem to produce ambiguity effects, one involving belief and the other involving preference. The former tends to moderate extreme beliefs when they are ambiguous. The latter inhibits people from choosing an option when they feel that information about its consequences is missing.

12.3 THE NORMATIVE CLAIM

Savage (1954) provided a rationale for a normative theory which implies that uncertain states of the world are all assigned personal probabilities and decisions are consistent with the maximization of expected utility based on these probabilities. An important implication of Savage's theory is that "... for a 'rational' man—all uncertainties can be reduced to risks" (Ellsberg, 1961; page 645). The ambiguity effect demonstrates that many people do make a distinction between different types of risk. Thus, people's intuitions are in conflict with a normative theory.

Central to Savage's theory is a form of the independence principle, which can itself be violated by people who are sensitive to ambiguity. Similar principles, along with the principle of transitivity, are used in later developments along the same lines (see Krantz *et al.*, 1971, Chapter 8).

12.3.1 Justification of the Independence Principle

Independence (in one form) requires an analysis of decisions into options, uncertain states of the world, and outcomes, which depend on the option and the state. According to the independence principle, if the option chosen does not affect the outcomes in some states of the world, then we can ignore the nature of these outcomes in those states. For example, in option A, you get a 1/1000 chance to win \$1000 if a coin flip comes up heads, and \$$Z$ for sure if it comes up tails. In option B, you get \$1 for sure if it comes up heads, and \$$Z$ for sure if it comes up tails. Z has the same value in both options. By the independence principle, you should make the same choice regardless of the value of *Z*, because in the state of the world "tails" the outcome is the same regardless of your choice. Your choice really comes down to whether you prefer the dollar or the chance to win $1000.

More generally, the independence principle can be described in terms of Table 12.3, in which the rows are the options and the columns are uncertain events or states of the world (as described by Jeffrey, 1983).

The entries in the table are the outcomes (*V*–*Z*). In the example just given, state 3 corresponds to tails, state 1 corresponds to heads and winning the lottery, state 2, to heads and losing. *W* and *Y* are both $1, *V* is $1000, and *X* is $0. By assumption, the entries in one column (state 3) are identical. The options therefore differ as a function of the choice only in the other columns. The independence principle states that the outcomes in the identical column (*Z*, here) should not affect the decision. The non-identical columns affect the decision in the same way, regardless of what is in the identical column.

If you follow independence and transitivity (plus other axioms that are less important), then you must make decisions as though you assigned probabilities to uncertain states of the world, assigned utilities to outcomes, multiplied the probability of each outcome by its utility, added up these products for the possible outcomes of each option, and chose the option with the highest sum (EU). If you accept the axioms as constraints on your decision, then, normatively, you should not violate this EU formula.

How can the independence axiom be justified? One line of justification may be based on the definition of utility in terms of goal achievement (or, equivalently for this purpose, desire satisfaction). Importantly, we take utility to be a real property of states of the world, not an intervening variable designed to explain preferences. Thus, as Kahneman and Snell (1992) argue, judgments of utility are more like predictions than reports of inner states. When we make a judgment of the utility of an outcome, we are predicting how much that outcome will achieve all of our goals taken together. Note that, by this view, to say that two entries in the table are the same (e.g. to label them with the same letter) is to say that they are equivalent in terms of achieving goals.

Now, given this kind of table, we have two possibilities. Either the identical state (state 3) occurs or one of the non-identical states (state 1 or 2). If the identical state occurs, then the nature of the identical outcome (*Z*) does not

Table 12.3

	state 1	state 2	state 3
option *A*	*V*	*X*	*Z*
option *B*	*W*	*Y*	*Z*

affect the achievement of goals as a function of the option chosen, since the outcome is the same regardless of the option chosen. If one of the non-identical states occurs, then the nature of the identical outcome does not affect the achievement of goals either, because the identical outcome (Z) does not occur. Achievement of a goal is a matter of fact, so it depends on what is true of the world after the decision is made. (Recall that we have assumed that no goals concern counterfactual outcomes.) In sum, the nature of Z, the identical outcome, does not matter if Z occurs, and it does not matter if Z does not occur, so it does not matter. Independence therefore follows from the idea that rational decisions should be determined by the extent to which their outcomes achieve goals. (The same kind of argument can be used to defend related principles, such as those involving dominance or independence of irrelevant alternatives.)

12.3.2 Why People might still want to violate Independence

The independence principle is usually illustrated with monetary outcomes, as in the Ellsberg paradox. When the entries in the table represent monetary outcomes, people may want to violate the principle for a couple of reasons. First, forgone or counterfactual outcomes affect their emotions, or more generally, the way in which consequences are experienced (before, during, or after they occur). For example, if Z is \$1 in Table 12.3 then X (\$0) could cause a feeling of regret, since you would realize that if you had chosen B you would have won something no matter what. If Z is \$0, however, it will be easier for you to tell yourself that you might have won nothing anyway. Your experience of X is therefore changed by your knowledge of Z. In terms of goal achievement, then, X is no longer the same outcome for different values of Z. It should be represented with different symbols depending on the value of Z. Because the independence principle for goal achievement requires that X be the same consequence, the premise of the independence principle is not true, and you have not violated it if you make different choices for different values of Z. In sum, violations of independence (or of EU itself) that depend on emotional experiences need not be violations at all once the experiences are included in the descriptions of consequences. The trouble comes from describing the consequences as amounts of money. (Frisch & Jones, 1993, make a similar point.) Evidence that subjects take such experiences into account in making decisions is summarized by Harless (1992).

In case it is difficult to imagine when the assumptions of the independence condition *are* met, consider the case of (what we shall call) Other decisions, in which each decision is made for another person, who does not know what the rejected options or counterfactual outcomes were, and in which we cannot assume that the Other has goals concerning the effect of these unknowns on choice (Baron, 1993). If the decision-maker truly took into account only the utilities of the recipient, not her own utilities connected with making the decision, the emotional effects of forgone or counterfactual outcomes would largely disappear. If the recipient's utilities concerned only the outcomes that he would know about, these effects would disappear completely.

For Self decisions (those made for the self), the fact that the same nominal outcome (e.g. "$1000") may lead to different real outcomes as a function of forgone options or counterfactual outcomes makes it difficult to test EU as a *descriptive* theory from behavior alone. To test the theory for Self decisions, we must measure the utility of outcomes *in the context of the decision itself*, using other methods than how people make decisions under risk (Baron, 1994, ch. 17). We can use the theory normatively and prescriptively in the same way, i.e., by describing the outcomes in the context of the whole decision and allowing its utility to depend on events that did not happen and options that were not chosen.

If we want the utilities of outcomes to be independent of the context, we do well to think about Other decisions. Our arguments in favor of the independence principle applied most clearly to this case. If we use Other decisions to test the theory descriptively, we will probably find all the same violations that have been found in Self decisions, such as the effect of certainty. Some experiments have used Other decisions (Baron & Hershey, 1988; Kahneman & Tversky, 1984; Ritov & Baron, 1990; Spranca, Minsk & Baron, 1991), finding that the theory still did not apply descriptively. In particular, the ambiguity effect found by Ritov and Baron (1990) was in the context of an Other decision, the vaccination of a child or (equally) a policy for vaccination of many children. When the conditions are met for the independence principle to apply, violations of that principle, such as the Ellsberg paradox, subvert the achievement of goals. In that sense, the pattern of choices observed in the Ellsberg paradox is nonnormative. We suggest that more research be done using Other decisions. Ambiguity effects in Self decisions are not clearly nonnormative. When these effects—and other effects—occur in Other decisions, they are more clearly nonnormative. They may be considered as overgeneralizations of heuristics that might be useful for Self decisions.

In sum, we have provided a defense of the independence principle in terms of goal achievement. This defense is intended as an answer to criticisms of the more traditional approach, which derives from the intuitive appeal of the axioms themselves (e.g. Slovic & Tversky, 1974).

12.4 THE ALLAIS PARADOX

The Allais paradox is another case in which the independence principle is violated (Allais, 1953). Consider the gambles shown in Table 12.4, in which

Table 12.4

	Number of ball drawn 1	2–11	12–100
Situation X			
Option 1	$1000	$1000	$1000
Option 2	0	5000	1000
Situation Y			
Option 3	$1000	$1000	$0
Option 4	0	5000	0

the outcome is decided by drawing a ball at random from an urn containing 100 balls with the numbers 1 through 100 written on them. Many people in this situation are tempted to choose Option 1 in Situation *X* and Option 4 in Situation *Y*. In situation *X*, they are not willing to give up the *certainty* of winning $1000 in option 1 for the chance of winning $5000 in option 2: This extra possible gain would expose them to the *risk* of winning nothing at all. (If you do not happen to feel this way, try replacing the $5000 with a lower figure, until you do. Then use that figure in choice 4 as well.) In situation *Y*, they reason that the difference between the two probabilities of winning is small, so they are willing to try for the larger amount.

This pattern of choices violates the independence principle. Balls 12–100 lead to the same outcome ($1000) regardless of whether we choose Option 1 or 2 in Situation *X*, and they lead to the same outcome ($0) whether we choose Options 3 or 4 in Situation *Y*. By the independence principle, you should choose Options 1 and 3, or Options 2 and 4, but you should not choose Options 1 and 4. Usually, the independence principle is intuitively attractive, but many people are prone to violate it by choosing Options 1 and 4.

Shafer (1986) argues that it is not necessarily irrational to choose Options 1 and 4. He says that the "constant" outcomes—those that are the same regardless of our choice—affect our goals or desires in the situation. (Lopes, 1987, makes a similar argument.) When we see that we can win a substantial sum of money for sure in Option 1, this reduces our desire for the larger sum. When we see that we are likely to lose no matter what, in Options 3 and 4, our desire to "win big" increases.

This argument is less relevant if we change the example. Instead of the decision-maker getting the money, it is donated anonymously and without explanation to his favorite nephew, or whoever. This is an Other decision. The nephew does not know what options were foregone or what states did not occur, so his experiences are unaffected by these things. Moreover (we assume), the decision-maker has no reason to think that the nephew has any particular goals concerning options that were not chosen, or states that did not occur, in having decisions made on his behalf. The nephew's utilities thus cannot be affected by counterfactual outcomes, so Shafer's argument does not apply. We see here how Other decisions are, in a sense, simpler than Self decisions. By Shafer's account, the assumptions of the independence principle are typically not met in Self decisions, but it is easy to imagine how they might be met in Other decisions.

The perspective of Other decisions also strengthens another argument for the independence principle. Raiffa (cited in McClennan, 1983) points out that we may view the original problem as a sequential decision, as follows:

> First, a ball will be drawn out of an urn with balls numbered 1–100. If the number drawn is between 12 and 100 inclusive, the outcome is $1000 (for Situation X) or $0 (for Situation Y). Otherwise, a second draw is made from a new urn with balls numbered 1 through 12. For Option 5, the outcome of the second draw is $1000, no matter what. For Option 6, the outcome is $5000 if the number is between 2 and 11 inclusive but $0 if the ball is 1.

If we get to the stage of making the choice between Options 5 and 6, then the outcome for number 12–100 is irrelevant, for it did not occur. Raiffa argues that it should be irrelevant whether we make the decision before we know whether we get to the second stage of the game (as in the original Situations X and Y) or after we know (as in this example). McClennan (1983), points out that Raiffa and others who make similar arguments give no reason why the timing of the decision should not matter; they simply assert it, or suggest that most people's intuition would agree. But, to answer McClennan, it is clear that the timing would not matter to some Other who simply experienced the consequences without knowing the sequence of events that led to it (assuming that the Other has no goals concerning these non-experienced events).

12.5 ISSUES IN APPLICATION

We have argued that ambiguity effects can result from overgeneralization of heuristics concerning the postponement of decision-making when information is perceived as missing. These effects can be nonnormative, that is, in opposition to the optimal achievement of our goals. But issues remain concerning the practical treatment of situations in which information *is* missing, for example, cases in which probability judgments disagree and we lack information about how to resolve the conflict. We discuss this problem here, as well as the problem of defining true probability in practical contexts, and the role of experts in decisions under ambiguity.

12.5.1 Conflct

Lindsey, Tversky, and Brown (1979) have discussed the problem of conflicting judgments from a Bayesian point of view. Theoretically, they assume that judgments are a function of some underlying probability that we might call "true". If assumptions are made about the probability of each judgment given each possible true probability, then Bayes' theorem can be used to derive a probability distribution over the possible true probabilities, and the mean of this distribution can be taken as the best estimate. In this way, judgments made by different methods or by different people can be reconciled. Lindsey et al. give several examples.

12.5.2 True Probability

But the true probability is still known only probabilistically. In fact, the concept of true probability requires explication. The very distinction that inspired Ellsberg, that between uncertainty and risk (Knight, 1921; Luce and Raiffa, 1957) implies that some probabilities can be known with certainty but others cannot be known, only judged. This distinction lies at the heart of a number of recent alternatives to EU theory, reviewed by Camerer & Weber (1992). Bayesians in the tradition of Savage are skeptical about this distinction, however. They see cases of "known risk" as merely convenient simplification, in which various judges and methods agree closely on the probability. In some cases, this agreement has resulted from overwhelming of priors by extensive data. In other cases, it results from ignoring relevant data, as when a judgment is made about the probability of a certain patient having a disease on the basis of population statistics, ignoring potentially relevant data about the individual. From this Bayesian point of view, the only possible "true" probabilities are zero and one, and these apply mostly after the fact. Everything else involves judgments based on incomplete information.

This sort of Bayesian stance runs into conflict with our way of talking about probabilities. We say things like, "I thought that the probability was *X*, but it was really *Y*." In some cases, laws and regulations are stated in terms of probabilities, such as limits on the probability of disease caused by exposure to a chemical. These regulations are written as if the probability were an objective fact.

Brown (1993) proposed a Bayesian analysis of the idea of true probability, an analysis that allows such ways of talking to make sense. The true probability is the judgment that experts would converge on, as further relevant information became too costly to collect. In each case, the specific information required would differ, and a true probability need not exist in every case. For example, in determining the cancer risk from a chemical, the true probability might be thought of as the estimate derived from epidemiological data concerning cancer rate as a function of yearly exposure in the whole population of interest. Experts would form their priors on the basis of animal studies and theoretical beliefs about the form of the dose–response function. As more data were collected, these beliefs would approach the same asymptote. In principle, given sufficient time, data like these could be collected for different groups of patients. But such data on the interaction between exposure and individual characteristics would presumably be too costly to collect, so the population asymptote would be the one that experts would have in mind as the "true probability". Thus, Brown's analysis assumes both that an intermediate asymptote exists and that expert judgments would converge. He argues for the plausibility of these assumptions in many cases.

This definition of true probability avoids the conclusion that "the true probability is always just 0 or 1". It assumes that there is some sort of standard body of evidence that people want in each case. To take another example, a doctor might sensibly say, "I can't assign a probability that the patient has cancer until we get back all the test results." Here, the standard tests constitute the standard evidence. Note that a biopsy would be definitive here, but that is not included in the standard tests because it is considerably more costly (and perhaps because it would not make sense to speak of probabilities at all if it were available).

Brown's account fits neatly with our own theory of ambiguity as missing information. When a standard body of evidence exists and has not been obtained, people will be aware that this information is missing, and they will desire to collect it before acting. In most cases, this hesitation will be justified. In some cases, however, the situation will be classified as one in which the standard information is easily available, when, in fact, the information is not available at reasonable cost. From this perspective, then, ambiguity effects arise in situations seen as similar to those in which additional information is available. In the Ellsberg urn, for example, the proportion of balls is seen as something that is usually given. The unique aspect of the problem is that the experimenter won't tell.

12.5.3 Expert Judgment vs. Democracy

People fear risks that are not well known (Slovic, Lichtenstein & Fischhoff, 1984). These risks include those of new technologies such as genetic engineering. Another example is the risks resulting from changes in legal standards: part of the US "liability crisis" of the 1980s was the unwillingness of insurance companies to write liability policies when court standards could change retroactively, as they did several times in recent years (Huber, 1988).

Hacking (1986) makes an argument with which many would probably sympathize. He is happy enough to have policy decisions made on his behalf by decision analysis when probabilities of relevant outcomes are well known, but not when probabilities are subjectively judged. Presumably, probabilities would be well known for things like the success rate of various medical therapies for various disorders. Probabilities would not be well known for events such as meltdowns of nuclear power plants (especially when their design is new). In cases of the latter type, we would have to rely more heavily on traditional methods of decision-making, which stress participation of those affected, or holistic subjective judgments by elected representatives.

On the other hand, we have argued that missing information is always present whenever probabilities are involved. What changes from case to case is its psychological salience. Normatively, we ought to make decisions on the basis of our best estimate of the probability, it would seem.

An exception to this argument occurs when the risk to one person is correlated with the risk to another and when the utility function for harm is nonlinear with the number of people. Correlated risks are found in the case of disasters, e.g. hurricanes or earthquakes, since harm to one person from such a source implies that others are more likely to be harmed as well. But the argument as stated here applies to individuals.

As we have noted already, the "best estimate" could be systematically biased against caution in the case of new technologies. Often, the best estimate is arrived at by trying to imagine all possible ways in which something could go wrong. Yet, as Fischhoff, Slovic & Lichtenstein (1978) have shown, we might tend to err on the side of leaving things out because of our inability to think of them, and therefore estimate on the low side. The public's intuition that experts underestimate risks ("You've been wrong so many times before, so why should we believe you now?") might be justified.

On the other hand, the public could be basing its judgment on a biased sample of cases that come to mind simply *because* the experts erred against caution, such as the Three Miles Island nuclear incident and the problems with some intrauterine devices. Perhaps as many, or more, cases could be found in which experts erred in favor of caution. Experts, too, could be sensitive to ambiguity effects. (The US Food and Drug Administration is said to routinely boost risk estimates when the data on which they are based are in any way inadequate.)

In principle, these problems are remediable. Enough experience exists with risk estimates to allow a direct test of the existence of bias. Such tests have not been done. In the meantime, risk analysts ought to do the best they can. Perhaps they should correct for various sources of error. Putting this another way, our true best estimate should include a correction—if needed—for under- or overestimation as a function of the amount of information available. Analysts can also use risk analysis to determine when more data will be helpful and when it will not.

Political factors are sometimes relevant. One of the purposes of risk analysis is to help reduce political friction. For this purpose, the risk analysis ought to be open to criticism by the public. Conceivably, such criticism can improve the accuracy of risk analysis, but even if it impairs accuracy it might be worth soliciting. In addition to soliciting public input, risk analysis should also consider educating the public about such matters as the ambiguity effect just described. The intuition that "we should not act until we know the probability" should be understood as one that has a legitimate basis only insofar as systematic bias enters the process of risk analysis or insofar as collection of additional data is worthwhile.

Intertwined with the ambiguity effect is also a bias toward the status quo, or toward inaction (Ritov & Baron, 1990). Ambiguity seems to exaggerate this bias (Ritov & Baron, 1990), but it is present in any case. The amount of money that people will pay to rid themselves of a risk they already have is far less than the amount that they will accept in order to take on the same risk (Thaler, 1980; Viscusi, Magat & Huber, 1987). If people could learn to overcome this bias—and it seems that they can to some extent (Larrick, Morgan & Nisbett, 1990)—we could take their resistance to new technology more seriously. The existence of this bias toward inaction therefore makes more plausible the claim that people subvert their own goals by favoring present risks over smaller risks that just happen to be new.

What of Hacking's argument? In cases in which the public has reason to distrust those in charge of a decision analysis, traditional methods of decision-making might be better. As noted, self-serving bias—the basis of distrust—can be minimized by precautions surrounding the analysis itself. If substantial self-serving bias is absent, however, or if adequate precautions are taken to avoid its effects, perhaps Hacking would do well to trust his fate to the best guess of experts rather than to the political process. The political process itself is hardly perfect.

REFERENCES

Allais, M. (1953) Le comportement de l'homme rationnel devant le risque; Critique des postulates et axioms de l'école américaine. *Econometrica*, **21**, 503–546.

Baron, J. (1993) *Morality and Rational Choice*. Kluwer, Dordrecht.

Baron, J. (1994) *Thinking and Deciding* (2nd edn). Cambridge University Press, New York.

Baron, J. & Hershey, J.C. (1988) Outcome bias in decision evaluation. *Journal of Personality and Social Psychology*, **54**, 569–79.

Becker, S.W. & Brownson, F.O. (1964) What price ambiguity? Or the role of ambiguity in decision making. *Journal of Political Economy*, **72**, 62–73.

Brown, R.V. (1993) Impersonal probability as an ideal assessment based on accessible evidence: a viable and practical construct? *Journal of Risk and Uncertainty*, **7**, 215–235.

Camerer, C. & Weber, M. (1992) Recent developments in modelling preferences: uncertainty and ambiguity. *Journal of Risk and Uncertainty*, **5**, 325–70.

Curley, S.P. & Yates, J.F. (1985) The center and range of the probability interval as factors affecting ambiguity preferences. *Organizational Behavior and Human Decision Processes*, **36**, 272–87.

Curley, S.P. & Yates, J.F. (1989) Am empirical evaluation of descriptive models of ambiguity reactions in choice situations. *Journal of Mathematical Psychology*, **33**, 397–427.

Curley, S.P., Yates, J.F. & Abrams, R.A. (1986) Psychological sources of ambiguity avoidance. *Organizational Behavior and Human Decision Processes*, **38**, 230–56.

Einhorn, H.J. & Hogarth, R.M. (1985) Ambiguity and uncertainty in probabilistic inference. *Psychological Review*, **92**, 433–61.

Einhorn, H.J. & Hogarth, R.M. (1986) Decision making under ambiguity. *Journal of Business*, **59**, S225–50.

Ellsberg, D. (1961) Risk, ambiguity, and the Savage axioms. *Quarterly Journal of Economics*, **75**, 643–99.

Fischhoff, B., Slovic, P. & Lichtenstein, S. (1978) Fault trees: Sensitivity of estimated failure probabilities to problem representation. *Journal of Experimental Psychology: Human Perception and Performance*, **4**, 330–4.

Frisch, D.G. (1988) The effect of ambiguity on judgment and choice. Doctoral dissertation, Department of Psychology, University of Pennsylvania.

Frisch, D. & Baron, J. (1988) Ambiguity and rationality. *Journal of Behavioral Decision Making*, **1**, 149–57.

Frisch, D. & Jones, S.K. (1993) Assessing the accuracy of decisions. *Theory and Psychology*, **3**, 115–135.

Gärdenfors, P. & Sahlin, N.-E. (1982) Unreliable probabilities, risk taking, and decision making. *Synthese*, **53**, 361–86.

Hacking, I. (1986) Culpable ignorance of interference effects. In D. MacLean (ed.), *Values at Risk*. Rowman & Allanheld, Totowa, NJ. pages 136–84.

Hammond, P.H. (1988) Consequentialist foundations for expected utility. *Theory and Decision*, **25**, 25–78.

Harless, D.W. (1992) Actions versus prospects: the effect of problem representation on regret. *American Economic Review*, **82**, 634–49.

Heath, C. & Tversky, A. (1991) Preference and belief: ambiguity and competence in choice under uncertainty. *Journal of Risk and Uncertainty*, **4**, 5–28.

Humber, P.W. (1988) *Liability: The Legal Revolution and its Consequences*. Basic Books, New York.

Jeffrey, R.C. (1983) *The Logic of Decision* (2nd revised edition). University of Chicago Press, Chicago.

Kahneman, D., Slovic, P. & Tversky, A. (eds.) (1982) *Judgment under Uncertainty: Heuristics and Biases*. Cambridge University Press, New York.

Kahneman, D. & Snell, J. (1992) Predicting changing taste: do people know what they will like? *Journal of Behavioral Decision Making*, **5**, 187–200.

Kahneman, D. & Tversky, A. (1984) Choices, values, and frames. *American Psychologist*, **39**, 341–50.

Kashima, Y. & Maher, P. (1994) Framing of decisions under ambiguity. Manuscript, Department of Psychology, La Trobe University.

Knight, F.H. (1921) *Risks, Uncertainty, and Profit*. Macmillan, London.

Krantz D.H., Luce, R.D., Suppes, P. & Tversky, A. (1971) *Foundations of Measurement*, Vol. 1. Academic Press, New York.

Kunreuther, H. & Hogarth, R.M. (1989) Risk, ambiguity, and insurance. *Journal of Risk and Uncertainty*, **2**, 5–35.

Larrick, R.P., Morgan, J.N. & Nisbett, R.E. (1990) Teaching the use of cost–benefit reasoning in everyday life. *Psychological Science*, **1**, 362–70.

Lindley, D.V., Tversky, A. & Brown, R.V. (1979) On the reconciliation of probability assessments. *Journal of the Royal Statistical Association A*. **142**, 146–80 (with commentary).

Loewenstein, G. & Mather, J. (1990) Dynamic processes in risk perception. *Journal of Risk and Uncertainty*, **3**, 155–75.

Lopes, L.L. (1987) Between hope and fear: The psychology of risk. In L. Berkowitz (ed.), *Advances in Experimental Social Psychology*, Vol. 20, Academic Press, New York, pages 255–95.

Luce, R.D. & Raiffa, H. (1957) *Games and Decisions*. Wiley, New York.

McClennan, E. (1983) Sure thing doubts. In B.P. Stigum & F. Wenstøp (Eds.), *Foundations of Utility and Risk Theory with Applications*, pages 117–36. Reidel, Dordrecht.

Maher, P. (1993) *Betting on Theories*. Cambridge University Press, Cambridge.

Marschak, J. (1975) Personal probabilities of probabilities. *Theory and Decision*, **6**, 121–53.

Murphy, A.H. & Winkler, R.L. (1977) Can weather forecasters formulate reliable probability forecasts of precipitation and temperature? *National Weather Digest*, **2**, 2–9.

Raiffa, H. (1961) Risk, ambiguity, and the Savage axioms: comment. *Quarterly Journal of Economics*, **75**, 690–4.

Ramsey, F.P. (1931) Truth and probability. In R.B. Braithwaite (ed.), *The Foundations of Mathematics and other Logical Essays by F.P. Ramsey*. Harcourt, Brace, New York.

Rawls, J. (1971) *A Theory of Justice*. Harvard University Press, Cambridge, MA.

Ritov, I. & Baron, J. (1990) Reluctance to vaccinate: omission bias and ambiguity. *Journal of Behavioral Decision Making*, **3**, 263–77.

Roberts, H.V. (1963) Risk, ambiguity, and the Savage axioms: comment. *Quarterly Journal of Economics*, **77**, 327–36.

Savage, L.J. (1954) *The Foundations of Statistics*. Wiley, New York.

Shafer, G. (1976) *A Mathematical Theory of Evidence*. Princeton University Press, Princeton.

Shafer, G. (1981) Constructive probability. *Synthese*, **48**, 1–60.

Shafer, G. (1986) Savage revisited. *Statistical Science*, **1**, 463–501 (with discussion).

Slovic P., Lichtenstein, S. & Fischhoff, B. (1984) Modeling the societal impact of fatal accidents. *Management Science*, **30**, 464–74.

Slovic, P. & Tversky, A. (1974) Who accepts Savage's axioms? *Behavioral Science*, **14**, 368–73.

Spranca, M., Minsk, E. & Baron, J. (1991) Omission and commission in judgment and choice. *Journal of Experimental Social Psychology*, **27**, 76–105.

Thaler, R.H. (1980) Toward a positive theory of consumer choice. *Journal of Economic Behavior and Organization*, **1**, 39–60.

Viscusi, W.K., Magat, W.A. & Huber, J. (1987) An investigation of the rationality of consumer valuation of multiple health risks. *Rand Journal of Economics*, **18**, 465–79.

Chapter 13

Risk Perception: Main Issues, Approaches and Findings

Wibecke Brun
University of Bergen

Risk is a fact of life. A fact we all face and act upon daily. Yet the ways we perceive and react to the fears and hazards surrounding us are poorly understood and hard to predict. Although researchers for years have tried to clarify and understand human responses to dangers, there are still controversies about basic issues as to how risk should be defined and which components constitute or compose an intuitive risk concept (Vlek & Stallen, 1980; Drottz-Sjöberg, 1991; Lopes, 1987; Hansson, 1989; Vlek & Keren, 1992; Jungerman & Slovic, 1992; Yates & Stone, 1992a).

This chapter will first present some definitions of risk and discuss different decompositions of the risk concept. Next we will give a short presentation of the main approaches to the study of risk perception, along with some of their main findings. We will then take a closer look at the psychological mechanisms or "cognitive components" that are hypothesized or found to be of importance as determinants for lay perceptions of risk. These components (or "dimensions of risk"), can be grouped according to the nature of the risks, whether these components have to do with the causes of the risks, the characteristics of the hazards and their consequences, or with our relationship to and reactions towards these hazards. Risk judgments also vary according to the degree and type of uncertainty involved and the way the risk information is obtained.

Subjective Probability. Edited by G. Wright and P. Ayton.
© 1994 John Wiley & Sons Ltd.

13.1 DEFINITIONS OF RISK

Different definitions of risk have been proposed. Some of them are closely linked to formal probability theory and aimed at providing a rule or a procedure for calculating risk in an "objective" way, for instance, by defining risk as a product of the probability of a loss and its magnitude. These definitions vary regarding the *basis* for the estimation of the probability and the severity components. What is, for instance, the "natural" unit of risk: reduced life expectancy? probability of death per hour of exposure? Furthermore, there exist different views as to the *relative importance* of the probability and the severity components (Drottz-Sjöberg, 1991). Should the uncertainty and severity components be treated as multiplicative in the sense that a small probability of a large loss is considered equivalent to a larger probability of a smaller loss? Research indicates that this does not necessarily reflect the way laypeople think (Bettmann, 1973; 1975; Slovic, 1967; Slovic, Fischhoff & Lichtenstein, 1980). If one takes the view that risk should be objectively measured, there is still the problem of which rule to follow, e.g. how these components should be merged into one. Should for instance the subjective significance of losses (their disutility) be taken into account? Should the value of possible gains from a risky product or activity be included? How should the question of ambiguity (uncertain probabilities) be handled? In one definition Allais (1953, cited in Vlek & Stallen, 1980) defines risk as the *variance* of the probability distribution over the utilities of all possible consequences. Other examples of formal definitions can be found in Vlek and Stallen (1980) and Kaplan & Garrick (1981).

Other definitions focus more closely upon lay perceptions and are aimed at reflecting what laypeople intuitively understand by the term. These more descriptive definitions are typically linked to a subjectivistic view of probability. Yates and Stone (1992a), considering risk as a subjective construct, give one such general definition. They state that risk has to do with *losses*, the *significance* of those losses and the *uncertainty* associated with them. According to Vlek and Hendrickx (1988) the experience of risk has to do with a lack of *perceived control* associated with a serious undesired consequence.

It is important to note that some definitions focus upon the uncertainty associated with *one* given outcome. Other definitions stress that risk is a joint estimate of *several* accident possibilities, and focus upon the "openness" of a situation where several possible accident scenarios exist. Both these categories refer to risk as some sort of product of the probability and severity of loss. A slightly different perspective is present in definitions that refer more directly to the *process or mechanism* underlying how an accident may come about. The definition of risk as "lack of controllability" could be seen as belonging to this last category. Here the uncertainty aspect of risk refers to the individual's own lack of control or competence, e.g. possibility or ability to master, alter or avoid a given outcome. Given complete control no accident would occur, hence risk refers to lack of control.

Apparently most definitions of risk include an estimate of *uncertainty* (a likelihood, possibility or judged subjective probability) for a *negative event* to happen (a possible loss or a negative consequence of an action). It follows that risk *perception* has a perceived probability/uncertainty aspect as well as a perceived severity aspect to it.

Since probability is one of the main components in most definitions of risk, many issues discussed within the area of risk estimation and risk perception parallel discussions within the area of probability theory and its applications. Here we find the main fronts represented by the *frequentists*, who interpret probability as relative frequency, and the *subjectivists*, who interpret probability as degree of belief (Shafer, 1993). In attempts to integrate these different views probabilities have been seen as possibilities of an event for which there is some sort of proof, reason or evidence. The strength of a belief may be more or less warranted by empirical data. One may let knowledge of long-run frequencies of relevant states of the world justify a feeling of confidence or belief in a given outcome, or the belief may be justified or influenced by more subjective aspects, such as how easy it is to construct mental scenarios representing a given outcome, or the availability of it (Kahneman & Tversky, 1982a; Tversky & Kahneman, 1982). It has been suggested that the difference between objective and subjective probabilities is related to the amount and/or the quality of the available information on which the probability judgment is based, for instance the perceived completeness of the given information (Frisch & Baron, 1988; Kaplan & Garrick, 1981). The same reasoning can readily be applied to risk estimates.

Is it possible to formulate a valid definition of risk that holds for *different* risky situations? It may be that the content or meaning of the term risk is so closely linked to the nature of the particular event in question, that its definition has to depend upon an interpretation of the total context where it is used. According to Vlek and Stallen (1980) it is quite plausible that "risk" is primarily associated with the *probability* of a loss whenever possible losses are small and of a similar magnitude and probabilities are well specified, but that "risk" refers to the *size* of a loss (e.g. the possible magnitude or severity of an accident) in contexts where negative consequences can be serious, but the probabilities are vague and hard to assess.

Another relevant question to be asked is whether the interpretation of the word risk differs between different *individuals* to such an extent that it is hard to speak of a common understanding of the term. One way to try to approach this topic is simply to ask what people mean by the term. Drottz-Sjöberg (1991) gave her subjects the choice between four alternative answers. Altogether 26.4% agreed that "the meaning of the risk concept is entirely based on the nature of the event", 21.8% agreed that "risk is mainly a question of the extent of the consequences of an event", 21.5% agreed that "risk is a combination of probability and consequence", while 45.8% agreed that "risk is mainly a question of the probability of an event."[1] The study further showed the subjects' risk definitions to be of consequence for their estimates of magnitude of a particular risk. Subjects who defined risk in terms of probabilities gave lower risk ratings than those subjects who defined risk in terms of adverse consequences. Unfortunately Drottz-Sjöberg did not ask those subjects who claimed that the meaning of risk is entirely based on the nature of the event, to explain more precisely what they had in mind.

A small-scale Norwegian study provides a somewhat different perspective to the lay concept of risk (Brun, unpublished data). After the subjects (university students, $n = 72$) had listened to a short introduction on different perspectives of risk (probability, consequences and exposure) they were told that different people seem to attach different meanings to the term and that we wanted them to give their opinion by completing the sentence: "When I'm using the word risk, I mean ...". After this introduction 62.5% of the subjects defined risk in a way that could be classified as a *combination* of the probability and consequence aspects (for instance: "... that the chances are great that I will be hurt or hit, or fail in a situation", or "... there is a chance for something negative to happen"). 34.7% of the answers could be classified as mainly or entirely focused on the consequences or the severity of the risk (e.g. "... something is dangerous"). Only two persons (2.8%) focused exclusively on the probability or uncertainty aspect, defining risk as "... the probability is great that something will happen" and "... something I am not sure of (feel uncertain about). If I'm to do something for the very first time". When the same subjects later were given the following three scales (1–7), their ratings again confirm the importance of probability and severity of consequences as risk determinants. Mean rating on Scale A (If something is to be called risky possible consequences must be serious) was 5.5 (SD = 1.34). Mean rating on Scale B (If something is to be called risky there is a great probability that something negative will happen) was 5.2 (SD = 1.86). The subjects did not, however, agree to the statement in Scale C (If something is to be called risky it has to be widespread, hit or have an impact on many people): mean rating 2.2, SD = 2.09. When the subjects later were randomly divided into three groups and asked to rate a set of 75 risks with regard to either probability for a negative outcome, the size of a negative outcome or the generality of the hazard, these estimates were found to correlate substantially with risk magnitude estimates for the same hazards, given by another group ($r = 0.81$, 0.68, 0.64 for probability, seriousness and exposure respectively).

The main result is that most of our subjects in general expressed that risk has to do with both the severity of an event and the uncertainty associated with it. Secondly the subjects expressed a very *individualistic* risk perspective. Although risk exposure (e.g. how many people are exposed to the risk) was specifically mentioned in the introduction of the study, only one person made any mention of risk exposure, and then by denying its relevance ("The probability of big (negative) consequences is great, but it does not necessarily hit many people"). That laypeople often take an individualistic perspective has been found earlier in studies of risk perception (Teigen, Brun & Slovic, 1988) and in the interpretation of probability phrases in general (Murphy et al., 1980; Brun & Teigen, 1988).

Though disagreement exists as to how the aspects of probability and severity/magnitude of risk should be estimated, combined or adressed, they are still important aspects in both laypeople and expert risk definitions. But laypeople differ from experts by also allowing other characteristics of the risks, such as the dread feeling they evoke or the catastrophic potential that is associated with them, to influence their judgments. The result is that some hazards are perceived as riskier and others as less risky than what is accounted for by statistical risk estimates like annual fatalities (Slovic, Fischhoff & Lichtenstein, 1980). These characteristics may influence risk perceptions in various ways. They may contribute *directly* to lay perceptions of risk magnitude (as when a new risk is perceived as riskier than an old one of the same "objective" magnitude) or probability of occurrence (a new risk may be seen as more uncertain and a low statistical death rate is harder to trust than equal fatality rates, given an old and well-known risk source). Or they may have a more *indirect* impact on risk perceptions, by influencing the way the public perceive the *significance* of a given loss. Further it may influence public attitudes towards *responsibility for risk management* (Brun, 1992), *acceptability* of risk (risk tolerance) and *willingness to pay for risk reductions* (McDaniels, Kamlet & Fischer, 1992). Some risk dimensions may also predict whether an accident is seen as a *warning*, signalizing future accidents (Slovic, Lichtenstein & Fischhoff, 1984).

Does a discrepancy between "actual" risk measures like statistical fatality estimates and the subjectively perceived risk really constitute a problem? The importance of having inter-subjective consensus and clear operational definitions of risk is obvious in the area of risk assessment and risk comparisons, but seems less important when it comes to capturing and understanding the lay perceptions themselves. Fischhoff (1989) suggests that both objective risk assessments (e.g. fatality rates) and subjective risk judgments can be seen as alternative risk perceptions—made by "the experts" or by the lay public, respectively.

Studies of how hazards are viewed in terms of more qualitative risk characteristics provide an opportunity to capture a lay conception of risk, but are also useful by providing insight into laypersons' general reactions towards hazards and hazard management. Studies within the psychometric tradition have especially adressed this problem. An important assumption in this tradition is that risk is inherently subjective. Risk does not exist "out there", waiting to be measured, but is a concept the human mind has invented to help us understand and cope with the dangers and uncertainties of life. From these studies it becomes obvious that the lay concept of risk is a complex and multidimensional one that is hard to reduce to a single figure (Slovic, 1987; 1992).

13.2 HOW CAN RISK PERCEPTIONS BE STUDIED?

The main body of research within the area of risk perception have been revolving around basically two issues. Several studies has been conducted where the main goal has been to gain knowledge of *what the major public concerns are, and whether these concerns differ from the concerns of the risk experts or the authorities*. Which kinds of risks do people fear and which do they tolerate? The interest in public preferences for, or attitudes towards, certain categories of risks has yielded a large body of applied research. Several studies have focused on individual hazards of special concern like means of transportation, energy production and chemical waste sites, and have been partly financed by relevant industries or governmental agencies.

The second, and more theoretically important question, is *what "psychological mechanisms" can explain public reactions (attitudes and actions) towards hazards?* If there is a discrepancy between expert and lay perceptions of risk, how should it be understood? What "psychological laws" do risk judgments follow? When answering these questions public reactions to specific hazards are not interesting in their own right, but can be taken as indicators of the more general cognitive or judgmental strategies that people use when facing uncertainty and dangers.

These two main questions/perspectives give answers that are not too easy to integrate within the same general framework. To complicate the picture further, some studies focus on laypeople's *general* reactions towards a given phenomenon, like means of transportation, and include both positive and negative aspects. Other studies are more focused directly on risky activities related to the use of the risk source, and still others are asking for lay reactions to a given hypothetical accident. For example, some focus upon *hazards* (threats to humans and what they value) while others are studies of *risk* (some quantitative measure of the probability of harm or loss) from a given hazard. Furthermore some ask for personal risk, and others for estimates of risk for the society as a whole. The different perspectives have given results that are not easy to compare across different studies. It has also been noted that risk perceptions and risky behavior are not necessarily directly related. Risk perception may be one—among several—determinants of risk behavior. Studies within the experimental approach to risk have primarily focused upon studying determinants for risk *behavior*, while the psychometric approach has been aimed at demonstrating cognitive components important for lay *perceptions* of risk. Thus, in the following, we will pay most attention to results gained from research within the psychometric approach.

As is well known from risk studies as well as other areas of decision-making, the way questions are asked, choice of response format or the given context may have an influence on the answers, through framing effects, anchoring processes and attention biases (Tversky & Kahneman, 1981; Fischhoff & MacGregor, 1983; Eiser & Hoepfner, 1991). When asked to rate or compare different risky activities, subjects' attention is focused towards these given risks, with the possibility of neglecting or underestimating other and maybe equally important characteristics. Several studies specifically address the question of whether different methods or judgmental tasks give rise to different representations of risk (Johnson & Tversky, 1984; Tyszka & Goszczynska, 1993).

13.2.1 The Experimental Approach to Risk

The "traditional" way to study risky decision-making in psychology and economics has been through gambling studies. Here subjects are presented with "lotteries", described in terms of possible outcomes (gains and losses) and the uncertainties (probabilities/relative frequencies) associated with these outcomes, and are asked to bet or to express their preferences for different lotteries, or asked to rate or compare their riskiness. Outcomes and probabilities are easily manipulated and the methodological stringency of the studies is appealing. These lotteries have been hypothesized to model risky choice in the real world, and models of risky behavior have been constructed on the basis of the results (Lopes, 1987). Many studies have focused on violations of utility models and have aimed at testing general rules for risky choice. People have been found to be "satisfizing" rather than "optimizing" when confronted with decision tasks, and to use risk strategies that do not necessarily maximize benefits, but assure positive payoffs and avoid major disasters. In addition to the basic risk elements of the gambles (probabilities and magnitude of consequences); individual, situational/contextual and social characteristics have been studied as predictors of risky behavior (Slovic, 1972; Lopes, 1987; Shurr, 1987; Lamm, 1988, Mann, 1992; Shoemaker, 1990; Bromiley & Curley, 1992; Davis, Kameda & Stasson, 1992).

It is generally suggested that the weights people use in evaluating gambles are not identical to the objective probabilities; they may for instance ignore small probabilities and treat them as nonexistent (as is also said to be the case when people judge traffic accident risk), or optimistically assign more weight to them than is warranted by the objective rules of the game (as when buying lottery tickets). According to Lopes (1987), most experimentalists explain risky choice by positing an internal process for evaluating gambles that is structurally similar to computing expected values, with the difference that objective values are replaced by subjective values (utilities) and objective probabilities by subjective uncertainty weights.

The methodological advantage of the gambling approach could also be viewed as a weakness. In real life people are not presented with well-defined decision options with all relevant parameters clearly laid out. Real-life gains and losses are often multidimensional (physical, psychological and economical losses, damage to nature and environment, etc.) and the associated uncertainties often ambiguous and vague. In addition, real-life choices are frequently made under time pressure. This makes risky life situations more complex and unclear than normal laboratory tasks. On the other hand real-life situations provide the subjects with contextual information and cues that normally facilitate the handling of the task.

Criticisms of the fruitfulness of the gambling/betting perspective of risk have also come from researchers within the paradigm. Lopes (1983; 1987) suggested that studying risks as static lotteries has limited our understanding of the psychological processes involved because it has not provided an opportunity for related and important issues to be investigated. Some recent experimental studies have tried to meet some of these criticisms by studying risk taking in real-life contexts like driving a car (see for instance Hendrickx & Vlek, 1991a).

13.2.2 The Psychometric Approach to Risk

Dissatisfaction with the gambling paradigm as being ecologically invalid and irrelevant for understanding everyday reactions to risk, led in the 1960s and 1970s to other—and apparently more valid—approaches.

Starr (1969) based his studies of risk acceptance upon the method of *revealed preferences*. His assumption was that society over time arrives at levels of risk for different areas that it finds acceptable due to a risk–benefit trade-off. By examining current levels of risk (number of fatalities, amount of damages etc.), one can estimate the levels of risk society is willing to accept also for other sources of risk. Although this view was found appealing and interesting, several limitations and problems could be pointed out. Does accepted risk mean the same as acceptable risk? Are laypeople rational in the sense that they are able to—and actually do estimate levels of risks and benefits? This position is based on a view of society and its members as well-informed decision makers who accept and reject risks according to a consistent set of criteria.

This assumption was especially challenged by the "Oregon group", Paul Slovic, Sarah Lichtenstein and Baruch Fischhoff, who "founded" the psychometric approach to risk perception (for a more thorough description of the historical development of the psychometric approach, see Slovic, 1992). Rather than focusing upon the revealed preferences they focused on the *expressed* preferences of the public. There were several indicators that laypeople held different conceptions of risk compared to the experts. When stating their perceived risk of an activity or a hazard, the experts relied on fatality statistics, while for the public these estimates seemed to be less important (Slovic, Fischhoff & Lichtenstein, 1980; von Winterfeldt, John & Borcherding, 1981). How can this be explained? The fact that laypeople are able to give relatively well-calibrated estimates of fatality rates showed that the discrepancies were not due to erroneous calculations, so explanations had to be sought elsewhere. Laypeople simply didn't have the same risk concept as risk experts. This led to a long-lasting search for "risk dimensions" or "risk characteristics" capable of explaining the "unexplained variance" of lay risk perception. Lowrance (1976) had originally presented a list of 10 risk considerations or dimensions that he hypothesized would influence safety judgments. These and other risk dimensions were later empirically studied as factors likely to influence lay perceptions of risk (Fischhoff *et al.*, 1978). Surveys were conducted to establish perceived magnitude of risk and benefit of several different risk sources, while other groups rated the same hazards on several other more qualitative risk characteristics, like dread and familiarity. Through factor-analytic procedures several different dimensions (from 9 up to 18) have been reduced to two (or sometimes three) more basic factors, explaining a substantial proportion of the total variance. From these two factors a "factor-analytic space" can be constructed where the given hazards are plotted and comparisons between them can be made (Fischhoff et al., 1978; Slovic, Fischhoff & Lichtenstein, 1980; 1985). Several studies have showed the same general pattern, with one factor labeled *dread risk* (characterized mainly by the dimensions dread, fatality, catastrophic potential and lack of control) and another factor that can be labeled *unknown risk* (characterized by dimensions like unfamiliarity, unknown to science and unknown to those exposed to the risk). The studies generally show that lay perceptions of magnitude of risk can be relatively well explained from ratings on the dread factor, but are relatively independent of the factor unknown risk. A similar study by Vlek and Stallen (1981) gave the same general pattern with two factors explaining most of the variance. The factors in this study were labeled *catastrophic risk* and *degree of organized safety control*, and have a substantial similarity with the factors found by Slovic and his associates. As in the former studies, the catastrophic risk was found to predict lay judgments of risk magnitude, whereas the second factor was unrelated to this measure. Teigen, Brun & Slovic (1988) and Brun (1992) labeled their two main factors the *potency factor* and the *active-passive factor*. The active–passive dimension could be seen as referring to uncertainty regarding the *cause* of a risk, while the potency factor to a larger extent refers to the *effects* of it.

The main criticism of the psychometric tradition has been raised on methodological issues and has especially concerned the correlational nature of the studies. When particular dimensions of risk are found to co-vary with perceived magnitude of risk, this may indicate a causal relationship, but could also be an artifact due to the particular stimulus set. Most studies within this tradition has been of an exploratory kind, and confirmatory testing of the theoretical framework developed by the psychometric tradition is now needed (Slovic, 1992). Experimental studies of the effect of various risk dimensions on risk perception or risk-taking propensity, based upon ratings of ecologically valid scenario descriptions, could be one way to go (see Hendrickx, Vlek & Oppewal, 1989; Hendrickx, Vlek & Caljé, 1992). But even in such designs it is hard to secure full experimental control and separate the effect of each risk dimension from the others, since risk dimensions tend to be confounded in real life. It is for instance hard to vary a given risk's "novelty" without affecting "knowledge of risk" at the same time.

Another criticism addresses the problem of inter-individual variability. The main body of studies within the psychometric approach have been based on correlations between mean ratings of various risks. These aggregated data may have concealed important individual differences, and the generalized risk perceptions presented may not be representative at an individual level. Vlek and Stallen (1981) and Kuyper and Vlek (1984) have for instance found that the average correlation between subjects (over stimuli) is quite low for estimates of catastrophic potential, and even lower for estimates of probability.

The Generality of Risk Perception

Over the past years numerous studies have been conducted within the psychometric paradigm using the same or similar methodology with respondents from different social and cultural groups and with different sets of hazards. The studies have commonly concluded that risk perceptions follow a remarkably similar pattern. A replication study conducted by Slovic and his associates after an interval of 10 years showed substantial stability of the factor structure (Slovic, 1992). Neither differences of *culture* (nationality), nor changes in the set of *hazards* in the studies have altered the general picture of the factory-analytic representations or "dimensionality" of risk (Engländer et al., 1986; Teigen, Brun & Slovic, 1988; Keown, 1989; Goszczynska, Tyszka & Slovic, 1991; Kleinhesselink & Rosa, 1991). These studies have normally focused upon man-made risks. Including a set of *natural* hazards led, however, in one study to the appearance of a third factor (a *novelty factor*) in addition to the two basic factors, one *potency factor* and one *active–passive factor* found previously (Brun, 1992).

Analysis of risk perceptions of various accident scenarios *within a single hazard domain* is another way of testing the generality of the risk dimensions.

Psychometric analyses of lay perceptions of a set of railroad accidents and a set of car accidents have provided the same "factor-analytic space" represented by the factors *dread risk* and *unknown risk* as found in analyses across different hazards (Slovic, MacGregor & Kraus, 1987; Kraus & Slovic, 1988). Gardner et al. (1982) studied perceptions of risk from nuclear power on an *individual* level and found basically the same pattern of correlations between ratings of magnitude of risk and the major risk dimensions (dread, catastrophic potential and environmental damage) as reported in the studies based on aggregated data. The similarities in perceptions of the main dimensions of risk across subjects and domains have made researchers ask whether these risk dimensions represent a "universal" way of perceiving risk, and answers the criticism that the "representations of risk" from a given study is heavily dependent upon the set of risks selected for the study.

But some striking differences between studies have also been found. These refer mainly to estimates of perceived *magnitude of risk*, e.g. what are perceived as the *major concerns* of the individual and the society. When this issue is raised the answers seem to vary for different social (Borcherding, Rohrmann & Eppel, 1986) and national groups (Engländer et al., 1986; Teigen, Brun & Slovic, 1988; Hoefer & Raju, 1991; Mechitov & Rebrik, 1990; Goszczynska, Tyszka & Slovic, 1991). Subjects from America and Hong Kong have generally expressed the highest levels of risk, and subjects from Hungary and what was the Soviet Union the lowest. Nor have the rankings of hazards been the same. Apart from a common fear of nuclear power, Hungarians have for instance expressed greatest concern over common and everyday hazards (motor vehicles, caffeine), Americans have shown greater concern over chemical substances and new technologies, Poles rate the dangers of warfare and nuclear weapons especially high, while Norwegians have shown greater concern over narcotics and psycho-active drugs than the other national samples. And *females* have generally been found to give higher risk ratings than men, especially with respect to risk from nuclear power (Slovic, Kraus, Letzel & Malmfors, 1989; and Slovic, Kraus, Lappe & Major, 1991, both cited in Slovic, 1992; Bastide et al., 1989; Hoefer & Raju, 1991; Sjöberg, 1993). To summarize: There seems to be a substantial amount of similarity in risk perception as long as we refer to the *main representations of risk* (risk dimensions), but less so if we focus upon the *expressed level of riskiness* associated with given hazards.

13.2.3 Qualitative Studies of Risk Perception

There have also been some attempts to study risk perception by more qualitative methods: By use of verbal protocols/reports (Tyszka & Goszczynska, 1993), repertory grid techniques (Green & Brown, 1980, cited in Slovic, Fischhoff & Lichtenstein, 1984), and interviews or "open-ended" questionnaires

(Fischer et al., 1991). Instead of making the subjects choose among, or rate risks from, ready-made lists constructed by the experimenter, subjects in these studies have normally been asked to focus upon and list the risks *they personally* think of as important. In some cases subjects are asked to produce unrestricted lists of risks, in others they are asked to select a given number of risks of major concern. The domains can be completely open ("whatever comes to your mind") or specified by the experimenter (e.g. "health, safety and environmental risks").

This approach has provided a different perspective upon what are the main public concerns from studies within the psychometric approach. From the latter it has been concluded that the public shows great concern and worry about technological hazards like nuclear power plants as opposed to "everyday hazards" like risks from transportation. But when subjects were asked to produce their own lists, Fischer et al. (1991) found that the category "accidents" was the most frequently mentioned (37% of the responses), with motor vehicle accidents alone accounting for 21.6% of the responses. Diseases were the second most common category of concern. When another group was asked to list the "health, safety and environmental risks" of major concern, various environmental risks were mentioned most often (44.1%), followed by health risks (23.8%) and safety risks (22.4%), with general societal risks (10.7%) ranking last. Similar results were found in a Norwegian interview study (Teigen, unpublished data). Here diseases and accidents accounted for 68% of the answers to questions about what the subjects personally were most worried about, while environmental damages were the most frequent category (28%) for perceived risks to society. In these open-ended risk studies one also finds topics like failures in self-realization, job and studies, interpersonal relationships, economical affairs etc.; risks rarely to be included in the traditional risk studies (Fischer et al., 1991; MacGregor, 1991; Teigen, unpublished data; Brun, unpublished data).

13.2.4 Alternative Perspectives

Studies within the so-called "cultural theory" of risk have recently offered a challenge to the traditional psychological risk studies. This approach is based on the work by the anthropologist Mary Douglas (1966) and later refined and developed by anthropologists, psychologists and sociologists (Douglas & Wildavsky, 1982; Thompson, 1980; Buss & Craik, 1983; Buss, Craik & Dake, 1986; Dake, 1991; Rayner, 1992). The cultural theory states that cultural biases and world-views shared by broader social groups shape the individuals' perceptions of risk, and that people select risks in order to defend their preferred lifestyles. The culture provides the individual with socially constructed systems of beliefs or "cultural biases", which the individual internalizes and which shape the individual perception of risk. Wildavsky and Dake (1990) have made an attempt to empirically test different explanations of risk perception, and find that cultural theory is a better predictor of lay perceptions of risk than political orientations, personality factors and people's knowledge of risk.

A recent development is the theory of *social amplification of risk* (Kasperson et al., 1988; Kasperson, 1992). This theory or model attempts to supply a framework for understanding individual perceptions of risk by placing them within a broader network of social processes. The assumption is that the occurrence of a hazardous event influences psychological, social, institutional and cultural processes in ways that can further heighten or attenuate perceptions of risk and shape risk behavior. Behavioral responses to risk generate secondary social or economic consequences extending beyond the direct harm to human health and environment, and these secondary effects will again trigger demands for additional institutional responses or, conversely, responses aimed at suppressing the risk. The theory has been criticized for being too general (Svenson, 1988) and impossible to falsify (Rayner, 1988), but a recent study used confirmatory analyses to test the assumptions in the model and found support for them (Renn et al., 1992).

Risk can be studied from different perspectives and with different methods. Most psychological studies have taken an individualistic perspective, trying to explain risk perception and risk-taking behavior in terms of general cognitive and judgmental strategies, personality traits and situational demands. Some studies have had a more social psychological perspective focusing on group processes in judgment, as seen for instance in the literature on group polarization and the "risky shift" phenomenon (Lamm, 1988), but this perspective also discusses risk on a micro-level. The cultural theory and the theory of social amplification of risk present a broader perspective claiming that individual risk perception must be related to social, institutional and cultural processes.

13.3 CHARACTERISTICS OF PERCEIVED RISK

Different classifications of risks have been suggested according to the nature of the risk in question. Johnson and Tversky (1984) suggest from similarity estimates and clustering techniques that risks can be grouped as violent acts, hazards, accidents, technological disasters and diseases. Others have classified risks as technological risks, substances, and activities as opposed to natural hazards. Others have found an even coarser categorization of risks in manmade versus natural type as fruitful for understanding public perceptions of—and reactions to—hazards (Baum, Fleming & Davidson, 1983; Baum, 1988; Kasperson & Pijawka, 1985; Pijawka, Cuthbertson & Olson, 1987–88; Brun, 1992). The assumptions behind these taxonomies are that people's reactions are similar for hazards that are perceived to share some common features, and that attitudes and reactions to a given hazard are influenced by features of the broader category to which the hazard belongs. Classifications of risks in broader categories are accordingly seen as useful tools for risk communication and risk management purposes.

Cvetkovich and Earle (1985) argue for a more process-oriented classification based on the hazard's "life history". Hazards could be differentiated according to their causes, their physical and psychosocial characteristics and typical individual and aggregate responses to hazards.

13.3.1 Causes of Risk

There are reasons to believe that the *origin* of a risk affects how it is perceived. For instance, a man-made as opposed to a natural risk implies that someone can be blamed for an accident. "Self-induced" accidents are perceived differently than "chance" events where nobody can be seen as immediately responsible for the outcome. For example, people have expressed reluctance to vaccinate a child when the vaccination itself can cause death, even when this is much less likely than death from the disease itself (Ritov & Baron, 1990). This is also reflected in the amount of media coverages allotted to human vs. non-human factors in accidents.

The degree to which a given technology is perceived as intended to harm living organisms has been found to be related to overall estimates of perceived risk, while no correspondence was found for estimates of annual mortality (Hohenemser, Kates & Slovic, 1983). Vlek and Keren (1992) suggest that the nature of the source or cause of a possible risk constitutes a separate dimension of risk which they have labeled "harmful intentionality". Some classes of risks are seen as generally dangerous if they are designed to harm living organisms, while others are perceived as benign even if they happen to have harmful "side-effects", because the harm is not intended.

13.3.2 Characteristics of the Hazards

One important information for societal risk estimates is the universality of a given risk source, e.g. the size of the population at risk. This can have a wide impact on policy strategies for risk communication and risk acceptance. From a societal perspective it is sometimes easier to accept an activity of high relative risk if very few people belong to the risk group, but it is also easier to take drastic measures against it, e.g. make it illegal. In some factor-analytic studies of perception of societal risk, *number of people exposed to a risk* has come out as a separate factor, independent of the dread and knowledge factors (Slovic, Fischhoff & Lichtenstein, 1980). But we have also seen that laypeople normally hold an individualistic perspective and do not pay sufficient attention to the number of people exposed to a risk when estimating risk magnitude (Teigen, Brun & Slovic, 1988).

Another important dimension of risk is *novelty*. The interpretation of an accident seems to be influenced by whether the hazard is seen as new or old.

If a small accident happens in an unfamiliar and poorly understood system (e.g. chemical production, nuclear technology, genetic engineering etc.) the accident may serve as a *signal*, confirming our suspicions that the technology or activity is dangerous, and hinting at the possibility for even greater accidents to happen (the "tip of the iceberg" phenomenon). The signal potential of a risk appears to be systematically related to dimensions loading on both the dread and the unknown factors (Slovic, Lichtenstein & Fischhoff, 1984). For new and dreadful events, it seems that it is the potential or *possibility*, not the numerical probability given from fatality rates, that matters. An increase of fatalities is generally related to perceived risk (Slovic, Fischhoff & Lichtenstein, 1980) and increases in new and unknown system will be taken as a sign of danger (Slovic, Fischhoff & Lichtenstein, 1985), while increases of fatalities in older systems are more likely to be seen as natural fluctuations.

13.3.3 Characteristics of the Consequences

This group of risk characteristics is primarily focused on the accident potential of a risk source and what negative consequences an accident may have. But relevant also are *other* consequences the risk source may have, primarily its *benefits*. Most of the common "everyday risks" like driving cars and eating unhealthy food are primarily thought of in terms of their benefits, not as potential sources of injuries or death. So risk is not seen as the main characteristic of such activities although we all know from statistics that they are the source of many fatalities.

Also the temporal and spatial distribution of the consequences seems important (Vlek & Stallen, 1980; Lindell & Earle, 1983). Immediate benefits for oneself or close persons count more than risks that will affect strangers, or will only happen at a later time. This has been referred to as the NIMBY ("not in my back yard") effect (Marks & von Winterfeldt, 1984).

Probably one of the most mentioned single aspects in this group is the "disaster potential" of a hazard (Green & Brown, 1978; Slovic, Fischhoff & Lichtenstein, 1980). The maximal size of a possible accident and the maximal mortality rate (maximum loss) have been found to predict lay estimates of overall perceived risk (Vlek & Stallen, 1981; Hohenemser, Kates & Slovic, 1983). This challenges the notion of risk as a unidimensional and additive concept, e.g. that several small losses should count the same as one or a few more severe accidents with the same total number of lives lost. Both on empirical and theoretical grounds it has been argued that this is not the only natural and rational way to perceive risk (Hansson, 1989). Most empirical studies show that a large catastrophic event is perceived to be worse than several smaller accidents (von Winterfeldt, John & Borcherding, 1981), but the opposite view has also been claimed: that large accidents should be given proportionally lower weights because they normally occur in geographically restricted areas, they often kill whole families and therefore will have an impact on fewer survivors altogether (Zeckhauser, 1975).

"Deaths" are not calculated as "cold numbers" by the public: both the question of "who" and "how" is important. Some ways of dying (painful or dreadful deaths) are perceived as worse than others. The existence of a special vulnerable and *innocent risk group* (e.g. children) also matters and may make a risk appear especially negative. There are also rational arguments for counting the death of a young and healthy child as more severe (in years lost) than the death of an elderly person. But one also finds arguments for the view that hazards that literally "hit out of blue" are perceived as more frightening than well-defined and predictable hazards. A quote from the Head of The Medical Advisary Board of the Norwegian Sport Federation illustrates the point. When asked to comment on the sudden deaths of some Nordic runners in orienteering races he told the newspaper: "Yes, I am frightened, what's horrible with the TWAR-bacteria is that it can hit anybody" (Verdens Gang 17 November, 1992).

More generally the distribution of risks and benefits in the population and the question of *equity* plays a major role in discussions of the acceptability of risks. It should be noted that how widespread a risk source is in the public and the actual distribution of the risky *consequences* are not necessarily the same thing. For instance, if the risk has *delayed and irreversible effects* with a potential for harmful consequences on future generations it is perceived as especially serious. Especially new environmental risks seem to be associated with a high *ambiguity* concerning the probability of harm and the possible consequences that may follow. High ambiguity is generally found to influence the evaluation of an event in a negative way (Curley, Yates & Abrahams, 1986). Subjects are, for instance, found to be willing to pay for reductions in chance ambiguity (Becker & Browson 1964 cited in Vlek & Stallen, 1980), and insurance experts have been found to suggest considerably higher prices for insurances against events when the probabilities are ambiguous compared to when they are well specified (Hogarth & Kunreuther, 1992). The potency factor found in the psychometric studies can be seen to reflect both psychological effects of a hazard, such as feelings of dread and lack of control, and the physical (actual) consequences like catastrophic potential, delayed and fatal consequences.

13.3.4 The Person's Relation to the Hazard

Risk is sometimes defined as insufficient controllability. Psychologically this aspect is then seen as the main determinant of the uncertainty a person feels when confronted with a possible negative event. The aspect of *control* is seen as important in various areas of risk perception and risky decision-making (Weinstein, 1984; Langer, 1975; DeJoy, 1989; Hendrickx & Vlek, 1991b).

When drivers perceive less risk for a car accident than their passengers do, this may be due to the greater perceived control on the part of the driver (Bragg & Finn, 1982 cited in Jonah, 1986). Cognitive strategies of "coping behavior" are also relevant here. Feelings of being in control, either by being able to ward off an accident or by being able to mitigate its effects, contribute to the person's psychological state of well-being.

Related to controllability is the frequently mentioned risk characteristic "voluntariness of exposure to risk" (Wildavsky, 1986; Lyng, 1990). According to Starr (1969; 1985), the public seems willing to accept higher risks from voluntary activities than involuntary ones that provide the same level of benefit, voluntariness of exposure of risk being the key mediator of risk acceptance. Some factor-analytic studies have found this risk characteristic to constitute a basic dimension in public perceptions of risk, but it seems unrelated to measures of *magnitude* of risk and more important for laypeople's views on risk management (Teigen, Brun & Slovic, 1988; Brun, 1992).

A third relevant aspect is the amount of accumulated *knowledge* of risk (Johnson, 1993). Studies have found the dimensions of amount of personal and scientific knowledge of risk to be related to measures of risk magnitude, although the way these measures are related seems to differ cross-culturally (Teigen, Brun & Slovic, 1988).

13.3.5 Responses to Hazards

There is a great variety of responses to hazards. These include possible actions for accident prevention and more intra-psychological effects as feelings of fear, dread and helplessness. Risks differ according to how much horror and dread they evoke in the public. Some risks have consequences that people have learned to live with and can think of relatively calmly, others evoke strong emotional reactions, and this is found to be a main predictor of perceived magnitude of risk. There is also reason to believe that the choice of *preferred agent for risk management* is related to several such qualitative characteristics of the hazards (Brun, 1992). Risks are also judged differently according to the possibilities of human intervention. Is the risk predictable? Is it possible for anybody to prevent the risk from occurring? And is it possible to escape the risk? Studies show the aspect of perceived control over risk to be an important factor which, together with perceived size of a hazard, predicts whether individuals will take protective measures (Beck, 1984). Preference for risk management agent is also related to type or kind of risk in focus. While societal agents are seen as responsible for managing new and dreadful hazards like damage to the ozone layer, biotechnology and acid rain, voluntary and "common killers", like smoking, consumption of alcoholic beverages and motor vehicle travel, are expected to be managed on an individual basis (Fischer et al., 1991; Brun, 1992).

13.4 UNDERSTANDING PERCEIVED RISK: RISK CHARACTERISTICS, JUDGMENTAL STRATEGIES AND VARIANTS OF UNCERTAINTY

What is the status of the risk dimensions? Some of them seem to be rather "objective" descriptions of relevant properties of the hazard, while others are determined by the relationship between the individual and the hazard, and hence of a more subjective or relative kind. These characteristics are found (or hypothesized) to be of importance when laypeople make risk judgments, but their explanatory power is still a matter of debate. Some of them have relatively strong empirical support, while others must be seen as more hypothetical and in need of further empirical testing.

How should these risk attributes be understood and how do they influence risk perceptions? Krimsky (1992) has suggested that they function as *heuristics* when laypeople perceive and evaluate risk. Risk attributes serve as filters and *a priori* categories through which individuals experience events as risky or not. From this perspective one would think that they provide cues or signals for how individuals should evaluate a given risk, and what the optimal strategies for risk management are. The risk dimensions may mediate risk perceptions differently for various classes of risks (e.g. the significance of perceived novelty differs for manmade and natural risks), and the significance of the various dimensions will most likely differ between individuals and groups of individuals, according to personal characteristics and prevailing social, cultural and political attitudes.

How are the risk characteristics related to the traditional, normative components of risk? According to Hendrickx (1991, p. 14) the "catastrophic potential" associated with the dread factor in the risk perception studies parallels the "amount to lose" in the the gambling paradigm, while a dimension like "personal control" is related to the "probability of loss". Yates and Stone (1992b) similarly suggest that the two main factors found in the psychometric studies of risk—dread risk and unknown risk—correspond to the two main components of the risk concept, loss significance and loss uncertainty. Both these components are multidimensional. When considering the aspect of loss significance people may be concerned with several aspects of the dread and horror the risk evokes, such as the catastrophic potential of the risk and the fatality of the effects. When concerned with the loss uncertainty there are again multiple aspects going beyond the mere probability of injury or death. These could be lack of knowledge of a risk, the novelty of it or the ambiguity of the risk information. Following these arguments the attributes "dread" and "catastrophic potential" most likely facilitate the perception of a hazard as risky by strengthening and amplifying the perceived negative consequences of the risk, while dimensions like "novelty" and "lack of knowledge" focus the attention towards the ambiguity and uncertainty aspect of a hazard, i.e. its probability component.

Teigen (1994, see Chapter 10) discusses different conceptualizations of uncertainty and suggests that there exist several different uncertainty (or intuitive probability) concepts. According to this framework some "erroneous" examples of probability judgments could be understood as expressions of some other type of uncertainty, such as plausibility, propensity, or lack of control rather than estimates of probability in a more traditional (e.g. frequentistic) sense of the word. This reasoning seems highly relevant for risk judgments. Seemingly "overestimated" risks could be explained by assuming that the probability component in the risk estimate is substituted by some other variant of uncertainty. If, for instance, a situation appears *open* and it is hard to come up with arguments favoring one above the other possible outcomes, they will all appear "probable" (Teigen, 1988). If the causal mechanisms of a disease are not known and there does not exist a known risk group, everyone in the population may appear at risk. Likewise when the situation seems *representative* of an accident situation it may be judged risky, even if the "true" probability is low. If an event or a substance is seen as having the *potentiality* of being fatal, this possibility may be all that matters—not how big or small the probability really is. As one of our subjects said when stating his own definition of risk: "It doesn't need to be a high probability, but I have to know that the threatening consequences *might* happen." Overestimations are also found when one is perceived to be *close to* an accident and where one easily could imagine small changes that would have fatal consequences (even if this rarely happens) or there is very little extra effort needed for the risk to become materialized (as when one is said to be "one second from death"). Such situations show a *tendency* or are pointing in the direction of an accident. This means that the risk estimate does not have to reflect a high probability for an accident, but can refer to a possibility, a disposition or a tendency for the accident to occur. In other words people may be judging the "realism" of a disastrous outcome, rather than its probability (Teigen, 1993). Several events appear highly realistic, convincing and plausible in spite of their low probability and will hence be judged as disproportionately risky.

Risk perception cannot be understood separately from the more general mental models and judgmental strategies people use when facing uncertainty (Thüring & Jungermann, 1986; Svenson, 1988; Bostrom, Fischhoff & Morgan, 1992). Hendrickx and Vlek (1991b) ask whether different types of risk elicit different cognitive strategies for risk judgments and state that the quality and type of probabilistic *information* provided are important determinants for the strategy used. The importance of the nature and the quality of the probabilistic information has been stressed by several authors in the area of risky decision-making (Howell & Burnett, 1978; Kahneman & Tversky, 1982b; Palmer & Sainfort, 1992). Hendrickx and Vlek (1991b) suggest that two fundamentally different types of information underlie probability judgments: *frequency* information based on the outcomes of similar situations in the past, and *process* information based on knowledge of the mechanisms that determine an activity's future course and outcome. Corresponding to these types of information there exist different types of mental strategies for likelihood judgments. Frequentistic information will normally facilitate traditional probability estimates. Process information in simple domains (like tossing a coin) will lead to logical deductions of the probability estimates, but will in complex domains serve as a basis for constructing mental scenarios that specify how future accidents might happen. Empirical studies (Hendrickx, Vlek & Oppewal, 1989; Hendrickx, Vlek & Caljé, 1992) have shown people to be sensitive to both types of information, but their relative importance varies with the type of risky activity. For small-scale personally controllable risks, scenario information dominated, while for large-scale and uncontrollable risks frequency information was more important.

Apparently one can arrive at risk estimates through several different mental strategies, some more intuitively (subjective) based and others more rule based. In very simple domains and when the risk is not related to one's own person and competence, the "objectivistic" or rule-based approaches can be used. But when people are personally involved (as is the case when risks are controllable) the natural way of evaluating is through one's own experience of the same or similar situations in the past and by judgments of one's own competence. Hence, it probably seems irrelevant to base judgments on statistical frequencies, since the importance of one's own behavior appears much more salient. When making risk estimations for new technologies with no history of accidents, one will simply be forced to rely on mental constructions of scenarios describing possible mechanisms or causes of an accident, since no reliable frequency information is available. In these situations probability estimates may be replaced by possibility judgments instead. Following this line of thought one can suggest that frequentistic information and process information in simple domains serve as basis for probability estimates (in the classical sense of the word), while process information, especially in new and complex domains, serves as a basis for estimates of the degree of "realism" (judged, for instance, by the possibility, plausibility or representativeness) of a given outcome rather than its probability.

In agreement with the psychometric tradition we will conclude that risk is multidimensional. We will further stress that this also applies to—and could be analyzed at the level of—the basic risk components. It has long been stated that the loss component of risk must be seen as multidimensional, including different types of losses, some more material and others psychological, some quantifiable and others more vague and harder to assess. Apparently, not only the loss component but also the probability/uncertainty component must be seen as multidimensional. There exists a diversity of different forms of losses, and likewise different variants of uncertainty. They may combine in different ways and constitute different lay concepts of risk. Whether people rely on rule-based or intuitive strategies for risk judgments may be dependent on the risk concept activated.

ACKNOWLEDGEMENT

The author would like to thank Professor Karl Halvor Teigen for comments on a previous draft of the manuscript.

NOTE

(1) As several of the respondents agreed to more than one definition the percentages sum up to more than 100%.

REFERENCES

Bastide, S., Moatti, J.P., Pages, J.P. & Fagnani, F. (1989) Risk perception and social acceptability of technologies: The French case, *Risk Analysis*, **9**, 215–23.
Baum, A. (1988) Disasters, natural and otherwise, *Psychology Today*, **4**, 57–60.
Baum, A., Fleming, R. & Davidson, L.M. (1983) Natural disaster and technological catastrophe, *Environment and Behavior*, **15**, 333–54.
Beck, K.H. (1984) The effects of risk probability, outcome severity, efficacy of protection and access to protection on decision making: A further test of protection motivation theory, *Social Behavior and Personality*, **12**, 121–5.
Bettman, J.R. (1973) Perceived risk and its components: A model and empirical test, *Journal of Marketing Research*, **10**, 184–90.
Bettman, J.R. (1975) Information integration in consumer risk perception: A comparison of two models of component conceptualization, *Journal of Applied Psychology*, **60**, 381–5.
Borcherding, K., Rohrmann, B. & Eppel, T. (1986) A psychological study on the cognitive structure of risk evaluations. In B. Brehmer, H. Jungermann, P. Lourens, and G. Sevón (eds.) *New Directions in Research on Decision Making*, Elsevier, Amsterdam, pages 245–62.
Bostrom, A., Fischhoff, B. & Morgan M.G. (1992) Characterizing mental models of hazardous processes: A methodology and an application to radon, *Journal of Social Issues*, **48**, 85–100.
Bromiley, P. & Curley, S.P. (1992) Individual differences in risk taking. In J.F. Yates (ed.) *Risk Taking Behavior*, Wiley, London, pages 87–132.
Brun, W. (1992) Cognitive components in risk perception: Natural versus manmade risks, *Journal of Behavioral Decision Making*, **5**, 117–32.
Brun, W. & Teigen, K.H. (1988) Verbal Probabilities: Ambiguous, context dependent or both? *Organizational Behavior and Human Decision Processes*, **41**, 390–404.
Buss, D.M., & Craik, K.H. (1983) Contemporary worldviews: Personal and policy implications, *Journal of Applied Social Psychology*, **13**, 259–80.
Buss, D.M., Craik, K.H. & Dake, K. (1986) Contemporary worldviews and perception of the technological system. In V.T. Covello, J. Menkes & J. Mumpower (eds.) *Risk Evaluation and Management*, Plenum Press, New York, pages 93–130.
Curley, S.P., Yates, F.J. & Abrams, R.A. (1986) Psychological sources of ambiguity avoidance, *Organizational Behavior and Human Decision Processes*, **38**, 230–56.
Cvetkovich, G. & Earle, T.C. (1985) Classifying hazardous events, *Journal of Environmental Psychology*, **5**, 5–35.

Dake, K. (1991) Orienting dispositions in the perception of risk. An analysis of contemporary world views and cultural biases, *Journal of Cross-Cultural Psychology*, **22**, 61–82.
Davis, J.H., Kameda, T. & Stasson, M.F. (1992) Group risk taking: Selected topics. In J.F. Yates (ed.) *Risk Taking Behavior*, Wiley, London, pages 163–99.
Defoy, D.M. (1989) The optimism bias and traffic accident risk perception, *Accident Analysis and Prevention*, **21**, 333–40.
Douglas, M. (1966) *Purity and Danger: Concepts of Pollution and Taboo*. Routledge and Kegan Paul, London.
Douglas, M. & Wildavsky, A. (1982) *Risk and Culture: An Essay on the Selection of Technological and Environmental Dangers*. University of California Press, Berkeley.
Drottz-Sjöberg, B.M. (1991) Perception of risk. Studies of risk attitudes, perceptions and definitions, Center for Risk Research, Stockholm School of Economics, Gotab, Stockholm.
Eiser, J.R. & Hoepfner, F. (1991) Accidents, disease, and the greenhouse effect: Effects of response categories on estimates of risk, *Basic and Applied Social Psychology*, **12**, 195–210.
Engländer, T., Farago, K., Slovic, P. & Fischhoff, B. (1986) A comparative analysis of risk perception in Hungary and the United States, *Social Behaviour*, **1**, 55–66.
Fischer, G.W., Morgan, M.G., Fischhoff, B., Nair, I. & Lave, L.B. (1991) What risks are people concerned about? *Risk Analysis*, **11**, 303–14.
Fischhoff, B. (1989) Risk: A guide to controversy. In *Improving Risk Communication*, National Academy Press, Washington, D.C. pages 211–308.
Fischhoff, B. & MacGregor, D. (1983) Judged Lethality: How much people seem to know depends upon how they are asked, *Risk Analysis*, **3**, 229–35.
Fischhoff, B., Slovic, P., Lichtenstein, S., Read, S. & Combs, B. (1978) How safe is safe enough? A psychometric study of attitudes towards technological risks and benefits, *Policy Sciences*, **9**, 127–52.
Frisch, D. & Baron, J. (1988) Ambiguity and rationality, *Journal of Behavioral Decision Making*, **1**, 149–57.
Gardner, G.T., Tiemann, A.R., Gould, L.C., DeLuca, D.R., Doob, L.W. & Stolwijk, J.A.J. (1982) Risk and benefit perceptions, acceptability judgements, and self-reported actions toward nuclear power, *The Journal of Social Psychology*, **116**, 179–97.
Goszczynska, M., Tyszka, T. & Slovic, P. (1991) Risk perception in Poland: A comparison with three other countries, *Journal of Behavioral Desision Making*, **4**, 179–93.
Green, C.H. & Brown, R.A. (1978) Counting lives, *Journal of Occupational Accidents*, **2**, 55–70.
Hansson, S.O. (1989) Dimensions of risk, *Risk Analysis*, **9**, 107–12.
Hendrickx, L. (1991) How versus how often. The role of scenario information and frequency information in risk judgment and risky decision making, Doctoral Thesis, Groningen, Holland.
Hendrickx, L. & Vlek, C. (1991a) Effects of risk information on speed choice in blind curves. In G.B. Grayson & J.F. Lester (eds.) *Behavioural Research in Road Safety*. Proceedings of a seminar at Nottingham University, 26–27 September 1990, Transport and Road Research Laboratory, report PA2038/91. TRRL, Crowthorne (UK).
Hendrickx, L. & Vlek, C. (1991b) Perceived control, nature of risk information and risk taking. An experimental test of a simple taxonomy of uncertainty, *Journal of Behavioral Decision Making*, **4**, 235–47.

Hendrickx, L., Vlek, C. & Caljé, H. (1992) Effects of frequency and scenario information on the evaluation of large-scale risks, *Organizational Behavior and Human Decision Processes*, **52**, 256–75.
Hendrickx, L., Vlek, C. & Oppewal, H. (1989) Relative importance of scenario information and frequency information in the judgment of risk, *Acta Psychologica*, **72**, 41–63.
Hoefer, M. P. & Raju, V.S. (1991) Technological risk perception: Similarities and dissimilarities in French and American samples. In B.J. Garrick & W.C. Gekler (eds.) *The Analysis, Communication, and Perception of risk*, Plenum Press, New York, pages 25–30.
Hogarth, R.M. & Kunreuther, H. (1992) Pricing insurance and warranties: Ambiguity and correlated risks, *The Geneva Papers on Risk and Insurance Theory*, **17**, 35–60.
Hohenemser, C., Kates, R.W. & Slovic, P. (1983) The nature of technological hazard, *Science*, **220**, 378–84.
Howell, W.C. & Burnett, S.A. (1978) Uncertainty measurement: A cognitive taxonomy, *Organizational Behavior and Human Performance*, **22**, 45–68.
Johnson, B.B. (1993) Advancing understanding of knowledge's role in lay risk perception. *RISK: Issues in Health and Safety*, **4**, 189–212.
Johnson, E.J. & Tversky, A. (1984) Representations of perceptions of risks, *Journal of Experimental Psychology: General*, **113**, 55–70.
Jonah, B.A. (1986) Accident risk and risk taking behavior among young drivers, *Accident Analysis and Prevention*, **18**, 255–71.
Jungermann, H. & Slovic, P. (1992) Characteristics of Individual Risk Perception. Ms.
Kahneman, D. & Tversky, A. (1982a) The simulation heuristic. In D. Kahneman, P. Slovic, & A. Tversky (eds.), *Judgement under Uncertainty: Heuristics and Biases*. Cambridge University Press, Cambridge, pages 201–8.
Kahneman, D. & Tversky, A. (1982b) Variants of uncertainty, *Cognition*, **11**, 143–57.
Kaplan, S. & Garrick, B.J. (1981) On the quantitative definition of risk, *Risk Analysis*, **1**, 11–27.
Kasperson, R.E. (1992) The social amplification of risk: Progress in developing an integrative framework. In S. Krimsky & D. Golding (eds.) *Social Theories of Risk*, Preager, Westport, Conn., pages 153–78.
Kasperson, R.E. & Pijawka, K.D. (1985) Societal response to hazards and major hazard events: Comparing natural and technological hazards, CENTED Research Report No. 45. MA.: Clark University, Worcester.
Kasperson, R.E., Renn, O., Slovic, P., Brown, H.S., Emel, J., Goble, R., Kasperson, J.X. & Ratick, S. (1988) The social construction of risk: A conceptual framework, *Risk Analysis*, **8**, 177–97.
Keown, C.F. (1989) Risk perceptions of Hong Kongese vs. Americans, *Risk Analysis*, **9**, 401–5.
Kleinhesselink, R.R. & Rosa, E.A. (1991) Cognitive representation of risk perceptions. A comparison of Japan and the United States, *Journal of Cross-cultural Psychology*, **22**, 11–28.
Kraus, N.N. & Slovic, P. (1988) Taxonomic Analysis of perceived risk: Modeling individual and group perceptions within homogeneous hazard domains, *Risk Analysis*, **8**, 435–55.
Krimsky, S. (1992) The role of theory in risk studies. In S. Krimsky & D. Golding (eds.) *Social Theories of Risk*, Preager, Westport, Conn., pages 3–22;
Kuyper, H. & Vlek, C. (1984) Contrasting risk judgments among interest groups, *Acta Psychologica*, **56**, 205–18.

Lamm, H. (1988) A review of our research on group polarization: Eleven experiments on the effects of group discussion on risk acceptance, probability estimation, and negotiation positions, *Psychological Reports*, **62**, 807–13.

Langer, E.J. (1975) The illusion of control, *Journal of Personality and Social Psychology*, **32**, 311–28.

Lindell, M.K. & Earle, T.C. (1983) How close is close enough: Public perceptions of the risk of industrial facilities, *Risk Analysis*, **3**, 245–53.

Lopes, L.L. (1983) Observations. Some thoughts on the psychological concept of risk, *Journal of Experimental Psychology: Human Perception and Performance*, **9**, 137–44.

Lopes, L.L. (1987) Between hope and fear: The psychology of Risk, *Advances in Experimental Social Psychology*, **20**, 255–95.

Lowrance, W.W. (1976) *Of acceptable risk: Science and the determination of safety*. William Kaufmann, Los Altos, Ca.

Lyng, S. (1990) Edgework: A social psychological analysis of voluntary risk taking, *American Journal of Sociology*, **95**, 851–86.

McDaniels, T.L., Kamlet, M.S. & Fischer, G.W. (1992) Risk perception and the value of safety, *Risk Analysis*, **12**, 495–503.

MacGregor, D. (1991) Worry over technological activities and life concerns, *Risk Analysis*, **11**, 315–24.

Mann, L. (1992) Stress, affect, and risk taking. In J.F. Yates (ed.) *Risk Taking Behavior*, Wiley, London, pages 202–30.

Marks, G. & von Winterfeldt, D. (1984) "Not in my back yard": Influence of motivational concerns on judgments about a risky technology, *Journal of Applied Psychology*, **69**, 408–15.

Mechitov, A.I. & Rebrik, S.B. (1990) Studies of risk and safety perception in the USSR. In K. Borcherding, D.I. Larichev, & D.M. Messick (eds.) *Contemporary Issues in Decision Making*, Elsevier, Amsterdam, pages 261–70.

Murphy, A.H., Lichtenstein, S., Fischhoff, B. & Winkler, R.L. (1980) Misinterpretation of precipitation probability forecasts, *Bulletin of the American Meteorological Society*, **6**, 695–701.

Palmer, C.G.S. & Sainfort, F. (1992) Risk perception: a theoretical framework and empirical results. Poster presented at the annual meeting of the J/DM Society in St Louis, November 15.

Pijawka, K.D., Cuthbertson, B.A. & Olson, R.S. (1987–88) Coping with extreme hazard events: Emerging themes in natural and technological disaster research, *OMEGA*, **18**, 281–97.

Rayner, S. (1988) Muddling through metaphors to maturity: A commentary on Kasperson et al., The Social Amplification of Risk, *Risk Analysis*, **8**, 201–4.

Rayner, S. (1992) Cultural theory and the risk analysis. In S. Krimsky & D. Golding (eds.) *Social Theories of Risk*. Praeger, Westport, Conn., pages 83–116.

Renn, O., Burns, W.J., Kasperson, J.K., Kasperson, R.E. & Slovic, P. (1992) The social amplification of risk: Theoretical foundations and empirical applications, *Journal of Social Issues*, **48**, 137–60.

Ritov, I. & Baron, J. (1990) Reluctance to vaccinate: Omission Bias and Ambiguity, *Journal of Behavioral Decision Making*, **3**, 263–77.

Shafer, G. (1993) Can the various meanings of probability be reconciled? In G. Keren & C. Lewis (eds.) *A Handbook for Data Analysis in the Behavioral Sciences. Methodological Issues*. Erlbaum, Hillsdale, NJ, pages 165–96.

Shoemaker, P.J.H. (1990) Are risk attitudes related across domains and response modes? *Management Science*, **36**, 1451–63.

Shurr, P.H. (1987) Effects of gain and loss decision frames on risky purchase negotiations, *Journal of Applied Psychology*, **72**, 351–8.
Sjöberg, L. (1993) Confidence in experts and perceived nuclear waste risk. Paper read at the Conference on Social Amplification of Risk, Center for Risk Research, Stockholm, February 22–23.
Slovic, P. (1967) The relative influence of probabilities and payoffs upon perceived risk of a gamble, *Psychonomic Science*, **9**, 223–4.
Slovic, P. (1972) Information processing, situation specificity, and the generality of risk taking behavior, *Journal of Personality and Social Psychology*, **79**, 139–45.
Slovic, P. (1987) Perception of risk. *Science*, **236**, 280–5.
Slovic, P. (1992) Perception of risk: Reflections on the psychometric paradigm. In S. Krimsky & D. Golding (eds.) *Social Theories of Risk*, Preager, Westport, Conn., pages 117–52.
Slovic, P., Fischhoff, B. & Lichtenstein, S. (1980) Facts and Fears. In R.C. Schwing & W.A. Albers, Jr. (eds.) *Societal Risk Assessment. How Safe is Safe Enough?* Plenum Press, New York, pages 181–216.
Slovic, P., Fischhoff, B. & Lichtenstein, S. (1984) Behavioral decision theory perspectives on risk and safety, *Acta Psychologica*, **56**, 183–203.
Slovic, P., Fischhoff, B. & Lichtenstein, S. (1985) Characterizing perceived risk. In R.W. Kates, C. Hohenemser, J.X. Kasperson (eds.) *Perilous Progress: Managing the hazards of technology*, Westview, Boulder, CO., pages 91–125.
Slovic, P., Lichtenstein, S. & Fischhoff, B. (1984) Modeling the societal impact of fatal accidents, *Management Science*, **30**, 464–74.
Slovic, P., MacGregor, D. & Kraus, N.N. (1987) Perception of risk from automobile safety defects, *Accident Analysis and Prevention*, **19**, 359–73.
Starr, C. (1969) Social benefit versus technological risk, *Science*, **165**, 1232–8.
Starr, C. (1985) Risk management, assessment, and acceptability, *Risk Analysis*, **5**, 97–102.
Svenson, O. (1988) Mental models of risk, communication, and action: Reflections on social amplification of risk, *Risk Analysis*, **8**, 199–200.
Teigen, K.H. (1988) When are low-probability events judged to be "probable"? Effects of outcome-set characteristics on verbal probabilities, *Acta Psychologica*, **68**, 157–74.
Teigen, K.H. (1994) Variants of subjective probabilities: Concepts, norms and biases. In G. Wright & P. Ayton (eds.) *Subjective Probability*, Wiley, London, pages 211–237.
Teigen, K.H. (1993) Risk perception and variants of subjective realism. Paper read at the Conference on Social Amplification of Risk, Center for Risk Research, Stockholm, February 22–23.
Teigen, K.H., Brun, W. & Slovic, P. (1988) Societal risk as seen by a Norwegian public. *Journal of Behavioral Decision Making*, **1**, 111–30.
Thompson, M. (1980) *An Outline of the Cultural Theory of Risk*. International Institute for Applied System Analysis, Laxenburg.
Thüring, M. & Jungermann, H. (1986) Constructing and running mental models for inferences about the future. In B. Brehmer, H. Jungermann, P. Lourens & G. Sevon (eds.) *New Directions in Research on Decision Making*, Elsevier, North Holland, pages 163–74.
Tversky, A. & Kahneman, D. (1981) The framing of decisions and the psychology of choice, *Science*, **211**, 453–8.
Tversky, A. & Kahneman, D. (1982) Availability: A heuristic for judging frequency and probability. In D. Kahneman, P. Slovic, & A. Tversky (eds.) *Judgement under*

uncertainty: Heuristics and biases, Cambridge University Press, Cambridge, pages 163–78.
Tyszka, T. & Goszczynska, M. (1993) What verbal reports say about risk perception, *Acta Psychologica*, **83**, 53–64.
Vlek, C. & Hendrickx, L. (1988) Statistical risk versus personal control as conceptual bases for evaluating (traffic) safety. In J.A. Rothengatter & R.A. deBruin (eds.) *Road User Behaviour: Theory and Research*, Van Gorcum, Assen.
Vlek, C. & Keren, G. (1992) Behavioral decision theory and environmental risk management: Assessment and resolution of four "survival" dilemmas, *Acta Psychologica*, **80**, 249–78.
Vlek, C. & Stallen, P.J. (1980) Rational and personal aspects of risk, *Acta Psychologica*, **45**, 273–300.
Vlek, C. & Stallen, P.J. (1981) Judging risk and benefits in the small and in the large, *Organizational Behavior and Human Performance*, **28**, 235–71.
von Winterfeldt, D., John, R.S. & Borcherding, K. (1981) Cognitive components of risk rating, *Risk Analysis*, **1**, 277–87.
Weinstein, N.D. (1984) Why it won't happen to me: Perceptions of risk factors and susceptibility, *Health Psychology*, **3**, 431–57.
Wildawsky, A. (1986) Defining risk, *Science*, **232**, 439.
Wildavsky, A. & Dake, K. (1990) Theories of risk perception: Who fears what and why? *DÆDALUS* Journal of the American Academy of Arts and Sciences, **119**, 41–60.
Yates, F.J. & Stone, E.R. (1992a) The risk construct. In F.J. Yates (ed.) *Risk Taking Behavior*, Wiley, London, pages 1–25.
Yates, F.J. & Stone, E.R. (1992b) Risk appraisal. In F.J. Yates (ed.) *Risk Taking Behavior*, Wiley, London, pages 49–86.
Zeckhauser, R.J. (1975) Procedures for valuing life, *Public Policy*, **23**, 419–64.

Chapter 14

Relations Between Confidence and Skilled Performance

Nigel Harvey
University College, London

Procedural knowledge is concerned with "knowing how" whereas declarative knowledge is concerned with "knowing that" (cf. Ryle, 1949). Various arguments have been put forward to support the psychological reality of this distinction. For example, amnesic patients have difficulty in using or acquiring declarative knowledge but are still able to use and acquire procedural knowledge (Cohen & Squire, 1980).

Performance of some tasks depends primarily on exercise of innate abilities. Tests of perceptual acuity come into this category. Performance of other tasks depends primarily (though not exclusively) on use of procedural knowledge that has been acquired through learning. These tasks are termed skilled tasks. They are characterized by individual differences in performance that are related to the amount of practice that the individuals have experienced.

Here I shall discuss the relationship between how well people perform skilled tasks and the confidence that they have in their performance. I shall not cover confidence in answers to general knowledge tests (e.g. Lichtenstein, Fischhoff & Phillips, 1982) or confidence in use of innate abilities in tests of perceptual acuity (e.g. Keren, 1988). These topics are covered in chapters by other authors (e.g. McClelland & Bolger, Chapter 18, this volume; Gigerenzer, Chapter 7, this volume).

Subjective Probability. Edited by G. Wright and P. Ayton.
© 1994 John Wiley & Sons Ltd.

Skilled tasks are often divided into those that are motor (e.g. surgery, driving), cognitive (e.g. clinical decision-making, judgemental forecasting), perceptual (e.g. reading X-rays, determining the sex of chickens) and social (e.g. bargaining, negotiation). Most of the work to which I shall refer focused on motor skills and cognitive skills. However, if similar principles underlie acquisition and performance of all different types of skill (Rosenbloom & Newell, 1986; Welford, 1980), the findings can be expected to generalize to other domains.

Few psychologists working on skilled behaviour have been interested in confidence *per se*. Instead they have used confidence measures as tools to test various theories. In some cases, this has led them to study whether confidence accurately reflects performance and whether changes in performance produce changes in confidence. In other words, performance is treated as the cause and confidence as the effect. Sometimes, however, their theoretical concerns have led them to examine how confidence influences performance. In this case, confidence is treated as the cause and performance as the effect.

In this chapter I shall review both types of study. As far as I am aware, they have not been treated together before. By doing this, I hope to show that people working on each one have produced findings relevant to, but ignored by, those working on the other. An integrative approach would pay theoretical dividends for both. Confidence and performance should be viewed as coupled together as a dynamical system; neither should be seen as just a cause nor as just an effect.

Both confidence and performance can be measured in various different ways. Performance can be measured by the proportion of times that a goal is reached ("she cleared the five-foot bar eight out of ten times"), by error along some performance-relevant dimension ("she was five inches short") or by measuring performance along that dimension ("she jumped four feet"). In some studies, more than one measure must be taken. For example, in goal-setting experiments, conditions that cause people to fail more often lead to performance improvement ("the proportion of times that she cleared the bar dropped from 0.8 to 0.6 when it was raised six inches but the average height of her jumps increased by four inches").

Confidence can be measured by asking people to provide a rating on a fixed-point scale, by asking them to estimate a probability that they are successful or by asking them to estimate the frequency with which they are successful. It can also be assessed by asking them to estimate their performance (or the error in it) along some relevant dimension or by asking them how much they will bet on being successful. No method is perfect; each one has problems that have been well documented (e.g. Cohen, Dearnaley & Hansel, 1956a, 1957; Keren, 1991; Poulton, 1989).

Confidence judgements also differ in another important way; they can be prospective or retrospective. Assessment is prospective when it is made before a performance starts (or without a performance ever taking place). For example, someone might judge that they have a 15% probability of clearing a six-foot bar. Assessment is retrospective when it is made after a performance has finished but before its effectiveness has been revealed. For example, consider a skater in an international competition. After she has finished performing but before the judges announce their scores, she estimates that she has a 90% probability of exceeding her previous best. These two types of judgement are used in different situations. Studies of risk perception use prospective assessment, whereas people studying motor learning and skill training use retrospective assessment.

In what follows, I shall deal first with studies of how well confidence reflects performance. Then I shall turn to studies of the influence that confidence has on performance. In both cases, I shall group them according to the various theoretical traditions to which they belong. I shall finish with some comments about the relationships between these different types of study and the findings that they have produced.

Many books have been written on most of the content areas that I shall review. My aims are necessarily modest: for each area, I want to outline the theoretical issues that prompted work on confidence and to give an idea of what has been found out and what problems remain.

14.1 DOES CONFIDENCE REFLECT PERFORMANCE?

In what follows, I shall cover work on motor skill acquisition; illusions of learning and control; self-assessment in skilled tasks; risk assessment and preference. I shall then discuss whether models designed to account for confidence in declarative knowledge (see McClelland & Bolger, Chapter 18, this volume) are useful for accounting for the results that have emerged from these areas of work.

14.1.1 Motor Skill Acquisition

Much of the work on the relationship between confidence and performance in motor skill was stimulated by Adams' (1971) closed-loop theory of skill acquisition. According to this theory, movements are determined by two internal representations: the memory trace and the perceptual trace. The memory trace is a simple motor program. It is responsible for initiating all movements and for terminating those that are rapid and ballistic (i.e. made without reference to feedback). The perceptual trace is a store of the sensory consequences associated with the correct response. It is a representation of the "central tendency" of past feedback states that have occurred when the response was designated as correct. Whenever knowledge of results (KR) specifies that a correct response has been made, both perceptual and memory traces are strengthened.

When movements are slow and controlled, the perceptual trace is continually compared with sensory information coming from the ongoing moment. Action is terminated only when a match is obtained. Because a match has been obtained, people cannot perceive an error in their own movement. However, they are more confident that a movement is correct when their criterion for the correctness of movement is better (i.e. when the perceptual trace is stronger). Consequently, for this type of movement, confidence increases with experience of the correct response but does not depend on the magnitude of the objective error in the movement.

When movements are rapid and ballistic, the perceptual trace is matched to the sensory consequences of the finished movement: the larger the difference, the greater the perceived error in the movement and the less confident the person is in having made the correct movement. Of course, people will feel more certain about both these judgements when they are based on better criteria (i.e. when the perceptual trace is stronger). Hence, for this type of movement, confidence will increase with experience of the correct response and will be directly related to the accuracy of the movement. Furthermore, the relationship between confidence and accuracy will increase with experience of the correct response.

Adams' theory in its original and in its modified form (Schmidt, 1975) led to many experiments. Researchers were primarily interested in testing the predictions for ballistic movements: does people's ability to monitor the accuracy of their fast movements improve with practice? In a typical experiment, subjects were required to move a slide over a fixed distance in a target time (T). After each attempt, they had to estimate the time that they took (E). The experimenter compared this with their actual movement time (A). There have been many experiments of this general type (e.g. Adams & Goetz, 1973; Adams, Goetz & Marshall, 1972; Harvey, Garwood & Palencia, 1987; Kantowitz, 1974; Koch & Dorfman, 1979; McCracken & Stelmach, 1977; Marshall, 1972; Newell, 1974; Newell & Boucher, 1974; Newell & Chew, 1974; Newell & Shapiro, 1976; Schmidt & White, 1972; Schmidt & Wrisberg, 1973; Turpin, 1980; Wallace, De Oreo & Roberts, 1976; Zelaznik & Spring, 1976). To test the predictions, experimenters compared actual and estimated movement times by correlating these two variables or by calculating the absolute difference between them (i.e. $|A - E|$). The main findings are easily summarized.

First, people do have some insight into their own individual performances. Correlations between actual and estimated movements have usually been found to be around 0.5. Second, with few exceptions, the size of this correlation has not been found to change with practice at the task. Third, when KR is provided, the absolute difference between actual and estimated movement decreases over trials. Fourth, when KR is withdrawn, this difference tends to increase. The size and speed of the increase depend on the number of trials with KR that were provided prior to KR withdrawal. When fewer have been provided, the increase is larger and faster.

The decrease in the size of the absolute difference between actual and estimated movement as people experience more trials with KR is consistent with a strengthening of the perceptual trace. The absence of any corresponding increase in the correlation between the two variables can be explained by assuming that practice also restricts the range of one or both of them (Koch & Dorfman, 1979). This is a reasonable assumption: it is well known that responses become less variable as people learn to perform a task (e.g. McCracken & Stelmach, 1977).

The increase in the size of the absolute difference between actual and estimated movement after KR withdrawal is consistent with forgetting caused by fading of the perceptual trace. Again, the lack of any corresponding decrease in the correlation between the two variables is to be expected if forgetting causes an increase in variability of one or both of them.

Work to test Adams' theory has shown that people have some insight into their own performance and that this insight improves with practice. However, it tells us nothing about whether people are underconfident or overconfident. This is because of the measure that the researchers used to test the theory (i.e. absolute difference between actual and estimated movement). To assess over/underconfidence they would have had to calculate the subjective–objective error difference: i.e. the difference between the absolute size of the actual movement time error and the absolute size of error derived from the estimates of movement time. For example, suppose that the target time for producing the movement is 200 ms and that its actual duration is 220 ms. If its estimated duration is 210 ms, the absolute difference between actual and estimated duration ($|A-E|$) is 10 ms and the subjective–objective error difference ($|T-A|-|T-E|$) shows 10 ms overconfidence. However, if its estimated duration is 230 ms, the absolute difference between actual and estimated duration is still 10 ms but the subjective–objective error difference now shows 10 ms underconfidence.

14.1.2 Illusions of Learning

In some of the experiments designed to test Adams' theory, people were asked to make an additional response on each trial. After estimating the duration of their movement, they had to use a fixed-point scale to rate their confidence that their movement was within a narrow range of the target. Analyses of this variable are not so easy to summarize. However, despite the inconsistencies, certain aspects of the results are intriguing.

Marshall (1972) found that practice improved performance but had no systematic effect on confidence. Correlations between these two variables were close to zero. Adams & Goetz (1973) found that practice improved performance without raising confidence in one of their experiments but obtained the reverse pattern of results in the other. They did not report correlations. Schmidt & White (1972) tested people for 120 trials on a first day and for 50 trials on a second day. On both days, performance improved and confidence increased. This is what Adams' theory predicts. However, on each day confidence reached an asymptote after only five trials. Furthermore, at the start of the second day it had dropped back to the same low level as at the start of the first day. Although this drop in confidence from one day to the next could have been due to forgetting of the perceptual trace, the authors felt that the transitory nature of the increase on each day suggested a more cognitive explanation. Subjects may have had a "set" that "dictated" to them that they should not be confident on the first few trials of any task. Ratings increased because confidence was released from the suppressive effects of this set rather than because of strengthening of the perceptual trace.

Other work tends to confirm Schmidt and White's suspicions about the importance of cognitive effects related to set and expectancy. Newell & Boucher's (1974) subjects moved a slide along a track until it hit a block. They could not see their movements. After 20 trials, they estimated the movement distance. One group of subjects made their estimates in inches and another made them in millimetres. The block was then removed and all subjects attempted to reproduce the movement. There was no difference in accuracy between the groups but ratings on a seven-point scale showed that confidence in accuracy was higher for the group that had earlier estimated the movement's length in millimetres than for the group that had estimated it in inches. People appear to expect precision to be associated with accuracy. Those that have estimated their movement precisely expect to perform it accurately!

On each trial of their experiment, Harvey, Garwood & Palencia (1987) played subjects a musical interval and asked them to sing it. After each attempt, subjects used a seven-point scale to rate their confidence in having reproduced the target interval correctly. Two of their findings are of interest to us here. First, correlations between confidence and performance were close to zero: people could not monitor their own performance. However, confidence did increase over a sequence of 24 test trials. As there was no corresponding improvement in performance, it appears that this increase occurred because people expected practice at the task to improve their performance. Second, subjects who had previously practised the task actively by singing musical intervals themselves were more confident on the test trials than those who had just listened to the experimenter perform the task. However, as there was no corresponding difference between how well the two groups performed on the test trials, this effect must have arisen because the expected improvement from active (singing) practice was greater than the expected improvement from passive (listening) practice.

A third example of a between-groups effect that can be attributed to the groups' having different expectations about the effects of practice is provided by Adams & Goetz (1973, Experiment 2). Subjects who had more experience with a criterion stimulus did not perform any better but did have higher confidence in their performance. Presumably both groups started off with the same level of confidence and this (but apparently not memory for the criterion) was incremented as more experience was gained. These increments must have been based on the expected rather than on the actual effects of experience.

We saw in the last section that people have some ability to estimate the duration of their own movements. This ability increases over trials if they are given post-estimate information about the actual duration of their movements. In contrast, the work that I reviewed in this section reveals no evidence of a relationship between confidence in the correctness of a movement and how close a movement is to being correct. Also the relationship between confidence in correctness and length and type of practice seems to depend largely on people's expectations about the effects of practice rather than on the actual effects of practice. In cases in which no learning occurs (e.g. Adams & Goetz, 1973, Experiment 2; Harvey, Garwood & Palencia, 1987), these expectations produce "illusions of learning". Presumably people are susceptible to them whenever they assume a skill orientation. These illusions can be thought of as biases arising from use of heuristics that normally serve people well. Practice usually serves to improve performance and active learning is usually more effective than passive observation. In situations where people cannot monitor their own chances of success, they can obtain an estimate of them by using these heuristics. It is only in situations in which practice does not have its usual effects that the illusions appear.

Before going on to discuss a similar but better known type of illusion (Langer, 1975), it is worth mentioning that illusions of learning are not restricted to motor skills studied in the laboratory. Marteau et al. (1989) found that nurses with considerable practice in performing resuscitation were more confident but no better than those who had less practice at the skill. Marteau et al. (1990) replicated this finding on doctors.

14.1.3 Illusions of Control

Langer (1975) examined people's confidence of success in situations in which actual success was governed by chance alone. She found that confidence was affected by factors that could be expected to influence performance only when people have real control over events. People assume a skill orientation when it is inappropriate: they have an illusion of control.

Among the phenomena that she interpreted as illusions of control were the following. First, people betting on who will draw the higher card from a shuffled deck bet more when their opponent appears incompetent than when their opponent appears competent. Second, people require a higher price to sell lottery tickets that they have selected themselves than lottery tickets that they have been given. Third, in a motor task in which the correctness of the response depends on purely random factors, people are more confident in the correctness of their own responses than in those of an experimenter. They are also more confident when they have familiarized themselves with the apparatus. Langer argues that assessment of the competition, personal control over which response is made or selected and familiarity with apparatus are all factors that should affect confidence of success in skilled tasks. However, they could only have influenced confidence in her studies if people inappropriately adopted a skill orientation.

Langer argued that people are poor at discriminating controllable from uncontrollable events because chance is involved in many skilled tasks. Hence, heuristics appropriate for determining confidence in success in skilled tasks are overgeneralized to produce illusions of control. Langer pointed out that illusions of control (overgeneralization of a learnt dependence between actions and events) could be regarded as the converse of learned helplessness (overgeneralization of a learnt independence between actions and events). This latter phenomenon occurs when long experience with uncontrollable events is followed by a failure to attempt to influence controllable ones. People suffering from depression are known to exhibit learned helplessness (Seligman, 1975). If Langer is correct, they should not also be subject to illusions of control. Many studies have been done to test this prediction. In the majority of them, it has been borne out (e.g. Alloy, Abramson & Viscusi, 1981; Golin, Terrell & Johnson, 1977; Golin et al., 1979).

It is important to be aware of a conflict between the implicit and explicit definitions of illusions of control. It is the implicit definition that is used as a criterion for their presence. Specifically, illusions of control are taken to be present when subjects' confidence in their own performance in uncontrollable situations is influenced by factors that experimenters think should influence performance in controllable ones. This is quite different from their explicit definition: "An illusion of control is defined as an expectancy of a personal success probability inappropriately higher than the objective probability would warrant" (Langer, 1975, page 313). In fact, subjects are not asked for their estimates of their probability of success in experiments carried out within the illusion of control paradigm. It is possible (though admittedly unlikely) that Langer's subjects could have produced all the effects cited above without ever showing the overconfidence that the explicit definition requires.

Behaviour in line with the implicit definition tells us something about the grounds on which people are making their judgements about performance: it indicates that they are assuming that events depend on their actions. What are the circumstances in which adoption of this assumption will result in the overconfidence demanded by the explicit definition? If adopting a skill orientation results in people believing that probability of success is increased by their intervention, then overconfidence can be expected when people know what the probability of success would be if they did not intervene or if they responded randomly. However, in many tasks, people are not told what this probability is and cannot find it out: the structure of the task provides few cues, there is little opportunity to make random or null responses and no feedback is given. When probability of success without intervention is relatively high, range-frequency effects may cause people to underestimate it (Poulton, 1989). For example, they may estimate it to be 75% when it is actually 90%. If they then judge that their intervention will raise probability of success by 10% and if their control is completely illusory, they will be 5% underconfident.

14.1.4 Self-assessment of Performance: Overconfidence or Underconfidence

None of the work that I have discussed so far has allowed determination of whether people are underconfident or overconfident in their performance. However, some studies do allow us to say something about this matter. In this section, I shall deal with retrospective judgements; in the next section I shall cover prospective ones.

If people are overconfident in their performance, they may fail to make an effort to attend to or seek the KR that would inform them of the true state of affairs and that would help them improve their skills. Conversely, if they are underconfident, they may expend effort needlessly acquiring costly KR. It is hardly surprising that applied psychologists are interested in the validity of self-assessment (Mabe & Wells, 1982).

The procedures used to measure retrospective confidence in procedural knowledge are directly analogous to those used to measure confidence in declarative knowledge (e.g. Lichtenstein et al., 1982). The results are very similar, too. Harvey (1990a) asked people to use their judgement to alter the parameters of a dynamical system to bring its output into a target range. After each control response, people estimated the probability that it had been effective. Calibration analyses showed that people were overconfident and that this overconfidence increased as difficulty of the task increased.

When frequency estimation rather than probability estimation is used to measure confidence, underconfidence rather than overconfidence appears. Harvey (1988) required people to intercept 1000 targets in a step-tracking task. After every block of 100 targets, subjects estimated how many they had actually succeeded in hitting. They were 20% underconfident after the first block of trials but less than 10% underconfident after each of the last three blocks. This decrease in overconfidence occurred as people became better at the tracking task: i.e. as it became easier for them.

These findings of difficulty-dependent overconfidence with probability estimation and of underconfidence with frequency estimation are directly analogous to many that have been obtained when declarative knowledge has been assessed by multiple-choice general knowledge tests (e.g. Gigerenzer, Hoffrage & Kleinbölting, 1991; Lichtenstein, Fischhoff & Phillips, 1982; May, 1986; Sniezek & Switzer, 1989). Later I shall consider whether they are open to the same sort of theoretical explanations.

Before moving on, it is worth mentioning one paradoxical result. Harvey (1990b) asked people to intercept targets. After each interception, they had to decide whether or not they had been faster than on the previous trial and then give a probability estimate that this decision was correct. As they became better at the interception task, the difference between successive interception times decreased. This meant that their decisions about whether they were faster than on the previous trial became harder and that their overconfidence in those decisions increased. Thus, as people became better at their task, they became more overconfident in judgements of their performance. It would be interesting to discover how well this finding generalizes to other performance criteria.

14.1.5 Risk Assessment and Risk Preference

There are many definitions of risk (Pidgeon et al., 1992). For present purposes, I shall adopt a commonly accepted one: risk is the probability of undesired consequences. Given that failure is an undesired consequence of performance, someone who overestimates their probability of success in a task underestimates risk. It is useful to distinguish risk assessment from risk preference. In risk assessment experiments, we ask people to estimate their probabilities of success (or failure) under various conditions. However, they do not have to act on the basis of those assessments. In risk preference studies, we decrease risk until people say that they are willing to act (or else increase risk until they say that they are no longer willing to act). From our studies of risk assessment, we can then identify the subjective probability of failure associated with this point. This tells us what degree of risk people prefer.

Cohen, Dearnaley & Hansel (1956b) performed an experiment at a bus drivers' training school. The task was to drive an eight-ton double-decker bus between two wooden posts that were six feet (1.8 m) high. First, prospective confidence was measured by frequency estimation. The gap between the posts was initially much too narrow for the bus. It was increased in small steps. At each point, the driver estimated the number of times out of five that he could drive through the posts without touching them. Performance was then assessed by telling the drivers to drive through gaps of various sizes five times each. Figure 14.1 shows the results. Various points are worth noting. First, the classic finding of overconfidence except when the task was very easy appears again. However, in this case, it was obtained with a frequency estimation rather than a probability estimation task.

The second point to note is that, although more experienced drivers were more skilled, they were not significantly better at assessing how good their

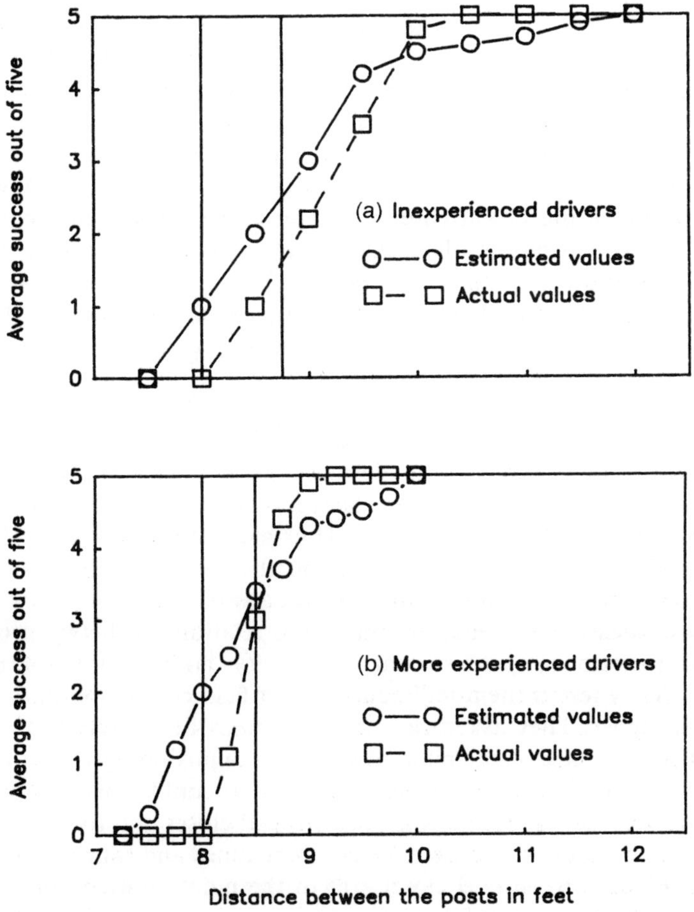

Figure 14.1 Estimated and actual frequencies of success (out of five attempts) at driving a bus through two posts placed different distances apart. The left-hand vertical lines indicate the width of the bus. The right-hand vertical lines indicate the average minimum distance between posts that drivers would voluntarily attempt when given a free choice. (Graphs are plotted from Cohen, Dearnaley & Hansel's (1956b) tabulated data. Reproduced by permission of the Operational Research Society.)

performance would be. Even an experienced driving instructor showed some degree of overconfidence. This aspect of Cohen, Dearnaley & Hansel's (1956b) results is consistent with the findings obtained for retrospective judgements by those testing Adams' theory (e.g. Marshall, 1972). Even when outcome feedback is given, practice has little effect (or acts very slowly) on the relationship between confidence and performance.

Cohen, Dearnaley & Hansel were interested in risk preference as well as risk assessment. What level of risk were drivers willing to adopt in practice? To find out, they determined the minimum gap through which drivers would voluntarily attempt to pass. The average size of these gaps is shown by the right-hand vertical lines in Figure 14.1. (The left-hand vertical lines show the actual width of the buses.) Overconfidence is again evident. Novice drivers were willing to attempt a gap that they later failed to drive through on an average of more than 50% of occasions. Actions of the more experienced drivers were still overconfident but the effect was reduced somewhat. The driving instructor did not attempt any gap that he did not later pass through on 100% of occasions. Cohen, Dearnaley & Hansel conclude that overconfidence is present in actions as well as in verbal estimates of performance. However, while practice has little effect on verbal estimates of riskiness, it reduces riskiness of actions. It affects risk preference rather than risk assessment.

In a similar experiment, Cohen, Dearnaley & Hansel (1958) found that drivers who had drunk alcohol attempted smaller gaps but required larger ones to succeed five times out of five. However, alcohol had no effect on the proportion of times that they estimated that they would successfully drive through a gap of the size that they attempted. Thus alcohol increased the riskiness of bus drivers' actions but had no effect on their verbal estimates of risk. It affected risk preference rather than risk assessment.

Cohen & Dearnaley (1962) examined the relationship between confidence and performance in professional, university and school football team members. They tested them individually on a field containing only the opposition goalkeeper. They asked each player to start in his half of the field and walk down the centre of the pitch towards the opposition goal until he was at the distance from it at which he first felt that he could score once out of 100 attempts. Next, he was told to move closer and closer and to specify when he first felt that he could score once, twice, three times and four times out of five attempts. Finally, he moved closer still to the point at which he felt that he could score 99 times out of 100 attempts.

Afterwards, each player made five attempts to score from five of the six estimated distances. (The authors do not report data for the distance from which subjects estimated that they had only a one in 100 chance of success.) Number of successes at each point was recorded. As Figure 14.2 shows, players were quite accurate in estimating their performance. However, they

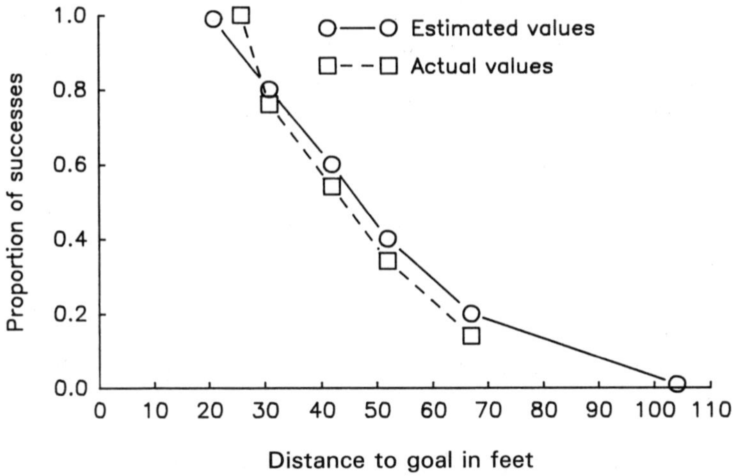

Figure 14.2 Estimated and actual proportions of success at shooting balls into an opponent's goal from different distances. No data were collected for actual performance at the distance estimated to produce success one in every 100 attempts. (Graphs are plotted from Cohen & Dearnaley's (1962) tabulated data. Reproduced from the *British Journal of Psychology* by permission of The British Psychological Society.)

were generally about 5% overconfident except when their task was very easy: again, the classic finding appears but with frequency rather than probability estimation.

As in their previous experiments, Cohen & Dearnaley (1962) were interested not just in verbal estimates of risk but also in the risks that people are actually willing to take. Hence, before the performance assessments just described, they asked each player to run with the ball at his feet towards the opposition goal and to shoot at the first appropriate moment. This was the longest shot that a player would voluntarily attempt. Cohen and Dearnaley examined two indices. The first, which they termed the margin of hazard, was the difference between the longest shot that a player would voluntarily attempt and the longest distance at which he succeeded in scoring on five out of five attempts in the later performance assessment. The second index, which they called the margin of safety, was the difference between the longest shot that a player would voluntarily attempt and the longest distance from which he estimated that he would succeed in scoring on 99 out of 100 attempts. They found that players' margins of hazard varied significantly across teams from which they were drawn but that margins of safety did not do so. Thus, again, a factor that affected riskiness of actions did not appear to influence verbal estimates of risk. Risk preference varied but risk assessment did not.

We have seen that certain variables (practice, drug dosage, group membership) affect riskiness of actions without influencing verbal estimates of risk.

These dissociations imply that the underlying psychological processes differ in some way. Verbal estimates just reflect ability to assess probabilities of success and failure under various conditions whereas the action-based measures reflect both this and decisions about the levels of acceptable risk under these conditions.

14.1.6 Theories of Overconfidence in Procedural Knowledge

Overconfidence except when tasks are very easy is not a phenomenon that is restricted to tests of procedural knowledge. It also occurs when declarative knowledge is tested in multiple-choice questions (Lichtenstein, Fischhoff & Phillips, 1982). Are theories that have been developed in that context relevant to self-assessment of skilled performance? Some appear to be more applicable than others.

Poulton (1989) has proposed that overconfidence with difficult items and underconfidence with easy ones arise because people are reluctant to use the extremes of probability scales. This suggests that the direction of the bias should switch around the centre of the scale that they use for their assessment. However, the switch generally seems to occur much closer to the top end of the scale than this (cf. Figures 14.1 and 14.2).

Although Poulton's basic model appears to be inadequate, Ferrell & McGoey (1980) have developed a signal detection model that is also based on the notion that people take insufficient account of differences in item difficulty when using the probability scale to express their feelings of certainty. Until recently (McClelland & Bolger, Chapter 18, this volume), relatively little attention has been paid to their model. However, taken in conjunction with Adams' (1971) proposal that people have an internal representation of the sensory feedback associated with correct performance, it appears to have considerable potential for being developed into an account of judgements of confidence in skilled behaviour. However, the result of any such theoretical confluence would be more appropriate for explaining retrospective than prospective judgements.

Gigerenzer, Hoffrage & Kleinbölting (1991) have proposed that people answer questions by searching through potentially relevant probabilistic cues until they discover one that can be activated. After using it to answer the question, they express their confidence as equal to the validity of the cue. For example, suppose that people must decide which of two German cities has the larger population. First they test the highest validity cue. This might specify that if one of the cities is in the area that used to be East Germany but the other is in what used to be West Germany, then there is an 85% probability that the latter city has the larger population. However, both cities in question are in West Germany. Consequently, this cue cannot be activated and the next most valid one must be tested. This might specify that if one city has a team in the German football league but the other does not, then there is an 80%

probability that the former has the larger population. The person answering the question knows that only one city does, in fact, have a team in the league. Consequently, the cue is activated: that city is given as the answer with an estimated probability of being correct of 80%.

Gigerenzer, Hoffrage & Kleinbölting argue that people are well adapted to their environments: internal cue validities approximate closely to their corresponding environmental probabilities. Overconfidence obtained in experiments is an artefact. It has arisen because experimenters have not selected alternatives in multiple-choice questions in a representative manner. Only this way of choosing will produce correct answers with probabilities equal to their cue validities. However, instead of doing this, experimenters have chosen alternatives in a way that causes activated cues to fail more often than would be expected on the basis of their validity. For example, alternatives in questions about German city populations have been chosen so that the football league cue fails on 40% rather than on 20% of occasions on which it is used. The resulting 20% overconfidence reflects a question-setting bias of the experimenter rather than a question-answering bias of the subject. Support for Gigerenzer et al.'s theory comes from experiments that have shown that selecting the alternatives in multiple-choice questions in a representative manner causes overconfidence to disappear (e.g. Juslin, 1993, 1994).

Can Gigerenzer et al.'s theory explain the overconfidence that Cohen and his colleagues found to be present in self-assessments made by bus drivers and football players? Although the experiments were performed outside the laboratory in contexts that had some degree of ecological validity, people had fewer cues to help them judge the difficulty of their task than they would have had when really practising their skill. Footballers made their estimates while stationary and with the opposition goalkeeper as the only other player on the field. Bus drivers made their judgements while stationary and in an off-street setting. However, a random reduction in the number of cues could only be expected to make self-assessment worse; there appears to be no reason to suppose that it would introduce a bias towards overconfidence. Subjects deprived of their highest validity cue would use one with a lower validity but the effectiveness of this substitute cue would not be distorted. At the very least, we must conclude that Gigerenzer et al.'s theory requires some elaboration if it is to account for overconfidence in skilled performance.

Various theories have been proposed that explain overconfidence in declarative knowledge in terms of a bias in the assessment of arguments for and against the chosen alternative. For example, Koriat, Lichtenstein & Fischhoff (1980) argued that people first select an alternative answer and then search for arguments in favour of it (or against the rejected one) in order to produce their confidence judgement. Because they seek to confirm but not disconfirm their own choices, overconfidence is produced. More recently, Griffin & Tversky (1992) have explained overconfidence in terms of a tendency to take into account strength of evidence (e.g. proportion of arguments in favour of the chosen alternative) while paying too little attention to its weight (e.g. total number of arguments being considered).

These ideas can be extended to account for overconfidence in procedural knowledge. For example, someone may feel that they have a high chance of successfully skiing down a black run because they emerged in one piece when they tried it previously. However, they may fail to take account of the fact that this judgement is based on just one previous experience that took place when they were younger and fitter. They may also ignore the many falls that they have suffered when skiing down easier runs. However, although this approach appears to have potential, it is not one that has been developed in the context of skilled performance. Perhaps formulating precise predictions from it or devising good tests of them is particularly difficult in this domain. For example, the implicit, non-verbalizable nature of procedural knowledge may frustrate people's attempts to express their reasons for having confidence in it.

Another explanation of overconfidence attributes it to unrealistic optimism (e.g. Weinstein, 1980, 1989) or wishful thinking (e.g. Babad, 1987; Harvey, 1992). In other words, it is caused by a generalized tendency to overestimate the probability of positive (favourable) events and to underestimate the probability of negative (unfavourable) ones. Proponents of this view do not see the bias as restricted to events over which people feel that they have control. Overconfidence in skilled performance is just an example of a much more general phenomenon that can also be observed in many situations in which people cannot expect to influence outcomes. For instance, Babad (1987) asked football supporters to predict outcomes of matches. The more strongly affiliated they felt to a team, the more likely they felt that it would win.

Attributing overconfidence to an unrealistic optimism bias begs questions about the mechanisms responsible for producing the bias. It could be just programmed into us by evolutionary processes because its advantages outweigh its disadvantages (Taylor & Brown, 1988). Alternatively, it could be generated by the sort of argument recruitment processes proposed by Koriat, Lichtenstein & Fischhoff (1980).

Finally, overconfidence may arise because of the control that people perceive they have over outcomes in tasks that they perform (Howell, 1971). We overestimate the power of our intellectual abilities (Dawes, 1980) and our skills (Svenson, 1981). This leads to overconfidence in tests of declarative and procedural knowledge, respectively. But why do we overestimate our intellectual capabilities and skills? Wright and Wishuda (1982) argued that intelligence and knowledge are highly regarded by others and that there is therefore a positive social utility in expressing certainty in their products.

Recently, McKenna (1993) has reported experiments designed to determine whether overconfidence in skilled performance is best explained by a general unrealistic optimism bias or by the perceived control that people have over outcomes. First, he found that drivers think that they are less likely than the average driver to experience an accident when they are driving themselves but as likely as the average driver to suffer one when they are passengers. Second, he found that this effect was strong and significant in high-control scenarios (e.g. "assess the likelihood of an accident in which the car that you are driving hits the car in front") but small and insignificant in low-control scenarios (e.g. "assess the likelihood of an accident caused by another vehicle hitting the car you are in from behind"). These findings suggest that overconfidence in skilled performance is better explained in terms of perceived control than in terms of a generalized bias towards unrealistic optimism.

Neither wishful thinking nor perceived control explains the underconfidence that is typically observed with very easy tasks. Jones (1977) argues that it can be interpreted as anticipatory face-saving; it would be humiliating to fail very easy tasks and so people try to view them as less easy than they are. As objective probability of success increases, the strength of this bias rises and that of opposite biases (e.g. wishful thinking) declines. At any given level of task difficulty, opposing biases summate algebraically to determine whether overconfidence or underconfidence is present.

14.2 DOES CONFIDENCE INFLUENCE PERFORMANCE?

Work relevant to this question emanates from research on levels of aspiration, work motivation and goal-setting. I shall discuss each of these issues in turn.

14.2.1 Levels of Aspiration

When different versions of a task vary in difficulty level, a performer must decide which version to attempt. The version that is chosen defines the performer's level of aspiration. What determines level of aspiration? How does it depend on confidence of success? Psychologists have studied this issue for over 50 years (e.g. Festinger, 1942; Lewin et al., 1944). Because of its relevance to entrepreneurship and economic development (e.g. McClelland, 1955), they have been primarily interested in whether consistent individual differences in level of aspiration exist and, if they do, whether they can be explained in terms of personality variables (Atkinson, 1957, 1964; Sorrentino, Hewitt & Raso-Knott, 1992).

As Feather (1959a) pointed out, early analyses of level of aspiration interpreted choice of task difficulty as a process akin to maximization of subjective expected utility (SEU). Lewin et al. (1944) suggested, not unreasonably, that subjective probability of success decreases as task difficulty increases. More controversially, they also argued that the valence (utility) of success increases as task difficulty increases. Because of this latter claim, their theory implies that utility and subjective probability of success are inversely related and, therefore, that their product (SEU) is at a maximum when subjective probability of success is 0.5. In the absence of other externally set rewards, people should be most likely to choose tasks in which they feel that they have an evens chance of success.

Various experiments confirmed that the attractiveness of succeeding in a task is inversely related to subjective probability of success (e.g. Feather, 1959b) and provided some general support for this SEU type of approach. For example, Litwin (cited in Whiting, 1979) performed the following experiment in 1958. People took part in a game in which they had to throw rings over a peg from different distances (up to a maximum 4.6 m (15 ft)). Before actually throwing the rings, they made two sets of judgements. First, they had to estimate how many times out of 100 attempts an average player would achieve success at each of the six distances. Second, they had to estimate how much money (out of a maximum of $1.00) would be suitable as a prize for a successful throw from each distance. After making these judgements, they threw the ring 400 times. The proportion of times that each distance was chosen was well predicted both by the product of subjective probability of success and its complement and by the product of subjective probability of success and the value of the prize that had been associated with success.

Litwin also reported the actual probability of success at each distance. As was to be expected from the work of Cohen and colleagues, people were highly overconfident; objective probability of success was about 25% lower than its subjective counterpart. If people modify their prospective confidence on the basis of feedback, this overestimation should lessen with practice in which KR is provided. In Litwin's task, a shorter distance should be associated with a perception of an evens chance of success. Practice should reduce the difficulty of the task that people decide to attempt.

Results from various experiments are consistent with this prediction. For example, Hamilton (1974) found that the average distance that people attempted in their first 10 throws in Litwin's task was 3.5 m (11.5 ft). He then gave them 10 trials' practice at each of 13 different distances. For each subject, he calculated the distances that were associated with different probabilities of success (0.1 to 1.0 in 0.1 steps). Markers containing this probability information were placed at the appropriate distances from the thrower. Each subject then selected distances for a final 10 attempts at the task. The average distance chosen for these attempts was 3 m (9.9 ft) significantly less than that for the first 10 attempts. So, in line with Cohen, Dearnaley & Hansel (1956b), practice reduced riskiness of actions.

Task performance can improve because skill learning occurs. How does this (and knowledge of it as a possibility) affect level of aspiration? This is an issue that has not been adequately researched. In practice, study of the effect of performance improvement on level of aspiration has proved difficult. One problem is in demonstrating that learning has occurred when there are simultaneous changes in the versions of the task that people attempt. For example, Hamilton (1974) found that probability of success increased from 0.19 at the start of his experiment up to 0.39 after practice. He argued that this improvement was produced not only because people selected simpler tasks but also because they had learnt to perform better. To support this assertion, he weighted each success by the distance at which it was obtained and then totalled up the results to obtain an overall score. He concluded that learning had occurred because this score was 17.5 before practice and 29.3 afterwards. The difficulty with this procedure is that it involves assuming that doubling the distance of a throw makes it twice as hard. In fact, Hamilton's own data suggest that, within the range 0.9–3.6 m (3–12 ft), doubling the distance makes the task four times as hard. Given this, it seems unlikely that any learning occurred in his study.

Most of the research on level of aspiration has focused on individual differences. Atkinson (1957, 1964) proposed that there is a subset of people who do not select from task variants by choosing one with a subjective probability of success close to 0.5 and that these people are motivated more by a fear of failure than by a desire for success. To avoid the anxiety and social humiliation associated with failing in a task in which there is a chance of success, they adopt one of two strategies. They either attempt very easy tasks that they have negligible chance of failing or else they tackle extremely difficult tasks. In the latter case, no humiliation is associated with failure because no-one can really be expected to succeed.

Many experiments set out to test whether people scoring higher in a test of fear of failure than in a test of need for achievement do respond in the way that Atkinson suggested. The results of this work have been reviewed many times (e.g. Meyer, Folkes & Weiner, 1976; Weiner, 1980; McClelland, 1987). The consensus is clear. Failure-oriented people (who are more concerned about failure than about success) respond in a similar way to success-oriented people (who are more concerned about success than about failure). In both cases, they prefer tasks that they perceive to be of intermediate difficulty.

There have been two broad classes of response to the failure of Atkinson's predictions. The first was based on observations from some of the experiments suggesting that levels of aspiration of success-oriented and failure-oriented individuals might still differ in certain consistent ways. For example, there were hints that the curve relating strength of preference to task difficulty might have less kurtosis or more skew in failure-oriented individuals (e.g. De Charms & Davé, 1965). As a consequence, the original theory was elaborated (Kuhl, 1978; Raynor, 1969) and reformulated (Atkinson & Birch, 1970; Kuhl and Blankenship, 1979a) to produce new predictions about how the two groups' levels of aspiration should differ. Unfortunately, these new predictions have proved to be much more difficult to subject to convincing experimental tests (e.g. Kuhl & Blankenship, 1979b, page 561).

The second type of response was more radical. Trope & Brickman (1975) challenged the view that accomplishing more difficult tasks is inherently more rewarding. They felt that the SEU style of analysis described above was inappropriate. Instead they argued that people select level of task difficulty to maximize the information that task outcomes provide about their level of ability. They show how a Bayesian analysis of this selection process can explain why people prefer tasks of intermediate difficulty.

Suppose that people with high (H) ability and people with low (L) ability perform a task that has only two possible outcomes, success (S) and failure (F). For one task variant, $P(S \mid H) = 0.97$ and $P(S \mid L) = 0.80$. For the other, $P(S \mid H) = 0.52$ and $P(S \mid L) = 0.48$. Here the first task variant provides a better means of distinguishing ability levels. It is more diagnostic.

Very easy and very difficult task variants cannot be so diagnostic as intermediate-difficulty task variants. In general, $P(S) = P(S \mid H)P(H) + P(S \mid L)P(L)$. Suppose that people with low and people with high ability are equally common and contribute equally to the overall determination of difficulty of task variants. Initial probabilities of high and low ability will then both be 0.5. This means that initial probabilities of success must be symmetrical about the overall difficulty level. Clearly $P(S \mid H)$ and $P(S \mid L)$ can (and usually would) be further apart when they have to average out at 0.5 than they can be when they have to average out at 0.9 or 0.1. People tend to select task variants in which they have an evens chance of success because these situations are most diagnostic of their ability.

To test their analysis, Trope & Brickman (1975) contrived a situation in which the hard and easy task variants were more diagnostic than the one of intermediate difficulty. In these conditions, the latter alternative was the least favoured rather than the most favoured task variant. They concluded that people were motivated by a desire to find out about their own abilities rather than by the reward inherent in success. Work by Trope (1975) and Meyer, Folkes & Weiner (1976) provided additional support for their view.

Given these findings, we might expect individuals who are motivated to discover new things about themselves and their environment (uncertainty-oriented people) to behave differently from those who ignore or avoid new or inconsistent information about themselves and their environment (certainty-oriented people). Sorrentino, Hewitt & Raso-Knott (1992) used personality tests to select subjects for their experiment. They accepted only those who scored at the extremes of the uncertainty-oriented/certainty-oriented and success-oriented/failure-oriented continua. They tested the resulting four groups on Litwin's ring-throwing task. Those who were uncertainty-oriented showed the classic preference for intermediate-difficulty task variants. This preference was somewhat weaker in failure-oriented than in success-oriented people. Those who were both certainty-oriented and failure-oriented displayed a complete reversal of the classic pattern; task variants of intermediate difficulty were the least preferred. Sorrentino et al. suggest that an effect of success-orientation/failure-orientation had not been previously found because researchers testing Atkinson's theory did not separate out certainty-oriented and uncertainty-oriented people and the majority of those whom they tested were uncertainty-oriented.

Sorrentino et al. found that the two personality variables that they examined had additive effects on performance. This suggests that people separately assess the difficulty of a task variant and the diagnosticity of its possible outcomes. Do they take anything else into account when selecting their level of aspiration? There may be situations in which high levels of aspiration are rewarded quite independently of the quality of the performances that follow them. For example, Damm (1968) argued that people may publicly set high levels of aspiration for themselves because it is socially desirable to appear competitive and ambitious.

To examine effects of rewarding level of aspiration rather than performance, Feather (1964) studied people in a card-sorting task. On each trial, they first had to estimate how many cards they would be able to sort within the time limit. Their performance was successful when the number of cards that they sorted equalled or exceeded their estimate. When points for success equalled their estimate but points for failure equalled the number of cards they actually sorted, level of aspiration was set high. It exceeded performance; people appeared to be overconfident. However, when the points that they received equalled the number of cards that they actually sorted irrespective of their success or failure, level of aspiration was much lower. Overconfidence appeared much reduced. (By giving people zero points when they failed, Feather could reduce level of aspiration still more; people appeared to be underconfident.)

Via its effects on judged difficulty and diagnosticity of task variants, subjective probability of success influences level of aspiration. However, it is clear from the work of Feather (1964) and others (e.g. Smith, 1963) that level of aspiration is affected by other factors as well. For a given subjective probability of success, it may be raised or lowered (Kuhl, 1978). We know that these factors relate to personalities of individual performers but they are also likely to relate to task characteristics. In some tasks, it is important to act even though risk is high; in others, it is better to withhold action in such conditions. To select an appropriate level of aspiration for a given task, people must learn both the risk levels associated with each task variant and the level of risk that it is most appropriate to adopt given the circumstances. Cohen, Dearnaley & Hansel's (1958) research on bus drivers (discussed above) suggests that these different types of information are acquired separately and that, in experts, the latter may compensate for errors in the former.

14.2.2 Expectancies in Work Motivation

Expectancy theories of work motivation adopt an SEU type of approach to explain the amount of effort that people put into occupational tasks (e.g. Campbell et al., 1970; Galbraith & Cummings, 1967; Georgopolous, Mahoney & Jones, 1957; Vroom, 1964). These theories and the empirical studies designed to test them have been the subject of frequent reviews (e.g. Campbell & Pritchard, 1976; Heneman & Schwab, 1972; Mitchell, 1974, 1979; Schwab, Olian-Gottlieb & Heneman, 1979).

People are generally assumed to have internal representations of the relationships between effort and performance, between performance and outcomes and between outcomes and utilities. They make use of them when deciding how much effort to put into their work. The effort–performance relationship is assumed to be probabilistic in nature; it is usually measured by asking subjects for their estimates of the probabilities that various different performance levels will be met at each effort level. Nowadays the performance –outcome relationship is often assumed to be probabilistic as well. Hence it can be measured in a similar way; subjects can be asked for their estimates of the probabilities that various different outcomes will arise from each of a number of levels of performance.

With, say, six levels of effort, six levels of performance and four levels of outcome, this procedure gives two subjective probability matrices of 36 and 24 cells for each subject. Given the amount of data, composite scores that reduce each matrix to a single value are often used to test expectancy theory predictions about the overall effects of critical independent variables. For example, Ilgen, Nebeker & Pritchard (1981) wanted to study the effects of switching from a flat-rate to an incentive-payment system on people's perceptions of the effort–performance and performance–outcome relationships in a clerical task. They extracted an expected value score (E) from each matrix, where

$$E = \frac{\sum_{i=1}^{6} R_i \sum_{j=1}^{m} r_{ij} C_j}{\sum_{i=1}^{6} \sum_{j=1}^{m} r_{ij} C_j}$$

Here R_i is the ith performance level (i.e. 11, 15, 19, 23, 27 and 31 blocks of work per hour); C_j is the jth effort level (i.e. 30, 36, 42, 48, 54 and 60 minutes of work per hour) or the jth outcome level (i.e. one of four levels of pay); m is the number of effort or outcome levels (six and four, respectively); and r_{ij} is the subjective probability that the ith level of performance would be associated with the jth level of effort or outcome.

Tests of predictions from expectancy theories have produced some findings that appear relevant to the issue of how confidence influences performance.

However, a number of factors make their interpretation difficult. First, actual performance is not always assessed. Instead, changes in performance due to learning or increased effort are often hypothesized as ways of explaining the effects of independent variables on subjective probabilities. Second, as we have seen, results are often reported as composite scores (e.g. Dachler & Mobley, 1973; Ilgen, Nebeker & Pritchard, 1981). Changes in these scores can result from different types of changes in the factors that make them up. Different types of underlying change have different implications for models of the interaction between confidence and performance. However, the additional analyses required to identify the nature of the underlying change in composite scores are seldom reported.

14.2.3 Goal-setting

Mace (1935) found that people told to reach specific scores on each trial of a task improved much faster than those who were just told to do their best. In an experimental bargaining task Siegal & Fouraker (1960) found that people with a high level of aspiration at the start of the bargaining process actually achieved higher profits in the end. Feather's (1964) manipulations, which changed level of aspiration, also appear to have moved performance in the same direction. In all these studies, people performed better when they had more difficult goals to meet.

The potential importance of these findings for work motivation in organizational settings has been appreciated. Over the past 25 years, a large corpus of research on goal-setting effects has accumulated. (For reviews, see Latham & Locke, 1991; Locke & Latham, 1990; and Locke et al., 1981.) Some broad generalizations can be extracted from it. Given that level of ability is controlled, performance generally increases linearly with task difficulty. Specific difficult goals lead to better performance than goals that are easier or vaguer. Such goals serve to direct attention, increase effort and motivate people to develop more efficient task strategies. In some domains (e.g. sport), evidence for goal-setting effects is currently equivocal; this may be because of methodological problems (Locke, 1991) or because motivational processes are idiosyncratic in these areas (Weinberg, 1992; Weinberg & Weigand, 1993).

Goal-setting theory does not address the issue of the relationship between confidence and performance directly. It predicts that "harder goals lead to better performance than easy goals, despite their lower probability of being reached" (Locke et al., 1981). The reference here is to the objective probability of success. (In goal-setting studies, people are usually provided with explicit information about the difficulty of their task. This is often done by telling them what proportion of their peers have succeeded in the past.) Confidence only enters the picture because the above prediction is not always fulfilled. For example, Motowidlo, Loehr & Dunnette (1978) found a curvilinear relationship between performance and objective probability of success. Maximum performance occurred at an intermediate rather than at the lowest probability level.

To account for such findings, goal-setting theory includes assumptions about goal commitment. Any downturn in performance at high difficulty levels occurs because people have decided that the goals that are very difficult to reach are not worth attempting. When an individual's subjective probability of success drops below some personal criterion value, goal commitment declines. Consequently, performance no longer increases with task difficulty but starts to decrease. Locke (1969) showed that the effort that people put into a task generally decreased when their subjective probability of success dropped below 0.5. However, for a subgroup of people who said that they were still trying to succeed despite low subjective probabilities of success, this decline in effort was not statistically significant.

Locke & Latham (1990, pages 83–5) point out that this analysis is consistent with level of aspiration studies that show that people prefer (i.e. are committed most to) task variants that they perceive to be of intermediate difficulty. Of course, it does assume that judgements of subjective probability of success are made before goal commitment decisions. In other words, these judgements represent people's estimates of the probability that they would succeed given that they were trying as hard as they could. This appears to have implications for the relationship between subjective and objective probabilities of success.

Suppose that people would be well calibrated if they actually tried as hard as they could at all difficulty levels. However, on the basis of their subjective probability judgements, they decide not to be fully committed to very difficult versions of the task. For these versions only, objective probabilities of success would be depressed below their subjective counterparts. (Subjective probabilities could not reflect the depression of objective ones because they are used as a basis for the commitment decisions causing the depression.) Consequently, overconfidence would be observed for these difficult versions of the task but for no others. If we assume that commitment is a matter of degree rather than all-or-none and that people become less committed as tasks become more difficult, then we can expect overconfidence to increase as difficulty increases. Unwittingly, Locke and Latham have provided us with another explanation of this classic finding (e.g. Cohen et al., 1956b).

When it has been assessed, the relationship between performance and subjective probability of success has been found to be either positive (Arvey, 1972; Motowidlo, Loehr & Dunnette, 1978) or absent (Mento, Cartledge & Locke, 1980). As Locke & Latham (1990, pages 66–7) point out, these findings do not have to be seen as conflicting with those supporting goal-setting. Higher-performing (higher-goal) groups have lower expectations of success than lower-performing (lower-goal) groups. However, within each goal-specific group, people with higher expectations could still perform better. The overall correlation between performance and subjective probability for all subjects in all groups would depend on the relative contribution of these between-group and within-group effects.

We have seen that those working on goal-setting have examined the relationship between performance and objective probability of success and between performance and subjective probability of success. However, the relationship between objective and subjective probability of success has not been one of their concerns. Only correlations between the base-rate probabilities (provided to subjects as difficulty information) and subjective probabilities of success are calculated as manipulation checks. I suggested above that overconfidence may be relevant to goal-setting theories. However, it has not been recognized as such and, therefore, it has not been measured.

14.3 CONCLUSION

We have seen that measures of confidence in research on skilled behaviour have been used to test theories of motor learning, risk-taking, work motivation and individual differences in levels of aspiration. The main point that I want to emphasize here is that people working in each of these areas have paid relatively little attention to the findings produced by colleagues researching the other domains. Those studying motor learning have rarely considered the relevance of overconfidence biases to their work; those interested in risk assessment and preference have seldom considered the importance of research on goal-setting and levels of aspiration; those working in these latter two areas do not appear to have taken full account of the possibility that their subjects are overconfident and learn during experiments.

The types of study that we have considered fall into two broad classes. The first is concerned with how well confidence reflects performance (past or future). This issue is similar to the one addressed by those working on calibration of confidence in declarative knowledge (e.g. Lichtenstein, Fischhoff & Phillips, 1982; McClelland & Bolger, Chapter 18, this volume; Keren, 1991). However, we have seen that it is not always easy to transfer theories from this domain to the procedural one. There are various reasons for this. Those working on confidence in declarative knowledge rarely take account of factors that are known to influence skilled behaviour: e.g. the possibility of learning during experimental sessions; the degree of awareness of any such learning; potential effects of fatigue.

The second type of study differs from the first type in two important ways. First, variations in confidence are seen as causes rather than effects of variations in performance level. Second, theoretical concerns focus on motivational factors (effort, aspiration, arousal, fatigue) rather than on cognitive ones (cue validities, memory traces, judgemental biases). This type of study includes the work that we have reviewed in the last three sections.

These two types of study remain conceptually separate. People working on each one make implicit but unreasonably strong assumptions about the other. For example, cognitive psychologists employing confidence as a dependent variable typically assume that people are as motivated to perform difficult tasks as easy ones. Conversely, motivational psychologists employing confidence as an independent variable typically assume that no learning occurs during experiments and that individuals' internal assessments of factors such as goal difficulty and goal acceptance are explicit (i.e. conscious and verbalizable).

Theoretical integration of cognitive and motivational substrates of behaviour is currently one of the major developments within psychology (e.g. Oatley & Johnson-Laird, 1987; Ortony, Clore & Foss, 1987; Sloman, 1987). In this chapter I have contributed little to that integration but I hope that I have been able to show that there is great scope for it within the area that I have covered. Some interaction across the cognitive-motivational divide is already taking place. For example, Henry & Sniezek (1993) have suggested that subjective probabilities given as prospective judgements of confidence in an information-retrieval task could themselves act as self-set goals. Horgan (1992) has argued that changes in these judgements in response to previous successes and failures can be used to characterize individual motivational styles.

Neither performance nor confidence should be thought of solely in terms of cause or solely in terms of effect. We should think of them as part of a unitary dynamical system that produces a pair of time series as output. It may be possible to model the system and produce computer simulations of its output series. These could be compared with real series produced by people. Given that previous dynamic models of the performance series alone have proved hard to test (Kuhl & Blankenship, 1979a), this more ambitious endeavour would certainly be a challenging one. However, the foundations for it have already been laid (e.g. Kuhl, 1986).

ACKNOWLEDGEMENTS

I would like to thank Peter Ayton and Fergus Bolger for their helpful comments on an earlier draft of this chapter.

REFERENCES

Adams, J.A. (1971) A closed-loop theory of motor learning. *Journal of Motor Behavior*, **3**, 111–50.

Adams, J.A. & Goetz, E.T. (1973) Feedback and practice as variables in error detection and correction. *Journal of Motor Behavior*, **5**, 217–24.
Adams, J.A., Goetz, E.T. & Marshall, P.H. (1972) Response feedback and motor learning. *Journal of Experimental Psychology*, **92**, 391–7.
Alloy, L.B., Abramson, L.Y. & Viscusi, D. (1981) Induced mood and the illusion of control. *Journal of Personality and Social Psychology*, **41**, 1129–40.
Arvey, R.D. (1972) Task performance as a function of perceived effort-performance and performance-reward contingencies. *Organizational Behavior and Human Performance*, **8**, 423–33.
Atkinson, J.W. (1957) Motivational determinants of risk-taking behavior. *Psychological Review*, **64**, 359–72.
Atkinson, J.W. (1964) *An Introduction to Motivation*. Van Nostrand, Princeton, NJ.
Atkinson, J.W. & Birch, D. (eds) (1970) *The Dynamics of Action*. Wiley, New York.
Babad, E. (1987) Wishful thinking and objectivity among sports fans. *Social Behavior: An International Journal of Applied Social Psychology*, **4**, 231–40.
Campbell, J.P. & Pritchard, R.D. (1976) Motivation theory in industrial and organizational psychology. In M.D. Dunnette (ed.), *Handbook of Industrial and Organizational Psychology*. Rand McNally, Chicago.
Campbell, J.P., Dunnette, M.D., Lawler, E.E. & Weick, K.E. (1970) *Managerial Behavior, Performance and Effectiveness*. McGraw-Hill, New York.
Cohen, J. & Dearnaley, E.J. (1962) Skill and judgement of footballers in attempting to score goals: A study of psychological probability. *British Journal of Psychology*, **53**, 71–86.
Cohen, J., Dearnaley, E.J. & Hansel, C.E.M. (1956a) The addition of subjective probabilities: The summation of estimates of success and failure. *Acta Psychologica*, **12**, 371–80.
Cohen, J., Dearnaley, E.J. & Hansel, C.E.M. (1956b) Risk and hazard: Influence of training on the performance of bus drivers. *Operational Research Quarterly*, **7**, 67–128.
Cohen, J., Dearnaley, E.J. & Hansel, C.E.M. (1957) Measures of subjective probability: Estimates of success in performance in relation to size of task. *British Journal of Psychology*, **48**, 271–5.
Cohen, J., Dearnaley, E.J. & Hansel, C.E.M. (1958) The risk taken in driving under the influence of alcohol. *British Medical Journal*, **1**, 1438–42.
Cohen, N.J. & Squire, L.R. (1980) Preserved learning and retention of pattern-analysing skills in amnesia: Dissociation of knowing how from knowing that. *Science*, **210**, 207–10.
Dachler, H.P. & Mobley, W.H. (1973) Construct validation of an instrumentality-expectancy-task-goal model of reward motivation: Some theoretical boundary conditions. *Journal of Applied Psychology*, **58**, 397–418.
Damm, J. (1968) Effects of interpersonal contexts on relationships between goal setting behaviour and achievement motivation. *Human Relations*, **21**, 213–26.
Dawes, R.M. (1980) Confidence in intelligence judgments vs confidence in perceptual judgments. In E.D. Latermann and H. Feger (eds.), *Similarity and Choice*. Hans Huber, Bern.
De Charms, R. & Davé, P.N. (1965) Hope of success, fear of failure, subjective probability and risk-taking behavior. *Journal of Personality and Social Psychology*, **1**, 558–68.
Feather, N.T. (1959a) Subjective probability and decision under uncertainty. *Psychological Review*, **66**, 150–64.

THE LIBRARY
BOLTON INSTITUTE OF HIGHER
EDUCATION-DEANE CAMPUS

Feather, N.T. (1959b) Success probability and choice behavior. *Journal of Experimental Psychology*, **58**, 257–66.
Feather, N.T. (1964) Level of aspiration behaviour in relation to payoffs and costs following success and failure. *Australian Journal of Psychology*, **16**, 175–84.
Ferrell, W.R. & McGoey, P.J. (1980) A model of calibration for subjective probabilities. *Organizational Behavior and Human Performance*, **26**, 32–53.
Festinger, L. (1942) Wish expectation and group standards as factors influencing the level of motivation. *Journal of Abnormal and Social Psychology*, **37**, 184–200.
Galbraith, J. & Cummings, L.L. (1967) An empirical investigation of the motivational determinants of task performance: Interactive effects between instrumentality-valence and motivation-ability. *Organizational Behavior and Human Performance*, **2**, 237–57.
Georgopolous, B.S., Mahoney, G.M. & Jones, N.W. (1957) A path-goal approach to productivity. *Journal of Applied Psychology*, **41**, 34–53.
Gigerenzer, G., Hoffrage, U. & Kleinbölting, H. (1991) Probabilistic mental models: A Brunswikian theory of confidence. *Psychological Review*, **98**, 506–28.
Golin, S., Terrell, T. & Johnson, B. (1977) Depression and the illusion of control. *Journal of Abnormal Psychology*, **86**, 440–24.
Golin, S., Terrell, T., Weitz, J. & Drost, P.L. (1979) The illusion of control among depressed patients. *Journal of Abnormal Psychology*, **88**, 454–7.
Griffin, D. & Tversky, A. (1992) The weighing of evidence and the determinants of confidence. *Cognitive Psychology*, **24**, 411–35.
Hamilton, J.O. (1974) Motivation and risk taking behavior: A test of Atkinson's theory. *Journal of Personality and Social Pychology*, **29**, 856–64.
Harvey, N. (1988) Are models of the future used to anticipate targets in tracking tasks? In A.M. Colley and J.R. Beech (eds), *Cognition and Action in Skilled Behaviour*. Amsterdam, North Holland.
Harvey, N. (1990a) Effects of difficulty on judgemental probability forecasting of control response efficacy. *Journal of Forecasting*, **9**, 373–87.
Harvey, N. (1990b) Overconfidence in self-assessment of motor skill performance. In J.-P. Caverni, J.-M. Fabre and M. Gonzalez (eds.), *Cognitive Biases*. North-Holland, Amsterdam.
Harvey, N. (1992) Wishful thinking impairs belief-desire reasoning. A case of decoupling failure in adults? *Cognition*, **45**, 141–62.
Harvey, N., Garwood, J. & Palencia, M. (1987) Vocal matching of pitch intervals: Learning and transfer effects. *Psychology of Music*, **15**, 90–106.
Heneman, H.G., III, & Schwab, D.P. (1972) An evaluation of research on expectancy theory predictions of employee performance. *Psychological Bulletin*, **78**, 1–9.
Henry, R.A. & Sniezek, J.A. (1993) Situational factors affecting judgments of future performance. *Organizational Behavior and Human Decision Processes*, **54**, 104–32.
Horgan, D.D. (1992) Children and chess expertise: The role of calibration. *Psychological Research*, **54**, 44–50.
Howell, W.C. (1971) Uncertainty from internal and external sources: A clear case of overconfidence. *Journal of Experimental Psychology*, **89**, 240–3.
Ilgen, D.R., Nebeker, D.M. & Pritchard, R.D. (1981) Expectancy theory measures: An empirical comparison in an experimental simulation. *Organizational Behavior and Human Performance*, **28**, 189–223.
Jones R.A. (1977) *Self-fulfilling Prophecies. Social, Psychological and Physiological Effects of Expectancies*. Erlbaum, Hillsdale, NJ.
Juslin, P. (1993) An explanation of the hard-easy effect in studies of realism of

confidence in one's general knowledge. *European Journal of Cognitive Psychology*, **5**, 55–71.

Juslin, P. (1994) The overconfidence phenomenon as a consequence of informal experimenter-guided selection of almanac items. *Organizational Behavior and Human Decision Processes*, **54**, 226–246.

Kantowitz, B.H. (1974) Modality effects in recognition short-term motor memory. *Journal of Experimental Psychology*, **103**, 522–9.

Keren, G. (1988) On the ability of monitoring non-veridical perceptions and uncertain knowledge: Some calibration studies. *Acta Psychologica*, **67**, 95–119.

Keren, G. (1991) Calibration and probability judgments: Conceptual and methodological issues. *Acta Psychologica*, **77**, 217–73.

Koch, C.G. & Dorfman P.W. (1979) Recall and recognition processes in motor memory. Effects of feedback and knowledge of results delay. *Journal of Motor Behavior*, **11**, 23–34.

Koriat, A., Lichtenstein, S. & Fischhoff, B. (1980) Reasons for confidence. *Journal of Experimental Psychology: Human Learning and Memory*, **6**, 107–18.

Kuhl, J. (1978) Standard setting and risk preference: An elaboration of the theory of achievement motivation and an empirical test. *Psychological Review*, **85**, 239–48.

Kuhl, J. (1986) Motivation and information processing: A new look at decision making, dynamic conflict and action control. In R.N. Sorrentino and E.T. Higgins (eds.), *The Handbook of Motivation and Cognition: Foundations of Social Behavior*. Guildford, New York.

Kuhl, J. & Blankenship, V. (1979a) The dynamic theory of achievement motivation: From episodic to dynamic thinking. *Psychological Review*, **86**, 141–51.

Kuhl, J. & Blankenship, V. (1979b) Behavioral change in a constant environment: Shift to more difficult tasks with constant probability of success. *Journal of Personality and Social Psychology*, **37**, 551–63.

Langer, E.J. (1975) The illusion of control. *Journal of Personality and Social Psychology*, **32**, 311–28.

Latham, G.P. & Locke, E.A. (1991) Self regulation through goal setting. *Organizational Behavior and Human Decision Processes*, **50**, 212–47.

Lewin, K., Dembo, T., Festinger, L. & Sears, P.S. (1944) Level of aspiration. In J.M. Hunt (ed.), *Personality and the Behavior Disorders*. Ronald Press, New York.

Lichtenstein, S., Fischhoff, B. & Phillips, L.D. (1982) Calibration of probabilities: The state of the art to 1980. In D. Kahneman, P. Slovic and A. Tversky (eds.), *Judgment under Uncertainty: Heuristics and Biases*. Cambridge University Press, Cambridge.

Locke, E.A. (1969) Goals, probability of success (expectancy), and anticipated effect as determinants of performance level. In E.A. Locke, N. Cartledge, C.S. Knerr and J.F. Bryan. *The Motivational Effects of Knowledge of Results*. Silver Spring, MD: American Institutes of Research Report No R69-2.

Locke, E.A. (1991) Problems with goal-setting research in sports—and their solution. *Journal of Sport and Exercise Psychology*, **13**, 311–16.

Locke, E.A. & Latham, G.P. (1990). *A Theory of Goal Setting and Task Performance*. Prentice-Hall, Englewood Cliffs, NJ.

Locke, E.A., Shaw, K.N., Saari, L.M. & Laltham, G.P. (1981) Goal setting and task performance: 1969–1980. *Psychological Bulletin*, **90**, 125–52.

Mabe, P. & Wells, S. (1982) Validity of self-evaluation of ability: A review and meta-analysis. *Journal of Applied Psychology*, **67**, 230–296.

McClelland, D.C. (1955) Some social consequences of achievement motivation. In M.R. Jones (ed.), *Nebraska Symposium on Motivation, 1955*. University of Nebraska Press, Lincoln, NE.

McClelland, D.C. (1987) *Human Motivation*. Cambridge University Press, Cambridge.
McCracken, H.D. & Stelmach, G.E. (1977) A test of the schema theory of discrete motor learning. *Journal of Motor Behavior*, **9**, 193–201.
Mace, C.A. (1935) *Incentives: some experimental studies*, Industrial Health Research Board (Great Britain) Report No. 72.
McKenna, F.P. (1993) It won't happen to me: Unrealistic optimism or illusion of control? *British Journal of Psychology*, **84**, 39–50
Marshall, P.H. (1972) Recognition and recall in short-term motor memory. *Journal of Experimental Psychology*, **95**, 147–53.
Marteau, T.M., Johnston, M., Wynne, G. & Evans, T.R. (1989) Cognitive factors in the explanation of the mismatch between confidence and competence in performing basic life support. *Psychology and Health*, **3**, 172–82.
Marteau, T.M., Wynne, G., Kaye, W. & Evans, T.R. (1990) Resuscitation: Experience without feedback increases confidence but not skill. *British Medical Journal,* **300**, 849–50.
May, R.S. (1986) Inferences, subjective probability and frequency of correct answers: A cognitive approach to the overconfidence phenomenon. In B. Brehmer, H. Jungermann, P. Lourens and G. Sevo'n (eds.), *New Directons in Research in Decision Making*. North Holland, Amsterdam.
Mento, A.J., Cartledge, N.D. & Locke, E.A. (1980) Maryland vs Michigan vs Minnesota: Another look at the relationship of expectancy and goal difficulty to task performance. *Organizational Behavior and Human Performance*, **25**, 419–40.
Meyer, W.-U., Folkes, V. & Weiner, B. (1976) The perceived informational value and affective consequences of choice behavior and intermediate difficulty task selection. *Journal of Research in Personality*, **10**, 410–23.
Mitchell, T.R. (1974) Expectancy models of job satisfaction, occupational preference and effort: A theoretical, methodological and empirical appraisal. *Psychological Bulletin*, **81**, 1053–77.
Mitchell, T.R. (1979) Organizational behavior. *Annual Review of Psychology*, **30**, 243–82.
Motowidlo, S.J., Loehr, V. & Dunnette, M.D. (1978) A laboratory study of the effects of goal specificity on the relationship between probability of success and performance. *Journal of Applied Psychology*, **63**, 172–9.
Newell, K.M. (1974) Knowledge of results and motor learning. *Journal of Motor Behavior*, **6**, 235–44.
Newell, K.M. & Boucher, J.-L.B. (1974) Motor response recognition: Two processes. *Journal of Motor Behavior*, **6**, 81–86.
Newell, K.M. & Chew, R A. (1974) Recall and recognition in motor learning. *Journal of Motor Behavior*, **6**, 245–53.
Newell, K.M. & Shapiro, D.C. (1976) Variability of practice and transfer of training; Some evidence toward a schema view of motor learning. *Journal of Motor Behavior*, **8**, 233–43.
Oatley, K. & Johnson-Laird, P.N. (1987) Toward a cognitive theory of emotion. *Cognition and Emotion*, **1**, 29–50.
Ortony, A., Clore, G. & Foss, M.A. (1987) The referential structure of the affective lexicon. *Cognitive Science*, **11**, 341–64.
Pidgeon, N., Hood, G., Jones, D., Turner, B. & Gibson, R. (1992) Risk Perception. In *Risk: Analysis, Perception and Measurement*. The Royal Society, London.
Poulton, E.C. (1989) *Bias in Quantifying Judgments*. Erlbaum, Hillsdale, NJ.
Raynor, J.O. (1969) Future orientation and motivation of immediate activity: An

elaboration of the theory of achievement motivation. *Psychological Review*, **76**, 606–10.
Rosenbloom, P.S. & Newell, A. (1986) The chunking of goal hierarchies: A generalized model of practice. In R.S. Michalsk, J.G. Carbonell and T.M. Mitchell (eds.), *Machine Learning: An Artificial Intelligence Approach*. Kaufmann, Los Altos, CA.
Ryle, G. (1949) *The Concept of Mind*. Hutchinson, London.
Schmidt, R.A. (1975) A schema theory of discrete motor skill learning. *Psychological Review*, **82**, 225–60.
Schmidt, R.A. & White, J.L. (1972) Evidence for an error detection mechanism in motor skills: A test of Adams' closed-loop theory. *Journal of Motor Behavior*, **4**, 143–53.
Schmidt, R.A. & Wrisberg, C.A (1973) Further tests of Adams' closed-loop theory: Response-produced feedback and the error detection mechanism. *Journal of Motor Behavior*, **5**, 155–64.
Schwab, D.P., Olian-Gottlieb, J.D. & Heneman, H.G., III. (1979) Between subjects expectancy theory research: A statistical review of studies predicting effort and performance. *Psychological Bulletin*, **86**, 139–47.
Seligman, M.E.P. (1975) *Helplessness: On Depression, Development and Death*. Freeman, San Francisco.
Siegal, S. & Fouraker, L.E. (1960) *Bargaining and Group Decision Making*. McGraw-Hill, New York.
Sloman, A. (1987) Motive, mechanisms and emotions. *Cognition and Emotion*, **1**, 217–33.
Smith, C.P. (1963) Achievement-related motives and goal setting under different conditions. *Journal of Personality*, **31**, 124–40.
Sniezek, J.A. & Switzer, F.S. (1989) The over-underconfidence paradox: High Pi's but poor unlucky me. Paper presented at the Judgment and Decision Making Society Annual Meeting. Atlanta, Georgia.
Sorrentino, R.M., Hewitt, E.C. & Raso-Knott, P.A. (1992) Risk-taking in games of chance and skill: Informational and affective influences on choice behavior. *Journal of Social and Personality Psychology*, **62**, 522–33.
Svenson, O. (1981) Are we all less risky and more skilful than our fellow drivers? *Acta Psychologica*, **47**, 143–8.
Taylor, S.E. & Brown, J.D. (1988) Illusion and well-being: A social psychological perspective on mental health. *Psychological Bulletin*, **103**, 193–210.
Trope, Y. (1975) Seeking information about one's own ability as a determinant of choice among tasks. *Journal of Personality and Social Psychology*, **32**, 1004–13.
Trope, Y. & Brickman, P. (1975) Difficulty and diagnosticity as determinants of choice among tasks. *Journal of Personality and Social Psychology*, **31**, 918–26.
Turpin, B.A.M. (1980) Response recognition of a rapid, discrete, gross movement: Implication of Adams closed-loop theory of motor learning. In P. Klavora and J. Flowers (eds.), *Motor Learning and Biomechanical Factors in Sport*. University of Toronto School of Physical and Health Education Publications Division, Toronto.
Vroom, V.H. (1964) *Work and Motivation*. Wiley, New York.
Wallace, S.A., De Oreo, K.L. & Roberts, G.L. (1976) Memory and perceptual trace development in ballistic timing. *Journal of Motor Behavior*, **8**, 133–7.
Weinberg, R.S. (1992) Goal-setting and motor performance: A review and critique. In G. Roberts (ed.), *Motivation in Sport and Exercise*. Human Kinetics, Champaign, IL.
Weinberg, R. & Weigand, D. (1993) Goal setting in sport and exercise: A reaction to Locke. *Journal of Sport and Exercise*, **15**, 88–96.

Weiner, B. (1980) *Human Motivation*. Holt, Rinehart and Winston, New York.
Weinstein, N.D. (1980) Unrealistic optimism about future life events. *Journal of Personality and Social Psychology*, **39**, 806–20.
Weinstein, N.D. (1989) Optimistic biases about personal risks. *Science*, **246**, 1232–3.
Welford, A.T. (1980) On the nature of higher order skills. *Journal of Occupational Psychology*, **53**, 107–10.
Whiting, H.T.A. (1979) Subjective probability in sport. In G.C. Roberts and K.M. Newell (eds.), *Psychology of Motor Behavior and Sport—1978*. Human Kinetics, Champaign, IL.
Wright, G. & Wishuda, A. (1982) Distribution of probability assessments for almanac and future event questions. *Scandinavian Journal of Psychology*, **23**, 219–24.
Zelaznik, H. & Spring, J. (1976) Feedback in response recognition and production. *Journal of Motor Behavior*, **8**, 309–12.

Chapter 15

The Ups and Downs of the Hope Function In a Fruitless Search

Ruma Falk
The Hebrew University, Jerusalem
Abigail Lipson
Harvard University
and
Clifford Konold
University of Massachusetts, Amherst

It is seldom, if ever, that human beings are not actively searching for something. They may be searching for the next correct turning in the road they travel; for a misplaced object of value; for a name to put to the familiar face that suddenly confronts them; or for a solution of tomorrow's problems. All such search is beset with uncertainties. (Bell, 1979, page 14)

Imagine searching for a paragraph that you read some time ago. You have a visual memory of that paragraph on a right-hand page of a book, toward the top. Though you think you remember the particular book, you are not absolutely certain. Systematically, you begin leafing through the book's 10 chapters. The paragraph does not turn up in the first chapter, or in the second, third . . . . As you proceed without success through the chapters, does your hope of finding the paragraph in the next chapter increase or decrease?

Subjective Probability. Edited by G. Wright and P. Ayton.
© 1994 John Wiley & Sons Ltd.

And what of your hope of finding it in the book at all? Imagining yourself in this familiar situation, you may feel that before you reached the end of the book, despair would set in ("this must be the wrong book"). On the other hand, the longer you search the more reluctant you may be to quit, not only because of the efforts invested up to now, but because of a persisting intuition that the chances of finding the paragraph in the next chapter increase after each successive disappointment.

We all too often find ourselves in this type of search process. Without a realistic assessment of the uncertainties involved, we may either overestimate our chance of success, thus wasting more time in a futile search, or underestimate our chances, giving up too early in frustration and unjustified despair (MacGregor, Fischhoff, & Blackshaw, 1987). Considering the simplicity of the search situation in question and everybody's familiarity with the experience, it has surprised us to find that studies analyzing probabilistic reasoning in such situations are scarce. The psychological studies concerning search that we found deal mostly with seeking *strategies*, not with the course of the searcher's optimism throughout a systematic search characterized by prior uncertainty. (We make this statement despite realizing that our own search strategies might have been suboptimal; we might have abandoned the search prematurely.)

Bell (1979) reviews investigations of several types of physical search, conducted mainly by John Cohen and his collaborators. In their studies, subjects (children) choose locations in which to search for an object which is known for sure to be in one of the available locations (see, e.g., Cohen & Meudell, 1968, Experiment 4). Thus, subjects' hope assessments (confidence ratings) in these studies confound probabilistic judgments with evaluations of the wisdom of their own choices. Another class of studies concerns search decisions and confidence assessments in complex hierarchical systems. These studies include investigations of locating general items of knowledge in a *Statistical Abstract*, and searching computerized databases (see, for example, MacGregor, Fischhoff & Blackshaw, 1987, and references therein).

The more typical real-world search process involves situations where initial uncertainty about the existence of a target object in a finite field of locations is followed by a systematic search of these locations, with a series of negative results. We have encountered variations of such situations in math-education journals, in popular scientific literature, in fiction, and in daily living. Consider the following four examples.

Example 1 *The Case of Sherlock Holmes.* In Arthur Conan Doyle's story, *The Six Napoleons* (cited by Jones, 1966), the great detective Sherlock Holmes deduces that one of six plaster busts of Napoleon conceals a priceless pearl. As the story unfolds, the busts are smashed one by one, until Sherlock finds and dramatically smashes the last one, recovering the pearl. As usual, the detective reveals his reasoning, noting that the numerical chances of finding the pearl in the next bust increased as their number dwindled, until with the last bust it reached certainty. Jones (1966) points out that the scientific viewpoint would doubt Sherlock's initial certainty, and would start with, say, only a 50% chance that Sherlock's theory is right: "As successive busts are smashed and no pearl is found, the rising chance of finding it in the next is balanced by the evidence of this growing succession of failures that Sherlock is wrong, and that there isn't any pearl at all." (page 466)

Example 2 *Doctor Fischer's Bomb Party.* Graham Greene's (1980) Dr Fischer wants to test the limits of greediness. He invites six wealthy guests to a party and shows them a barrel in a corner of his garden in which are six Christmas crackers. Five of the crackers, he explains, contain a cheque for two million Swiss Francs. The sixth contains enough explosive so primed as to end the life of whoever pulls the cracker. The guests are challenged to approach the barrel one by one and try their luck. Dr Fischer assures them that the cheques are there, but the matter is complicated by the possibility that the presence of the bomb might be a hoax. While one of the guests prepares (hesitantly) to make his move, he is preempted by Mrs Montgomery who pushes ahead of him to the barrel, explaining that "the odds would never be as favorable again" (Greene, 1980, page 127). Is she right? (See Ayton & McClelland's, 1987, delightful paper on that ghastly party.)

Example 3 *The Key Problem.* A man comes home at night during a blackout. He has two similar bunches of keys in his pocket; one for home, one for work. In the darkness, he picks one bunch from his pocket. The bunch comprises n keys of which only one will fit his door; if, that is, he has picked the right bunch. He tries the keys successively (sampling without replacement). We are interested in his confidence that he's got the right bunch, and in his immediate expectancy of unlocking the door when key after key fails to do the job (L.V. Glickman, personal communication, 1984. Adapted from a problem in Feller, 1957, page 54).

Example 4 *Let Sleeping Flies Lie.* Raphael Falk, a Hebrew University geneticist, told us about his experience of expecting a phone call from the Dean of his faculty. The Dean had told him the previous day that he *might* call him in his lab between 10 and 11 a.m. Raphael spent that morning examining successive bottles inhabited by Drosophila flies, looking for a certain rare mutant. His routine was to etherize the flies in each bottle for a few minutes and then inspect them under the microscope. If the inspection were to be interrupted, the flies would wake up and fly away. He kept working calmly until about 10:30 a.m., by which time the Dean had still not called. Raphael reported feeling that the chances of the Dean calling were dropping steadily as time went on. However, he became increasingly nervous about etherizing the flies in each successive bottle, fearing that the Dean's impending call would disrupt the inspection

In order to investigate the nature of probabilistic reasoning in situations like those described above, we devised two experimental problems, each of which involved two hope questions (long- and short-term). Problem 1 (inspired by Meshalkin, 1963/1973, page 21) concerns a standard *search* situation (similar to Example 3). Problem 2 involves an equivalent *wait* situation (similar to Example 4). The two situations are structurally analogous, although the first describes an active search process while the second describes an extended wait for a target event to occur.

We will present the two standard problems along with the Bayesian solution. Then we will discuss a number of features of the solution by applying it to a variety of situations including the four examples just cited. After describing how our subjects reasoned about the standard problems, we will present a didactic device we developed to make the search problem more conducive to resolution. Finally, we will explore subjects' ability to transfer the lesson learned from the didactic device to the analogous wait problem.

15.1 STANDARD PROBLEMS AND THEIR SOLUTION

Problem 1 *The Standard Search Problem.* The Desk: Seek and you Shall Find?

Long-term probability version (Desk-Long—DL). Imagine that you are searching for an important letter that you received some time ago. Usually your assistant puts your letters in the drawers of your desk after you have read them. He remembers to do this in 80% of the cases, and in 20% of the cases he leaves them somewhere else.

There are eight drawers in your desk. If indeed your assistant *has* placed the letter in your desk, you know from past experience that it is *equally likely* to be in any of the eight drawers.

You start a thorough and systematic search of your desk.

(A) You search the first drawer, and the letter *is not* there. How would you *now* evaluate the probability that the letter is *in the desk*?
(B) You continue to search the next three drawers, until altogether you have searched four drawers. The letter *is not* there.
How would you *now* evaluate the probability that the letter is *in the desk*?
(C) You continue to search three more drawers, until altogether you have searched seven drawers. The letter *is not* there.

How would you *now* evaluate the probability that the letter is *in the desk*?

Short-term probability version (Desk-Short—DS). Same problem-stem as DL, but the three questions are:

(A) You search the first drawer, and the letter *is not* there. How would you *now* evaluate the probability that the letter is *in the next drawer* (i.e., in the second drawer)?
(B) You continue to search the next three drawers, until altogether you have searched four drawers. The letter *is not* there.
How would you *now* evaluate the probability that the letter *is in the next drawer* (i.e., in the fifth drawer)?
(C) You continue to search three more drawers, until altogether you have searched seven drawers. The letter *is not* there.
How would you *now* evaluate the probability that the letter is *in the next drawer* (i.e., in the eighth drawer)?

Problem 2 *The Standard Wait Problem.* At the Bus Stop.

Long-term probability version (Bus-Long—BL). Imagine that you and your friend are tourists in a big foreign city. You find yourself late in the evening looking for transportation back to your hotel. You approach a bus stop that doesn't display any timetable. You know, however, that the buses in this city run punctually *each half hour* during the evening, only it is now so late that you are somewhat worried that they might have already stopped running.

You know that 60% of the bus routes in the city operate this late, and 40% do not, but you don't know whether this particular bus is still running or not. It is now 11:30 p.m., and you decide to wait until either the bus arrives or midnight, whichever happens first.

Since you have no idea about the bus's exact schedule, you figure that the bus is *equally likely* to arrive in any of the six five-minute intervals during the coming half-hour (if indeed it is still running).

(A) The bus *does not arrive* in the first five minutes. It is now 11:35.
How would you *now* evaluate the probability that the bus will arrive *sometime before midnight*?
(B) Another ten minutes elapse. The time is now 11:45, and the bus *has not arrived*.
How would you *now* evaluate the probability that the bus will arrive *sometime before midnight*?
(C) Ten more minutes go by. The time is now 11:55, and the bus *has not arrived*.
How would you *now* evaluate the probability that the bus will arrive *sometime before midnight*?

Short-term probability version (Bus-Short—BS). Same problem-stem as BL, but the three questions are:

(A) The bus *does not arrive* in the first five minutes. It is now 11:35.
How would you *now* evaluate the probability that the bus will arrive *during the next five minutes* (i.e., between 11:35 and 11:40)?

(B) Another ten minutes elapse. The time is now 11:45, and the bus *has not arrived.*
How would you *now* evaluate the probability that the bus will arrive *during the next five minutes* (i.e., between 11:45 and 11:50)?

(C) Ten more minutes go by. The time is now 11:55, and the bus *has not arrived.*
How would you *now* evaluate the probability that the bus will arrive *during the next five minutes* (i.e., between 11:55 and midnight)?

15.1.1 The Mathematical Long- and Short-run Functions

We solve the Standard Search Problem (Problem 1) for the *general case* of n equally likely drawers and prior probability L_0 that the letter is in the desk. The solution applies as well to the isomorphic Wait Problem (Problem 2). If the letter *is in* the desk, the conditional probability of *not* finding it when searching the first i drawers is $(n-i)/n$; if the letter *is out* of the desk, *not* finding it in the first i drawers is a certainty. Let's denote the respective long- and short-term posterior probabilities we wish to find by $L_i = P$ (letter is in desk | letter was not in first i drawers), $S_i = P$ (letter is in next drawer | letter was not in first i drawers). Clearly, $S_0 = L_0/n$, and $S_i = L_i/(n-i)$. By Bayes' rule,

$$L_i = \frac{[(n-i)/n]L_0}{[(n-i)/n]L_0 + (1-L_0)}$$

A few algebraic manipulations yield

$$L_i = \frac{(n-i)L_0}{n-iL_0} \qquad i = 0, 1, 2, \ldots, n-1, n \qquad (15.1)$$

$$S_i = \frac{L_0}{n-iL_0} \qquad i = 0, 1, 2, \ldots, n-1 \qquad (15.2)$$

Formulas (15.1) and (15.2) describe the *hope functions* for the long run (L_i) and the short run (S_i), given i initial failures.

Table 15.1 presents the specific forms which L_i and S_i assume in the case of Problems 1 and 2, along with the answers to the questions posed in the Problems. The numbers in Table 15.1, as well as formulas (15.1) and (15.2), indicate that the long-term hope function, L_i, decreases as i grows, whereas the short-term hope function, S_i, increases with i, until $L_{n-1} = S_{n-1}$.

Table 15.1 Long- and short-term hope functions for the desk and bus problems (Problems 1 & 2)

		Long Run	Short Run
		DL	DS
Problem 1		$L_i = \dfrac{8-i}{10-i}$	$S_i = \dfrac{1}{10-i}$
Desk			
$L_0 = 0.80$	(A) $i = 1$	$7/9 = 77.8\%$	$1/9 = 11.1\%$
$n = 8$	(B) $i = 4$	$4/6 = 66.7\%$	$1/6 = 16.7\%$
	(C) $i = 7$	$1/3 = 33.3\%$	$1/3 = 33.3\%$
		BL	BS
Problem 2		$L_i = \dfrac{6-i}{10-i}$	$S_i = \dfrac{1}{10-i}$
Bus			
$L_0 = 0.60$	(A) $i = 1$	$5/9 = 55.6\%$	$1/9 = 11.1\%$
$n = 6$	(B) $i = 3$	$3/7 = 42.9\%$	$1/7 = 14.3\%$
	(C) $i = 5$	$1/5 = 20.0\%$	$1/5 = 20.0\%$

15.1.2 Further Explorations

A number of issues surface as we extend our formal analysis to the examples cited earlier. Suppose $L_0 = 1$, as in the case of Sherlock's absolute confidence that the pearl *is* hidden in one of the busts (Example 1). If indeed there is no doubt whatsoever about the existence of the target object in one of the available locations, no initial sequence of failures, long as it may be, will shatter that (long-term) certainty. L_i will equal 1 for all values of i. The short-term probability of success in the next unit (location or time slot) will equal the inverse of the number of remaining units and will thus rise to 1 when only one unit remains (i.e., for $i = n - 1$). The results for the case of initial certainty may also be obtained from formulas (15.1) and (15.2) by substituting 1 for L_0. These formulas are, in fact, valid for the entire range of possible values of L_0, including the end points 1 and 0.

Figure 15.1 presents the long- and the short-term hope functions for the data of the standard search problem (Problem 1). Sherlock's short-term hope function (Example 1), in which $L_0 = 1$ and $n = 8$, is added for comparison (inspired by Jones, 1966).

Dr Fischer's bomb party (Example 2) raises a third question, in addition to our long- and short-run questions: Which (if any) is the safest serial position *beforehand*? The *a priori* probability of blowing the bomb (finding the object) in ordinal position (location) i, denoted A_i, can be successively computed,

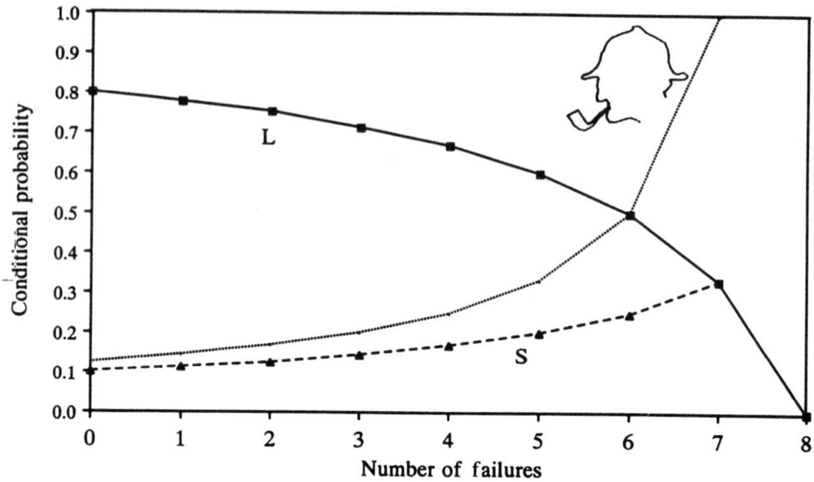

Figure 15.1 Long- and short-term hope functions for $L_0 = 0.80$ and $n = 8$ ($L_0 = 1$ in Sherlock's case)

given L_0. Suppose $L_0 = \frac{1}{2}$. Let n be 6, as in Greene's (1980) story. The conditional probability L_i that a bomb exists in the barrel, given that i crackers have been safely pulled, is obtained by applying (15.1) to the present case:

$$L_i = \frac{6-i}{12-i} \qquad i = 0, 1, 2, \ldots, 6$$

For player i, we multiply the probability of the previous $i-1$ players *not* detonating the bomb by the conditional probability of the presence of a bomb given that information (i.e., L_{i-1}). We then multiply that result by the probability of player i pulling the bomb-cracker out of the remaining $6-i+1$ crackers. These three factors are listed, in turn, in each row of Table 15.2. Computing these products, we see that the *a priori* probabilities of pulling the bomb are the same for all the ordinal positions (Ayton & McClelland, 1987). The function A_i is thus *constant* over all the values of i. There was no reason for Mrs Montgomery to rush to play first.

In hindsight, it should have been obvious that, prior to starting the game, all the participants are equally likely to detonate the bomb (just as the *a priori* probabilities of finding the letter in any of the drawers of the desk are equal). Without loss of generality, we can imagine that instead of going in turn, the six players are assigned a cracker at random, and they all pull simultaneously. The modified version is evidently symmetric with respect to all players. Consequently, their chances of detonating the bomb are *equal* (see Falk, 1993, Problems 2.3.3, 2.4.12, and 2.4.13).

Table 15.2 *A priori* probabilities of blowing the bomb as a function of serial position. $L_0 = 0.50$; $n = 6$ (Example 2)

i	A_i
1	$1 \times \frac{1}{2} \times \frac{1}{6} = \frac{1}{12}$
2	$\frac{11}{12} \times \frac{5}{11} \times \frac{1}{5} = \frac{1}{12}$
3	$\frac{10}{12} \times \frac{4}{10} \times \frac{1}{4} = \frac{1}{12}$
4	$\frac{9}{12} \times \frac{3}{9} \times \frac{1}{3} = \frac{1}{12}$
5	$\frac{8}{12} \times \frac{2}{8} \times \frac{1}{2} = \frac{1}{12}$
6	$\frac{7}{12} \times \frac{1}{7} \times 1 = \frac{1}{12}$

As we saw, our hope functions, which are defined as *conditional* probabilities given an initial sequence of i negative outcomes, are generally *not* constant (see Figure 15.1). This is true for all cases, barring L_i when $L_0 = 1$ (Example 1). In terms of Dr Fischer's bomb party (Example 2), the course of the function L implies that "if we entertain any degree of doubt concerning the presence of a bomb in *any* of the crackers then that doubt will be fuelled the more crackers that are pulled without a bomb exploding" (Ayton & McClelland, 1987, page 180). At the same time, the course of the function δ indicates that the risk of the next cracker blowing up increases with the number of innocuous crackers that have been pulled.

By the same token, the man who tries consecutive keys in the bunch and fails to unlock the door (Example 3) should realize that the possibility he holds the wrong bunch is becoming more and more probable. On the other hand, he is not to blame for persisting in his attempts with the same bunch, because in each successive trial he is slightly more likely to succeed.

Suppose the police are scanning house after house in a given neighborhood in search of an escaped prisoner. The information that the runaway might be in the neighborhood was received from a source that is usually reliable. The police are right to become increasingly *alert* when moving from one house to the next. Their mounting apprehension, however, does not contradict the assessment that the overall chances of finding the escapee in the neighborhood keep dropping as the search progresses unsuccessfully. These two apparently conflicting tendencies characterize all situations where we sequentially search for an object in a given space, provided we lack complete certainty that it is there and the object is equally likely at the beginning to be in each unit of the space.

When waiting for an initially uncertain event to happen in consecutive time units, the long- and short-range conditional probabilities of occurrence behave precisely as the respective hope functions in search situations. Thus, the geneticist (Example 4) was justified as time elapsed both in losing confidence that the Dean would call, and in hesitating to anesthetize another batch of flies. His feelings matched the course of the actual long- and short-term probabilities of receiving the phone call.

Finally, the search (or wait) for Mr Right is roughly subject to the same apparently paradoxical rules. Patterns of nuptiality in several societies from about ages 18 to 30 indicate that although individuals who do *not* marry for several years are less likely ever to do so, their short-term conditional probabilities of marrying within a year keep rising for a while (Gabriel, 1960). The long- and short-term functions describe the two faces of our optimism, or pessimism, depending on the desirability of the target event.

15.2 SUBJECTIVE HOPE

> The ordinary person looking for some lost object instinctively holds to the scientific viewpoint . . . . He is neither philosophically unmoved by the progress of the search, nor does his optimism rise increasingly as successive possibilities are eliminated. His initial cautious hope is increasingly balanced by the growing conviction, born of successive failures, that it's not there, that it's not *anywhere*: and when he regards this as adequately proven, he gives up. (Jones, 1966, page 466)

To find out whether Jones' evaluation of the "ordinary person" is true, we asked subjects to answer the questions posed in Problems 1 and 2. The general question of whether people intuitively grasp the Bayesian solution can be decomposed into several more specific questions. To what extent is base-rate information (prior probability) taken into account? How is the ongoing failure to find the object incorporated into the reasoning? Do people correctly assess the *direction* of the two functions, namely, the simultaneous descent of the long-term hope (L) and ascent of the short-term hope (S)? Do they experience an intuitive conflict when trying to evaluate S_i, sensing that the general hope is decreasing but the diminishing number of remaining possibilities suggests that success in the next trial becomes more likely?

In addition to the *numerical* versions of Problems 1 and 2 given above, we composed *directional* versions of these examples which differed only in asking about directions instead of numbers. Thus, for example, question (A) in directional DL version asked whether the probability that the letter is in the desk is now greater than, equal to, or less than 80%. Question (B) asked whether the same probability is now greater than, equal to, or less than what it was in (A), and (C) asked to compare the target probability with what it was in (B). The same was true for the directional DS version which asked in (A) whether the probability that the letter is *in the next drawer* is now less than, equal to, or greater than what it was for the first drawer. Question (B) asked for a comparison of the short-run probability with that of (A), and so on. Equivalent changes were introduced into the directional versions of BL and BS.

The design included eight kinds of problems made up of all combinations of three binary variables: (1) *story* (desk or bus), (2) *range* (long or short), (3) *question type* (numerical or directional). Sixty-one subjects—36 undergraduate students of psychology from the University of Massachusetts, Amherst, and 25 senior high-school students (of ages 17 & 18) from Massachusetts— answered two problems. The two forms each subject got differed on all three dimensions. Thus, a subject who first got directional BL would then receive numerical DS. Order of administration and all other aspects of design were counterbalanced. Subjects were instructed at the head of the form to read the problem carefully and trust their *common sense* in answering the questions. They were asked at the end to *explain* their reasoning.

15.2.1 Directional and Numerical Assessments

We first analyzed the data *ordinally*. Ignoring the exact values in the numerical versions, we sorted responses into three main types: strictly *increasing*, strictly *decreasing*, or a *constant* function. A fourth category (*other*) included functions which changed directions or were weakly monotonic. (The undergraduate and senior high-school students' responses were pooled since the patterns of responses of the two groups were very similar.) Table 15.3 shows the two-dimensional distribution, pooled across story types, of the 61 subjects according to the kind of L and S functions which they produced.

The results in Table 15.3 show that a majority of the subjects (35) intuitively sensed the decline of the L function. The modal group of subjects (27) produced an increasing S function. Yet, only about one fifth of the subjects (12) generated the correct *combination* of a decreasing L and an increasing S function. It is noteworthy that in a pilot study with 42 undergraduate law students at the Hebrew University of Jerusalem, about one fifth (8) produced the correct combination. The pilot study used different but isomorphic stories

Table 15.3 Subjects classified according to the long- and short-term hope functions they produced

	Long-run hope				
Short-run hope	Increasing	Decreasing	Constant	Other	Total
Increasing	—	12	11	4	27
Decreasing	—	7	1	1	9
Constant	—	6	7	1	14
Other	—	10	—	1	11
Total	—	35	19	7	61

(searching an escaped prisoner in successive houses, and waiting for a forgetful professor to come to an appointment).

None of the 61 subjects responded with a correct triplet of numerical probabilities to *any* of the L or S forms. This was true for all the numerical versions and for many of the directional versions in which subjects gave numerical answers while explaining their choices. Overall, it is clear that students of fairly high ability are *incapable* of correctly assessing the L and S hope probabilities, but they have a rudimentary conception of the correct directions of the two functions.

15.2.2 Principal Assumptions Underlying Solution Strategies

Solution strategies are suggested by the pattern of subjects' numerical responses and the explanations they provided. In examining these, a few heuristics appear to us to be guiding a substantial number of responses. In particular, in many cases assumptions of *constancy* underlie the choice of the three answers.

Suppose one assumes that the given L_0 of 0.80 in Problem 1 (desk) stays unchanged despite failing to find the letter in the first i drawers. That assumption, which we label *constant L*, entails an identical response of 0.80 to all questions of DL and an increasing triplet of answers to DS—(A) 0.114 (i.e., 0.80/7), (B) 0.20, (C) 0.80 (see the correct set of answers in Table 15.1). One may, however, assume that the probability of success per drawer (unit) stays unchanged. We label that assumption *constant S*. It entails an identical response of 0.10 (i.e., S_0) to all the questions in DS and a decreasing triplet of answers to DL—(A) 0.70, (B) 0.40, (C) 0.10 (cf. Table 15.1). The corresponding predictions of responses to BL and BS under the two constancy assumptions can be easily obtained.

The responses of twelve subjects to the two forms were compatible with the constant L assumption. Six subjects assumed constant S across both forms, and another 13 assumed constant L in answering one form and constant S in the other. Among the remaining 30 subjects, 19 assumed constancy in only one of the forms (6 constant L, and 13 constant S). Overall, of the 122 forms answered by 61 subjects, 81 (i.e., 66.4%) were based on constancy assumptions: 43 constant L, and 38 constant S.

The heuristic of adhering to one constant parameter of the setup (whether L_0 or S_0) reduces the complexity of the hope problems. But it may also reflect subjects' conception of probability as an unchanging *propensity* of the situation at hand. Kahneman and Tversky (1982) draw a distinction between two loci to which uncertainty can be attributed: the external world or our state of knowledge. Real-world systems are frequently perceived as having dispositions to produce different events, and the probabilities of these events are judged by assessing the strength of these dispositions. The propensity of the desk (or drawer) to produce the missing letter (or, for that matter, of the transportation system to produce the bus) may have been considered a fixed parameter of the setup by many of our subjects. This would explain why they refused to update that parameter in light of the accumulating search results. They did not interpret the question as addressing their state of knowledge, and were consequently impervious to the effect of new evidence.

Subjects often explicitly expressed the idea that constant probability was a characteristic disposition of the chance setup. The following statements were made by subjects who responded invariably with an answer of 80% to all the questions in numerical DL: "The probability that the letter is in the desk is 80%, and that's it!" A deliberate attempt to *ignore* the information about successive failures (as if the subject is wary of falling prey to the gambler's fallacy) is notable in another subject's words: "Like the lottery, no matter how many times you play or what number you use, you have the same probability in winning. So each desk has an 80% chance of having the letter." Similar insistence on the irrelevance of the reported outcomes is found in: "Finding empty drawers doesn't change probability that letter is in desk," and "The letter is equally likely (80%) to be in any of the drawers—so the fact that x number of drawers was checked does not lower the probability."

The constancy of the long-run hope for the arrival of the bus (Problem 2) was justified by "I figure that the exact time between 11:30 and 12:00 (11:35, 11:45, 11:55) doesn't really matter—since 60% of the buses operate this late I think there is still a 60% chance that a bus will come." However, the same subject assumed constancy per unit when asked about short-term probabilities: "Since there are 8 drawers and the letter, if it is in any of the drawers, is equally likely to be in any of the drawers, the probability that the letter is in any one drawer is 10%. *This doesn't change if the letter is not in one or more of the other drawers* [italics added]." Had this discussion taken place in class, the teacher could have asked at that point, "and what if the letter *is* found in the *i*th drawer, would you still think the probability doesn't change?"

Assuming constant S when answering numerical BL, results in a decreasing L function ((A) 50%, (B) 30%, (C) 10%). This was typically justified by answers such as: "I estimated that since there was 60% chance that the bus was still running ... the chance of it arriving decreased by 10% as each 5 minute (of 6) passed." Another subject's explanation repeats the same rationale for DL: "There is 80 percent chance of letter in desk and 20% not. Checking one drawer with an unsuccessful try drops your chances of it being there by 10%, to 70%, and so on."

In terms of the issue of "Evidential Impact of Base Rates" (the title of a paper by Tversky & Kahneman, 1982), an assumption of constant L represents an extreme point of "conservatism" on the continuum of use versus neglect of base-rate data. In fact, constant L is the reverse of the "base-rate fallacy" according to which subjects typically ignore the base rate and consider only the specific evidence about the case at hand (as in Tversky & Kahneman's well-known cab problem). The constant S assumption, although resulting in exaggerated decrease of the L function, keeps the base-rate *unit* unchanged instead of duly increasing it in light of the evidence. In this sense, constant S is conservative as well.

Our impression is that subjects' conservatism, as revealed by the prevalence of the constancy assumptions, is a consequence of their *external* attribution of uncertainty (Kahneman & Tversky, 1982). The parameters L_0 and/or S_0 are apparently perceived as properties that belong to the desk, like color, size and texture. Subjects think of these parameters in terms of "the probabilities of the desk", whereas the Bayesian view would imply expressions like "*my* probability of the target event". Thus, subjects fail to incorporate the additional knowledge they acquire when given successive search results.

15.2.3 Other Strategies

Several subjects denied the presence of chance altogether and acted as if it were *certain* that the letter was in the desk (the bus is going to come), and others embraced the historic position of *equal ignorance* and responded with "fifty–fifty," in apparent disregard of the givens of the problem.

Eleven subjects relied on assumption of certainty in response to one of the problems they answered. Another three subjects assumed certainty in both problems. Most of the certainty-based responses were made by subjects assuming either constant L or constant S. Thus, for example, assuming initial certainty and constant S when responding to numerical BL means that $S_0 = 17\% = \frac{1}{6}$, and that S_0 is subtracted from the L function (starting with $L_0 = 100\%$) for every five-minute interval in which the bus does not arrive. This results in: (A) 83%, (B) 50%, (C) 17%. One subject justified this triplet as follows: "I made a time table of 30 minutes. I take a fraction of how much time has elapsed, then divide by 100%, giving the answer." Note that this subject was not disturbed by the fact that probability of 83% following a five-minute wait for the bus was higher than the given initial probability of 60%.

The double assumption of initial certainty and constant L means that $L_0 = 100\%$ stays unchanged. Thus, when answering numerical DS, these 100 percents are divided each time in equal shares among the remaining drawers, resulting in (A) $14\% = \frac{1}{7}$, (B) 25%, (C) 100%. We quote one subject's elaborate justification of the above triplet: "If you didn't eliminate drawers and randomly pointed to any drawer there would be still $\frac{1}{8}$ probability because there is replacement. But here we don't have replacement and each drawer is equally likely of containing the letter, so however many drawers you have its $1/N$ probability."

No less surprising than the responses that converted the initial probability of 80% (or 60%) into certainty were those that assumed total ignorance and concluded therefore that the target probability should be one half. Ten subjects appeared to invoke the maxim of "insufficient reason" assigning *equal probabilities* to the two possible outcomes. Consider the explanation of a subject who gave a constant 50% answer to all three BS questions: "Because since you don't have any idea what time the bus arrives and you don't even know if the bus is coming, then it is equally likely to arrive at any time." Another subject, who responded similarly, wrote: "It doesn't matter that the bus didn't arrive in the last 5 minutes. There is always a 50% chance it will come and a 50% chance it will not come." A uniform 50% response to the three BL questions was explained by: "There is no ↑ in probability it will come because there's only 5 min left—there's still a 50/50 chance it will either come or its doesn't." One subject's "ingenious" reasoning with respect to BL resulted in: (A) 41.6%, (B) 25%, (C) 8.3%. His telegraphic-style explanation ran as follows:

> 6 5 min intervals from 11:30–12:00
> —said it was = ly likely at 11:30 (50%)
> 50% : 6 = 8.3
> each 5 min interval decreases probability by 8.3%

We see here an interesting combination of the equal-ignorance and constant-S heuristics.

The human tendency to remove chance from our considerations has been observed in various judgmental contexts (several examples are reviewed by Falk & Konold, 1992). The same is true for people's inclination to assume equal likelihood once uncertainty is acknowledged. The tendency to identify randomness with equiprobability and thus assign equal chances to the available options has been widely documented in empirical investigations (e.g., Konold et al., 1991; Shimojo & Ichikawa, 1989). The primacy of the equiprobability intuition has been described in studies of the historical development of probability theory. Uniformity was the first presumption on which probability calculations were based (Gigerenzer et al., 1989; Hacking, 1975, Chapter 14). Converging evidence thus testifies to the genuine power of the intuitive bent toward symmetry (Falk, 1992; Zabell, 1988).

Paradoxically, subjects' assumptions of certainty and of equal ignorance, although diametrically at odds with each other, might be viewed as the two Janus-faces of the same orientation. Konold (1989) has referred to that orientation as the *outcome approach*. People reasoning via the outcome approach tend to interpret a request for a probability of some event as a request to *predict* whether or not that event will occur on the next trial. Contrary to current scientific thinking, these reasoners do not view probability as a measure of one's uncertainty, nor as answering the question about the relative frequency of occurrence of the target event in many repeated trials. According to Konold's (1989) description, outcome-oriented subjects translate probability values into yes/no decisions, transforming their probability evaluations into certainty. Thus, a probability of 20% means "it *won't* happen," a probability of 80% means "it *will* happen." When they sense a total lack of knowledge about the outcome, however, they express it by the 50/50 numerical probability, which means "it either will happen or won't happen—don't know which." Konold found in several studies that a certain subgroup of the subjects (not necessarily a majority) was fairly consistent in responding according to this outcome-oriented perspective.

Although we cannot predict whether an outcome-oriented subject would convert the probabilities given in our problems into certainty or into equal ignorance, it stands to reason that the former would occur more often when the probabilities are close to 100% (or to zero) and the latter when the probabilities are close to 50%. Our data show roughly this pattern. Of the 17 forms which elicited certainty-based responses, 10 were desk problems ($L_0 = 80\%$) and 7 bus problems ($L_0 = 60\%$). In contrast, of the 10 equally likely answers, 3 were given in response to the desk story and 7 to the bus. These include two subjects who responded by certainty to the desk *and* by equal ignorance to the bus. Overall, the conjecture that the outcome approach has played some role in answering the hope problems is weakly supported. It remains a possibility that should be further explored.

15.2.4 Toward a Solution

It was somewhat surprising that we did *not* find among the explanations of the *S* problems an explication of the *conflict* between the diminishing long-term hope and the increasing immediate hope implied by the fewer remaining units. One subject who produced a constant *S* function in response to directional BS did describe another conflict: "The probability that the bus will arrive in any given time slot is the same. Although my intuition would urge me to expect to see the bus more (meaning—I would assume the probability would be greater) as time elapsed, I believe that the 'laws' of probability would have it otherwise. But—as I think about it more, this could be argued against, saying that the probability changes as each *unknown* 5-minute segment became known." Several subjects, who produced a decreasing L function in response to directional versions (without giving numbers), gave a correct Bayesian-like explanation (e.g. "Well, if it is not in a drawer, then it could fall in the 20% zone and the more drawers you open without it being in there the lower the probability that it's in there").

Only one subject (No. 62), a precollege student enrolled at the Hebrew University of Jerusalem, responded correctly to both problems, in this case to numerical BL and directional DS. These were his explanations:

BL: At 11:30 the probability of the bus arriving by midnight was $\frac{6}{10}$, and of not: $\frac{4}{10}$
At 11:35 the probability of the bus arriving by midnight was $\frac{5}{9}$, and of not: $\frac{4}{9}$
At 11:45 the probability of the bus arriving by midnight was $\frac{3}{7}$, and of not: $\frac{4}{7}$
At 11:55 the probability of the bus arriving by midnight was $\frac{1}{5}$, and of not: $\frac{4}{5}$

DS: At the beginning of the search the letter could be in one of 10 "locations": 8 drawers and 2 "others." The 2 "others" stay in constant amount, whereas the number of drawers keeps decreasing. Therefore, the chance of finding the letter in the first drawer was only 10% (i.e., 1/10), in the second 1/9, in the fifth 1/6 and in the eighth 1/3.

These considerations yielded the same results (for each i) as the Bayesian computations. Note, however, that whenever several units are eliminated, the posterior probability distribution over the remaining units (including the imaginary "other" locations) stays uniform. That is why this subject's reasoning matched the Bayesian results. The same method would not work if applied to problems like that of the three prisoners, or Monty's notorious TV game "Let's make a deal" (Falk, 1992).

Inspired by that subject's method of solution, we devised a simplified version of the desk problem. The main change in the modified version was the addition of a concrete representation of the sample space that includes the two "locations" out of the desk.

15.3 THE HOPE PROBLEM—SIMPLIFIED

The simplified desk problem, presented below, is isomorphic to Problem 1:

Problem 1R *The Revised Desk Problem.* (Revised-Desk-Long—RDL; Revised-Desk-Short—RDS). The problem stem of both RDL and RDS reads as follows:

Imagine that you are searching for an important letter that you received some time ago. Your assistant always puts your letters in the drawers of your desk after you have read them.

There are ten drawers in your desk. You know that the letter is *equally likely* to be in any of the ten drawers. You notice, however, that drawers #9 and #10 are locked (see figure), and your assistant has gone home with the keys. You realize the chances that the letter is in one of the unlocked drawers is 80%. So you start a thorough and systematic search of the eight unlocked drawers.

Figure 15.2 presents the drawing which appeared in each form. The three questions in RDL were the same as in DL of Problem 1, except they asked for an evaluation of the probability that the letter is *in one of the unlocked drawers*. RDS included exactly the same questions as DS of Problem 1. Only *numerical* revised forms were prepared.

A pilot test was run at the University of Massachusetts, Amherst with 13 subjects (including graduate and postgraduate students). Each subject responded to only one form: 6 to RDL and 7 to RDS. Three of the responses to RDL and 6 of the responses to RDS were perfectly correct. Of the other

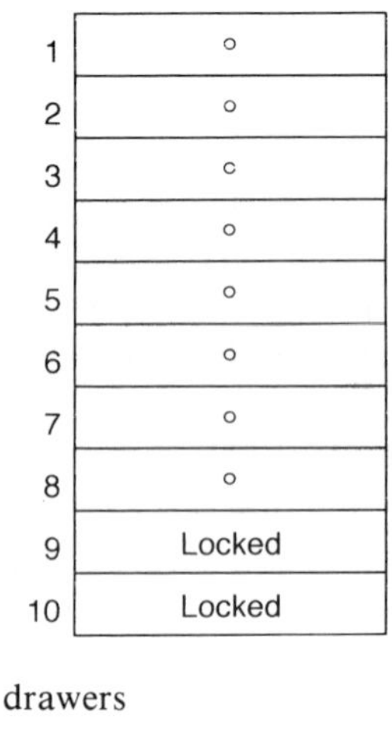

Figure 15.2 A desk with 10 drawers

3 RDLs, subjects gave 2 constant *L* responses and 1 constant *S* response. The seventh RDS answer assumed constant *L*.

Based on the results of this pretest, we conducted larger-scale surveys. Our aim was both to confirm the indications that the revised versions facilitate reaching the correct solution and to test whether subjects who succeed in solving Problem 1R would transfer the solution principle to the Standard Wait Problem (Problem 2) as originally phrased.

Fifty four subjects—34 undergraduate psychology students from the University of Massachusetts, Amherst and 20 senior high-school students from Massachusetts—participated in the first survey. Each subject was asked to answer two problems: either RDL and numerical BL, or RDS and numerical BS. The revised desk problem was always given first; 26 subjects received two *L* versions and 28 received two *S* versions.

Eighteen of the 54 revised forms were answered correctly (9 RDLs, and 9 RDSs). Compared with no correct answers to numerical DL and DS in the original group of 61 subjects, the rise to 33.3% correct represents a "dramatic" improvement. The 36 incorrect responses to the revised forms included 14 based on constancy assumptions (12 constant *S* and 2 constant *L*), 2 based on certainty and 1 on equal ignorance (i.e., 50/50).

None of the 54 bus problems was answered correctly, indicating no transfer of the solution strategy by those 18 subjects who have just solved a search problem (desk). Incorrect responses included 26 constancy-based answers (20 constant *S* and 6 constant *L*), 10 certainty and 5 equal ignorance.

Correct answers to the revised desk problem were often accompanied by lucid explanations of the underlying reasoning. Here is one example given in response to RDL: "The probability that I gave is the number of unlocked drawers remaining (unsearched) divided by the *total* number of drawers remaining (unlocked + locked)."

Similar to the explanations of incorrect answers to Problem 1, a constant 80% answer to RDL was justified by: "The overall probability doesn't change no matter how many drawers are searched," and, as maintained by another subject: "regardless of whether I looked in them or not." Some subjects who responded 80% throughout seemed to work hard not to be swayed by the given results: "It's like the boy/girl baby problem, even if you get BBBBBBBBBG the probability still remains at chance—50/50." Constant *S* responses to RDL were justified by, "I was almost fooled, but upon further thought I decided that as drawers are searched and found empty the statistics of the problem do not change. Same as if weather person says 50% chance of rain & it rains. Then is probability of rain 50% or 100%? It's still 50%." One subject, who gave 50% answers to RDS, explained: "The possibility of finding the letter was 50%, just like yes or no."

In a second survey, each subject received RDL *and* RDS, with order of presentation counterbalanced, *and* a bus problem of the same range (*L* or *S*) as the second of the two revised problems. The 109 subjects were undergraduate students of psychology or graduate students of education at the Hebrew University of Jerusalem. Fifty three got RDS, RDL, BL, in that order, and 56 got RDL, RDS, BS. Because of the extra length of this assignment subjects were not asked to explain their reasoning.

Correct responses were given to 48 of the 109 RDLs (44%), and to 73 out of 109 RDSs (67%), which is 56% correct overall. Every subject who correctly solved RDL correctly solved RDS as well, but not vice versa, suggesting that the revised short-term problem is more transparent. This makes sense if one notes that answering the *S* versions involves adjustment of only the denominator (the total number of remaining units) since the numerator is always one, whereas answering the *L* versions requires adjustment of both numerator and denominator. Assumptions of constancy, certainty, and equal ignorance were observed among the incorrectly answered forms. However, the absence of supporting explanations prevented a determination of subjects' underlying reasoning.

No single correct triplet of answers was given to any of the 109 bus problems. This was true despite the high rate of correct solutions of the immediately preceding revised desk problems. In particular, 58 of 109 subjects solved their second revised desk problem but not the equivalent bus problem of the same range. In conclusion, while the revised desk problems elicited more than half correct solutions, transfer of the method of solution to the bus problem failed to occur.

15.4 DISCUSSION

On the whole, subjects were unable to solve the numerical long- and short-term hope problems the way they were originally presented. To summarize our findings, we list several solution methods that subjects employed and beliefs they expressed. To be sure, this list is not exhaustive.

The load of processing the various details given in Problems 1 and 2 is eased if one of the givens (either L_0 or S_0, which is inferred from L_0) is held constant. Many subjects indeed based their answers on one of these constancy assumptions, solving the problem by reducing the number of variables involved. In so doing, they ignored one type of evidence, namely the search results, and considered only the *a priori* success probability and sometimes also the number of units.

Subjects' choice of the type of evidence may be linked to an external attribution of uncertainty. Many of the explanations cited above indicate that the prior was viewed as an inherent and unalterable characteristic of the setup. It may seem more "objective" than the information about the subsequent fruitless search (wait), and may therefore come to dominate subjects' reasoning.

Some subjects clearly resisted the urge to use epistemic considerations. The burden of providing the required probability, they insisted, should lie with the desk (bus system). We should note that all subjects had had some kind of introductory statistics course. Their cursory statistical knowledge apparently alerted them to the gambler's fallacy. The first examples of random processes usually given in class (successive coin tosses, childbirths, lotteries, etc.) are typically characterized by statistical independence. Students learn that they should *not* learn from experience since a coin has no memory. This lesson may be overgeneralized to the case of the hope problems, where successive failures do have a diagnostic value.

Those subjects who were aware of the need to consider the changes in their state of knowledge usually sensed the direction of the hope functions but did not know how to update their probabilities arithmetically. The concrete aid offered in the revised desk problem helped many of these subjects to simultaneously see the whole sample space and the subspace in which success may occur.

Failures in responding to the revised versions occurred when subjects were strongly committed to constancy assumptions. Whoever believes that the probability of finding the letter is 10% per drawer, regardless of how many drawers have been searched, will fail to adjust for the changing total number of drawers and will simply obtain the L function by multiplying 10% by the number of unlocked drawers that have not yet been eliminated.

In addition, a certain subgroup of subjects who answered the original problems, and the revised desk problem, was apparently outcome oriented. They resorted either to certainty or to complete indifference, both of which resulted in incorrect answers.

15.4.1 Why Didn't the Transfer Work?

We were puzzled by the failure of all the subjects who had solved the revised desk problem to transfer the solution's rationale to the bus problem. However, on second thought, and as a result of postexperimental discussions with some of the subjects, we have one possible reason for this failure of transfer.

The solution in the revised version was suggested by extending the dimension along which the search was carried out: two units (drawers) were added so that subjects could visualize the whole sample space and see the reason for the *a priori* L_0 of 80%. As they eliminated drawers, they could see the remaining "favorable" (unlocked) and "unfavorable" (locked) units of the changing space. When facing the bus problem, however, one cannot apply the same trick without changing the nature of the story. Extending the units of wait beyond midnight would not help to see the reason why L_0 is 60%. That prior reflects the fact that 60% of the bus routes operate this late, and 40% do not. A revision, equivalent to that of the desk problem, would have the bus *certain* to arrive sometime between 11:30 and 12:20, with equal probabilities for all the 10 five-minute intervals. The tourists, however, decide to wait until either the bus arrives or midnight, whichever happens first. Viewing the original bus problem as isomorphic to the revised desk problem was apparently too much to expect of subjects in an experimental situation.

The locked-drawers device can easily be applied to Dr Fischer's bomb situation (Example 2). Without loss of generality, we can change the story so that there are 12 Christmas crackers: one contains a bomb for sure, 11 contain checks. Only six, however, are at the guests' disposal for this party. The other six are kept for the next party. It is now easy to see that L_0 is 50% and to assess the L and S probabilities of pulling the bomb throughout the game's progress. We didn't pose this problem to our subjects. Our guess, however, is that transfer from the revised desk problem to this particular problem would have been within reach of some subjects.

15.4.2 Possible Extensions

Several subjects who viewed the search/wait process as analogous to coin flipping, incidentally raised an interesting question: *what if the search were to be conducted* **with** *replacement?* Suppose the man who comes home in the darkness with two bunches of keys (Example 3) is drunk. He would not be able to remove keys that have failed to unlock the door (see Feller, 1957, page 46). Or, imagine that an absent-minded professor is looking for a misplaced letter through the drawers of her desk (Problem 1) while her mind is deeply engaged in some other problem, thus forgetting instantaneously which drawers have been already searched. A "with replacement search" thus describes the case in which a key may be tried again after being found not to work or a drawer may be searched again after being found empty.

Does it make sense to think of waiting for the bus "with replacement"? Ennis (1985) describes a situation perfectly suited for our case. He imagines

> waiting for a bus on a route which offers a "15-minute service". Because of heavy traffic, the buses do not arrive at exact 15-minute intervals but randomly. The operators (Poisson Motor Services) do, however, provide a service which *averages out* at 15 minutes between buses. (page 27)

We only need to change 15 to 30 minutes, add the qualification that the chances are 60% that the bus is running that late, and Poisson Motor Services provide us a wait problem *with* replacement.

The computation of the long-term (L') and the short-term (S') hope functions for sampling with replacement requires a minor adjustment of formulas (15.1) and (15.2). One easily obtains, for the case of sampling *with* replacement:

$$L_i' = \frac{(n-1)^i L_0}{(n-1)^i L_0 + n^i(1 - L_0)} \qquad (15.1')$$

$$S_i' = \frac{1}{n} L_i' \qquad (15.2')$$

In both formulas $i = 0, 1, 2, \ldots, n-1, n, \ldots$

In contrast to the case of sampling without replacement, where the function L_i decreases with i and S_i increases (Figure 15.1), in the case of sampling with replacement, both L_i' and S_i' decrease. The rate of their decline, however, is slower than that of L_i. Figure 15.3 presents the course of the functions L_i' and S_i' compared with that of L_i and S_i, for the desk problem. In the limit, as i grows indefinitely, both L_i' and S_i' tend to zero. This means that despair creeps in justifiably in an extended fruitless search (wait) with replacement. In a without-replacement search, the rising S function may boost our morale to some degree. It is probably the short-range increase in hope that keeps most of us going.

Another extension of the original probabilistic model is obtained if in Problem 1 we allow for a less than perfect search. One may assume, for instance, that the conditional probability of finding the letter in a drawer, given it *is* there, is always p (such that $0 < p < 1$). This would, in fact, describe more realistically the state of affairs in desks of people like ourselves. In

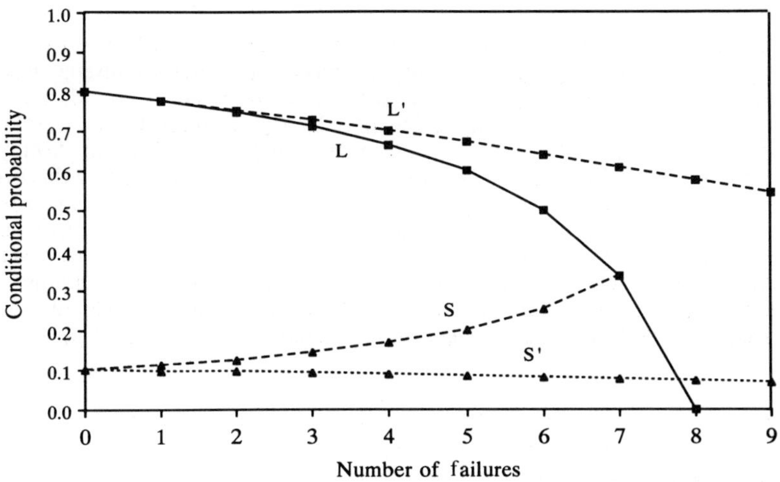

Figure 15.3 Long-term (L') and short-term (S') hope functions "with replacement" (compared with L and S)

addition to that, one may monitor the course of the hope functions while searching *within* each drawer. This will amount to extending the problem from the discrete to the continuous case.

Decisions of whether to continue or end a search (wait) depend not only on the long- and short-term success probabilities. The considerations should include the costs and benefits associated with each decision. These, however may change as the search proceeds.

Note that our search and wait problems involved desirable outcomes. The desirability of the target event makes no difference formally. It would be interesting to see, however, whether people's subjective functions, based on the same objective statistics, differ in any way when viewing the target event as "success" versus "failure." Clearly, picking one of two bunches of keys (Example 3) and trying them successively with or without replacement, is isomorphic to picking one of two guns, knowing that there is a bullet in one of them, and playing Russian Roulette with or without replacement. The "target" event, however, is so dramatically different in these two cases, that finding differences in probabilistic assessments would not be surprising. Likewise, waiting for malignant symptoms to reappear during a five-year period after treatment, although structurally equivalent to waiting for your loved one to come to a date during five successive short-time intervals, may be evaluated differently in probabilistic terms. Clearly, continued study of these phenomena is required.

ACKNOWLEDGEMENTS

Part of the research described in this chapter was carried out while the first two authors were at the Department of Psychology at the University of Massachusetts, Amherst. The study was supported in part by the National Science Foundation grant MDR-8954626 to Clifford Konold, and in part by the Sturman Center for Human Development, The Hebrew University, Jerusalem. We are grateful to Arnie Well, Sandy Pollatsek, and Jill Lohmeier for preliminary discussions concerning this project, to Dorit Rivkin and Oren Gilady for conducting pilot tests and raising suggestions, to Peter Ayton for contributing engaging examples, and to Oren Falk for being the first to solve the hope problems by common sense. Many of our students, who acted as subjects, have enriched our understanding by their comments and discussions.

Special thanks are due to Raphael Falk, who has devotedly helped us in all the stages of the study: developing concepts, identifying examples, and writing it up.

REFERENCES

Ayton, P. & McClelland, A. (1987) The despicable Doctor Fischer's (Bayesian) bomb party. *Teaching Mathematics and Its Applications*, **6**, 179–83.

Bell, C.R. (1979) Psychological aspects of probability and uncertainty. In C.R. Bell (ed.), *Uncertain Outcomes*, MTP Press, Lancaster, England, pages 5–21.
Cohen, J. & Meudell, P. (1968) Strategies of search. *Acta Psychologica*, **28**, 322–43.
Ennis, J. (1985) Statistics, St Petersburg and Sellafield. *New Scientist*, May, 26–28.
Falk, R. (1992) A closer look at the probabilities of the notorious three prisoners. *Cognition*, **43**, 197–223.
Falk, R. (1993) *Understanding Probability and Statistics: A Book of Problems*. AK Peters, Wellesley, MA.
Falk, R. & Konold, C. (1992) The psychology of learning probability. In F.S. Gordon & S.P. Gordon (eds.), *Statistics for the Twenty-First Century*. The Mathematical Association of America, USA, pages 151–64.
Feller, W. (1957) *An Introduction to Probability Theory and its Applications*, Vol. 1 (2nd edn). Wiley, New York.
Gabriel, K.R. (1960) Nuptiality and Fertility in Israel. Doctoral dissertation (in Hebrew with English summary). The Hebrew University, Jerusalem.
Gigerenzer, G., Swijtink, Z., Porter, T., Daston, L., Beatty, J. & Krüger, L. (1989) *The Empire of Chance: How Probability Changed Science and Everyday Life*. Cambridge University Press, Cambridge.
Greene, G. (1980) *Doctor Fischer of Geneva or the Bomb Party*. Penguin, London.
Hacking, I. (1975) *The Emergence of Probability*. Cambridge University Press, Cambridge.
Jones, D.E.H. (1966) On being blinded with science. *New Scientist*, November, 465–7.
Kahneman, D. & Tversky, A. (1982) Variants of uncertainty. *Cognition*, **11**, 143–57.
Konold, C. (1989) Informal conceptions of probability. *Cognition and Instruction*, **6**, 59–98.
Konold, C., Lohmeier, J., Pollatsek, A., Well, A., Falk, R. & Lipson, A. (1991) Novices views on randomness. *Proceedings of the Thirteenth Annual Meeting of the International Group for the Psychology of Mathematics Education—North American Chapter*, **1**, 167–73.
MacGregor, D., Fischhoff, B. & Blackshaw, L. (1987) Search success and expectations with a computer interface. *Information Processing and Management*, **23**, 419–32.
Meshalkin, L.D. (1973) *Collection of Problems in Probability Theory* (L.F. Boron & B.A. Haworth, trans.) Noordhoff, Leyden, The Netherlands (Original work published 1963).
Shimojo, S. & Ichikawa, S. (1989) Intuitive reasoning about probability: Theoretical and experimental analyses of the "problem of three prisoners". *Cognition*, **32**, 1–24.
Tversky, A. & Kahneman, D. (1982) Evidential impact of base rates. In D. Kahneman, P. Slovic & A. Tversky (eds.), *Judgment under Uncertainty: Heuristics and Biases*. Cambridge University Press, Cambridge, pages 153–60.
Zabell, S.L. (1988) Symmetry and its discontents. In B. Skyrms & W.L. Harper (eds.), *Causation, Chance, and Credence*. Vol. 1. Kluwer, Dordrecht, pages 155–90.

Part Three

Accuracy of Probability Judgments

Chapter 16

Subjective Probability Accuracy Analysis

J. Frank Yates
The University of Michigan

Sally Miller belongs to a parole board. So does Bill Coleman. They are participating in a parole hearing for Charles Starks. Ms Miller votes to deny parole, contending that "Mr Starks is clearly a bad risk." She expects that, if released, Mr Starks would soon commit other serious crimes and, if society is lucky, end up right back in the penitentiary. In contrast, Mr Coleman thinks that Mr Starks has responded well to the prison's rehabilitation efforts. In his view, if Mr Starks were allowed to leave, he would be a "productive, law-abiding citizen." Such disagreements between Sally Miller and Bill Coleman are not uncommon. In fact, Ms Miller and Mr Coleman seldom concur on the prospects of potential parolees returning to lives of crime. Thus, more often than not, their votes on the parole board offset each other, forcing the sentiments of Irene Thomas, the third board member, to carry the day.

The scene described above provides a good illustration of the issues addressed in this chapter. Whenever a prisoner is brought before the parole board, Sally Miller forms an opinion of how likely it is that the prisoner would get in trouble if released from confinement early. If that opinion is sufficiently pessimistic, she votes to deny parole. Bill Coleman and Irene Thomas use similar decision rules. But, as we have seen, all the board members tend to differ in their judgments of the chances that particular inmates would resume

Subjective Probability. Edited by G. Wright and P. Ayton.
© 1994 John Wiley & Sons Ltd.

their criminal activities. The following are sensible questions we might then ask:

- *Overall accuracy*. How good are the opinions of the parole board members as predictions of inmates' actual post-prison criminal behavior? Which board members' opinions are best, and whose are worst? Hence, whose opinions should be relied on most heavily and whose should be given less weight?
- *Incentives*. We can imagine various reasons why board members might be disinclined to offer their "best" opinions about parolees' recidivism chances. How might they be induced to do otherwise?
- *Accuracy elements and contributors*. In what specific ways are board members' forecasts of post-release activity especially good or poor—and why? That is, what are distinguishable components of judgment accuracy, and what conditions and skills affect those elements?
- *Training*. Suppose the parole board decides that it wants to improve its members' skills at predicting parolees' post-release behavior. How could such training be effected?
- *Selection*. Eventually, new parole board members will be required. How could the prison commission select board members who are good at anticipating prisoners' chances of "going straight" if released from prison early?

As they are articulated here, these questions apply to our fictional parole board. But situations formally equivalent to these circumstances are ubiquitous. That is, people must often form opinions about the chances of various events occurring. Sometimes they must act on, or even publicly report, those opinions. And in each of these situations questions similar to the present ones arise. This chapter surveys how such questions can be approached generally, in all kinds of contexts.

What do these issues have to do with subjective probability? In actual practice, only occasionally are people required to render their opinions in the form of subjective probability judgments per se. But in principle they could be asked to do so at any time. For instance, our parole board members could be requested to state P'(No trouble) for every prospective parolee who appears before them. In this notation, P' denotes a subjective probability judgment and "No trouble" describes the event that a parolee remains free of criminal activity for a specified time period. The approaches to answering the questions reviewed here all assume that the focal individuals have in fact indicated their opinions as subjective probability judgments. Indeed, the availability of these analytic techniques is a major advantage of having opinions expressed as probability statements. Most of these methods apply to judgments for categorical events, e.g. "No trouble," as in the current example. But some work concerns judgments for quantities, e.g. the number of weeks a parolee stays out of trouble (e.g. Matheson & Winkler, 1976; Schaefer & Borcherding, 1973; Seaver, von Winterfeldt, & Edwards, 1978). Because of space limitations, we can discuss only the simplest discrete event methods. One consolation, however, is that the essential ideas generalize.

16.1 OVERALL ACCURACY MEASURES: A PERSPECTIVE

Conceptually, probability judgments are said to be accurate overall if events that actually occur tend to be assigned high probabilities while events that do not happen receive low ones. In the extreme, probability judgments of 1 and 0 are given to events that do and do not occur, respectively. For instance, an ideal, clairvoyant recidivism forecaster would report $P'(\text{No trouble}) = 1$ for each parolee who stays free of criminal activity and $P'(\text{No trouble}) = 0$ for every one who does not. Implicit is the general strategy commonly employed for characterizing the overall accuracy of subjective probability assessments.

Why should overall accuracy measures interest us? First of all, they could be useful in the selection of judges. A hallowed principle in all selection procedures is that a good predictor of future performance is past performance. Thus, a judgment "consumer" should seek the assessments of judges who have earned good accuracy scores in the past. Accuracy measures have been expected to have incentive value, too. Since the judgment consumer's aim is accuracy, then it only makes sense that judges should be compensated on the accuracy scores they achieve. If they are, then this should encourage judges to devote appropriate effort. This leads naturally to a third potential use for accuracy scores—training. Such measures could be useful as feedback in improvement efforts. They could guide the course of those activities, e.g. the retention of training operations that tend to improve scores and the abandonment of ones that do not.

A variety of overall accuracy measures have been proposed (cf. Winkler & Murphy, 1968; Yates, 1984, 1990). However, we focus here on only one of them, the quadratic score. It is by far the most frequently employed of these measures. The quadratic score has several important features in common with those other measures and thus can be used to illustrate those features. But it also has useful properties that the others do not (cast against, of course, the strengths of those other approaches).

16.2 THE QUADRATIC SCORE

16.2.1 Notation

We represent a probability judgment for a given target event A, e.g. "No trouble," by

$$f = P'(A)$$

Typically, studies of probability judgment accuracy do not require that—or even examine whether—such judgments conform to probability theory axioms. Thus, it is an open question whether, strictly speaking, such statements constitute true subjective "probabilities." Instead, the focus is solely on what is called the "substantive goodness" (Winkler & Murphy, 1968), the "external correspondence" (Yates, 1982), or, as here, simply the "accuracy" of those judgments as predictions of real-world occurrences (Yates, 1990). Nevertheless, analysts normally impose on their data the minimal constraint that judgments for complementary events sum to 1, i.e. that $P'(A^c) = 1 - f$, where A^c represents the complement of event A.

An indicator variable called the *outcome index* (d) can be defined as follows:

$$d = 1 \text{ when } A \text{ occurs}$$

and

$$d = 0 \text{ otherwise, i.e., when } A^c \text{ occurs}$$

It sometimes helps our intuitions to think of d metaphorically as the judgment of a clairvoyant. Such an individual reports $P'(A) = 1$ whenever target event A occurs and $P'(A) = 0$ when it does not.

16.2.2 The Rule

The *quadratic score* essentially says that a real person's judgments are accurate to the degree that they are close to the judgments of our fictional clairvoyant. And it entails a specific interpretation of what "close" means. One particular form of the *quadratic scoring rule* is the following:

$$Q(f, d) = 1 - (f - d)^2 \qquad (16.1)$$

Thus, larger scores indicate greater accuracy. Perfect accuracy is indicated when $Q = 1$, for in that instance there is a perfect match ($f = d$) between the assessments of the clairvoyant and the real judge. Values of Q decrease quadratically with increasing discrepancies between f and d. The extreme is observed when the judge is absolutely sure that A will occur ($f = P'(A) = 1$), but it does not ($d = 0$), or vice versa. Of course, analysts normally are interested in mean values of Q for representative collections of cases, i.e.

$$\bar{Q} = (1/N)\Sigma Q \qquad (16.2)$$

for samples of size N.

In practice, the most commonly used version of the quadratic score is a linear transformation known as the *probability score* (PS):

$$PS = (f - d)^2 \qquad (16.3)$$

Clearly, the ideal value of PS is 0, and the worst possible accuracy is indicated when $PS = 1$. The mean value of PS,

$$\overline{PS} = (1/N)\Sigma\, PS \tag{16.4}$$

for a sample of N observations, describes typical judgmental accuracy. $\overline{PS}$ is also often described as the *Brier score*, after the meteorological statistician who introduced it (Brier, 1950).

16.2.3 Standards

Suppose a judge's accuracy is being graded and perhaps rewarded according to a scoring rule, say, $\overline{PS}$. Then we might expect the judge to do his or her best to optimize that score. But motivation research suggests that this might well not occur. Indeed, a sizable literature indicates distinct advantages of specific performance goals over "do-your-best" appeals (e.g. Locke et al., 1981). If a judge (or the consumer of his assessments) wants to set such concrete targets, which ones are reasonable? This implies the need for accuracy score standards, for ways of answering the question, "Is this good enough?" Here we discuss several answers to this question.

Human judge standards. One reasonable standard that might be set for a given judge is defined by the performance of other human judges:

- *The best judge*. Suppose that records have been kept on lots of judges who have considered similar if not identical cases. Then the best score of those competing judges is one standard one might set.
- *The typical expert judge*. Especially for a beginning judge, the performance of the best judge ever observed might be too exacting a standard. The implied performance level might be so high that it is out of reach and hence useless as a realistic target. But suppose there are judges who, for various reasons, are considered experts. Then the average performance of those individuals might constitute a more suitable goal.
- *The average peer judge*. Suppose a more modest but perhaps more effective goal is sought. Then the standard set by the judge's peers could make sense. For instance, a judge might find it especially compelling to try to achieve a value of $\overline{PS}$ better than those earned by half the other individuals assessing similar cases.

Artificial judge standards. In some judgment situations there exist artificial judges designed to accomplish the same task as human judges. Typically, these are computer-driven models. They might be ones instantiating optimal statistical procedures (as illustrated below). Or they could be production rule models intended to mimic acknowledged human experts.

Constant judge standards. A constant judge is an idealized individual who reports the same judgment for each case that comes along. That is, no distinctions are made among the situations that arise; $f = P'(A) = c$, a constant, is reported in every instance, regardless of the particulars of the case. For example, a constant judge in a medical diagnosis situation might report P'(Hepatitis) = 0.18 for every patient considered, no matter what the patient's signs and symptoms are. On the face of it, it seems that any judge "worth his salt" ought to be able to outperform such a constant judge. Nevertheless, as described here, several constant judges are commonly used as meaningful standards:

- *The uniform judge.* Suppose an individual believes that he knows so little that he can say nothing defensible about whether target event A will or will not occur. Then that judge—in the spirit of the Laplace (1951) decision rule for conditions of ignorance—might then consider all possibilities equally likely. When only A and its complement A^c are distinguished, behaving like such a "uniform judge" would imply reporting $f = P'(A) = 0.5$ in every case. It is easy to show that such a uniform judge would earn a mean probability score of $\overline{PS} = 0.25$, regardless of how often the target event actually occurs. It might seem that the uniform judge is such a modest performance standard that no one would ever fail to achieve it. Not so; in certain contexts, people commonly fall short of the uniform judge's standard (e.g. Staël von Holstein, 1972; Yates, McDaniel, & Brown, 1991; Yates et al., 1989). In fact, it is of considerable psychological interest how and why this happens as often as it does.
- *The historical base rate judge.* Suppose that over some specified collection of past cases, target event A has occurred at a *historical base rate* of h. For instances suppose that for each of a given number of years, stocks of a certain class have risen in price 35% of the time. Then the pertinent historical base rate is $h = P'$(Price increase) = 0.35. Now, suppose that for a collection of current cases 0.35 is reported for every stock. That is, the judge indicates P'(Price increase) = h = 0.35 for each new security, regardless of the company involved. Then this individual is behaving like a "historical base rate judge." The performance level of this particular constant judge is another standard that might be applied to the assessments of real, human judges. Such a standard is indeed used in the US National Weather Service, where it is described as the "climatological" forecaster (e.g., Murphy et al., 1985).
- *The sample base rate judge.* Suppose a judge considers N cases. In the notation introduced previously, the relative frequency or *sample base rate* for the target event in that collection of cases is

$$\bar{d} = (1/N)\Sigma d \qquad (16.5)$$

  Imagine that there is a judge (a semi-clairvoyant?) who can somehow anticipate the sample base rate and reports $f = P'(A) = \bar{d}$ for each of the N

cases considered. Such a fictional individual is described as a "(sample) base rate judge." As can be shown with decompositions of $\overline{PS}$ described below, the mean probability score earned by an arbitrary constant judge who reports $f = P'(A) = c$ for every case is given by

$$\overline{PS}_{\text{constant}} = \bar{d}(1 - \bar{d}) + (c - \bar{d})^2 \tag{16.6}$$

It is thus apparent that the base rate judge is the best possible constant judge, one that earns a mean probability score of

$$\overline{PS}_{\text{base rate}} = \bar{d}(1 - \bar{d}) \tag{16.7}$$

16.2.4 The Properness Concept: Expected Compensation for Candor

We can easily imagine reasons a judge might withhold his or her true opinion about the chances of a target event's occurrence, to hedge in one direction or the other. For instance, a physician making diagnoses might want to report probabilities of disease that are somewhat higher than the physician really believes, feeling that this is in the patient's best interests (Wallsten, 1981). A certain class of accuracy scores—which includes the quadratic rule—has a property called "properness" that in principle should discourage such hedging.

Simply put, the properness idea is the following (see Yates, 1990, Chapter 8): Suppose a judge is being compensated for accuracy according to score S, such that high scores are good (e.g. $Q = 1 - PS$). On a given judgment occasion, the judge has a true opinion $f_T = P'(A)$. On the other hand, the judge might consider reporting any number of judgments f_R, all but one of them different from f_T. From the judge's perspective, the subjective expected score (SEV) for reporting f_R is thus:

$$\begin{aligned} SEV(\text{Report } f_R) = {} & P'(A)\text{Score(When } f_R \text{ is reported and } A \text{ occurs)} \\ & + P'(A^c)\text{Score(When } f_R \text{ is reported and } A^c \text{ occurs)} \end{aligned}$$

In our current notation, we have

$$SEV(\text{Report } f_R) = f_T S(f_R,\ d = 1) + (1 - f_T) S(f_R,\ d = 0)$$

We say that the scoring rule S is *proper* (Winkler & Murphy, 1968) or *reproducing* (Shuford, Albert, & Massengill, 1966) if SEV is maximized when $f_R = f_T$. That is, a scoring rule is proper if the judge maximizes his or her subjective expected score by candidly reporting his or her actual opinion. Thus, since the quadratic score is proper, this constitutes something of a "fringe benefit" of using it as a means of compensating judges for their accuracy.

16.2.5 Implementation

Have scoring rules like *PS* been applied as means of compensating judges? Although not common, applications do exist. For instance, they are not unusual in experimental situations as a means of encouraging subjects to take tasks seriously and to be forthright (e.g., Yates, McDaniel, & Brown, 1991).

US National Weather Service meteorologists make various kinds of forecasts in probability form, e.g. their precipitation predictions. The accuracy of these forecasts is indexed by a form of the quadratic score. It is unclear whether the meteorologists are financially rewarded according to their individual scores (Murphy & Winkler, 1984). However, the mere fact that their performance is graded on that basis implies a form of compensation, to the extent that the forecasters are concerned about how their supervisors and peers regard their work.

Academic testing is another context in which judges can, in essence, be compensated on the quality of their assessments as indexed by accuracy scores. Take, for instance, the kinds of multiple choice tests medical students at the University of Connecticut have taken sometimes (Rippey & Voytovich, 1983). For each item in these tests, the student does not state categorically which alternative answer is correct. Instead, the student indicates a subjective probability that each option is the right one. The student's grade for the test is a function of the logarithmic scores for all the items. The logarithmic score is an accuracy measure in the class of proper scoring rules that includes the quadratic score (Yates, 1984).

16.2.6 Effectiveness

Are scoring rules effective as a tool for inducing people to produce and report better judgments? Curiously, despite the fact that the scoring rule idea has been around for a long time, no one appears to have attempted a definitive study of the issue. What is required, of course, is a controlled experiment. In one group subjects would make judgments given the promise of rewards or at least feedback according to the quality of those judgments as evaluated via the kinds of accuracy scores we have discussed. Control subjects would be given no such promise. Several studies have approached this ideal, but have not quite arrived there.

A study by Fischer (1982) in some respects was perhaps the closest to having a suitable design. In that investigation, subjects made probabilistic postdictions of other students' grade point averages. Some subjects were promised payoffs according to a so-called truncated logarithmic score. Other subjects not only were given this promise, but also learned the score they earned on each trial. Fischer found that the promise of compensation based on scores had a positive effect on judgment accuracy, but that case-by-case scoring rule feedback did not. Fischer did not attribute the observed incentive effect to scoring rules *per se*, however. Instead, subjects' response patterns suggested that the effect was probably due to the fact that the log rule yields very large negative scores for extreme misjudgments (e.g. reporting $f = 99\%$ for an event that does not occur), a property that is not shared by other rules, such as the quadratic. Another feature of the experiment further limits the conclusions it permits: the truncated log score is not proper (Shuford, Albert & Massengill, 1966).

Staël von Holstein's (1972) study of probabilistic stock market forecasting came close to having a sufficient design, too. A stated major purpose of that study was to determine whether providing periodic quadratic score feedback would allow subjects to improve the accuracy of their predictions of stock price movements. It did not. Several considerations suggest that accepting this negative result as definitive might be premature. There was no distinct control group as such in the experiment. Instead, given the within-subjects design, each subject served as his or her own control. The expectation was that, if feedback were effective, then as the experiment proceeded, accuracy should have improved. It is conceivable that the performance of a true control group would have been consistently worse than that of all the subjects, from the very beginning of the experiment; all groups in the study were promised feedback from the outset. Another potential problem is that subjects were required to forecast stock price changes over successive two-week periods. Some finance experts could contend that such short time horizons make the prediction task so difficult that realistically no one could be expected to improve.

Echternacht (1972) has reviewed the use of scoring rules in academic testing situations. He concluded that there is little evidence of their effectiveness. However, the scoring rules used in academic settings almost always have been truncated logarithmic rules, which have several problems associated with them, e.g. that they are not proper.

16.3 OVERALL ACCURACY MEASURE DECOMPOSITIONS: A PERSPECTIVE

It is possible that scoring rule feedback "works" in the sense that—especially when tied to compensation—it encourages great effort by the judge. But such feedback might be too coarse to inform such a motivated individual about precisely *how* to improve his or her judgments. This is consistent with the results of Staël von Holstein's (1972) stock market forecasting experiment. Detailed information about specific aspects of judgment performance that are particularly good or poor might be required.

More generally, decomposition efforts have been envisioned as a tool for explaining how and why judges achieve the levels of performance indicated by measures of the overall accuracy of their assessments. These analyses typically do not definitively provide detailed explanations directly. Instead, they narrow the range of possibilities, directing the analyst toward plausible hypotheses that can be tested more pointedly by other methods, such as experimentation. Here we describe and illustrate two of the more popular approaches. Other methods exist that apply to the same types of judgments as well as others (e.g. Blattenberger & Lad, 1985; Hsu & Murphy, 1986; Murphy & Winkler, 1992; Shapiro, 1977; Stewart, 1990). Space limitations preclude discussion of them, however.

The present analysis differs from most previous ones. Earlier discussions have been limited mainly to statistical issues. The emphasis here is on the judgment procedures and other forces that might underlie various statistical indicators. This is important to both scientific and practical aims. There have been numerous studies in which researchers have provided subjects with feedback about their performance in the form of the measures identified below. These efforts have met with only mixed success. There is reason to suspect that this disappointing record is at least partly due to subjects not knowing specifically what they can do to improve their performance measures. The requisite concrete advice is implicit in the conceptual analyses offered here.

16.4 THE MURPHY DECOMPOSITION

16.4.1 The Decomposition

The most frequently used $\overline{PS}$ decomposition is due to Murphy (1973). That decomposition can be expressed as follows (Yates, 1990, Chapter 3):

$$\overline{PS} = \bar{d}(1 - \bar{d}) + CI - DI \tag{16.8}$$

In this equation, note the following:

- $\bar{d}(1 - \bar{d})$ is the variance of the outcome index, Var(d), since d is an indicator variable.
- CI is called the *calibration index* and is defined by

$$CI = (1/N) \sum_{j=1}^{J} N_j(f_j - \bar{d}_j) \tag{16.9}$$

Here the judge's probability assessments are restricted (or rounded after the fact) to J categories. For instance, if the judge reports only deciles, we would have $J = 11$, with $f_1 = 0\%$, $f_2 = 10\%$, . . . , $f_{11} = 100\%$. The term N_j indicates that judgment category f_j is used that many times, e.g. for each of $N_2 = 35$

stocks, an analyst might report a probability of $f_2 = 10\%$ that the stock will increase in price, with

$$N = \sum_{j=1}^{J} N_j.$$

The term $\bar{d}_j$ can be thought of as a *conditional base rate* or *conditional observed proportion*, the proportion of times the target event actually occurs given that judgment f_j is reported. For instance, if the prices really do increase for 7 of 35 stocks judged to have 10% chances of increasing, then $\bar{d}_j = 7/35 = 20\%$. A judge is said to exhibit good *calibration* to the extent that that individual's judgments match the corresponding relative frequencies with which the target event actually happens. Given its form, *CI* clearly measures the judge's calibration skill, with small values indicating better calibration.

- DI is the *discrimination index* and is defined thus:

$$DI = (1/N) \sum_{j=1}^{J} N_j(\bar{d}_j - \bar{d}) \tag{16.10}$$

In this equation, all the terms are as described previously. The discrimination index measures the degree to which the judge's assessments exhibit the *discrimination* or *resolution* of events. The discrimination concept entails significant subtleties, so some discussion is in order.

Two kinds of judgment occasions exist, those on which target event A actually happens and those on which event A fails to happen. Suppose a judge is capable of placing these two kinds of occasions into distinct categories; the judge never puts into the same class two different occasions, one of which results in event A's occurrence, the other of which does not. Then clearly this judge is capable of perfectly discriminating instances when event A is destined to happen from those when it is not. The discrimination index *DI* measures one's ability to attain this kind of discriminative categorization.

In the probability judgment situations of interest here, the judge can apply any one of labels $f_1, f_2, \ldots, f_J$ to each case that comes along. Suppose that our perfectly discriminative judge decides—perhaps even arbitrarily—to put some of those labels, say, 0%, 30%, 70%, and 90%, into Group 1. The judge then forms another, disjoint collection of labels we can call Group 2, including, say, 10% and 80%. (Note that not all categories must be used.) Since our judge has perfect discrimination skill, she can attach any of the labels in Group 1 to instances when she "knows" that event A will occur, and any label in Group 2 to instances when she realizes that A will *not* happen. We can show that this perfectly discriminative judge's probability assignments will attain the best possible value of $DI = \text{Var}(d) = \bar{d}(1 - \bar{d})$.

At the opposite extreme, the worst possible value of $DI = 0$ would be achieved when there is no connection at all between a judge's tendency to select a judgment category and the tendency for the target event to occur or not occur. For under those conditions, the relative frequency of the target event's occurrence when an arbitrary category j is used ($\bar{d}_j$) is the same as it is when any other category j^* is selected ($\bar{d}_{j^*}$). That is, all conditional proportions must be identical to one another (i.e., $\bar{d}_1 = \bar{d}_2 = \cdots = \bar{d}_J$) and hence to the overall base rate of the target event ($\bar{d}$).

Note that discrimination performance has nothing to do with *which* labels the judge assigns to event A on particular occasions. All that matters is that *different* labels be given when event A is going to occur than when it is not. In contrast, calibration, as reflected in *CI*, essentially reflects a label-selection activity, the assignment of "appropriate" numerical labels to cases in given categories.

16.4.1 Contributors

In terms of underlying mechanisms, the Murphy decomposition is most easily understood with the aid of schematics like that shown in Figure 16.1. At the highest level of the tree is the overall accuracy construct, as indexed by the measure $\overline{PS}$. That construct is derived from three others, as isolated in the

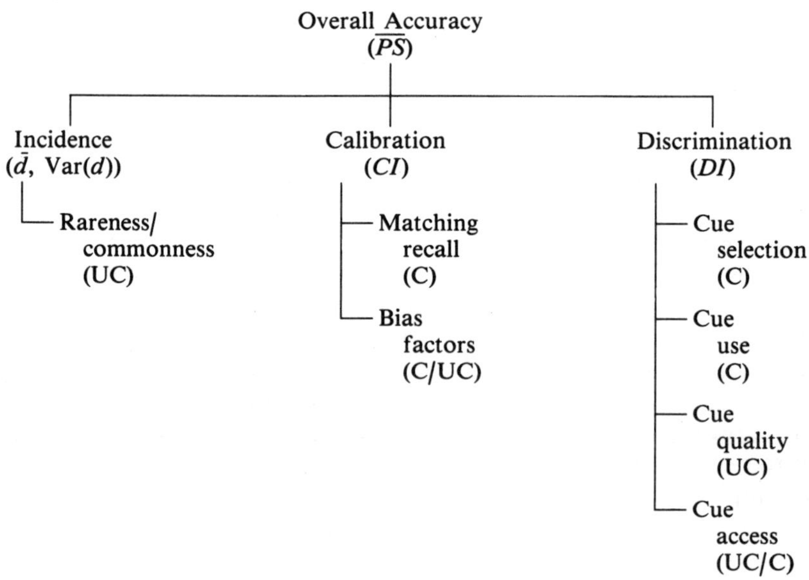

Figure 16.1 Schematic representation of overall accuracy in terms of its aspects as distinguished in the Murphy decomposition of the mean probability score, along with controllable (C) and uncontrollable (UC) factors that plausibly contribute to those aspects.

Murphy decomposition. First is the incidence for the target event, as reflected in the base rate $\bar{d}$ or, equivalently, the outcome index variance Var(d). Then there is the calibration construct measured by the calibration index CI. And, finally, there is discrimination, evaluated by the discrimination index DI. Each of these constructs in turn rests on one or more contributing factors. Figure 16.1 identifies a non-exhaustive collection of plausible ones.

As indicated in the figure, some contributing factors are controllable by the judge while others are not. The distinction is useful in evaluating judges. Suppose that Judges A and B tell a prospective consumer that their overall accuracy scores are the same. The consumer would be inclined to think that the judges are equally skilled. But suppose that Judge A accumulated his accuracy record in a context where accuracy is largely dictated by uncontrollable factors. In contrast, Judge B established his credentials under conditions in which uncontrollable factors played little part. Then obviously Judge B is the better judge. The controllable/uncontrollable distinction is pertinent to improvement efforts, too. In seeking means of enhancing accuracy, we should only pay attention to those performance dimensions that are subject to what the judge actually does. We now review various contributors, according to the Murphy decomposition elements they affect.

Incidence. The incidence of the target event entails a particular kind of judgment difficulty. As such, it is, of course, outside the judge's control. Specifically, it is the type of difficulty implicit in the target event's rareness or commonness, what we might call *incidence difficulty*. An example illustrates the concept.

Imagine a country that has dry and rainy seasons. In the dry season, it almost never rains; precipitation is observed on only about 3% of the days during that period. In contrast, during the rainy season, rain is recorded on roughly half the days. Many people would say that, intuitively, it seems quite easy to predict the weather on any given day during the dry season because rain virtually never occurs then. If categorical, nonprobabilistic judgments were required, they would advise, "Just say, 'No, it's not going to rain.'" In our scenario, this would guarantee a 97% hit rate. But such a simple strategy works poorly during the rainy season, when on any given day one is just as likely to see rain as not. The "No rain," constant prediction policy would yield a hit rate of only 50%. The probabilistic analog to such a policy is the strategy of a sample base rate judge, the best possible constant probabilistic judge. As indicated in Equation (16.7), in the dry season (assuming the current season is representative) the base rate judge would achieve a mean probability score of $\overline{PS}_{\text{base rate}} = (0.30)(0.97) = 0.0291$. But in the rainy season, the base rate judge would earn a much worse $\overline{PS}_{\text{base rate}} = (0.5)(0.5) = 0.25$.

Calibration. The calibration construct should be largely though not completely controllable. As indicated in Figure 16.1, it can be affected by detailed recall about the match or mismatch of the judge's assessments and the corresponding relative frequencies of the target event's occurrence. Indeed, there have been numerous demonstrations that feedback about *CI* (including graphical depictions of it) can be an effective means of improving calibration (e.g. Benson & Önkal, 1992; Lichtenstein & Fischhoff, 1980; Murphy et al., 1985). Such feedback can be seen as a kind of structured recall support. "Bias" is a common and special form of miscalibration which can affect miscalibration more generally, as measured by *CI*. Since bias is more easily understood in the context of the covariance decomposition of $\overline{PS}$, the discussion of it is deferred until that decomposition is introduced below.

Discrimination. In order to exhibit any discrimination beyond a chance level, the judge must rely on diagnostic information or "cues" presented by the judgment situation. "Diagnostic" means that there is a statistical association between target event *A* and the given cue, e.g. between a firm's eventual stock price change and some characteristic of the firm. As suggested in Figure 16.1, there are at least two controllable contributors to discrimination. The first is the judge's selection of cues. Out of the myriad items of information typically available in a judgment situation, the judge must choose to attend to some and ignore the rest. To the extent that the selected cues really are diagnostic and the ignored ones are not, then the chances of good discrimination are enhanced. These chances will not be realized, though, unless the other controllable contributor is favorable, that the judge uses the selected cues appropriately. For instance, the judge should respond sharply to highly diagnostic cues and more regressively to weaker cues. And there is little reason to expect that real people will always behave this way (e.g. Ganzach & Krantz, 1990; Kahneman & Tversky, 1973).

Certain uncontrollable discrimination contributors imply another kind of judgment difficulty (cf. Yaniv, Yates, & Smith, 1991), what we might call *cue-based difficulty*. Once more, an example highlights the idea. Consider predicting rain during the rainy season in the country described above. Also consider predicting whether a head will appear on a given toss of a fair coin. In both situations, the target event base rate is about 0.5; hence the incidence difficulty is the same. But most of us would say that predicting coin toss outcomes is far more difficult, bordering on "impossible." Why? The difference lies in the quality of accessible cues. In the case of rain, the appearance of the sky has some association with subsequent precipitation. And as long as we can look out the window, we can exploit that association. However, in the coin tossing case, such diagnostic cues either do not exist or are beyond current powers of discernment. More generally, situations can be expected to differ substantially in their possession of high quality, diagnostic cues. Those situations will also differ in the access they allow to such cues. Sometimes that access can be improved (e.g. by the development and deployment of weather satellites), but sometimes not.

16.4.2 An Example: Pneumonia Diagnoses

Tape et al. (1991) asked physicians at medical centers in Virginia, Illinois, and Nebraska to make probabilistic pneumonia diagnoses, i.e P'(Pneumonia), for all patients with complaints suggesting pneumonia as a possibility. Figure 16.2 shows the results of a Murphy decomposition analysis of those diagnoses.[1] The displays in Figure 16.2 are called *calibration graphs*. Potential judgment categories (f_j) are shown on the abscissa, and the corresponding proportions of times the target event actually occurred ($\bar{d}_j$) appear on the ordinate. For instance, the Virginia graph shows a point at about (0.65, 0.42), with the number 9 next to it. This indicates that there were 9 instances when the Virginia physicians indicated that patients had 65% chances of pneumonia and that about 42% of those patients ultimately were confirmed to have that disease. The calibration graphs shown in Figure 16.2 are enhanced in that the areas of the points are proportionate to the associated frequencies.[2] This allows the viewer to immediately and perceptually appreciate the significance of those points; recall (Equations (16.9) and (16.10)) that the calibration and discrimination measures CI and DI essentially weight points by the numbers of cases they entail (N_j). Also observe that each graph contains a horizontal dotted line at an elevation corresponding to the sample base rate $\bar{d}$. Pertinent decomposition statistics are shown in the body of each graph as well.

Consider the Virginia graph. There we see that the overall accuracy of the physicians' judgments, as indexed by $\overline{PS}$, was superior to that of any constant judge, since it surpassed that of the base rate judge. The same was true for the Nebraska diagnoses, but not for the ones in Illinois. In other words, in terms of $\overline{PS}$, the Illinois physicians would have done better had they reported P'(Pneumonia) = base rate = 11% as a diagnosis for each of their patients. It is significant that the base rates for pneumonia were radically different in the three locations, implying that in the sense of incidence difficulty, the diagnostic task was far harder in Nebraska than in either Virginia or Illinois. Nevertheless the Nebraska diagnoses were much better. The rest of the analysis offers at least some clues about how and why this happened.

The calibration indexes, CI, indicate that, although the calibration performance levels of the Virginia and Illinois physicians were comparable, that of the Nebraska physicians was markedly superior. This is apparent graphically, too. A calibration graph indicates good calibration to the degree that the points in the graph lie close to the 1 : 1 diagonal. The fit is distinctly better in the Nebraska graph.

Recall that large values of the discrimination index DI imply good discrimination. We see that, as in the case of calibration, there were substantial location differences in discrimination, although the pattern of strengths and weaknesses was somewhat different than was observed for calibration. The

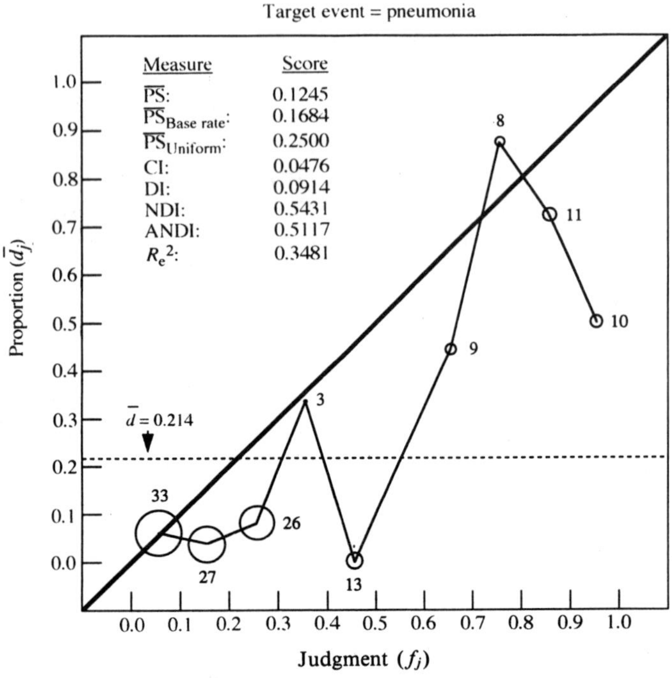
Target event = pneumonia
Measure
Score
PS:
0.1245
PS Base rate:
0.1684
PS Uniform:
0.2500
CI:
0.0476
DI:
0.0914
NDI:
0.5431
ANDI:
0.5117
R_e^2:
0.3481
$\bar{d}$ = 0.214
Proportion ($\bar{d}_j$)
Judgment (f_j)
33
27
26
3
13
9
8
11
10

(a)

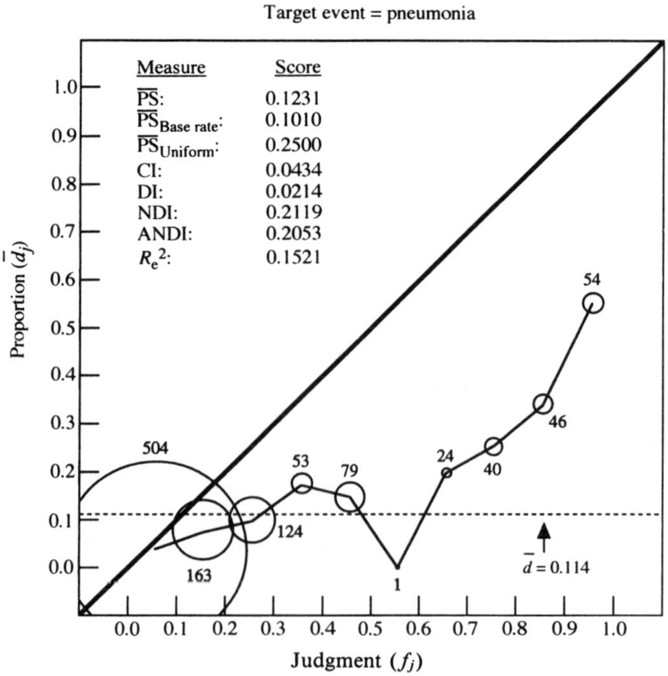
Target event = pneumonia
Measure
Score
PS:
0.1231
PS Base rate:
0.1010
PS Uniform:
0.2500
CI:
0.0434
DI:
0.0214
NDI:
0.2119
ANDI:
0.2053
R_e^2:
0.1521
$\bar{d}$ = 0.114
Proportion ($\bar{d}_j$)
Judgment (f_j)
504
163
124
53
79
1
24
40
46
54

(b)

Target event = pneumonia

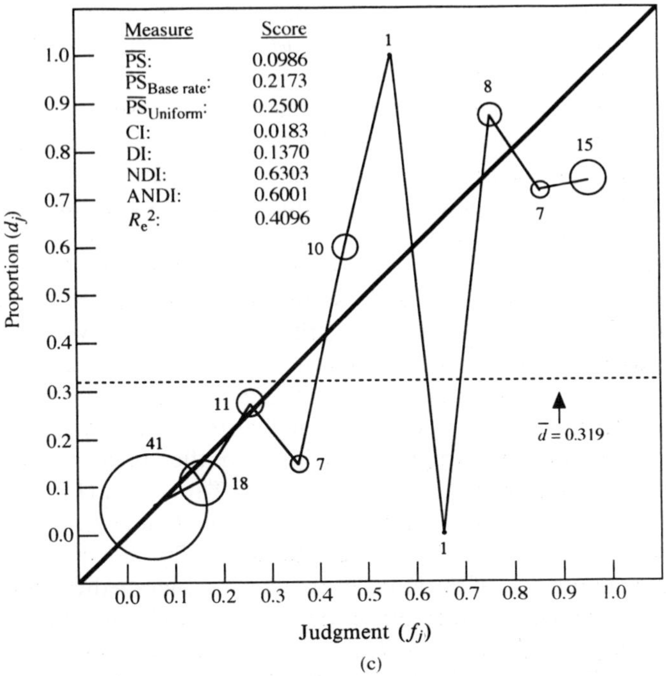

Figure 16.2 Calibration graphs of probabilistic diagnoses of pneumonia in (a) Virginia, (b) Illinois, and (c) Nebraska, as reported by Tape et al. (1991).

Illinois *DI* measure was less than a quarter that of the Virginia value, and more than 6 times smaller than the Nebraska statistic. Pictorially, good discrimination is evidenced by points that are vertically distant from the horizontal $\bar{d}$ line. Once again, the impressions created by the graphs are consistent with the numerical measures.

Since the best possible value of the discrimination index is $\text{Var}(d) = \bar{d}(1 - \bar{d})$, this implies different bounds on *DI*, depending on the base rate. A particular *DI* score, say, 0.12, means something quite different when the base rate is 0.45 than when it is 0.85. In the former case, the observed value of *DI* is only about half as large as it could possibly be, while in the latter, it has nearly reached its limit. Such observations led Sharp, Cutler, and Penrod (1988) to recommend reporting an η^2 statistic, which Yaniv et al. (1991) describe as the *normalized discrimination index* (NDI), defined as follows:

$$NDI = DI/[\bar{d}(1 - \bar{d})] \tag{16.11}$$

In essence, *NDI* measures relative discrimination, indicating how good is the obtained degree of discrimination in comparison to the best that is possible given the base rate. Yaniv et al. (1991) showed that *NDI* is a biased estimator of the corresponding population statistic. They also demonstrated that a particular transformation of *NDI*, the *adjusted normalized discrimination index*, *ANDI*, is unbiased.

Across-location comparisons of *NDI* and *ANDI* suggest less dramatic location differences in discrimination than did the *DI* comparisons. This is a good illustration of the effects of base rate and sample size on discrimination measures. Nevertheless, the discrimination differences are still quite remarkable: The Virginia and Nebraska physicians were able to get more than halfway to the maximum possible discrimination, while the Illinois physicians failed to reach even a quarter. The R_e^2 statistics shown in the calibration graphs implicate an intriguing partial answer to the mystery posed by these variations.

Before making a diagnosis for a given patient, each physician who participated in the study was required to note and record a standard set of facts about each patient, including demographic features as well as various medical signs and symptoms that might be helpful in the diagnosis, e.g. gender, wheezing history, and bronchial breath sounds. Tape et al. (1991) constructed optimal statistical models for predicting pneumonia from these various patient characteristics in each location. The values of R_e^2 shown in Figure 16.2 indicate the proportions of variance in patients' actual medical states predictable from the available information via those models. As we see, the proportions are surprisingly different from one location to another. In particular, pneumonia in Illinois was far less predictable than in either other location, but especially in comparison to Nebraska. This seems to be a particularly compelling demonstration of the notion of cue-based judgment difficulty. Apparently, the available cues for diagnosing pneumonia are simply less valid in Illinois than in Virginia and Nebraska. Why this should be so awaits further study.

16.4.3 Another Example: Intensive Care Prognoses

The physicians who participated in a study by McClish and Powell (1989) made a probabilistic prognostic judgment for each patient admitted to their intensive care unit. Specifically, the attending physician provided a response coded as P'(Die), where "Die" meant that the patient would never be discharged from the hospital alive. McClish and Powell constructed a logistic regression model that provided such judgments, too. The primary variable in this model is an APACHE II score. The APACHE (Acute Physiology Score and Chronic Health Evaluation) system is a commonly used method for evaluating the condition of intensive care patients. APACHE scores can be calculated easily by nursing staff through the completion of a simple scoring sheet based on readily available patient characteristics, such as heart rate, white blood count, and hematocrit level. McClish and Powell also formed a composite statistical model that combined the judgments of the physicians and the APACHE II model in an optimal way.

Table 16.1 shows the results of Murphy decomposition analyses derived from the statistics reported by McClish and Powell. Several results are noteworthy. Observe that all the "real judges" outperformed the best constant judge, the judge who would always report the base rate, i.e. $P'(\text{Die}) = 0.25$ for every patient. Table 16.1 also indicates the necessary fact that the entire value of $\overline{PS}_{\text{base rate}}$ is due to the target event incidence, i.e. that 25% of the patients died in the hospital. Further, since all the judges considered the same cases, the incidence portion of $\overline{PS}$ had to be the same for each of them. We are also reminded that the base rate judge has perfect calibration but nil discrimination.

Note that, in terms of overall accuracy, the physicians outperformed the APACHE II model. At first glance, this seems inconsistent with the apparent consensus of the literature that actuarial models almost always surpass human judges (e.g. Dawes, Faust, & Meehl, 1989). However, the decomposition analysis reveals how this superiority was achieved in this instance and highlights an important caveat that is often neglected when we read the literature on model vs. human comparisons.

Table 16.1 shows that the calibration of the APACHE II model was far superior to that of the physicians.[3] But the physicians had an advantage in discrimination. The calibration advantage of the model is, in fact, consistent with previous indications that the strong suit of models is their ability to optimally extract the statistical import of available evidence (the matching recall requirement for good calibration mentioned in Figure 16.1) and to do so reliably. On the other hand, the physicians' discrimination advantage most plausibly

Table 16.1 Murphy decomposition analysis statistics for intensive care unit mortality prognostic judgements [$P'(\text{Die})$] reported by McClish and Powell (1989)

Judgment source	Overall accuracy: $\overline{PS}$	Incidence: $\text{Var}(d)$	Calibration: CI	Discrimination: DI
Base rate judge ($\bar{d} = 0.25$)	0.1875	0.1875	0.0000	0.0000
Physicians	0.1240	0.1875	0.0210	0.0845
APACHE II model	0.1330	0.1875	0.0030	0.0575
Composite model	0.1015	0.1875	0.0005	0.0865

resulted from the physicians' access to more information than the APACHE system; the physicians were free to use any information sources they liked whereas the APACHE protocol is rigid and limited in scope. In most model vs. human comparisons, statistical and human judges have relied on identical information.

The final significant conclusion brought out by Table 16.1 concerns the composite model. We see that it did better than both the physicians and the APACHE model. As we would hope, it seemed to take the best of each, resulting in both better calibration and better discrimination than either.

16.5 THE COVARIANCE DECOMPOSITION

16.5.1 The Decomposition

The covariance decomposition of $\overline{PS}$ can be described as follows (Yates, 1982; 1988; Yates & Curley, 1985):

$$\overline{PS} = \text{Var}(d) + \text{MinVar}(f) + \text{Scat} + \text{Bias}^2 - 2\,[\text{Slope}]\,[\text{Var}(d)] \qquad (16.12)$$

We have already seen that $\text{Var}(d) = \bar{d}(1 - \bar{d})$, the first term in the Murphy decomposition. Thus, the covariance decomposition can be viewed as providing an alternative partition of the calibration and discrimination portions of the Murphy decomposition. The new terms in Equation (16.12) are defined and interpreted as follows:

- $\text{Bias} = \bar{f} - \bar{d}$, where $\bar{f}$ is the overall mean judgment for the target event A. So the bias statistic reflects a gross form of calibration, the tendency for the judge to effectively over- or underestimate the incidence of the target event. Ideally, we should observe $\text{Bias} = 0$.
- $\text{Slope} = \bar{f}_1 - \bar{f}_0$, where $\bar{f}_1$ is the mean judgment for the target event A reported on occasions when it actually happens, and $\bar{f}_0$ is the corresponding average for those occasions when that event does *not* occur. In the best of circumstances, $\text{Slope} = 1$, for in that instance the judge always reports $f = P'(A) = 1$ when event A occurs ($d = 1$) and $f = 0$ when it does not ($d = 0$).
- $\text{MinVar}(f) = \text{Slope}^2\text{Var}(d)$. This is the variability in the judge's assessments that is statistically required by the given value of Slope and the given base rate.
- $\text{Scat} = [N_1\text{Var}(f_1) + N_0\text{Var}(f_0)]\,/\,[N_1 + N_0]$. This "scatter" statistic reflects the variability in judgments that is *not* required by the given slope and base rate. As such, it indexes the amount of variability in judgment that is unrelated to the event in question. It is essentially "noise" as far as the judge's aims are concerned.

16.5.2 Contributors

The schematic in Figure 16.3 offers a useful means of understanding the covariance decomposition. As in the schematic for the Murphy decomposition (Figure 16.1), the second level of the tree describes the accuracy elements that yield overall accuracy. At the bottom level are controllable and uncontrollable factors that in turn contribute to each of these elements. Since we have already discussed incidence, we can proceed to the other constructs. At the outset, it is worth acknowledging that the covariance decomposition partitions overall accuracy more finely than does the Murphy decomposition. Thus, as we might expect, the contributing factors tend to be more molecular, too.

Bias. One controllable bias contributor is the extent to which the judge can recall the correspondence between the typical value of his or her judgments

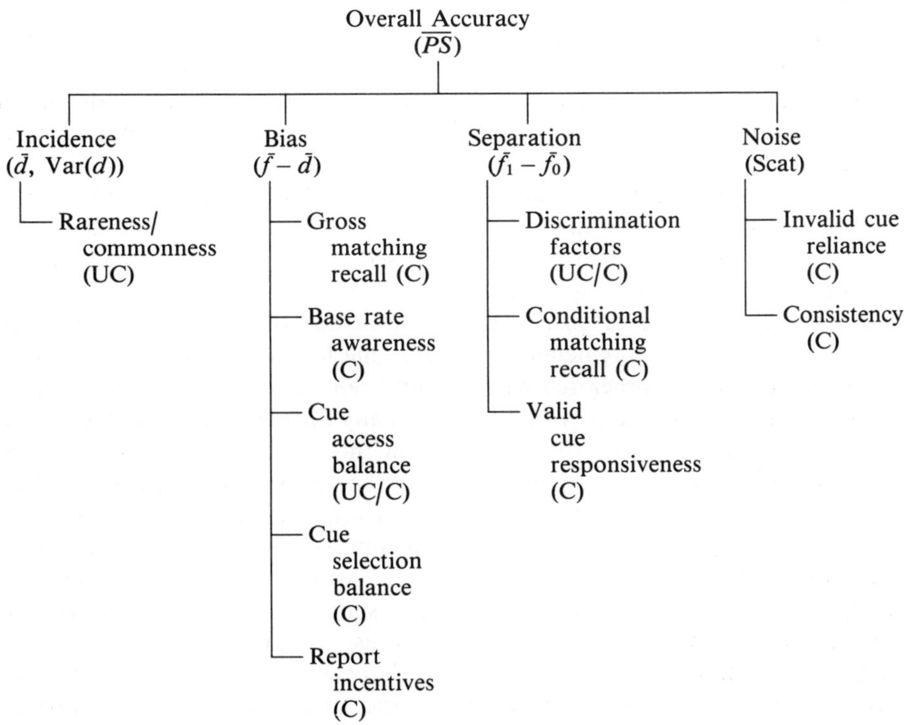

Figure 16.3 Schematic representation of overall accuracy in terms of its aspects as distinguished in the covariance decomposition of the mean probability score, along with controllable (C) and uncontrollable (UC) factors that plausibly contribute to those aspects.

and the incidence of the target event. This gross matching recall is a coarser version of the kind of recall that supports calibration more generally.

A closely related controllable bias contributor is awareness of the pertinent historical base rate. Depending on the nature of the underlying judgment process, if the judge believes that rate is very different from what it actually is, this almost guarantees significant bias. For instance, imagine a securities analyst who has been working in one industry and then moves to another. She might simply assume that the base rate for price increases in her new industry is the same as in her old one when in fact it is much lower. Then we should expect positive biases in her predictions.

We can imagine situations in which the cues available to a judge are directionally unbalanced, inappropriately favoring either event A or its complement. For example, it seems plausible that the tendency for sports fans to over predict the success of their local teams is partly due to such access imbalances. Because readers like them, local newspapers tend to contain more stories applauding and elaborating on the successes of the local team than similar stories about the team's opponents.

We might expect similar imbalances in how judges select the cues they use. The confirmation bias hypothesis (cf. Klayman & Ha, 1987) suggests as much. The claim is that, for various reasons, we are inclined to seek out information sources we have reason to expect will confirm our initial hypotheses about propositions.

The final controllable bias contributor we consider consists of incentives to report assessments either higher or lower than the judge believes to be the correct ones. Such incentives were discussed in our previous treatment of scoring rule properness.

Separation. The term "separation" implicates the fact that the judge produces two sets of judgments $f = P'(A)$, one for instances when the target event A occurs, the other for when it does not. Under the best of circumstances, these distributions would be maximally different or "separate" from each other. In particular, the difference in the conditional means of those distributions, Slope $= \bar{f}_1 - \bar{f}_0$, would be 1. Separation represents a blend of certain aspects of the discrimination and calibration constructs. Thus, in Figure 16.3, one class of separation contributors consists of the "discrimination factors" discussed in the context of the Murphy decomposition (Figure 16.1). The calibration aspect of separation should be especially sensitive to conditional matching recall, whereby the judge attempts to recollect average judgments conditional on the target event's occurrence and nonoccurrence. Finally, separation should be enhanced to the extent that the judge responds to cues in an appropriately regressive manner, i.e. in relation to their validity.

Noise. Two major factors can be expected to affect the noisiness of the judge's assessments. First, scatter should increase when the judge selects and uses cues that are invalid rather than valid. That is because those cues will induce the judge to vary judgments independently of the target event's occurrence. Judgments should also become noisier to the degree that the judge executes his or her judgment procedure inconsistently, responding to the same cues in different ways on different occasions. Such inconsistency appears to be the primary reason human judges are often outperformed by models of various kinds (Dawes et al., 1989).

16.5.3 An Example: Pneumonia Diagnoses Revisited

The previously discussed pneumonia diagnoses by the physicians studied by Tape et al. (1991) were analyzed using covariance decomposition techniques. Figure 16.4 shows the *covariance graphs* entailed in the analysis. Observe that such graphs consist mainly of two histograms of judgments $f = P'(A)$, one for the N_1 instances when the target event occurs (on the right), another for the N_0 cases when it does not (on the left). (The scale is indicated by the frequencies in parentheses on the longest bars.) The abscissa of a covariance graph has a dual interpretation. The endpoints represent the alternative values of the outcome index d, while the intermediate values mark off potential locations of the overall base rate, the mean of the outcome index. Thus, the line connecting the conditional means $\bar{f}_1$ and $\bar{f}_0$ is a regression line, with slope $\bar{f}_1 - \bar{f}_0$.

The overall accuracy measures in the study are necessarily the same as before. So we begin with bias, indexed by $\bar{f} - \bar{d}$. Bias is readily seen pictorially in a covariance graph. The vertical and horizontal dotted lines identify the base rate $\bar{d}$ and overall mean judgment $\bar{f}$, respectively. When judgments are unbiased, the intersection of these lines lies on the 1 : 1 diagonal. The magnitude of "miscalibration in the large" is thus indicated by the distance of that intersection from the diagonal. We immediately see that while the physicians in Virginia and Illinois grossly overpredicted pneumonia, by comparison, the bias of the Nebraska physicians was very slight. Factors related to base rate awareness are plausibly a contributor to the observed bias differences. Note that the base rates for pneumonia differed markedly across the locations, by about 20 percentage points. To some extent the physicians' judgments were sensitive to these differences. But the variability in average judgments seemed to lag. It is possible that the physicians in all the locations started with the same expectation of pneumonia incidence rates. However, consistent with other instances of anchoring and insufficient adjustment (Tversky & Kahneman, 1974), perhaps they did not alter those expectations to the degree they should have in response to local conditions.

Bias is often the major contributor to more general miscalibration, as revealed in Murphy decomposition analyses. That seems to have been the case here. Although bias is more clearly evident in covariance graphs, it is also

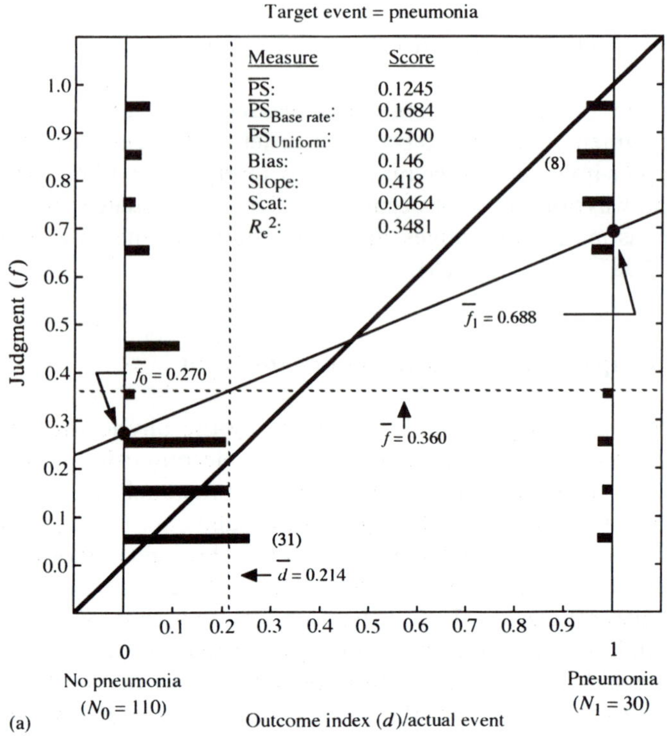
Target event = pneumonia
Measure
Score
PS: 0.1245
PS Base rate: 0.1684
PS Uniform: 0.2500
Bias: 0.146
Slope: 0.418
Scat: 0.0464
Re2: 0.3481
(8)
f1 = 0.688
f0 = 0.270
f = 0.360
(31)
d = 0.214
Judgment (f)
1.0
0.9
0.8
0.7
0.6
0.5
0.4
0.3
0.2
0.1
0.0
0.1 0.2 0.3 0.4 0.5 0.6 0.7 0.8 0.9
0
1
No pneumonia
(N0 = 110)
Pneumonia
(N1 = 30)
Outcome index (d)/actual event

(a)

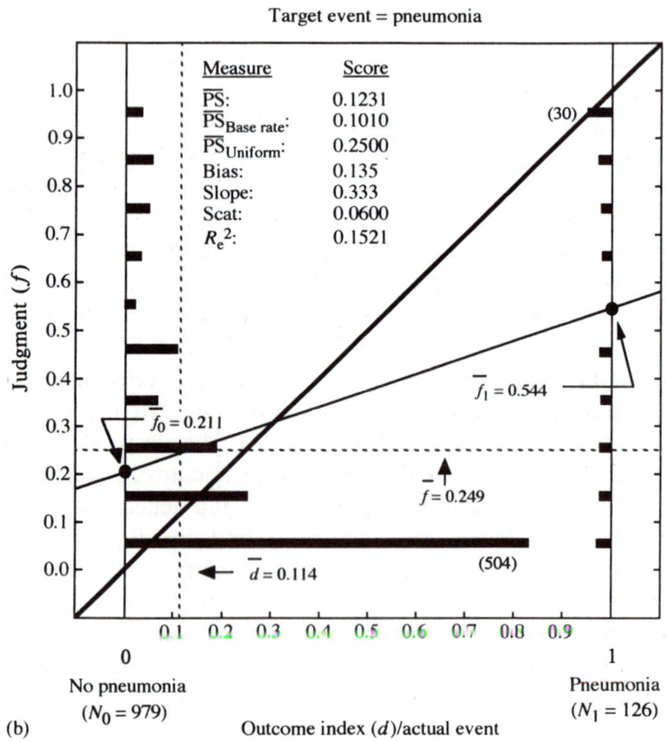
Target event = pneumonia
Measure
Score
PS: 0.1231
PS Base rate: 0.1010
PS Uniform: 0.2500
Bias: 0.135
Slope: 0.333
Scat: 0.0600
Re2: 0.1521
(30)
f1 = 0.544
f0 = 0.211
f = 0.249
(504)
d = 0.114
Judgment (f)
1.0
0.9
0.8
0.7
0.6
0.5
0.4
0.3
0.2
0.1
0.0
0.1 0.2 0.3 0.4 0.5 0.6 0.7 0.8 0.9
0
1
No pneumonia
(N0 = 979)
Pneumonia
(N1 = 126)
Outcome index (d)/actual event

(b)

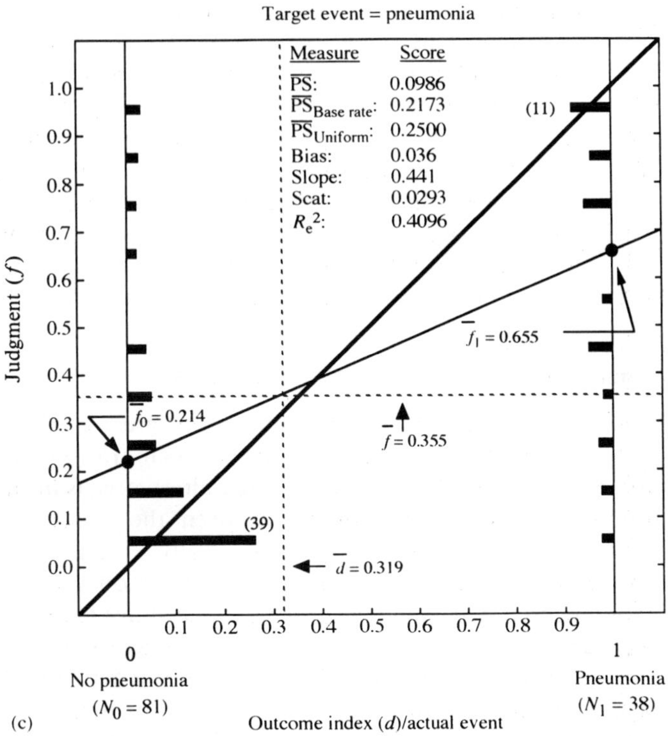

Figure 16.4 Covariance graphs of probabilistic diagnoses of pneumonia in (a) Virginia, (b) Illinois, and (c) Nebraska, as reported by Tape et al. (1991)

discernible in calibration graphs as a horizontal displacement of the plot of points to the left or right of the diagonal (see Figure 16.2).

Separation, as measured by the slope statistics (and as evident by the slopes of the regression lines), was comparable for the Virginia and Nebraska diagnoses, although the opinions from Nebraska were slightly better. But separation was markedly worse for the Illinois assessments. As suggested by Figure 16.3, there are several potential reasons for the differences. However, the most plausible is that it is due to the cue-based difficulty of diagnosing pneumonia at the Illinois site.

As in most comparisons, noise was worst (i.e. greatest) in Illinois and best in Nebraska. Noise, as measured by the scatter statistic, can be apprehended visually by the dispersion of the conditional judgment distributions on both sides of a covariance graph. (The Scat statistic is simply a weighted mean of the conditional variances.) The more dispersed the distributions, the greater the noise. The previous discussion suggested two reasonable potential explanations for the noise differences. The first is that, relatively speaking, the

Illinois physicians might have been especially inconsistent in applying their judgment policies. Analyses by Tape et al. (1991) yielded some evidence consonant with this reliability hypothesis. The other possibility is that the Illinois physicians relied on cues that were only weakly associated with pneumonia. Additional analyses by Tape et al. suggested that the judgment policy used by the Illinois physicians was very close to the best that one could use under the circumstances. But recall that the circumstances were actually quite different in the three locations. Standard predictors of pneumonia were simply not very good in an absolute sense in Illinois. So reliance on those cues necessarily would imply greater scatter, as was in fact observed.

16.5.4 A Final Example: Cross-National Variations

Consider the following question: Which country has greater oil reserves: (a) Venezuela or (b) Nigeria? After picking (a) or (b), you are to report a probability between 50% and 100% that your selected alternative is in fact correct. General knowledge questions like this have been the focus of the most common use of $\overline{PS}$ decomposition analyses. Typically, emphasis has been on what is called the bias statistic in covariance decomposition analysis. Imagine a subject responding to a large number of items in the above format. The mean judgment $\bar{f}$ is usually taken as an index of the respondent's confidence. This seems appropriate since the target event in this instance is actually "I chose the correct alternative." The base rate $\bar{d}$ is the proportion of items the subject in fact answered correctly. Hence, Bias $= \bar{f} - \bar{d}$ is often considered a measure of over- or underconfidence. For some time, it has been known that overconfidence in such general knowledge tasks is widespread (Lichtenstein, Fischhoff, & Phillips, 1982). The origins and limits of the phenomenon are the subject of intense study today (e.g. Gigerenzer, Hoffrage, & Kleinbölting, 1991).

In the late 1970s, in a series of studies, a group led by Wright and Phillips documented reliable cross-national variation in general knowledge overconfidence (e.g. Wright et al., 1978). They repeatedly found what most people consider to be a very surprising result, that overconfidence was more pronounced among various groups of southeast Asian subjects than among British subjects. Yates et al. (1989) tested the generality of those results to other Western and Asian countries and to other aspects of accuracy besides bias or overconfidence. Table 16.2 shows the results of a covariance decomposition analysis of judgments made by subjects in mainland China and the United States. It provides a good illustration of the technique.

The bias column shows that the previous differences did indeed generalize, with the overconfidence measure for the Chinese subjects being almost twice the size of that for the American subjects. But the story was actually much more interesting. Observe, for instance, that overall accuracy was essentially

Table 16.2 Median values of covariance decomposition statistics for Chinese and American subjects' general knowledge question judgments [P'(Correct)] reported by Yates et al. (1989).

Country	Overall accuracy: $\overline{PS}$	Incidence: $\bar{d}$	Bias: $\bar{f}-\bar{d}$	Separation: $\bar{f}_1-\bar{f}_0$	Noise: scat
China	0.2214	0.690	0.134	0.117	0.0323
United States	0.2121	0.655	0.072	0.089	0.0252
Signif. level	ns	ns	<0.005	<0.03	<0.001

identical for the groups. Also note that the Chinese judgments were significantly noisier than the American assessments. Given the compensatory character of the covariance decomposition, the pattern of comparisons among $\overline{PS}$, Bias, and Scat necessarily required that there was some accuracy dimension on which the Chinese subjects excelled. As we see, it was separation, as indexed by Slope $=\bar{f}_1-\bar{f}_0$.

Why do there exist such cross-national variations in probability judgment accuracy dimensions? This, too, is the subject of current study. Evidence is accumulating that the overconfidence difference is not mediated by affective processes, e.g. that Asian subjects have even more inflated opinions of their personal abilities than do Western subjects (e.g., Yates et al., 1991). It appears that, instead, such differences implicate fundamental differences in how people in different cultures approach judgment tasks cognitively (Yates, Lee, & Shinotsuka, 1992).

ACKNOWLEDGEMENTS

Part of the work discussed in this chapter was presented at the 25th International Congress of Psychology, Brussels, July 1992. Preparation of this chapter was supported in part by US National Science Foundation grant number SES 92-10027. The author thanks Andrew Parker for his computing assistance and other members of the Judgment and Decision Laboratory for their insights about the issues addressed.

NOTES

(1) The author gratefully acknowledges the generosity shown by Thomas Tape and Robert Wigton in making their data available for analysis.

(2) There exist computer programs that construct calibration graphs like these, as well as covariance graphs. One is *Probability Analyzer*, available through the University of Michigan's Judgment and Decision Laboratory and Information Technology Division.

(3) As is typical in medicine, McClish and Powell also performed an ROC analysis, the major products of which are measures of discrimination, such as the area under the ROC curve. The discrimination index *DI* provides an alternative to such measures and appears to be consistent with them generally (e.g. Yates *et al.*, 1990).

REFERENCES

Benson, P.G. & Önkal, D. (1992) The effects of feedback and training on the performance of probability forecasters. *International Journal of Forcasting*, **8**, 559–73.

Blattenberger, G. & Lad, F. (1985) Separating the Brier score into calibration and refinement components: A graphical exposition. *American Statistician*, **39**, 26–32.

Brier, G.W. (1950) Verification of forecasts expressed in terms of probability. *Monthly Weather Review*, **78**, 1–3.

Dawes, R.M., Faust, D. & Meehl, P.E. (1989) Clinical versus actuarial judgment. *Science*, **243**, 1668–74.

Echternacht, G.J. (1972) The use of confidence testing in objective tests. *Review of Educational Research*, **42**, 217–36.

Fischer, G.W. (1982) Scoring-rule feedback and the overconfidence syndrome in subjective probability forecasting. *Organizational Behavior and Human Performance*, **29**, 352–69.

Ganzach, Y. & Krantz, D.H. (1990) The psychology of moderate prediction: I. Experience with multiple determination. *Organizational Behavior and Human Decision Processes*, **47**, 177–204.

Gigerenzer, G., Hoffrage, U. & Kleinbölting, H. (1991) Probabilistic mental models: A Brunswikian theory of confidence. *Psychological Review*, **98**, 506–28.

Hsu, W.-R. & Murphy, A.H. (1986) The attributes diagram: A geometrical framework for assessing the quality of probability forecasts. *International Journal of Forecasting*, **2**, 285–93.

Kahneman, D. & Tversky, A. (1973) On the psychology of prediction. *Psychological Review*, **80**, 237–51.

Klayman, J. & Ha, U.-W. (1987). Confirmation, disconfirmation, and information in hypothesis testing. *Psychological Review*, **94**, 211–28.

Laplace, P.S. (1951) *A Philosophical Essay on Probabilities*. Dover, New York. (Originally published in French in 1796.)

Lichtenstein, S. & Fischhoff, B. (1980) Training for calibration. *Organizational Behavior and Human Performance*, **26**, 149–71.

Lichtenstein, S., Fischhoff, B. & Phillips, L.D. (1982) Calibration of probabilities: The state of the art to 1980. In D. Kahneman, P. Slovic & A. Tversky (eds.), *Judgment Under Uncertainty: Heuristics and Biases*, Cambridge University Press, New York, pages 306–34.

Locke, E.A., Shaw, K.N., Saari, L.M. & Latham, G.P. (1981) Goal setting and task performance: 1969–1980. *Psychological Bulletin*, **90**, 125–52.

McClish, D.K. & Powell, S.H. (1989) How well can physicians estimate mortality in a medical intensive care unit? *Medical Decision Making*, **9**, 125–32.

Matheson, J.E. & Winkler, R.L. (1976) Scoring rules for continuous probability distributions. *Management Science*, **22**, 1087–96.
Murphy, A.H. (1973) A new vector partition of the probability score. *Journal of Applied Meteorology*, **12**, 595–600.
Murphy, A.H. & Winkler, R.L. (1984) Probability forecasting in meteorology. *Journal of the American Statistical Association*, **79**, 489–500.
Murphy, A.H. & Winkler, R.L. (1992) Diagnostic verification of probability forecasts. *International Journal of Forecasting*, **7**, 435–55.
Murphy, A.H., Hsu, W.R., Winkler, R.L. & Wilks, D. S. (1985) The use of probabilities in subjective quantitative precipitation forecasts: Some experimental results. *Monthly Weather Review*, **113**, 2075–89.
Rippey, R.M. & Voytovich, A.E. (1983) Linking knowledge, realism and diagnostic reasoning by computer-assisted confidence testing. *Journal of Computer-Based Instruction*, **9**, 88–97.
Schaefer, R.E. & Borcherding, K. (1973) The assessment of subjective probability distributions: A training experiment. *Acta Psychologica*, **37**, 117–29.
Seaver, D. A., von Winterfeldt, D. & Edwards, W. (1978) Eliciting subjective probability distributions on continuous variables. *Organizational Behavior and Human Performance*, **21**, 379–91.
Shapiro, A.R. (1977) The evaluation of clinical prediction. *New England Journal of Medicine*, **296**, 1509–14.
Sharp, G.L., Cutler, B.L. & Penrod, S.D. (1988) Performance feedback improves the resolution of confidence judgments. *Organizational Behavior and Human Decision Processes*, **42**, 217–83.
Shuford, E.H., Albert, A. & Massengill, H.E. (1966) Admissible probability measurement procedures. *Psychometrika*, **31**, 125–45.
Staël von Holstein, C.-A.S. (1972) Probabilistic forecasting: An experiment related to the stock market. *Organizational Behavior and Human Performance*, **8**, 139–58.
Stewart, T.R. (1990) A decomposition of the correlation coefficient and its use in analyzing forecasting skill. *Weather and Forecasting*, **5**, 661–6.
Tape, T.G., Heckerling, P.S., Ornato, J.P. & Wigton, R.S. (1991) Use of clinical judgment analysis to explain regional variations in physicians' accuracies in diagnosing pneumonia. *Medical Decision Making*, **11**, 189–97.
Tversky, A. & Kahneman, D. (1974) Judgment under uncertainty: Heuristics and biases. *Science*, **185**, 1124–31.
Wallsten, T. S. (1981) Physician and medical student bias in evaluating diagnostic information. *Medical Decision Making*, **1**, 145–64.
Winkler, R.L. & Murphy, A.H. (1968) "Good" probability assessors. *Journal of Applied Meteorology*, **7**, 751–8.
Wright, G.N., Phillips, L.D., Whalley, P.C., Choo, G.T., Ng, K.O., Tan, I. & Wisudha, A. (1978) Cultural differences in probabilistic thinking. *Journal of Cross-Cultural Psychology*, **9**, 285–99.
Yaniv, I., Yates, J.F. & Smith, J.E.K. (1991) Measures of discrimination skill in probabilistic judgment. *Psychological Bulletin*, **110**, 611–17.
Yates, J.F. (1982) External correspondence: Decompositions of the mean probability score. *Organizational Behavior and Human Performance*, **30**, 132–56.
Yates, J.F. (1984) *Evaluating and analyzing probabilistic forecasts* (UMAP Instructional Module No. 572). Consortium for Mathematics and Its Applications, Lexington, MA.
Yates, J.F. (1988) Analyzing the accuracy of probability judgments for multiple events: An extension of the covariance decomposition. *Organizational Behavior and Human Decision Processes*, **41**, 281–99.

Yates, J.F. (1990) *Judgment and Decision Making*. Prentice Hall, Englewood Cliffs, NJ.

Yates, J.F. & Curley, S.P. (1985) Conditional distribution analyses of probabilistic forecasts. *Journal of Forecasting*, **4**, 61–73.

Yates, J.F., Lee, J. & Shinotsuka, H. (1992, November) Cross-national variation in probability judgment. Paper presented at the Annual Meeting of the Psychonomic Society, St. Louis.

Yates, J.F., McDaniel, L.S. & Brown, E.S. (1991) Probabilistic forecasts of stock prices and earnings: The hazards of nascent expertise. *Organizational Behavior and Human Decision Processes*, **49**, 60–79.

Yates, J.F., Zhu, Y., Ronis, D.L., Wang, D.-F., Shinotsuka, H. & Toda, M. (1989) Probability judgment accuracy: China, Japan, and the United States. *Organizational Behavior and Human Decision Processes*, **43**, 145–71.

Yates, J.F., Lee, J., Levi, K.R. & Curley, S.P. (1990) Measuring and analyzing probability judgment accuracy in medicine. *Philippine Journal of Internal Medicine*, **28** (Suppl. 1), 21–32.

Yates, J.F., Lee, J., Shinotsuka, H., Yen, N., Singh, R., Onglatco, M.L.U., Gupta, M. & Bhatnager, D. (1991, November) The origins of Asian versus Western differences in probability judgment accuracy patterns. Paper presented at the Annual Meeting of the Society for Judgment and Decision Making, San Francisco.

Chapter 17

Discrete Subjective Probabilities and Decision Analysis: Elicitation, Calibration and Combination

William R. Ferrell
The University of Arizona

17.1 INTRODUCTION

Subjective probability has found practical applications in industry for control, optimization and scheduling, in medicine for diagnosis and prediction, in government for assessing and communicating risks, and in decision analysis generally for evaluation and selection of strategies, alternatives and projects. Moreover, subjective probability is finding its way increasingly into the day-to-day routine of business, management and engineering through widely marketed software for decision trees and influence diagrams, Monte Carlo simulation with spreadsheets, and multiattribute utility analysis, all of which depend on probability judgments. This chapter focuses on several practical issues in subjective probability for discrete events from the standpoint of decision analysis. Decision analysis sets the standards for the use of subjective probability and points the way for other applications. It has become not only a technical field with different "schools" and exponents, but a small industry. As such it has a high stake in the success of subjective probability methods and

Subjective Probability. Edited by G. Wright and P. Ayton.
© 1994 John Wiley & Sons Ltd.

a high commitment to ensuring their reliability and validity. Of course, it also has a high stake in fostering and maintaining confidence in its product and methods. Hence, it should be made clear at the outset that the author is not a decision analyst and has no stake, either way.

The objective in this chapter is to examine *elicitation, calibration* and *combination* of discrete subjective probabilities in the light of a model that explains and brings order to a considerable amount of confusing experimental data. Calibration, the extent to which the observed proportions of events that occur agree with the assigned probability values, directly affects the quality of decision analysis and is the central issue. Elicitation, the process by which judgments are obtained, and combination, the process by which probabilities of the same event from different judges are aggregated, are intimately related to calibration and are considered from that standpoint. Although extremely important in decision analysis, probabilities for continuous quantities will not be addressed because much less is known about subjective probability density functions and because models explaining their calibration properties are less well developed (Curtis, Ferrell & Solomon, 1985). First, the general nature of decision analysis will be outlined and then subjective probability elicitation, calibration and combination will be considered in turn.

17.2 DECISION ANALYSIS AND SUBJECTIVE PROBABILITY

17.2.1 Structure of Decision Analysis

"Decision making" in Ronald Howard's classic definition, "is what you do when you don't know what to do" (Howard, 1980). Decision analysis (a term coined by Howard) applies to decision making (1) the formal tools of decision theory, probability theory, and mathematical modelling, (2) the accumulated research findings in the area of behavioral judgment and decision-making, and (3) the skilled judgment of analysts and of subject matter experts. It is costly in time and resources, so it is applied mainly to decision situations that are both important and complex. Keeney (1982) lists twelve typical sources of complexity:

(1) Multiple objectives, not all of which can be achieved
(2) Difficulty of identifying good alternatives
(3) The importance of intangible factors such as "morale" or "goodwill"
(4) Long time horizons with effects extending far into the future
(5) Many groups being affected and concerns for equity
(6) Risk and uncertainty from many sources including the actions of others, changes in priorities over time and lack of data or inherent unpredictability

(7) Risks to life and limb and other potentially dire consequences
(8) Need for expert knowledge from different disciplines
(9) Multiple decision makers and stakeholders
(10) Significant value trade-offs
(11) Attitudes toward risk taking must be considered
(12) Decisions being sequential, earlier ones conditioning those that follow

To deal with these complexities, decision analysis takes as its scope not just the comparative evaluation of decision alternatives, but the entire process leading up to it of structuring the problem, generating alternatives, modeling their probable impact, and assessing the preferences of the decision-makers. The steps are shown in Figure 17.1 along with the complexities that are addressed at each step. The objective of the analysis is not to select an optimum alternative which must be chosen, but to provide insight about the problem and to promote creativity in dealing with it and commitment to the alternative finally selected (Keeney, 1982).

17.2.2 Subjective Probabilities in Decision Analysis

Subjective probability can enter at any stage of the decision analysis process, implicitly or explicitly as a way of dealing with uncertainty. In generating alternatives, candidates may be rejected or accepted for further analysis on the basis of the subjective probability of instrumental efficacy or of side effects. Parameters of the models linking alternatives to consequences, such as disease detection probabilities, may be judged quantities for lack of measured values, and the determination of preferences may involve subjective probabilities when utility is measured or attribute weights determined using lotteries.

Subjective probabilities enter most explicitly, however, as the means of quantifying the uncertainties in the models that relate the alternatives to possible consequences. Quantification enables the computation of a probability distribution over those consequences for each alternative. In the evaluation stage, the alternatives are usually compared on the basis of expected utility. A weighted average of the utilities of the possible outcomes of each alternative is computed, using the outcome probabilities as weights, to obtain a single figure of merit for each alternative. Other criteria, such as the utility of the worst possible outcome, may be considered, but among feasible alternatives, expected utility is by far the most widely used basis for over-all comparison (von Winterfeldt & Edwards, 1986).

17.2.3 Quality of Decision Analysis

Decision analysis is not always highly dependent on probabilities, other aspects of the problem being more critical. But, in many cases subjective

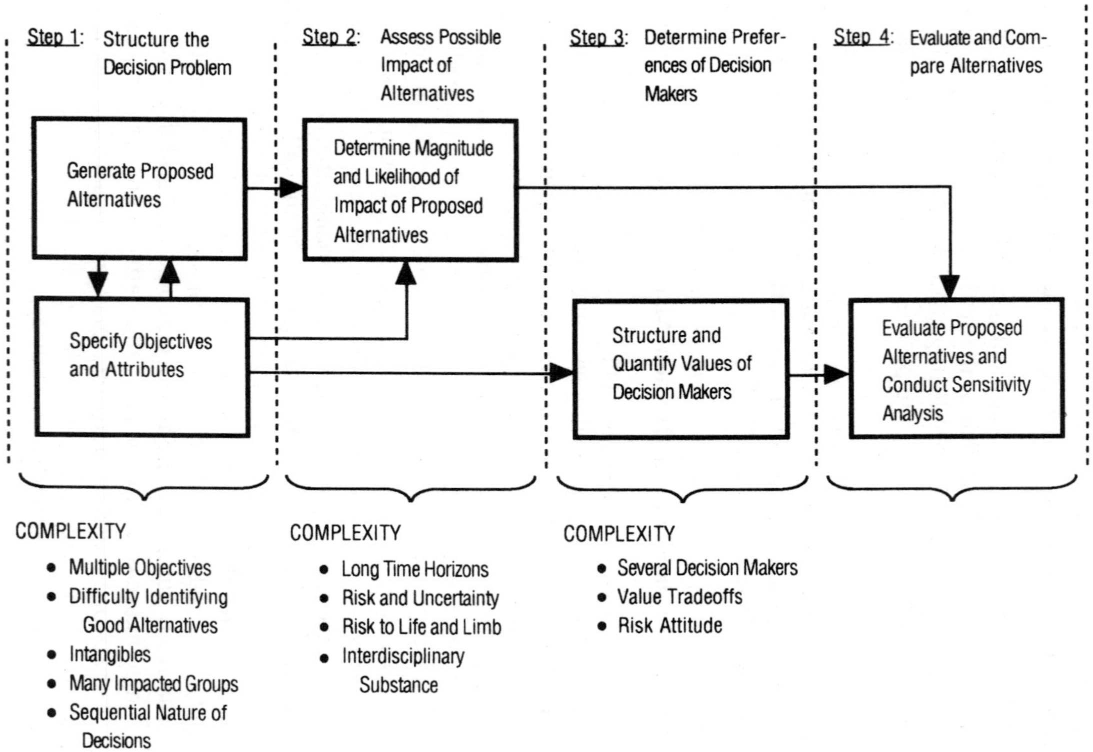

Figure 17.1 Steps in decision analysis. (Reproduced from Keeney, 1982, by permission of Operations Research Society and the author.)

probability judgments and their quality are extremely important. First, one can ask about the quality of decision analysis itself. Is it a reliable, valid, and effective approach to decision-making? Unfortunately, this question is not easily answered. Decisions with the complexity of those that merit the cost of decision analysis are usually quite different from each other and are seldom repetitive, preventing direct comparison of outcomes when decision analysis is used with outcomes when it is not. One might still expect that, on the whole, decisions aided by decision analysis have better results than those that are not thus aided. Although a generous sample of decision analyses have been reported in the literature (e.g. Keeney, 1982), they have not been reported with the detail, the temporal perspective, or the comparative data from decisions taken without it that would allow evaluation based on results.

The only alternative is to rely on the quality of the process by which decision analysis is carried out, under the assumption that, if the procedure is sound, one can have confidence that it will be conducive to the best decision given the resources devoted to it. Howard has suggested a format for appraisal of decision quality that explicitly considers the quality and completeness of elements of the successive steps represented in Figure 17.1 (Howard, 1988). Procedural quality depends upon the process being comprehensive, having a sound theoretical basis and being carefully and systematically applied. In general, decision analysis aims to be comprehensive, to address all aspects of the decision problem to the extent that they are relevant to choice. Its basic principles, as normative theory, are exceptionally convincing (Keeney, 1982; Howard, 1983) and rarely subject to serious attack, although there are significant differences in style and emphasis among practitioners (Phillips, 1989). This is not to say that all the procedures used in analyzing decisions are theoretically sound. The widely applied "analytical hierarchy process", for example, is on shaky ground (Dyer, 1990). Finally, although the care and attention with which the techniques are used differ according to the individual practitioners, the application standards set by the principal decision analysis groups are strikingly high.

This same reliance on the procedural guarantee of quality carries over to subjective probability within decision analysis. Concerning the theoretical basis, probability, as a mathematical construct, is well grounded, but there is considerable debate about the philosophical and psychological status of subjective probability, as is evident throughout this volume. As a practical matter, however, its use is not in doubt. Scarcely anyone would reject it and forego probabilistic modeling if no other source of information about uncertainty were available. Nonetheless, because of its slipperiness there should be an especially strong emphasis in decision analysis on the careful and systematic application of a comprehensive subjective probability elicitation process.

17.3 SUBJECTIVE PROBABILITY ELICITATION

Decision analysis requires a model relating the choice of an alternative to its relevant consequences. Almost invariably there is uncertainty about these, and the usual method of dealing with uncertainty is for the model to provide a probability distribution over the possible consequences for each alternative. The model may be highly aggregated, perhaps just the probability distribution itself or it may be highly disaggregated, with the relevant probabilities computed according to an elaborate physical or logical model of the situation. Although only the final distribution over the outcomes is important for the ultimate comparison of alternatives on the basis of expected utility, a model has the important objective of relating the probabilities that are needed for the distribution to ones that can be obtained with accuracy and at low cost, and of enabling computation of such useful quantities as the value of greater accuracy or of more information.

17.3.1 Steps in Elicitation

The usual practice for obtaining the subjective probabilities needed for the model is for a decision analyst, a person knowledgeable about the tools and practices of decision analysis, to interview the client or an expert, a person with substantive knowledge about the events whose uncertainty is to be assessed. The analyst applies knowledge of probability theory, of modeling, and of behavioral judgment to attempt to ensure that the encoding of the uncertainty into subjective probability is free of bias, consistent, and valid. The procedure recommended by the decision analysis group at Stanford Research Institute (Staël von Holstein & Matheson, 1979; Merkhofer, 1987) has five stages: (1) *motivating*, informing the expert and assessing the expert's motivation, (2) *structuring*, defining the events to be encoded (3) *conditioning*, making the expert aware of sources of bias, (4) *encoding*, obtaining the numerical judgments, and (5) *verifying*. Much of the following summary description is drawn from Merkhofer (1987), von Winterfeldt & Edwards (1986) and Spetzler & Staël von Holstein (1975).

(1) In *motivating* the expert properly, the decision analyst explains the decision model, the elicitation process, the importance of the probabilities to be elicited, and how they will be used in the analysis. The analyst also tries to determine, through discussion, whether there may be potential bias due to the expert's relationship to the problem. Several types of bias are possible. If the expert or client is in a position to affect the uncertain events, there may be a bias toward the outcomes that are intended to be produced. If the occurrence of an event is a goal of the expert, optimism can be expected. If the expert thinks that expertise entails being highly confident about predictions, then too

little uncertainty about the events can be expected. In addition, the analyst should attempt to determine the extent to which the expert will benefit personally from the events in question or from the outcome of the analysis. Bringing such factors to light may be sufficient, or it may be possible to restructure the model so that the relation of the events being assessed to the outcomes in which the expert has a personal stake are not apparent.

(2) *Structuring* refers to defining the specific events being considered and to the way in which the expert thinks about them, the cognitive structure in which information about them is stored. The events must be so clearly defined that there would be no ambiguity at all about whether or not they have occurred and they must be ones for which the expert's knowledge is adequate and in an appropriate form. In addition, they should also be ones that are minimally susceptible to cognitive biases. The event of "installation of the system within a month" will not do, a specific starting state and ending state and a specified number of working days is necessary to avoid any misunderstanding or miscommunication. The analyst must consider whether the model needs to be modified to accommodate the expert's knowledge. It might be more accurate to ask for probabilities associated with existing systems the expert is familiar with and use the model to extrapolate to a new type of system. The model may also need to be disaggregated further to minimize various biases associated with judgment under uncertainty. For example, the probability of the system installation taking less than a certain time might require the occurrence of a sequence of several independent events, a combination that, by being a causal scenario, may be judged more probable than it is in fact. Such a bias may be avoided if the probabilities of the events can be assessed separately and combined using probability theory. When the events to be assessed have been selected and are well specified, the analyst and expert need to be clear about the assumptions the expert is making when considering them. The expert may implicitly ignore some conditioning events, e.g. a labor strike or bad weather. If circumstances such as these are significant, it may be helpful to make assessments for each such scenario separately.

(3) In *conditioning* the judgments, the analyst attempts to stimulate the expert's awareness of all relevant information and to draw the expert's attention to possible problems in using it. The literature on subjective probability of the past 20 years; is filled with examples of bias and error in subjective judgments when compared with normative standards (Kahneman, Slovic, & Tversky, 1982). Although the literature itself may be biased toward calling attention to bias (Christensen-Szalanski & Beach, 1984), it is prudent to make the expert aware of the possibilities for it and of the ways it can come about. There is, as yet, no clear understanding of how instruction and guidance, i.e. stimulating awareness and giving advice, can be used to reduce bias, but there is a fair amount of evidence that establishing appropriate expectations and providing relevant information can have a beneficial, though

often insufficient, effect (Fischhoff, 1982a; Koriat, Lichtenstein, & Fischhoff, 1980; Kahneman & Tversky, 1979). The analyst tries to get the expert to become aware of all the kinds of information that bear upon the judgments, and through this discussion can assess what sorts of bias are especially likely. The analyst can then point out the bias and illustrate how it can come about. Typical biases would be neglect of base rate or distributional information in favour of case-specific evidence (Bar-Hillel, 1980; Kahneman & Tversky, 1973), overconfidence (Cooper, Woo, & Dunkelberg, 1988), over-reliance on recent or vivid instances or information (Tversky & Kahneman, 1973), the inappropriate application of judgment heuristics (Tversky & Kahneman, 1974), etc. Sometimes exercises or demonstrations, in which the expert makes judgments about things that are known and is later scored, are used to emphasize the possibility for bias. In a few cases, where there is sufficient information, it may be possible to adjust the expert's judgments to allow for bias, with the expert's understanding and concurrence (Merkhofer, 1987).

(4) *Encoding* is giving numerical expression to the expert's considered opinions. The most obvious source of bias in this for probabilities of discrete events is the type of response requested of the expert. Someone familiar with horse racing or gambling may feel more comfortable expressing uncertainty in odds, the ratio of the probability of the event to the probability of its complement, or a technically trained person may feel more comfortable with the probability scale. In the event that very large or small probabilities are involved, log-odds or log-probability scales have been found to reduce the rather common tendency to hedge extreme probabilities (Phillips & Edwards, 1966; Goodman, 1972). For those who to not feel comfortable using either probability or odds, the analyst can use a standard event for comparison, asking whether the expert would rather bet on the standard event or on the event in question to win a prize. When the probability of the standard event has been adjusted so that the expert is indifferent, that probability is taken as the subjective probability of the event. A wheel that can be spun and that has an adjustable segment is sometimes used. The nominal probability that the wheel will stop spinning with the segment opposite a fixed pointer is the relative size of the segment. However, since the relative size of the segment is so easily judged visually, this method may, in effect, be no different from requesting a numerical estimate (von Winterfeldt & Edwards, 1986). There are other encoding methods that involve inferring subjective probability from amounts bet. They require knowing the expert's utility for those amounts, and if a sure thing is compared with a gamble, they involve the expert's attitude toward risk. These restrictions may not matter; for small, hypothetical amounts, one can suppose utility linear with amount and a risk neutral attitude. An important use of such methods can be to test for consistency. In their survey of encoding, Walsten & Budescu (1983) stated, "The interesting conclusion emerges from the literature that high agreement exists among

various encoding methods, and that generally none conforms to the rules of additive probability theory" (page 167).

(5) *Verifying* is the process whereby the analyst attempts to ensure that the expert has given his or her true opinion and that these opinions are coherent, i.e., consistent with probability theory. This is done by such techniques as reformulating questions in logically equivalent ways to see if the results are consistent and by asking questions about events whose probabilities can be inferred from previous answers. Almost invariably there are inconsistencies, and these are brought to the attention of the expert and resolved by discussion and reconsideration. The question of whether this enforced coherence distorts experts' judgments has been asked (Walsten & Budescu, 1983) but not studied. However it is essential that there be coherence among the probabilities in the decision model.

17.3.2 Improvement of the Elicitation Process

The elicitation process described above relies largely on *ad hoc* efforts to forestall biases that have been found empirically. The process is well conceived and informed by practice, but it is not based on a clear representation of what the expert actually *does* in producing a subjective probability judgment. Lopes (1987) has shown that precise knowledge of the judgment process can lead to showing respondents how to organize their behavior to make essentially correct judgments, i.e. up to the Bayesian standard, in a sequential probability judgment task where other debiasing efforts, directed at compensating for poor performance by changing the response scale (Phillips & Edwards, 1966) or altering the nature of the judgment itself (Eils, Seaver, & Edwards, 1977), have been only partially successful. This level of understanding of the (far more complex) general probability judgment task leading to specification of appropriate judgment processes, is needed if one is to have confidence that elicited subjective probabilities are free of bias and well calibrated.

Following elicitation, a further step is often undertaken, one in which the subjective probabilities obtained from different experts are combined in an effort to improve judgment quality. The results of combination will be examined, but first it is necessary to consider an important aspect of the quality that is sought, calibration.

17.4 CALIBRATION

17.4.1 Definition and Implications for Decision Analysis

Probability judgments should agree with actual relative frequencies of the event judged or with normatively computed probabilities. This kind of agreement is called *external validity* (Walsten & Budescu, 1983). When events are unique, however, and there are no relevant relative frequencies, or when evidence is such that no normative computations can serve as a standard, subjective probabilities should still have good *calibration*. Good calibration means a relative frequency of occurrence x, in the long run, for *all* those events given a subjective probability of x. Calibration is typically represented by a graph of the observed relative frequency of occurrence *vs.* subjective probability. With good calibration, subjective probabilities can be taken at face value—e.g. an event given a probability of 0.3 considered equivalent in uncertainty to a random draw of a red ball from an urn with 3 red and 7 white balls. High external validity guarantees good *calibration*, but good calibration does not guarantee high external validity. Low external validity or poor calibration, or even the possibility of poor calibration, pose problems for decision analysis and can result in seriously suboptimal choices.

Clemen & Murphy (1990) use the example of weather forecasting to show that small improvements to the already extremely good calibration of weather forecasters could produce significant savings for those who use the forecast probabilities to compute whether or not to adopt measures to mitigate the effects of bad weather, i.e. who use them in decision analysis. If adequate information is available, they show that it is possible to recalibrate forecasts which are expected to be miscalibrated. It is unlikely that sufficient information would be available for most of the types of subjective probability used in decision analysis.

Even the possibility of miscalibration of subjective probabilities, as opposed to the possibility of their just being in error, raises potentially serious problems. This has been clearly shown by Harrison (1977). It is of prime importance in decision analysis to attempt to structure a model so that events whose probability is to be assessed are independent, as it simplifies both the elicitation task and the computational complexity of the model. From the standpoint of the decision analysis or of an expert whose judgments are used, however, the possibility of miscalibration implies that events whose subjective probabilities are given can not be assumed independent even though they are quite unrelated, and those considered dependent have their dependencies altered. It is easy to see why (and Harrison gives a good numerical example in illustration). If there were miscalibration, knowing the outcomes for some of the events would lead one to revise the subjective probabilities of the others. But the independence of two events implies that knowing the outcome for one event has no effect on the probability of the other. Hence the expectation of having to revise in this way implies that one cannot consider the events independent. The problem is not because of known miscalibration, where a correction can be made, as discussed by Clemen & Murphy (1990), or just because of uncertainty about the correct values of the probabilities of the events, but because of potential, but uncertain, miscalibration—the potential for a systematic error.

17.4.2 Results of Studies of Calibration

The elicitation procedures of decision analysis are designed to reduce the impact of possible causes of miscalibration that have been inferred from experiments. Important reviews of calibration are Lichtenstein, Fischhoff, & Phillips (1982), Keren (1991), and Walsten & Budescu (1983). In brief, the principal facts about calibration of discrete probabilities that have good empirical support are:

(1) Overconfidence is most frequently observed.
(2) There are systematic effects: (a) due to the difficulty of discriminating event occurrence from non-occurrence, and (b) due to the relative frequency of occurrence, the base rate of the events in question.
(3) Experts do not necessarily have better calibration.
(4) Calibration is rather easily changed with training, but improvement doesn't appear to generalize to other probability judgment tasks.

Overconfidence means subjective judgments are more extreme, closer to zero or one away from the default value for maximum uncertainty, than their corresponding relative frequencies. Underconfidence is the reverse. In effect, respondents give too much or too little weight to the evidence they have for responding with a value different from the default or base rate value. Over- and underconfidence should be distinguished from over- and underestimation, in which responses over the whole range tend to be greater or smaller than they should be. Most studies of calibration, for reasons of efficiency and statistical power, pool the judgments of many people, each of whom makes a limited number of judgments. The reported effects, however, have all been observed for individuals. A very large number of calibration studies have been reported in the literature and the results described in the following sections are quite robust to different circumstances. However, it is important to point out that in none of these studies, as far as the author knows, were the full techniques of probability elicitation used that are recommended for decision analysis.

Overconfidence and the effects of discriminability and base rates

The *half-range* task of selecting one of two answers to a question and giving a subjective probability on the interval [0.5–1] of having chosen correctly is rarely relevant to decision analysis, but is widely used in research. The discrimination involved is between true and false answers. Discriminability is directly measured by the proportion of correct choices $p(C)$. When the proportion correct is increased, the curve shifts in the direction of underconfidence and when it is decreased, in the direction of overconfidence. For questions that draw upon general knowledge, there is usually fairly good calibration when the proportion correct is about 75%. Typical examples of calibration for general knowledge questions are shown in Figure 17.2. (For the time being, references in the figures to the calibration model should be ignored. The model will be discussed further on.)

For the less frequently studied, but more practically relevant, *full-range* tasks of giving a subjective probability on [0–1] that a proposition is true or

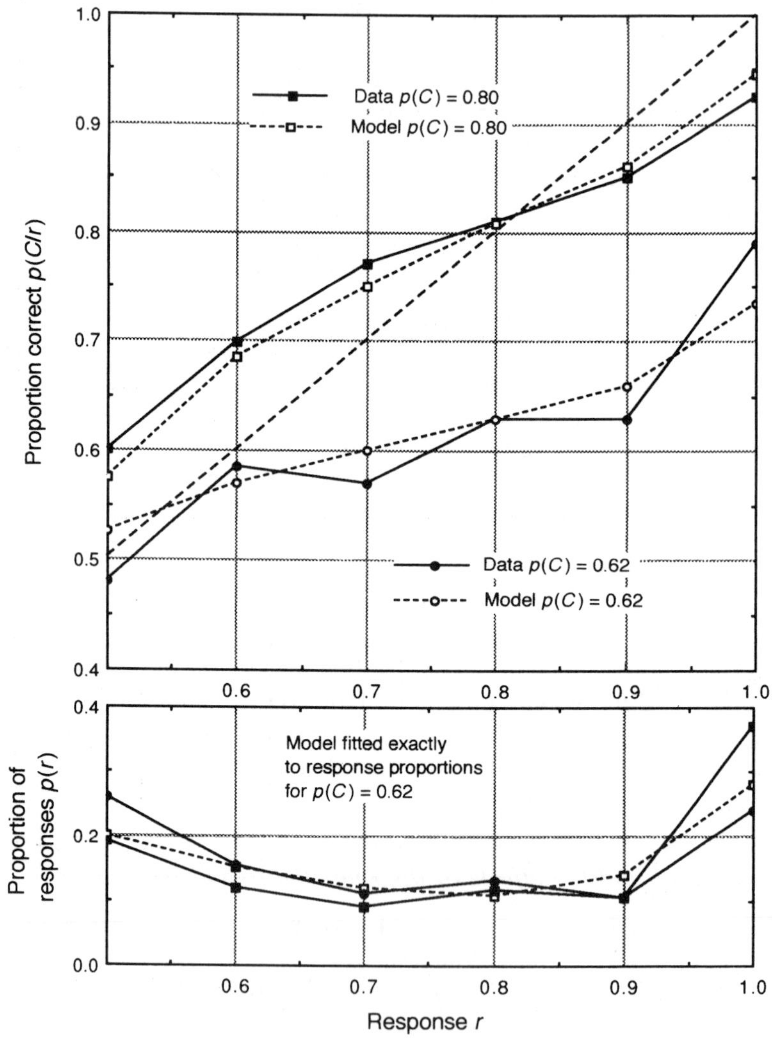

Figure 17.2 The discriminability (or "hard–easy") effect for half-range judgments. Data from Lichtenstein & Fischhoff (1977), experiment 5, for separate tests with general knowledge questions (Adapted from Ferrell & McGoey, 1980, by permission of Academic Press and the authors.)
Model: Normal, {0.347, 0.570, 0.744, 0.982, 1.219}, $p(C)$ as shown

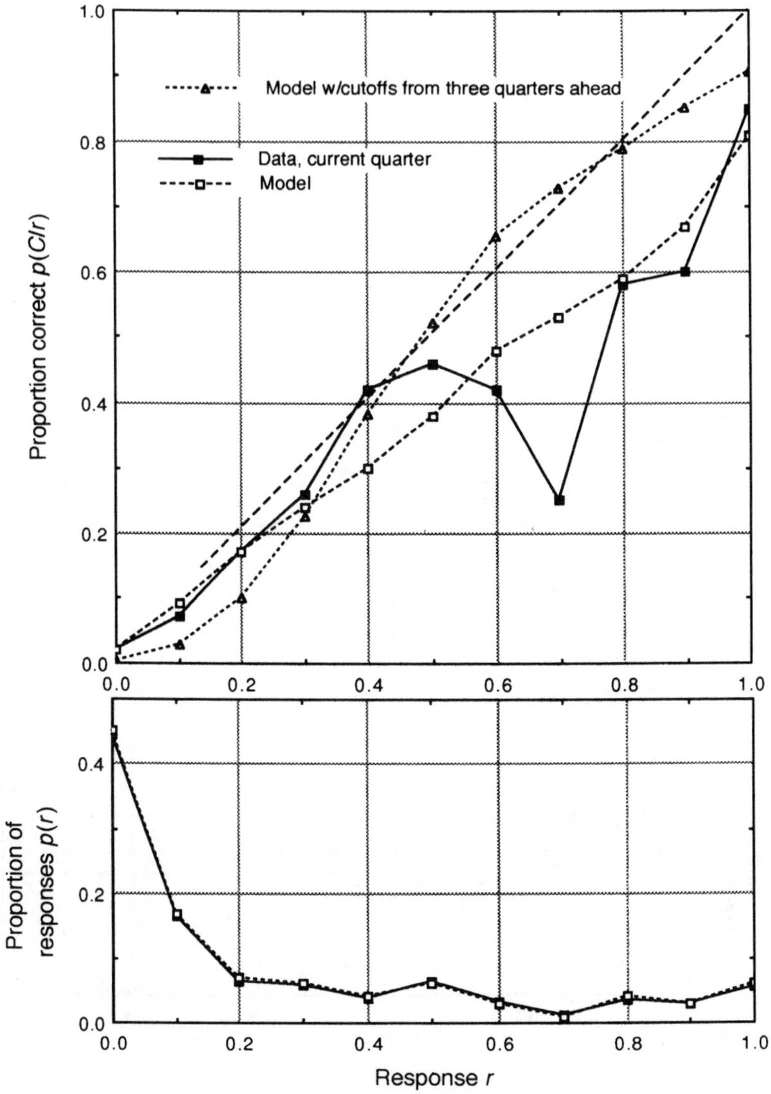

Figure 17.3(a) The discriminability effect for full-range judgments (Data from Braun & Yaniv, 1992, for expert forecasts of recession in the current quarter. Reproduced by permission.)
Model: Normal, $A = 1.87$, $B = 1.16$, {0.10, 0.61, 0.83, 1.05, 1.19, 1.47, 1.65, 1.72, 1.96, 2.22} $p(C) = 0.19$

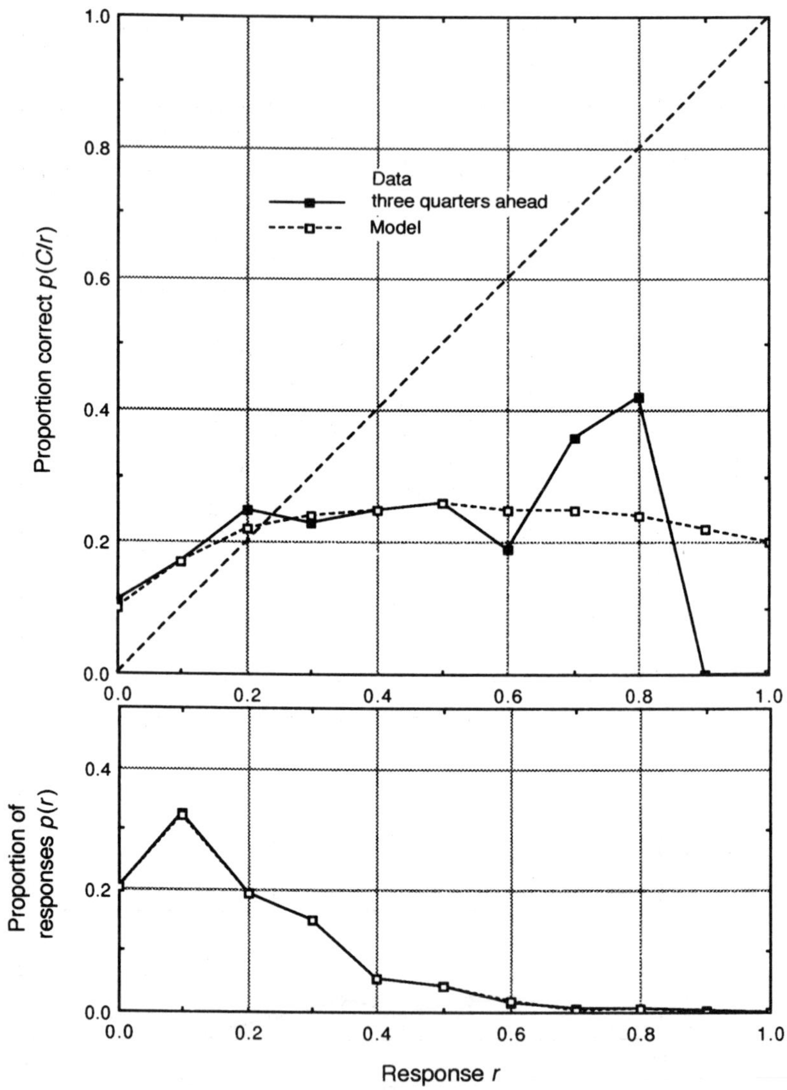

Figure 17.3(b) The discriminability effect for full-range judgments. (Data from Braun & Yaniv, 1992 for expert forecasts of recession three quarters ahead. Reproduced by permission.)
Model: Normal, $A = 0.37$, $B = 1.12$, $\{-0.76, 0.13, 364, 1.17, 1.48, 1.88, 2.2, 2.35, 2.75, 3.04\}$ $p(C) = 0.19$ (Reported as 0.2, but there is inaccuracy in reading data from the graph)

that an event will occur or has occurred, discriminability is also between truth and falsity, or between occurrence and non-occurrence. It can be measured, without bias, by, for example, signal detection methods (Green & Swets, 1974). Full-range tasks also show relative underconfidence for higher discriminability and overconfidence for lower discriminability. Figures 17.3a and b demonstrate this for expert forecasts of economic recession for the current calendar quarter and for three quarters ahead. Measured discriminability for three quarters ahead is about one fifth that for the current quarter and calibration becomes extremely overconfident for forecasts greater than the base frequency of about 0.2.

What is here called the discriminability effect is usually called the "hard–easy" effect in the calibration literature, but that is something of a misnomer. The effect is observed for the full-range task of responding to an open-ended question by giving an answer and a probability that it is correct.

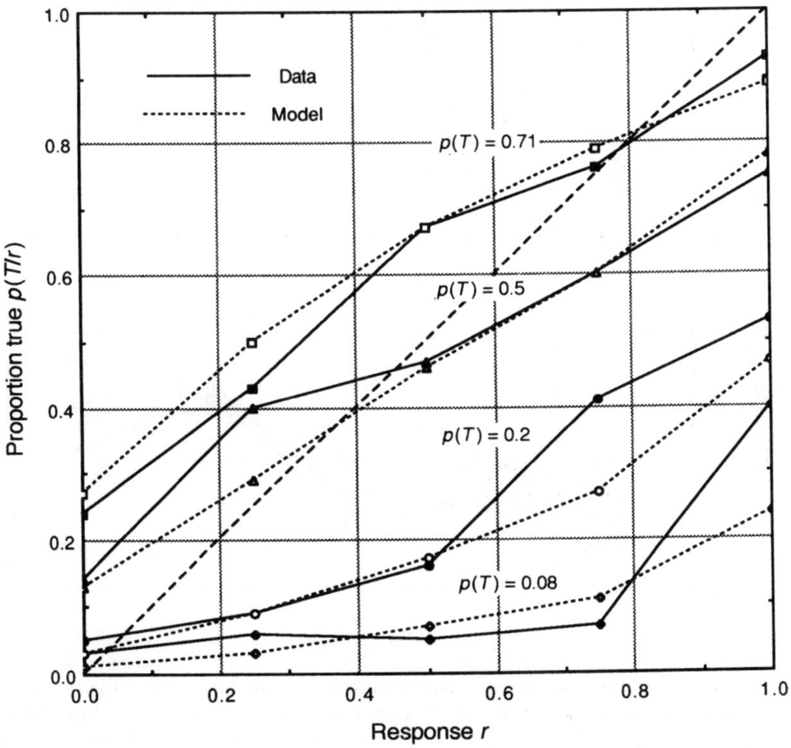

Figure 17.4 The base-rate effect (Data from Lichtenstein & Fischhoff as published in Lichtenstein, Fischhoff & Phillips, 1982, for judgments of the truth of propositions with different percentages of true test items. Adapted by permission of Cambridge University Press.)
Model: Normal, $A = 1$, $B = 1$, $\{-0.9, 0, 0.7, 1.2\}$ $p(C)$ as shown

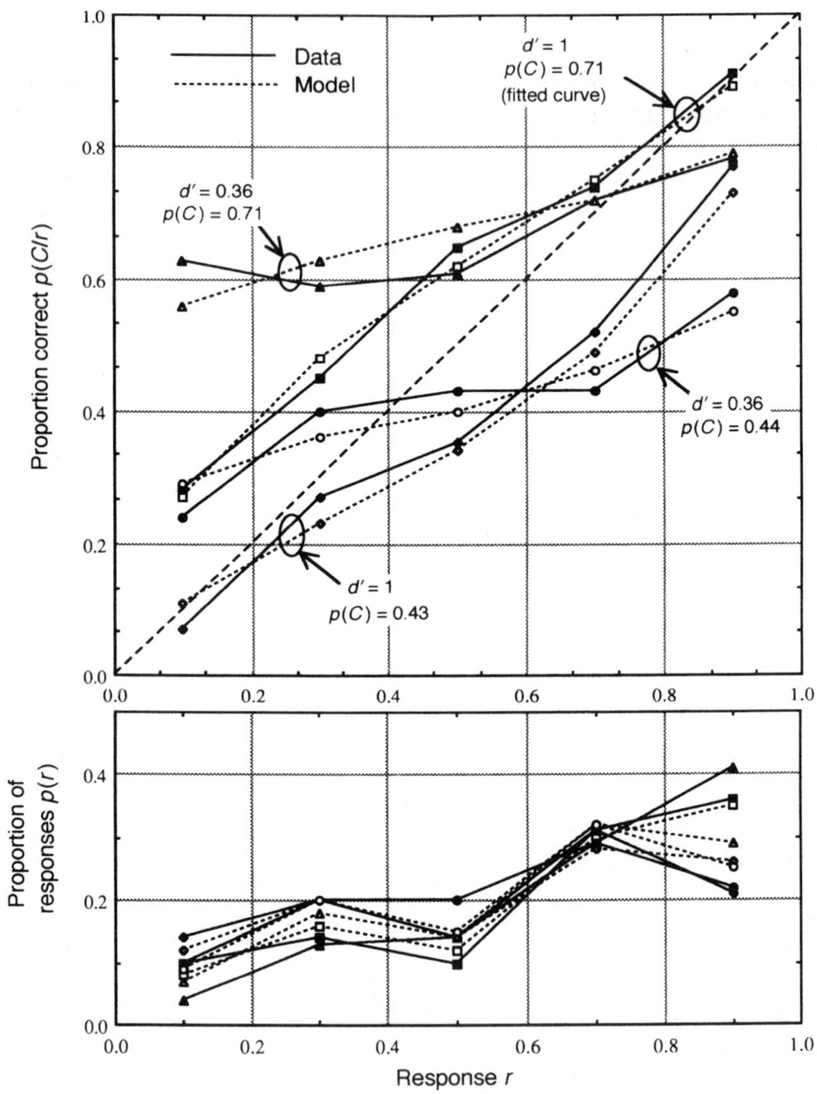

Figure 17.5 The combination of discriminability and base-rate effects. Data for students judging the probability their exam answers are correct. Students were separated into groups on the basis of their scores $P(C)$ and discriminability d' (Adapted from Ferrell & McGoey, 1980, by permission of Academic Press.)
Model: Normal, $A = d'$, $B = 1$, $\{-1.4, -0.6, -0.2, 0.65\}$ $p(C)$ and d' as shown

The effect then does not depend on whether the questions are hard or easy, but on the discriminability of correct from incorrect answers (Ferrell & McGoey, 1980).

For full-range tasks, there is an additional effect due to the the base rate, the overall frequency of the events. Increasing the base rate moves the whole calibration curve upward on the graph, giving general underestimation, and decreasing it moves the curve downward, giving general overestimation. This can be observed in Figure 17.4 for judgments of the probability of the truth of general knowledge statements with different proportions of true statements in the set.

The effects of discriminability and base rate combine to produce the resultant calibration curve. Figure 17.5 shows four curves representing the pooled judgments of four groups of students who gave subjective probabilities that their examination answers would be rated correct. The groups were separated by individual discriminability d' between correct and incorrect answers and by base rate, the proportion of correct answers $p(C)$. Individual d' and $p(C)$ values were uncorrelated. Students with high proportion correct and low discriminability have a combination of overestimation and overconfidence which shifts their curve upward and tilts it to the right. The curves from the other groups show comparable combined effects of their discriminability and base rate.

Expertise

Experts, those with much experience and domain-specific knowledge, might be expected to be better calibrated than others. The excellent calibration of weather forecasters (Murphy & Winkler, 1977) is usually cited. As another example, Keren (1987) found that expert bridge players were better calibrated than amateurs when judging the subjective probabilities of making the contracts they had bid even though they made a slightly smaller proportion of them. The calibration curves are shown in Figures 17.6a and b. Others have also found experts better calibrated (Dowie, 1976), but it is clear from a number of studies with negative results, especially with physicians (Lusted, 1977) and with economic forecasters (Braun & Yaniv, 1992), that expertise is not a sufficient condition for good calibration.

The conditions that are believed necessary to foster improvement in judgment skills, and presumably in calibration, are quite demanding (Fischhoff, 1989):

(1) Abundant practice with a set of reasonably homogeneous tasks
(2) Clear-cut criterion events for outcome feedback
(3) Task-specific reinforcement
(4) Explicit admission of the need for learning.

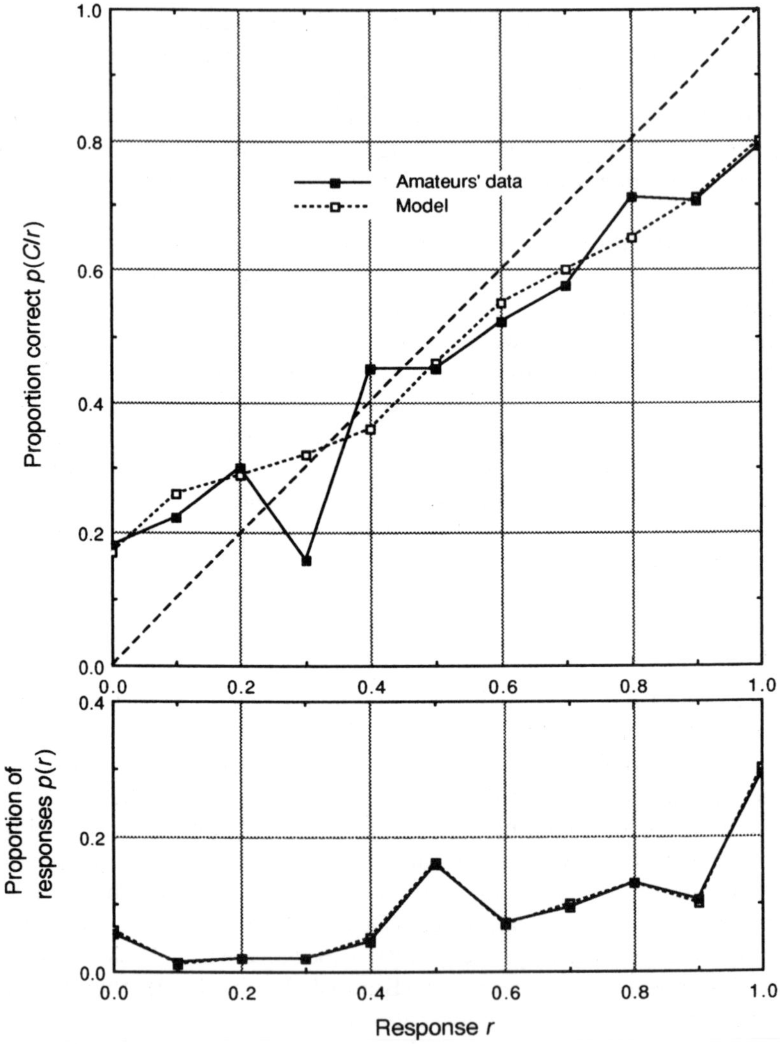

Figure 17.6(a) Amateur calibration. (Adapted from Keren, 1987, for judgments by amateur bridge players of the probability of making their contracts, by permission of Academic Press and the author.)
Model: Normal, $A = 0.85$, $B = 1.08$, $\{-1.2, -1.1, -0.95, -0.8, 0.2, 0.4, 0.76, 1.08\}$ $p(C) = 0.6$

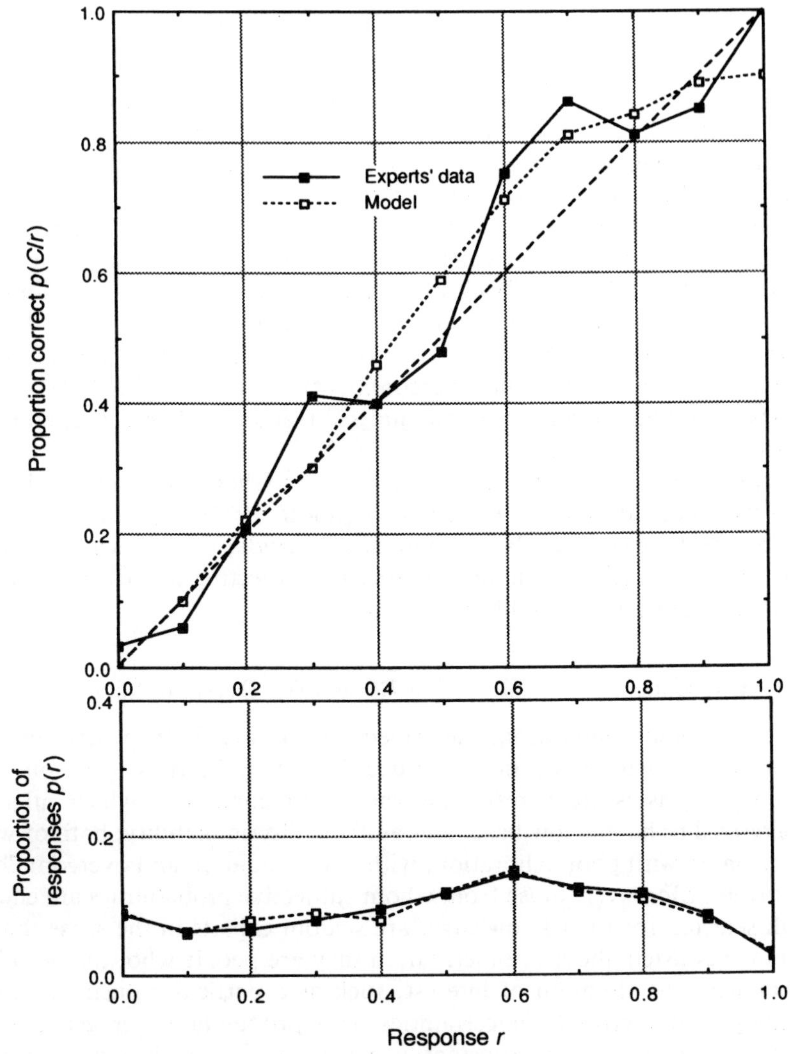

Figure 17.6(b) Expert calibration. (Adapted from Keren, 1987, for judgments by expert bridge players of the probability of making their contracts, by permission of Academic Press and the author.)
Model: Normal, $A = 1.56$, $B = 1.31$, $\{-0.9, -0.6, -0.2, 0.2, 0.4, 0.8, 1.2, 1.5, 1.9, 2.5\}$, $p(C) = 0.56$

There is ample evidence that these conditions are often not fulfilled for experts (Wright & Bolger, 1992).

Training

The final fact about calibration that seems to be well established is that it can be improved, relatively easily, but the change in performance does not tend to generalize to judgments about other types of event (Lichtenstein & Fishchhoff, 1980). A modest amount of training for calibration in which judgments are made and calibration feedback is given, e.g. 200 judgments about general knowledge, clearly improves calibration for the same task (Lichtenstein & Fishchhoff, 1980). Just feedback of outcomes does not necessarily improve it (Sharp, 1988). A simple manipulation is to preface the questions to be used for determining calibration with a few (e.g. 5) that are deceptively difficult (assuming calibration will be overconfident) (Arkes et al., 1987). These methods would all seem to affect the judges' response criteria, i.e., choice of numerical values.

A different method is to train the respondents so they can better discriminate the events being judged (Lichtenstein & Fischhoff, 1977), or, though it is much less effective, to encourage the respondent to consider ways in which the event could fail to occur or to supply reasons for expecting a different outcome (Koriat, Lichtenstein & Fischhoff, 1982b).

17.4.3 Consequences of Calibration Results for Decision Analysis

The facts about calibration, described above are both worrisome and comforting in their implications for decision analysis. It is worrisome that decision analysis is frequently concerned with conditions which might be characterized as having low base rates and low discriminability, both of which are associated with poor calibration, with overestimation and overconfidence, respectively. Moreover, those from whom subjective probabilities are elicited, though selected for their knowledge, are seldom experts in the sense that the requirements listed above are met. Often they are people who, though highly familiar with the situation of interest, such as a particular business, do not habitually characterize its uncertainties with probabilities, or get adequate feedback on the outcomes. Substantive expertise by itself does not guarantee good calibration.

On the other hand, it is comforting that calibration can be improved relatively easily by suggestion and by training. It supports the hope that the procedural methods used to ensure the quality of subjective probability elicitation by decision analysts are effective. This optimism must be tempered by the findings that debiasing effects are seldom large and that training to improve calibration does not seem to generalize very well, so that, for example, demonstration exercises to illustrate overconfidence on general knowledge questions, as is sometimes advocated, may not affect calibration in the area of substantive knowledge relevant to the analysis.

17.4.4 Explanations

No general consensus has emerged among judgment and decision making researchers on a unified account of the effects described above . General models of probability judgment and judgmental forecasting (Beach, Barnes, & Christensen-Szalanski, 1986; Smith, Benson, & Curley, 1991) have not attempted to address calibration specifically. The base-rate effect is largely a consequence of the mathematical fact that the mean value of the ordinate of the calibration curve, the average over r of $p(C \mid r)$, must be equal to $p(C)$. There are, however, several accounts of overconfidence and the discriminability effect. First, it is widely observed that people often are overconfident about what they believe in the ordinary (as opposed to calibration) meaning of overconfident, (e.g. Cooper, Woo & Dunkelberg, 1988). Assuming people use numbers accurately to represent their feelings, it seems reasonable that calibration will surely reflect any biases of this kind. But no detailed account has been offered that would show how this by itself results in an effect that is so systematic with discriminability.

It has been proposed that the discriminability effect is due to an artifact of the experimental questions used in calibration experiments. Gigerenzer, Hoffrage, & Kleinbölting (1991) have developed a model in which the subjective probabilities are the experienced validities of the cues used to decide the question. They attribute overconfidence or underconfidence to the selection of questions for which the available cues are unrepresentatively effective (underconfidence) or ineffective (overconfidence). On this account, if the selection of questions and the respondents' experience are both unbiased with respect to a body of knowledge, then calibration is expected to be good, whatever the level of knowledge, i.e., value of $p(C)$. This, of course, substitutes calibration of cue validity for calibration of subjective probability.

Griffin & Tversky (1992) have proposed that the discriminability effect is due to people's giving too much emphasis to what the evidence considered indicates and not enough to its quality, as, for example, attending too much to the proportion of events and insufficiently to the sample size with a sample from a Bernoulli sequence. This can lead to certain patterns of underconfidence and overconfidence. Their arguments are based almost entirely on external validity experiments, and, although they have considerable merit, they don't go very far toward explaining observed calibration effects. When the authors do give an example of how their idea might apply to a calibration task, they give an instance of the model to be described below.

17.4.5 A Model of Calibration

von Winterfeldt & Edwards (1986) observe, "Research evidence about calibration is abundant but singularly hard to make sense of" (page 127). In what follows, a model that has been proposed for certain types of probability judgment (Ferrell & McGoey, 1980) is examined as a way to bring order to much of that evidence and as a basis for assessing the impact of the calibration problem on decision analysis. First, it is necessary to consider the type of judgment to which the model applies.

It is possible to distinguish at least two different ways in which numerical subjective probability responses might be produced, or perhaps there is a continuum with two extremes. At one extreme, the entire cognitive procedure may be based on mental computations or manipulations that involve numerical values, as, for example, when a test of external validity is made by giving the respondent numerical information about base rates and evidence reliability. In such a case the reasoning can be done with numbers so that a numerical result follows naturally. This is not to say that non-numerical information does not influence the process. The numerical result will reflect cognitive and attentional biases through the selection of computations that are performed and the heuristics used and through errors, approximations, etc. And the final reasoned result may be adjusted under the influence of a non-numerical impression or disposition. This description would appear to apply to many external validity studies, such as those of Griffin & Tversky (1992).

At the other extreme, the evidential basis for a subjective probability judgment may have no numerical information whatever. It seems reasonable to suppose that judgments of whether a visual or auditory signal is present are of this kind. A response number is assigned to the magnitude of some internal variable that is determined by processes that are not consciously numerical. Of course one may be told how likely it is that a signal is present, or how many possibilities there are to consider, etc., so the number finally assigned would be influenced by these numbers, but the basic evidence, the result of cognitive processes that attempt to discriminate signal events from non-signal events and on which one would make a decision, is an internal feeling or impression that has magnitude but no intrinsic numerical labels or natural metric that implies numerical values. There is a substantial body of experimental and theoretical work in signal detection that supports this conception (Green & Swets, 1974; Swets, Tanner, & Birdsall, 1961), and there is a model of the way in which the internal variable is categorized to produce responses, the rating model (Egan, Schulman & Greenberg, 1959).

The conceptual task in probability judgment, as in detection, is discrimination between the occurrence of an "event" and its non-occurrence. One seeks a basis for deciding which is the case, even though perfect discrimination may be impossible. It seems likely that probability judgment tasks besides those involving just sensory discriminations, tasks that are essentially cognitive, may also produce an internal magnitude, without an implicit scale, to which a number must be assigned to produce a subjective probability response, and to which the rating model of signal detection is appropriate. The fact that signal detection theory has been applied so widely in psychology is evidence for its applicability beyond sensory psychophysics.

If this is so it can be argued that the judgments required in decision analysis are frequently of this sort. The decision analysis process attempts to model the probabilistic features of the problem explicitly, in the interests of an accurate and objective representation of the decision situation. The subjective inputs tend to be those that *cannot* easily be modeled or derived by normatively inspired, computationally based methods of reasoning such as those described by Shafer & Tversky (1985).

A model of calibration, based on signal-detection theory, was proposed by Ferrell. It is described in detail in Ferrell & McGoey (1980) and presented tutorially in Smith & Ferrell, (1983). The model offers a straightforward explanation of the discriminability effect, the base rate effect and the ease with which training improves calibration but fails to generalize. In addition it reveals conditions under which perfect (or even good) calibration is impossible, it fits a large amount of published data, and in many instances it makes specific, quantitative predictions of calibration curves and response distributions. Moreover, since it describes mainly the numerical assignment process, it is quite compatible with much of the theoretical and empirical cognitive psychological research on calibration, which can be interpreted as dealing, for the most part, with determinants of the magnitude of the internal decision variable that is partitioned in the model.

The model assumes that consideration of a proposition about which a subjective probability is to be given results in a value of an internal decision variable, i.e. a variable with the property that it is larger, on average, when the proposition in question is true than when it is not. The decision variable range is partitioned into intervals in the manner of the signal detection rating model. Responses, which may be numerical or verbal or even arbitrary, but having some ordinal property, are assigned in increasing order to successive intervals on the decision variable. For subjective probabilities, an appropriate set of probability numbers is used. When a question is considered, a value of the decision variable is generated and the response that corresponds to the interval in which the value lies is given.

A calibration experiment produces two distributions of values of the internal decision variable X, one when the event in question E occurs $f(X \mid E)$, and one when it does not $f(X \mid -E)$. The calibration value $p(E \mid r)$ associated with a given response category r is just the proportion of times that the event occurred when the internal variable took on a value within that category. If one can estimate the two distributions, the base rate $p(E)$, and the values of

$\{x_i\}$ that partition the internal variable, then the model produces the calibration curve and the proportion of responses in each category. The model is not described in greater detail here for lack of space; a fuller account is available in Ferrell & McGoey (1980). A comprehensive technical review is currently in preparation.

In what follows, the way the model explains the salient experimental findings about calibration will be illustrated by showing its fit to data representing a range of subjective probability judgment tasks. Finally, implications for decision analysis will be drawn.

17.4.6 Results from the Model of Calibration

The discriminability effect

The discriminability effect—relative overconfidence for events that are harder to discriminate and underconfidence for those that are easier to discriminate—is explained by the model as the respondents' maintaining the same criteria, the same cutoff values $\{x_i\}$ on the decision variable, or adjusting them insufficiently, when discriminability changes. If clues are inadequate to indicate the change, the respondent has no basis for adjusting the cutoffs.

Figure 17.2[1] shows the calibration curves and response proportions for two tests using two-alternative general knowledge questions, given to different groups, from Lichtenstein & Fischhoff (1977). The proportions of correct responses $p(C)$ for the two sets of questions were different by design. The questions were selected from a larger set used previously. For each question on the easy test there was one on the harder test that scored 20 percentage points lower. The model was fitted to the data for the hard test by matching the response proportions exactly. The same cutoffs $\{x_i\}$ were then assumed for the selected set. Using only *one* parameter estimated from the data, $p(C) = 0.80$, the model predicts all the calibration proportions and the response proportions for the easy set. The difference in $p(C)$ accounts for the difference in calibration. In this model, as in all those presented here, normal distributions are assumed on the internal variable. That they fit so well is probably due in part to the group nature of the calibration data. The form of the distributions, however is a parameter of the model, and can be expected to be different with different tasks.

Clues can indicate that discriminability is worse and lead respondents to adjust their response criteria to attempt to maintain good calibration in spite of the change. An example is forecasts with a longer time horizon, as reported by Braun & Yaniv (1992). Subjective probabilities on the full-range [0,1] of future recession were given by members of a survey panel of professional economic forecasters. Figure 17.3 shows the data and the fitted model for the current quarter and for three quarters ahead, respectively. The model was fitted to the data using the maximum likelihood method (Dorfman & Alf, 1969; Grey & Morgan, 1972)[2] with results shown as "model" values on the graphs.

For both forecasts, the proportion of actual recessions was approximately 0.2. The discriminability, measured by the separation of the underlying normal distributions, is about one fifth as great for the longer time horizon. Us a result, the judgments are extremely overconfident. The cutoffs differ just as one would expect; those for three quarters ahead are much more strict than those for the current quarter. A forecaster must have stronger evidence, a more extreme value (extreme relative to the value of the internal variable associated with the base rate) before responding with the same probability number. But the experts' adjustment of their cutoffs is not sufficient. Indeed, manipulation of the model shows that it is *impossible* to have good calibration for responses above about 0.25 when the base rate and discriminability are so low; no amount of adjustment of the criteria will suffice. However, excellent calibration can be achieved with a restricted response set, e.g. {0, 0.1, 0.2, 0.25}. Analysis also reveals that the forecasters could have been much better calibrated at one quarter ahead for responses of 0.4 and up had they used the cutoffs they adopted for the longer time horizon, instead, as is shown by the indicated calibration curve. This suggests that training for calibration would have been of benefit for the shorter horizon even though it could not possibly produce good performance for the longer one at the higher response values.

The base rate effect

For full-range judgments, relative overestimation with lower base rates and relative underestimation with higher ones is the base-rate effect. It is explained by the model as the respondents' maintaining the same cutoffs, or adjusting them insufficiently, when the base rate changes. If clues to the base rate are inadequate, the respondent has no basis for adjusting the cutoff. With true–false questions, one expects, perhaps from experience, that about half of them will be true. If it is hard to discriminate true from false, then it will be hard to learn the actual base rate of true statements and one will tend to adopt criteria appropriate to a default base rate of about 0.5.

Figure 17.4 illustrates the base-rate effect with data from Lichtenstein and Fischhoff given in Lichtenstein, Fischhoff & Phillips (1982). The model was fitted by eye to the data for probability true $p(T) = 50\%$. Unfortunately, the response proportions were not available. Then, keeping the same criteria, the value of $p(T)$ *only* was changed to equal the values for the other curves on the graph. The model then predicts those curves, as shown in the figure. In contrast, the expert economic forecasters of Figure 17.3, who presumably know the base rate of recessions, 0.2, very well, do not show any effect of its being different from expected.

The discriminability effect and base-rate effect combine, and the model reflects this. The model was applied to the student judgments of Figure 17.5 by fitting those for $d' = 1$, $p(C) = 0.71$ to obtain the cutoffs. The model, with only the d' and $p(C)$ values for each of the other groups, then predicts their calibration curves and response proportions quite accurately.

Expertise

Experts are not always better calibrated, but in some cases they are much superior to non-experts and the model helps to show why. Keren (1987) compared expert and amateur contract bridge players' calibration for the probability of making a contract that they had bid, i.e. the probability of doing at least as well as promised in the subsequent round of card playing. Figures 6a and 6b show the calibration curves and response proportions for experts and amateurs, with the model fitted by maximum likelihood. Although they had a slightly smaller base-rate of contracts actually made (0.56 as

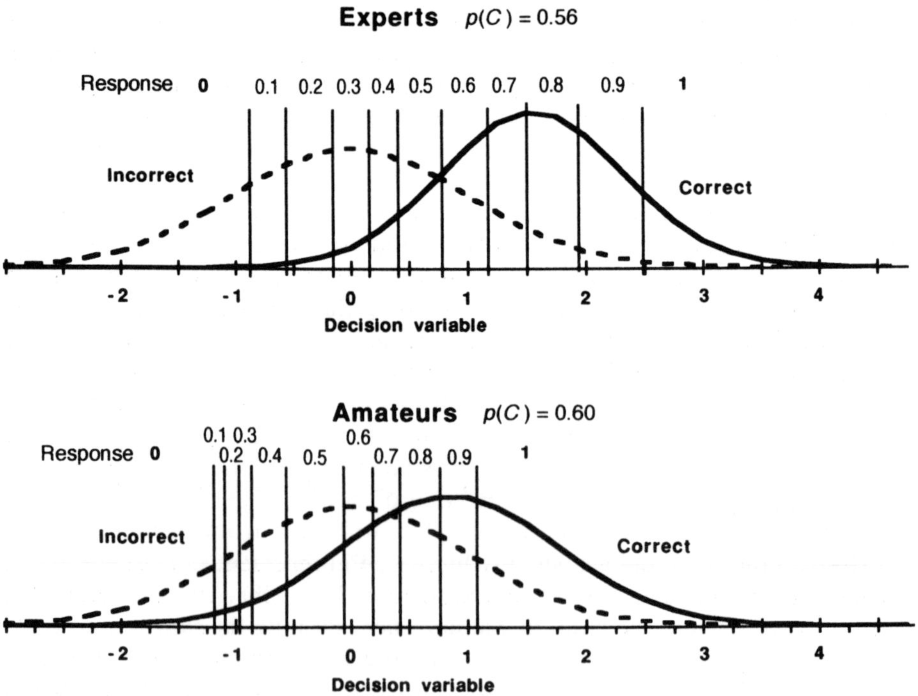

Figure 17.7 Graphical representation of the models for Keren's (1987) amateur and expert bridge players, showing the better discrimination and more widely spaced cutoffs of the experts.

opposed to 0.60), the experts are much more able to discriminate contracts they can make from ones they cannot. This can be seen in Figure 17.7. In particular, experts' criteria are more uniformly spread, especially over the distribution for contracts made, and the relative variability of that distribution is significantly smaller.

Training

In the model, the response cutoffs are, in principle, independent of the underlying distributions on the internal decision variable. To what extent, if the model is correct, they actually are independent of the cognitive processes that generate instances of the internal variable is an empirical question. The ability to change calibration through a modest amount of training (Lichtenstein & Fishchhoff, 1980) or cueing by a preliminary set of questions that establish an expectation of difficulty (Arkes et al., 1987) suggests that the criteria are easily changed with appropriate information, cues or feedback.

Weather forecasters are, by now, famously well calibrated (Murphy & Winkler, 1977). To be so they have to adjust their calibration to match weather conditions that vary from season to season and from year to year. Since they observe weather outcomes and their calibration results they have a basis for doing this. An example is a comparison of the precipitation forecasts for San Francisco for the years 1956–57 and 1957–58. There was a substantial increase from the first year to the second in the probability of rain and some increase in its discriminability. Figure 17.8 shows the calibration curves and response proportions with the model fitted to them (by eye). The forecasters modified their cutoffs as a result of the change. Had they remained the same in the second year, the model's calibration curve would have been as shown. Instead, the calibration is quite good. The difference in calibration between the two years is not due just to the different conditions, but required an adjustment of response criteria in reaction to those changed conditions, which is to be expected of experts with suitable feedback.

17.4.7 Implications of the Decision Variable Partition Model for Decision Analysis

If the calibration model is a valid theory, a good representation of the underlying structure of how some subjective probabilities are generated and given quantitative expression, then those probabilities must be incoherent, in the sense of not satisfying the axioms of probability when taken as a whole. In theory, probabilities are measures on an absolute scale. There is no general permissible transformation, no function of probabilities, that preserves all their information, just as there is no transformation of counts of discrete items that does not distort their meaning. The partitioning and response assignment

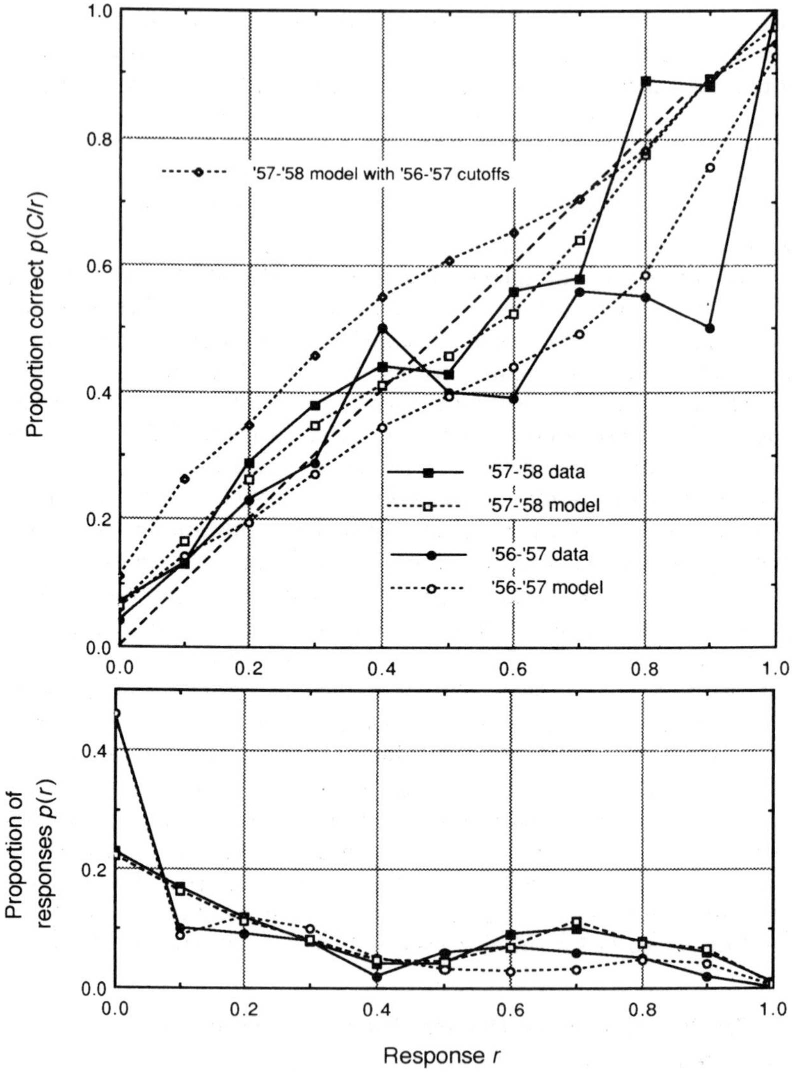

Figure 17.8 Precipitation forecast calibration curves for San Francisco from Root (1962) showing a change in calibration from one year to the next as the detectability of rain and the base rate of rain both increase. Forecasters changed their criteria to avoid overconfidence under the new conditions.
Model: 56–57: Normal, $A = 1.24$, $B = 1$, $\{-0.5, -2.5, 0.1, 0.45, 0.65, 0.8, 0.95, 1.15, 1.6, 2.8\}$ $p(\text{rain}) = 0.21$.
57–58: Normal, $A = 1.34$, $B = 1$, $\{-1.1, -0.55, -0.2, 0.05, 0.2, 0.35, 0.6, 1.1, 1.6, 2.8\}$ $p(\text{rain}) = 0.37$

processes described by the model preserve only the ordinal properties of the internal decision variable, whatever properties it may have before being expressed as a number. This is not good news, but, paradoxically, it argues overwhelmingly in favor of the decision analysis process which seeks to give coherence to the entire choice situation by means of modeling. The opportunity for the evaluation of uncertain alternatives to be misdirected by incoherent probability judgments is far greater if the judgments are holistic than if they are disaggregated and are knitted together by a careful analysis of the situation.

The model suggests that the number assignment aspect of subjective probability judgment, in certain cases, may not necessarily be tightly coupled to the process by which the non-numerical part of the judgment is produced. This is why the model is compatible with other cognitive research on reasoning and judgment, both numerical assignment and reasoning can be sources of over- or underconfidence. Clearly, this is a matter for empirical determination, but if it is the case, it appears to offer new opportunities for training for calibration and for guidance of the subjective probability elicitation process.

Since the model considers probability judgments themselves to be the result of decisions, an explicit consideration of this in decision analysis as in Levi (1985) may provide a firmer basis for, and give even more weight to, efforts in elicitation to avoid the judge's having any stake in the outcome of the event being judged.

In the presentation of their general framework for judgment under uncertainty Smith, Benson & Curley (1991) conclude "In our view, while the evidence to number translation is poorly understood, it is unlikely to be the critical factor in assessment. We anticipate that judgment is less variable and influential than reasoning in determining numerical outcomes" (page 314). This view certainly requires re-examination.

17.5 SUBJECTIVE PROBABILITY COMBINATION

17.5.1 Quality of Aggregated Judgments

The practice of combining subjective probabilities from several experts is intended to improve the quality of the probabilities used in a decision analysis. Aggregation methods can be classified as mathematical, of which there are many, simple averaging being a good example, behavioral, in which the experts agree on a value, and mixed, in which there is controlled interaction usually followed by mathematical combination (Ferrell, 1985). Mixed methods are generally designed to inhibit the negative effects in interacting groups of such things as dominating personalities, or status differences, but to retain the useful sharing of information.

It is not obvious how subjective probability quality and calibration in particular are affected by combination. *Coherence* is violated by many mathematical methods, such as averaging, but it is enforced by the analyst with respect to the total set of probabilities used in a decision model. Subjective probabilities should represent the *true opinion* of the judge, and this is ensured for individuals, as far as possible, by the elicitation process, but it is not a relevant consideration for aggregated values except, perhaps, for group consensus. *Reliability* is almost invariably increased by combination methods, since they reduce variability. Validity has at least two aspects, *discriminability* and *calibration*. Discriminability is likely to be increased. If those whose judgments reflect greater knowledge have more influence by being weighted more heavily, by being in a majority, or by influencing others, the group output will be more discriminating between the occurrence and non-occurrence of the events in question by a kind of voting effect. Discriminability affects the calibration of individuals' judgments, as was shown in the previous section, with increases leading to shifts in the direction of underconfidence. This might be expected for aggregated judgments, as well. There are other features of combination methods that potentially can affect calibration, so the over-all effects need to be determined empirically for different combination methods.

17.5.2 Seaver's Experiments

The main experimental work on calibration of combinations of subjective probabilities was done for his dissertation by Seaver (1979). The general structure of his experiment for discrete probabilities is diagrammed in Figure 17.9. Groups of four people, acquainted with each other, responded individually to 20 general knowledge questions by choosing one of two

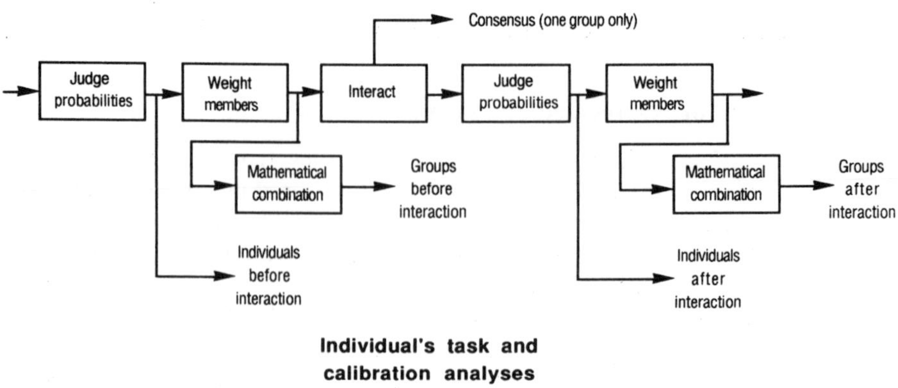

Figure 17.9 The structure of the calibration experiments of Seaver (1979).

answers and giving a probability of being correct on the interval [0.5,1]. For each question, they also assigned weights to members. Next, they considered each question as a group according to one of a number of structured methods such as Delphi, and finally they again responded individually to the set of questions and assigned weights to members. In one interaction condition, they discussed each question until reaching a consensus. Three different mathematical combination methods were then applied to the individual judgments before and after each interaction condition and using each of three weightings, equal, deGroot (1974), and normalized self-weighting. The mathematical combination methods, ones commonly used or proposed, were as follows:

- linear

$$p_G = \sum_{i=1}^{4} w_i p_i \tag{1}$$

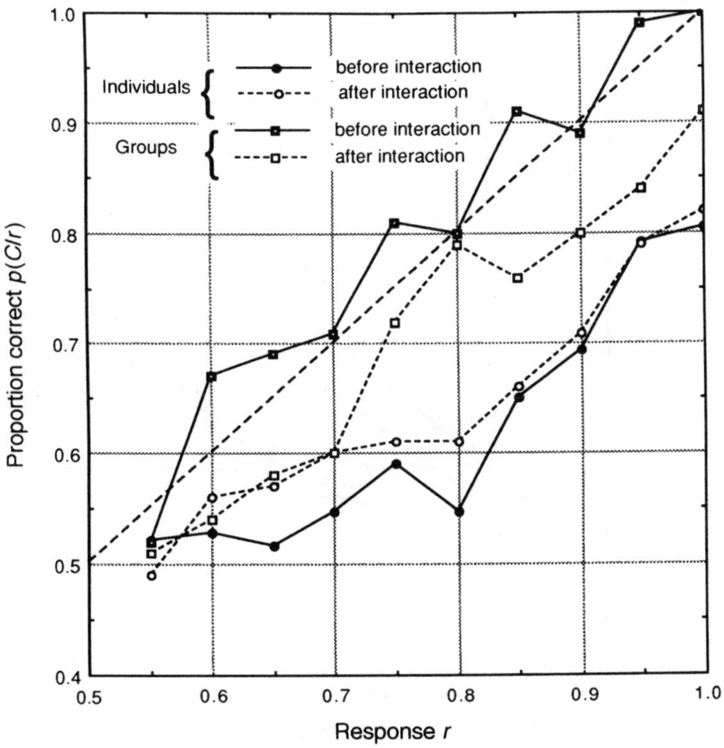

(a)

Figure 17.10(a) Data from Seaver (1979) for individuals and equally weighted average before and after interaction, pooled over interaction type (Reproduced from Ferrell & Rehm, 1980. Adapted from Seaver, 1979, by permission of the author.)

- weighted geometric mean

$$p_G = \frac{\prod_{i=1}^{4} p_i^{w_i}}{\prod_{i=1}^{4} p_i^{w_i} + \prod_{i=1}^{4} (1 - p_i)^{w_i}} \tag{2}$$

- and likelihood ratios

$$\frac{p_G}{(1 - p_G)} = \prod_{i=1}^{4} \frac{p_i}{(1 - p_i)} \tag{3}$$

These produce a group probability for the correct hypothesis from individual probabilities p_i where the value of p_i is the individual's half-range response r_i if the correct choice was made and $(1 - r_i)$ if not. The group choice and its half-range response r_G are similarly related to the group probability P_G.

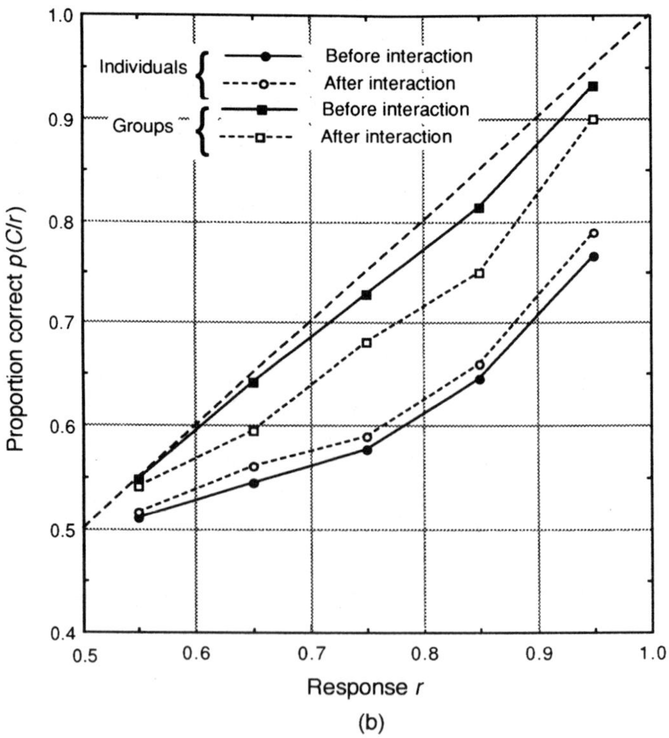

Figure 17.10(b) Model results corresponding to the conditions of (a) showing the same effects of aggregation and interaction. (Reproduced from Ferrell & Rehm, 1980.) Model: Normal, {0.2, 0.36, 0.53, 1.045}, $p(C) = 0.65$

17.5.2 Seaver's Results

In keeping with most other research on the subject (Wainer, 1976 and for a review, Ferrell, 1985), Seaver found that the three weighting methods did not differ significantly and did not have a significant effect on the results. He also found that the interaction methods did not significantly differ among themselves, but they did have a significant effect on both the quadratic probability score and on calibration.

Seaver's main results were that mathematical combination, except for the likelihood ratio method, substantially reduced the overconfidence of individuals' judgments, and that interaction reduced overconfidence for individuals, but increased it for the mathematically combined group judgments. Figure 17.10a shows the results for individuals compared to the equally weighted average and Figure 17.11a shows the results for the different combination methods. Calibration for consensus is not shown; that curve is

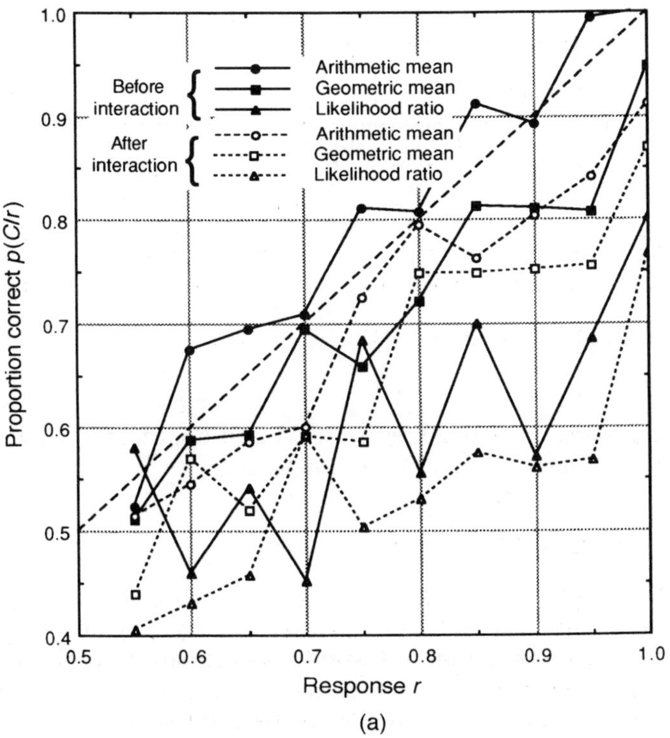

Figure 17.11(a) Data from Seaver (1979) for different types of mathematical aggregation before and after interaction, pooled over interaction type. Only the case of equal weights is included (Reproduced from Ferrell & Rehm, 1980. Adapted from Seaver, 1979, by permission of the author.)

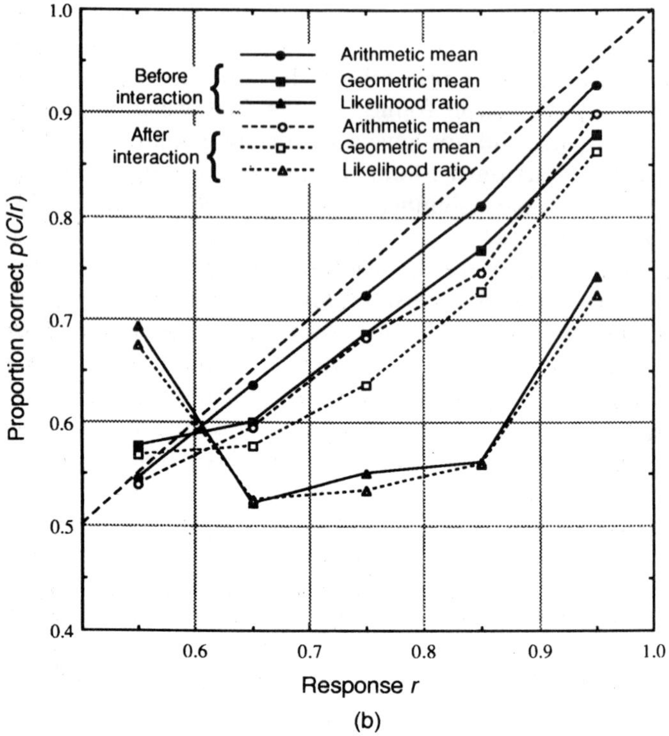

Figure 17.11(b) Model results corresponding to the conditions of Figure 17.11(a), showing similar effects of aggregation and interaction (Reproduced from Ferrell & Rehm, 1980).
Model: Normal, {0.2, 0.36, 0.53, 1.045}, $p(C) = 0.65$

rough, being based on only 20 responses, but it appears to fall about on the curve for the geometric mean before interaction.

17.5.3 Explanation

These results can be fully explained by two competing effects, both of which are induced by mathematical combination and by the interaction process, *discriminability* and *extremeness*. Consider mathematical combination first. If individuals are correct more often than not, then when their responses are mathematically combined, the group response will be correct even more often, just as in voting. This will increase discriminability, which, as was discussed in the previous section, shifts the calibration curve toward underconfidence. Mathematical combination can also result in somewhat more extreme responses, closer to zero or one, shifting the calibration curve in the opposite direction toward overconfidence. The likelihood ratio method produces the most extremeness. For example, if all members were to give a subjective probability of 0.6, the group probability using the likelihood ratio method is 0.835. The average produces no extremeness at all and the geometric mean is in between. The net effect with the likelihood ratio method is that the increase in discriminability and the extremeness cancel each other out so that the calibration has the same overconfidence as that of individuals. The average, with essentially no extremeness, is much less overconfident than individuals, and the geometric mean is in between the other two.

These same two factors, discriminability and extremeness, also result from interaction as a consequence of two very well-documented effects of group process, *conformity* (Insko & Schopler, 1972) and *polarization* (Lamm & Myers, 1978), respectively. Conformity produces a movement of individual opinion toward the group mean and polarization moves it away from the neutral point in the direction favored by the group mean. Seaver concluded, "Overall, interaction did produce a convergence of judgments" (page 42), and ". . . assessments tended to become more extreme after interaction" (page 39). Conformity, being an averaging process, induces increased discriminability, whereas polarization increases extremeness. With individuals, in this case, at least, conformity dominates, so that their calibration becomes somewhat less overconfident after interaction. However, conformity uses up part of the potential of combination to increase discriminability, so the effect of interaction plus mathematical combination is that of combination with additional extremeness added from polarization. Hence, combination after interaction shifts the calibration less in the direction of underconfidence than it does without interaction.

17.5.4 Application of the Model

The interpretation of Seaver's results given above is substantiated by a Monte Carlo simulation of his experiment using the calibration model described in the previous section. This model has the appropriate distinction between the underlying confidence dimension which is affected by interaction and the numerical expression of that confidence which is manipulated by combination. Ferrell & Rehm (1980) fitted the model to the calibration for individuals in Seaver's experiment, and using the partition $\{x_i\}$ and proportion correct $p(C)$ estimated from this, along with an addition to the model necessary to represent the effects of groups, determined calibration curves for interaction and the different types of combination. The addition was that judgments were assumed to be intercorrelated and interaction was simulated by a transformation x' of individual confidence values x based on the average of the other three in the group $\bar{x}_3$.

$$x' = x + k_1(\overline{x_3} - x) + k_2\overline{x_3} \tag{17.4}$$

where k_1 and k_2 are constants representing conformity and polarization, respectively.

Four unit-normal random numbers (with $\rho = 0.4$ to represent the commonality of shared information and a mean appropriate to $p(C)$ of 0.65 from the fitted model) were generated for each question. These represent the individuals' initial feelings of confidence. Negative values are for choice of the wrong answer. These initial values of x were transformed into subjective probability response values r by the partition from the fitted model (assumed symmetrical about zero) and the results are those for individuals before interaction. The response values r were then combined according to equations (17.1)–(17.3) using equal weights, to give the results for groups before interaction. The initial x values were then transformed by equation (17.4) (with $k_1 = 0.1$ and $k_2 = 0.2$) and responses r determined from the partition to give results for individuals after interaction. When these are combined by equations (17.1)–(17.3) they give results for groups after interaction.

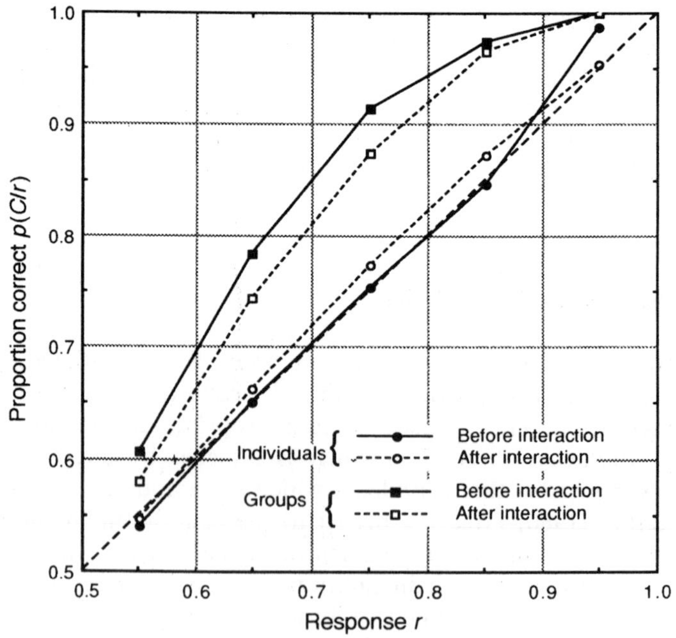

Figure 17.12 The effect on group calibration, predicted by the model when individuals are well calibrated. Groups are then underconfident (Reproduced from Ferrell & Rehm, 1980).
Model: Normal, {0.5, 1.1, 1.85, 3.55}, $p(C) = 0.65$

The model output is shown in Figures 17.10b and 17.11b. Comparison with Seaver's data in Figures 17.10a and 17.11a shows that the model captures the overall structure of Seaver's results, and shows the same effects of both combination and interaction and of the two together. In all, seven calibration curves are "predicted" from a total of eight parameter values. The lack of smoothness of Seaver's calibration curves is, presumably, due to the relatively small number of responses represented by each one.

A major implication of Seaver's results as interpreted through the calibration model is that interaction and/or combination do not *improve* the calibration of individuals, they just *change* it in a systematic way. From the standpoint of decision analysis there is a serious problem if this is so. A decision analyst who has carefully elicited probabilities from an expert and considers them to be well calibrated, cannot combine them either behaviorally or mathematically with those from another expert without expecting the result to be miscalibrated, to be underconfident. If underconfidence is not expected, then the original probabilities cannot be considered well calibrated, they must be overconfident. To illustrate, if the calibration model cutoffs that produce good calibration are used in the Monte Carlo simulation instead of those that fit the results for Seaver's individuals, the calibration curves of Figure 17.12 are obtained for individuals after interaction and for combination by averaging. Averaging produces *extremely* underconfident group results from otherwise well-calibrated individual judgments.

17.6 CONCLUSIONS

Decision analysis relies for its validity on the quality of the process it represents and on the soundness of the procedures by which the process is implemented. The decision theory basis of the over-all process is excellent. The procedures for elicitation of subjective probability judgments for reducing bias and error are sophisticated and well grounded in empirical observation. They are not, however, based on a theory of judgment, nor on a detailed knowledge of the process by which a subjective probability is produced.

Empirical results of calibration studies indicate that there are at least three sources of systematic miscalibration to which subjective probabilities in decision analysis are susceptible—low discriminability, base rates and combination. One tends to use judgment to assess what is difficult to know otherwise, and so events whose occurrence is hard to discriminate are likely to be appraised by subjective probability judgments. An example is the prediction of recession shown in Figure 17.3b. It might be observed that the relative frequency of the high probability judgments is quite small, so that their contribution to a probability score is small and the calibration is not as bad as the graph makes it look. But probability scores do not measure the importance of decisions taken with the probabilities, nor their costs. Base rates pose a problem, too. If the base rate of a critical phenomenon in an analysis is unknown, or wrongly assumed, all the probabilities contingent on it may be miscalibrated. And even when the base rate is known, it is difficult to know how sensitive judgment should be to it. Combination doesn't solve the problems. It almost certainly produces a change in calibration. Averaging, the most commonly recommended approach to combining subjective probabilities, shifts calibration toward underconfidence, a shift for the better only if the judgments being combined are appropriately overconfident.

It is concluded that subjective probability elicitation and combination needs a firmer theoretical structure if elicitation methods to prevent bias and combination methods to improve quality are to be used with confidence. This chapter has drawn attention to, and provided new evidence in support of, a model of subjective probability judgment and calibration that may be capable of providing a framework for that theoretical structure. The model organizes many of the empirical results that have been found puzzling and, in particular, gives a coherent picture of how the discriminability and base rate effects come about. It can represent a wide range of calibration data and provides a basis for quantitative prediction of calibration. Although it was not explained here, the model provides a unifying account of how responses to probability questions of different types (one, many or no-alternatives, probability true or probability correct) relate to each other (Ferrell & McGoey, 1980). And it relates cognitive inference and reasoning processes to numerical assignment, showing how they each can contribute separately to calibration. Consequently, if due allowance is made for types of judgment task, it is consistent with other, more cognitively focused, research on judgment under uncertainty.

This model is not a solution to major problems of subjective probability in decision analysis, but it is a possible step in that direction. It has, of course, its share of difficulties, puzzles and, perhaps, inconsistencies. Presenting challenges that are closer to psychophysics or engineering than to cognitive science, it may be considered unfashionable in psychological circles. Perhaps that is why it has been neglected by researchers over the past 10 years. Its consideration in Chapter 18 may be indicative of a change.

NOTES

(1) The model shown is fully specified in the following way in each figure caption: *Half-range models*: form of the distribution, cutoff values, proportion correct. *Full range models*: form of the distributions, A (the distance between means in units of sigma for the signal distribution), B (the ratio of noise to signal standard deviation), cutoff values (zero at mean of the noise distribution), base rate.

(2) The program, ROCFIT, was obtained courtesy of Dr Charles Metz, Dept. of Radiology, University of Illinois, in whose laboratory it was developed.

REFERENCES

Arkes, H.R., Christensen, C., Lai, C. & Blumer, C. (1987) Two methods of reducing overconfidence. *Organizational Behavior and Human Decision Processes*, **39**, 133–44.

Bar-Hillel, M. (1980). The base rate fallacy in probability judgments. *Acta Psychologica*, **44**, 211–33.

Beach, L.R., Barnes, V.E. & Christensen-Szalanski, J.J.J. (1986) Beyond heuristics and biases: a contingency model of judgmental forecasting. *Journal of Forecasting*, **5**, 143–57.

Braun, P.A. & Yaniv, I. (1992) A case study of expert judgment: Economists' probabilities versus base-rate model forecasts. *Journal of Behavioral Decision Making*, **5**, 217–31.

Christensen-Szalanski, J.J.J. & Beach, L.R. (1984) The citation bias: Fad and fashion in the judgment and decision literature. *American Psychologist*, **39**, 75–8.

Clemen, R.T. & Murphy, A.H. (1990) The expected value of frequency calibration. *Organizational Behavior and Human Decision Processes*, **46**, 102–17.

Cooper, A.C., Woo, C.Y. & Dunkelberg, W.C. (1988) Entrepreneurs' perceived chances for success. *Journal of Business Venturing*, **3**, 97–108.

Curtis, P.E., Ferrell, W.R. & Solomon, I. (1985) A model of calibration for subjective probability distributions and its application to aggregation of judgments. *Journal of Forecasting*, **4**, 349–61.

deGroot, M. (1974) Reaching a consensus. *Journal of the American Statistical Association*, **69**, 118–21.

Dorfman, D.D. & Alf, E. (1969) Maximum likelihood estimation of parameters of signal detection theory and determination of confidence intervals—Rating method data. *Journal of Mathematical Psychology*, **6**, 487–96.

Dowie, J. (1976) On the efficiency and equity of betting markets. *Economica*, **43**, 139–50.

Dyer, J.S. (1990) Remarks on the analytic hierarchy process. *Management Science*, **36**, 249–58.

Egan, J.P., Schulman, A.F. & Greenberg, G. Z. (1959) Operating characteristics determined by binary decisions and by ratings. *Journal of the Acoustical Society of America*, **31**, 768–73.

Eils, L.C., Seaver, D.A. & Edwards, W. (1977) Developing the technology of probabilistic inference: Aggregating by averaging reduces conservatism. (Research Report No. 77-3). Social Science Research Institute, University of Southern California.

Ferrell, W.R. (1985) Combining individual judgments. In G. Wright (ed.), *Behavioral Decision Making*. Plenum, New York, pages 111–45.

Ferrell, W.R. & McGoey, P.J. (1980) A model of calibration for subjective probabilities. *Organizational Behavior and Human Performance*, **26**, 32–53.

Ferrell, W.R. & Rehm, K. (1980) A model of subjective probabilities from small groups. In *Sixteenth Annual Conference on Manual Control*, Massachusetts Institute of Technology, Cambridge, MA, pages 271–84.

Fischhoff, B. (1982a) Debiasing. In D. Kahneman, P. Slovic & A. Tversky (eds.), *Judgment under Uncertainty: Heuristics and Biases*, Cambridge University Press, New York, pages 422–44.

Fischhoff, B. & MacGregor, D. (1982b) Subjective confidence in forecasts. *Journal of Forecasting*, **1**, 155–72.

Fischhoff, B. (1989) Eliciting knowledge for analytical representation. *IEEE Transactions on Systems, Man, and Cybernetics*, **19**, 448–61.

Gigerenzer, G., Hoffrage, U. & Kleinbölting, H. (1991) Probabilistic mental models: A Brunswikian theory of confidence. *Pychological Review*, **98**, 506–28.
Goodman, B. (1972) Action selection and likelihood estimation by individuals and groups. *Organizational Behavior and Human Performance*, **7**, 121–41.
Green, D.M. & Swets, J.A. (1974) *Signal Detection Theory and Psychophysics* (2nd edn). Wiley, New York.
Grey, D.R. & Morgan, B.J.T. (1972) Some aspects of ROC curve fitting: Normal and logistic models. *Journal of Mathematical Pychology*, **9**, 128–39.
Griffin, D. & Tversky, A. (1992) The weighing of evidence and the determinants of confidence. *Cognitive Psychology*, **24**, 411–35.
Harrison, J.M. (1977) Independence and calibration in decision analysis. *Management Science*, **24**, 320–8.
Howard, R.A. (1980) An assessment of decision analysis. *Operations Research*, **28**, 4–27.
Howard, R.A. (1983) Risk preference. In *Readings on the Principles and Applications of Decision Analysis*, Strategic Decisions Group, Menlo Park, CA.
Howard, R.A. (1988) Decision analysis: Practice and promise. *Management Science*, **34**, 679–95.
Insko, C.A. & Schopler, J. (1972) *Experimental Social Psychology*. Academic Press, New York.
Kahneman, D., Slovic, P. & Tversky, A. (eds.) (1982). *Judgment Under Uncertainty: Heuristics and Biases*. Cambridge University Press, New York.
Kahneman, D. & Tversky, A. (1973) On the psychology of prediction. *Psychological Review*, **80**, 237–57.
Kahneman, D. & Tversky, A. (1979) Intuitive predictions: Biases and corrective procedures. *TIMS Studies in Management Science*, **12**, 313–27.
Keeney, R.L. (1982) Decision analysis: An overview. *Operations Research*, **30**, 803–38.
Keren, G. (1987) Facing uncertainty in the game of bridge: A calibration study. *Organizational Behavior and Human Decision Processes*, **39**, 98–114.
Keren, G. (1991) Calibration and probability judgments: Conceptual and methodological issues. *Acta Psychologica*, **77**, 217–73.
Koriat, A., Lichtenstein, S. & Fischhoff, B. (1980) Reasons for confidence. *Journal of Experimental Psychology: Human Learning and Memory*, **6**, 107–18.
Lamm, H. & Myers, D.G. (1978) Group induced polarization of attitudes and behavior. *Advances in Experimental Social Psychology*, **11**, 145–95.
Levi, K. (1985) A signal detection framework for the evaluation of probabilistic forecasts. *Organizational Behavior and Human Decision Processes*, **36**, 143–66.
Lichtenstein, S. & Fischhoff, B. (1977) Do those who know also know more about how much they know? *Organizational Behavior and Human Performance*, **20**, 159–83.
Lichtenstein, S. & Fischhoff, B. (1980) Training for Calibration. *Organizational Behavior and Human Performance*, **26**, 149–71.
Lichtenstein, S., Fischhoff, B. & Phillips, L.D. (1982) Calibration of probabilities: The state of the art to 1980. In P. Kahneman, P. Slovic & A. Tversky (eds.), *Judgment under Uncertainty: Heuristics and Biases*, Cambridge University Press, Cambridge, pages 306–34.
Lopes, L.L. (1987) Procedural debiasing. *Acta Psychologica*, **64**, 167–85.
Lusted, L.B. (1977) *A study of the efficacy of diagnostic radiologic procedures: Final report on diagnostic efficacy*. Chicago: Efficacy Study Committee of the American College of Radiology.
Merkhofer, M.W. (1987) Quantifying judgmental uncertainty: Methodology,

experiences, and insights. *IEEE Transactions on Systems, Man, and Cybernetics*, **SMC-17**, 741–52.
Murphy, A.H. & Winkler, R.L. (1977) Can weather forecasters formulate reliable forecasts of precipitation and temperature? *National Weather Digest*, **2**, 2–9.
Phillips, L.D. (1989) *Decision analysis in the 1990's* Decision Analysis Unit, London School of Economics.
Phillips, L.D. & Edwards, W. (1966) Conservatism in a simple probability inference task. *Journal of Experimental Psychology*, **72**, 346–54.
Root, H. (1962) Probability statements in weather forecasting, *Journal of Applied Meteorology*, **1**, 163–8.
Seaver, D.A. (1979) Assessing probability with multiple individuals: group interaction versus mathematical aggregation. Dissertation, The University of Michigan.
Shafer, G. & Tversky, A. (1985) Languages and designs for probability judgment. *Cognitive Science*, **9**, 309–39.
Sharp, G.L. (1988) Performance feedback improves the resolution of confidence judgments. *Organizational Behavior and Human Decision Processes*, **42**, 271–83.
Smith, G.F., Benson, P.G. & Curley, S.P. (1991) Belief, knowledge, and uncertainty: A cognitive perspective on subjective probability. *Organizational Behavior and Human Decision Processes*, **48**, 291–321.
Smith, M. & Ferrell, W.R. (1983) The effect of base rate on calibration of subjective probability for true-false questions: Model and experiment. In P. Humphreys, O. Svenson & A. Vari (eds.), *Analyzing and Aiding Decisions*. North Holland, Amsterdam.
Spetzler, C.S. & Staël von Holstein, C.A.S. (1975) Probability encoding in decision analysis. *Management Science*, **22**, 340–58.
Staël von Holstein, C.A.S. & Matheson, J.E. (1979) *A manual for encoding probability distributions*, SRI International.
Swets, J.A., Tanner, W.P., Jr. & Birdsall, T. (1961) Decision processes in perception. *Psychological Review*, **68**, 301–40.
Tversky, A. & Kahneman, D. (1973) Availability: a heuristic for judging frequency and probability. *Cognitive Psychology*, **4**, 207–32.
Tversky, A. & Kahneman, D. (1974) Judgment under uncertainty: Heuristics and biases. *Science*, **185**, 1124–31.
von Winterfeldt, D. & Edwards, W. (1986) *Decision Analysis and Behavioral Research*. Cambridge University Press, Cambridge.
Wainer, H. (1976) Estimating coefficients in linear models: It don't make no nevermind. *Psychological Bulletin*, **83**, 213–17.
Walsten, T.S. & Budescu, D.V. (1983) Encoding subjective probabilities: A psychological and psychometric review. *Management Science*, **29**, 151–73.
Wright, G. & Bolger, F. (eds.). (1992) *Expertise and Decision Support*. Plenum, New York.

Chapter 18

The Calibration of Subjective Probabilities: Theories and Models 1980–94

Alastair G.R. McClelland and Fergus Bolger
University College London

18.1 INTRODUCTION

Why are individuals so often badly calibrated when making subjective probability judgements? In particular, why is overconfidence so frequently observed (the "overconfidence effect") and why does the degree of miscalibration seem to vary systematically with task difficulty (the "hard–easy" effect)? In the conclusion to their well-known review of research up to 1980 on the calibration of probabilities, Lichtenstein, Fischhoff & Phillips (1982, page 333) noted that

> . . . a striking aspect of much of the literature reviewed here is its "dust-bowl empiricism". Psychological theory is often absent, either as motivation for research or as an explanation of the results.

The aim of this chapter is to review the current situation, by providing a critical analysis of a number of substantive theories and models of subjective probability judgement for discrete propositions which have appeared in the last 14 years. Little will be said regarding the empirical research which has been

Subjective Probability. Edited by G. Wright and P. Ayton.
© 1994 John Wiley & Sons Ltd.

reported since Lichtenstein et al.'s review, excepting that which, in our view, provides either strong support for, or militates against, a particular model.

18.2 THE OVERCONFIDENCE EFFECT AND THE HARD–EASY EFFECT

Errors in probability judgements are not randomly distributed around the "target" value (e.g. the normative answer derived from the application of Bayes' theorem, or the proportion of correct answers associated with a particular subjective probability judgement). Rather, systematic errors or "biases" are frequently observed in a variety of tasks requiring individuals to produce probability judgements. So robust and compelling are these effects that they have been christened "cognitive illusions" by a number of authors (e.g. Kahneman & Tversky, 1982; von Winterfeldt & Edwards, 1986). In the calibration literature, the most commonly observed bias is the overconfidence effect. Subjects deliver probability estimates which are too high when measured against either the relative frequency of occurrence of an event assigned a particular probability estimate, or the proportion correct of answers which have been assigned a particular probability value.

A closely related phenomenon is the hard–easy effect; this is the observation that overconfidence decreases as task difficulty (usually indexed by the overall proportion correct) decreases. With easy tasks (over about 80% correct answers on a half-range probability scale, in a *two-alternative forced-choice* [2AFC] task[1]) the overconfidence effect disappears, and underconfidence is often observed. The main challenge facing theories and models of confidence is to explain these two effects.[2] It should be noted however, that there are clear anomalies in the literature. Good calibration has been found for "difficult" tasks (e.g. Keren, 1988); marked differences in calibration performance have been observed at the *same* level of task difficulty (e.g. McClelland, Coulson & Icke, 1990; Wright, 1982) as has good calibration at *different* levels of item difficulty (Juslin, 1993). Reversals of the hard–easy effect have also been noted, where the degree of overconfidence for a harder task is less than for an easier task (Keren, 1988; Ronis & Yates, 1987). Any principled account of how individuals make subjective probability judgements has to provide an explanation for these results, as well as the overconfidence effect and the hard–easy effect.

18.3 THE LOCUS OF BIAS IN PROBABILITY JUDGEMENTS

Over the last twenty years or so, two rival schools have developed, each with a radically different view as to the locus of the observed biases in calibration and other probability tasks. Jungermann (1983) termed one camp "the pessimists" and the other "the optimists". The pessimists believe that biases are in people—the optimists believe that biases are in research. The most representative members of the pessimist school are Daniel Kahneman and Amos Tversky. In their "heuristics and biases program" (Gigerenzer, 1991, page 85) they have argued that the locus of the bias is within the cognitive system, and have provided many demonstrations of the apparent irrationality of individuals when engaged in probabilistic reasoning (e.g. Kahneman, Slovic & Tversky, 1982; Tversky & Kahneman, 1974, 1983). Kahneman and Tversky claim that an explanation for this irrationality is that individuals use a variety of heuristics when reasoning probabilistically. Several of these heuristics have been cited as explanations (or partial explanations) for miscalibration, most notably the "anchor-and-adjust" heuristic. For example, Keren (1991) proposed that in laboratory-based 2AFC tasks, the expectation of the subjects regarding task difficulty might act as an anchor, and explain the relationship between difficulty and over/underconfidence. He suggested that subjects might anchor on a probability estimate reflecting intermediate difficulty (75%). When confronted with an item perceived to be either very easy or very difficult, they would adjust accordingly, but not sufficiently, and this would lead to under- or overconfidence respectively. Ferrell & McGoey (1986) made a similar suggestion. Wright (1982) also appealed to the anchor-and-adjust heuristic in order to explain the difference in calibration for past-event questions (e.g. has at least one member of the British Parliament died within the last fourteen days? (a) yes, (b) no) and future-event questions (e.g. will at least one member of the British Parliament die within the next fourteen days? (a) yes, (b) no). He suggested that the response anchor for past event questions might be 1.0 (reflecting certainty) whereas for future event questions it might be 0.5 (reflecting uncertainty). A failure to adjust sufficiently from these anchors would lead to the observed overconfidence for past-event questions and underconfidence for future-event questions.

Other theorists have sought explanations for miscalibration (and particularly overconfidence) in terms of cognitive style (Wright & Phillips, 1984), ignorance of processing limitations (Pitz, 1974), motivation (Milburn, 1978; Zakay, 1983), cognitive optimism (Dawes, 1980), and response-scale effects (Poulton, 1989). However, as Keren (1991) noted, many of these explanations are *post hoc* in nature, and whilst most of them are consistent with the finding of overconfidence, they cannot explain observations of underconfidence, good calibration and the hard–easy effect. The feature they have in common is that they all attribute miscalibration to human failing.

The most vigorous champion of the optimist school is Gerd Gigerenzer. In a number of papers (Gigerenzer 1991, this volume; Gigerenzer, Hoffrage & Kleinbölting, 1991) he and his colleagues have argued strongly against the pessimist school, and in particular the heuristics and biases approach. They have provided both a theoretical framework and empirical evidence to back the claim that biases in probabilistic reasoning are essentially artifacts, encouraged by the use of artificial and sometimes misleading tasks, and the nonrepresentative sampling of stimulus materials. In addition to suggesting that the locus of bias is in the main outside the cognitive system, Gigerenzer has also questioned the nature of probabilistic representations *within* the cognitive system. He has argued that the "intuitive statistician" within us is a frequentist, and not a Bayesian. Thus for Gigerenzer, probabilities are represented in terms of frequencies—and not as beliefs. This, Gigerenzer argues, has profound consequences both for the interpretation of the empirical evidence and for the nature of the theories and models required to explain human probability judgement, as we will discuss in a later section.

18.4 THEORIES AND MODELS

In this section, each of the theories and models we have chosen to review is briefly described, and in the next section critically evaluated. Although not exhaustive, we hope that we have included most of the major theoretical work from 1980 to date. Some of the models are clearly within the pessimist camp (locating the bias within the individual) and others the optimist camp (locating the bias within the experimental procedure, and in particular the nature of the stimulus materials). In addition, it is clear that some are domain specific (e.g. restricted to general-knowledge tasks), whereas others are presented as quite general models. Some of the models seem more applicable to situations in which the stimulus items are *essentially similar* (Ronis & Yates, 1987) or *related* (Keren, 1987, 1991) that is, they share common characteristics, whereas others are applicable to situations in which the stimulus items are *essentially unique* or *unrelated* (Keren, 1991). Despite these differences, we attempt a comparison of the models in a later section. The models are presented in chronological order.

18.4.1 The Stage Model

Koriat, Lichtenstein & Fischhoff (1980) proposed a three-stage model of the cognitive processes involved in answering a two-alternative general-knowledge question, and giving an associated confidence rating. In the first stage, memory is searched for relevant information and an answer chosen; in the second stage the evidence is assessed to arrive at a feeling of certainty, and in the third stage this feeling is translated into a numerical response. Koriat et al. suggested that unwarranted certainty (overconfidence) might be linked to one or more of the three stages

They proposed that in the first stage, individuals might be biased in the way they elicit knowledge, favouring positive rather than negative evidence. In the second stage, they suggested that individuals might have a tendency to disregard evidence inconsistent with the chosen answer. This tendency to elicit positive evidence, and disregard evidence contrary to the chosen alternative, would lead to overconfidence. Finally, they noted that in addition to the cognitive biases operating at the first two stages, there might be an inappropriate translation of feelings of certainty into a probability value. If this mistranslation were such that individuals gave values that were generally too high, this would also contribute to the overconfidence effect.

18.4.2 The Detection Model

Ferrell & McGoey (1980) proposed a model for calibration based on signal detection theory (also see Ferrell, this volume, and Smith & Ferrell, 1983). These authors not only provided an explanation of how perceived truth of propositions might be translated into numerical judgements of confidence (corresponding to the third stage in Koriat, Lichtenstein and Fischhoff's model), but also sought to explain the overconfidence effect, the hard–easy effect, and the effects of base-rate change on calibration performance.

Ferrell and McGoey suggested that the task facing subjects in a calibration study can be broken down into two parts; the first being a detection process, described by a signal detection model, and the second the assignment of a numerical probability value on the basis of the result from the first stage. The decision variable used is partitioned by a set of criterion values, one interval for each possible probability response (r). Each question generates a particular value on the decision variable, and the interval into which that value falls then determines the numerical response. In a 2AFC task, it is assumed that each alternative produces a value of apparent truth, and the subject chooses the alternative with the higher value. For simplicity, the distributions of apparent truth for the two alternatives are assumed to be normally distributed with equal variance. The decision variable is then taken to be the *absolute difference* between the truth values for the two alternatives, the larger the difference the greater the confidence that the correct alternative has been selected. The distributions of absolute difference when the correct answer produced the larger truth value, and when the incorrect answer produced it, are normal distributions truncated below zero. Calibration can then be determined from (1) the probability of a correct response [p(C)], (2) the cumulative distribution functions of the decision variable when the response is correct and not correct, and (3) the partition of the decision variable.

In order to be perfectly calibrated, subjects must choose a partitioning such that for each interval, $p(C \mid r) = r$. However, Ferrell and McGoey assume that the partition is determined by information obtained prior to the task (subjects appear to set their criteria for a task of intermediate difficulty, i.e. about 75% correct—see Ferrell & McGoey, 1980, page 40) and that it will *not change* unless feedback about performance is provided. In a two-alternative task, the partitioning should be determined solely by discriminability (as base rate is fixed at 50%). If subjects have set their partitioning for a task which is *easier* than the task actually presented, they will exhibit overconfidence—if on the other hand, the partitioning is set for a task *harder* than the one presented they will exhibit underconfidence. Ferrell and McGoey also showed that the model is not restricted to 2AFC tasks, but can be applied quite generally to any calibration task format. Finally, with the assumption that subjects are insensitive to changes in base rate as well as discriminability, they also provided predictions concerning the effects of base rate change on calibration performance in full-range tasks (see also Smith & Ferrell, 1983, and Ferrell, this volume for further details of the model).

18.4.3 The Process Model

May (1986a, 1986b) proposed a process model of subjective probability judgements which she claimed would allow the degree and direction of miscalibration to be predicted. In common with the later ecological models (see below) she argued that miscalibration should neither be seen as a bias in inferential reasoning, nor as a result of mistranslation of a feeling of uncertainty into a numerical response (cf. Ferrell & McGoey, 1980). Instead, she suggested that it was a consequence of the specific background knowledge possessed by subjects, of the tasks given to subjects, and the selection of items within the tasks.

Following Koriat, Lichtenstein & Fischhoff (1980) her model has three stages, which she labelled *problem-solving*, *emergence of subjective certainty* and *quantification* respectively, but she placed the origin of miscalibration at the first stage. She also identified two sources of difficulty which would affect the proportion of correct responses in a calibration task. The first source (Difficulty 1) was seen as a characteristic of the task such as the objective distance between two stimuli in a psychophysical task. The second source (Difficulty 2) was attributed to the subject having "wrong knowledge" (such as a distorted cognitive map when making a latitude judgement).

In May (1986a) two possible internal representations based on a subject's knowledge are presented. The first *mental model* is in the form of a syllogism, which May proposed might be used to answer a question such as "Which city has more inhabitants? (a) Hyderabad, (b) Islamabad." The second is in the form of a cognitive map, and she suggested that this type of representation might be used to answer a question such as "Which city is further north? (a) Rome, (b) New York."

With the syllogistic representation, May speculated that a subject might reason in the following way; Capital cities tend to have many inhabitants, Islamabad is a capital city, and therefore Islamabad is likely to have many inhabitants. On this basis, the subject would choose Islamabad as the answer.[3] May proposed that the confidence expressed by the subject would be a function of the perceived extent of the intersection between the set of capital cities and the set of cities with a large population, and possibly other relevant background knowledge. She argued that if the items (pairs of cities) were *randomly* sampled from the population of cities, good calibration would be expected in the long run, but if a set of items were selected so as to include a large number of "misleading" items (i.e. items for which the inference produced the wrong answer—as in the example above), overconfidence would result.

With the cognitive-map representation, May proposed that the difficulty of an item (e.g. deciding which of two cities was further north) would depend upon the *subjective* distance between them, and this would be reflected directly in the confidence given. However, distortions in subjects' cognitive maps could lead them to pick the wrong alternative, and depending on the extent of the distortion, to pick the wrong alternative with considerable confidence. Again, a large number of such "misleading" items in a set would produce overconfidence. In the cognitive-map representation, the probability of a correct answer (the *reaction* probability in May's terminology) is determined by the *subjective relationship* between the cities (i.e. which is subjectively further north) and the confidence (or subjective probability) by the *subjective distance* between the cities.

Finally, May draws a distinction between *populationwise* calibration and *itemwise* calibration. The former defines calibration for the universe of items that could be constructed within a certain knowledge domain. She suggests that defined this way, perfect calibration is impossible when misleading items are present. Itemwise calibration is defined with respect to single items, so that an item is calibrated when the mean reaction probability is identical to the mean subjective probability. Thus, by definition, only non-misleading items could be well calibrated in this sense.

18.4.4 The Memory Trace Model

Albert & Sponsler (1989) presented a mathematical model for the calibration of subjective probabilities. They proposed that the brain subconsciously makes subjective probability estimates based upon memories of similar prior experiences.

The basic principles underlying the model are fairly simple. Albert & Sponsler (1989) assume that when confronted with a new "fact pattern", the brain abstracts cues which permit it to identify a set of prior experiences characterized by a similar set of cues. For example, a weather forecaster might, on the basis of current weather conditions, identify days in the past upon which a similar pattern of weather conditions prevailed. The memory trace which is retrieved is considered to be composed of the results of *predictions* of a series of binary events (e.g. rain, no rain) which either did or did not occur. Successful predictions are imagined to be encoded as 1s and unsuccessful predictions as 0s. The "true" subjective probability is the relative frequency of successful predictions in the past for the entire set of events.

Albert and Sponsler suggested that an individual may not be able to retrieve the entire set of previous predictions, but rather a subset of the entire memory trace. The particular subjective probability estimate is then taken to be the relative frequency of successful predictions within the subset identified. An expert[4] estimator, according to the authors, will be able to identify the full set, and thus his or her estimate will match the true subjective probability. A less expert estimator will be able to retrieve only a subset of the full set of prior predictions, and thus his or her prediction will not necessarily correspond to the true subjective probability (the actual estimate depending upon the relative frequency of successes in the subset).

From this basic model, the authors derive the permissible range of subjective probability estimates allowed by the model for various values of the true subjective probability, and for various proportions of the memory trace retrieved. The range of possible subjective probability estimates a subject *could* produce is constrained (according to the model) by both the proportion of successful predictions in the full memory trace, and by the proportion of the trace retrieved on a given occasion. The authors assumed that, on average, the subjective probability estimate given is the midpoint of the range of permitted values for a given value of the true subjective probability and a given proportion retrieved. The rationale for this hypothesis is simply that when the midpoints are plotted against the true subjective probability values for various proportions retrieved, *overconfidence* is observed in that the midpoint values are greater than the corresponding true probabilities. The authors also conclude that the choice of subsets cannot be random (such that all subsets of a particular size are equally likely to be retrieved) as they show that this leads to an expected value of the subjective probability estimates which is equal to the true subjective probability value for all sample sizes. In other words, if the choice was random, subjects would in the long run be perfectly calibrated, and not demonstrate overconfidence.

In the remainder of their paper the authors speculated as to the possible shape of the distribution of the probability estimates in the various permitted ranges, and suggested that a transformation of the Beta distribution was particularly promising. They discussed the problems with attempting to estimate empirically the parameters in their model, and this leads to further speculation concerning the possible shape of the distribution of estimated true subjective probabilities. Two equations are given to calculate the expected value of the true probability estimates, one assuming a uniform distribution of values over the permitted range, and the other for a non-uniform distribution. The authors also note that an estimate of the true subjective probability could be obtained directly from an individual's calibration curve. Albert & Sponsler (1989) conclude that "... the entire theory demands, and it is hoped will receive, experimental verification" (page 308).

18.4.5 The Ecological Models

Working independently, Gigerenzer, Hoffrage & Kleinbölting (1991), and Juslin (1993, 1994) produced two models of remarkable similarity. We have termed these the "ecological" models. Both of these models are founded on the simple but powerful notion that, as a result of interaction with the natural environment, individuals encode the frequencies of co-occurrences of events in the environment, and use this information in a very direct fashion when making judgements about discrete propositions and attaching confidence ratings to those judgements.

The theory of probabilistic mental models (PMM theory) proposed by Gigerenzer, Hoffrage & Kleinbölting (1991) was developed to explain performance in 2AFC general-knowledge tasks, but the authors do suggest that the model could also be applied to 2AFC perceptual tasks. The authors outline the circumstances under which good calibration is to be expected, and provide explanations for the overconfidence effect, the hard–easy effect, observed reversals of the hard–easy effect, and a hitherto unobserved third phenomenon, termed the confidence–frequency effect (described below).

Underlying the model are the following assumptions;

(1) individuals are well adapted to their environments (see Brunswick, 1943, 1955),
(2) individuals are able to extract and store accurately, information regarding the frequency of occurrence of events in the environment—and do so with little if any conscious effort (see Hasher and Zacks, 1984 for a review),
(3) the basis for probability judgements are these stored frequencies—the "intuitive statistician" is a frequentist.

Gigerenzer, Hoffrage & Kleinbölting (1991) propose that if a solution to a given general-knowledge item cannot be obtained directly (e.g. via direct retrieval from memory, or by use of an elementary logical operation such as the method of exclusion) the subject will set up a *probabilistic mental model* (PMM). To take an example used by Gigerenzer *et al.*, imagine that the task consists of deciding which of two German cities with more than 100 000 inhabitants (*a* or *b*) is the larger. The PMM will contain the reference class (all cities in Germany with more than 100 000 inhabitants), a target variable (city size), probability cues (other variables related to the reference class) and cue validities. Gigerenzer et al. suggested that potential probability cues might include a soccer team cue (one city has a team in the Bundesliga and the other does not), an industrial cue (one city is located in an industrial area and the other a rural area), and a state capital cue (one city is a state capital and the other is not). A variable is a probability cue for the target variable in the reference class if the conditional probability of alternative *a* being the correct answer is different from the conditional probability of *b* being correct. For this example, a subject might use a soccer-team cue; that is choose the city which has a team in the German soccer Bundesliga (note that this assumes the cue can be activated; if both or neither of the cities had a team in the Bundesliga it could not be used). The *ecological validity* of this cue is 0.91, in that if all pairs in which one city has a team in the Bundesliga and the other does not are checked, one would find that in 91% of cases the city with the team in the Bundesliga has more inhabitants.

Cues are assumed to be generated, and if possible activated, in a hierarchical fashion. The probability cue with the highest validity is generated and tested first; if it can be activated it is used, if not, a further cue is generated and tested. If no cue can be activated, it is assumed that the subject chooses randomly and gives a 50% confidence rating.

Gigerenzer, Hoffrage and Kleinbölting argued that through interaction with the natural environment the ecological validities of cues become internalized through a process of observing the frequencies of co-occurrences of environmental events, and become the *cue validities* in the PMM. An individual uses a probability cue to both select an answer, and as the source of the confidence; once a choice is made, the cue validity is given as the confidence rating. If individuals have had repeated experience with a particular reference class, a target variable, and cues in the environment, it is assumed that the cue validities correspond well to the ecological validities. However, if a subject in a calibration experiment is given a set of items which are not *representative* of the reference class in the environment, performance will be systematically biased, as the cue validities used will not be appropriate.

Gigerenzer, Hoffrage & Kleinbölting (1991) made a number of predictions based on PMM theory. The first was that typical general-knowledge items (which have been used extensively in calibration studies) will produce both overconfidence and accurate judgements concerning the number of items correctly answered (frequency judgements). Overconfidence is attributed to a biased selection of items, with difficult, and importantly, "misleading" items, being over-represented. The use of cues and cue validities which would produce good calibration for a representative set of items leads to overconfidence with a selected "difficult" set. Imagine that a subject uses a cue with an ecological validity of 0.90 to answer ten questions. If these have been sampled randomly, the set of ten questions would be expected to contain one misleading item—an item where the cue fails to deliver the correct answer. The subject would be expected to get 9 correct, and give a confidence of 0.90 for each answer. If however the set has been informally selected, there may be, say, 3 such misleading items present, in which case the subject would only get 7 correct, but would still give a 0.90 confidence rating for each of the 10 questions. Hence the subject would be overconfident. However, the authors also predicted that if subjects were asked "how many items do you think you answered correctly?" they should give accurate estimates, as the reference class is now past general-knowledge tests that they have taken. If the current test is typical of the general-knowledge tests they have experienced in the past (i.e. representative of the reference class of general-knowledge tests), good calibration with respect to frequency estimates is to be expected. A corollary to the first prediction is that if subjects are given a set of items randomly selected from a particular reference class, they will exhibit good calibration with respect to confidence judgements, but should now *underestimate* the number of items correctly answered. Gigerenzer, Hoffrage & Kleinbölting (1991) referred to this as the *confidence–frequency effect*. They further predict that if two sets of items, hard and easy, are generated by the *same* sampling process (be it random or biased) the hard–easy effect should disappear. Finally, they predicted that if a set of items is representative of a "hard" reference class, and a second set is selected to be "difficult" but from an "easy" reference class, a reversal of the hard–easy effect should be observed. Empirical evidence is presented (and results in the literature reinterpreted) which, in the main, support the predictions from the model.

The arguments developed by Juslin (1993, 1994) to explain the overconfidence effect and the hard–easy effect are, in all essentials, the same as those of Gigerenzer, Hoffrage & Kleinbölting (1991). However, Juslin did not make the further predictions concerning the calibration of frequency judgements, nor did he predict the possible reversal of the hard–easy effect.

18.4.6 The Strength and Weight Model

Griffin & Tversky (1992) have provided the most complete model within the heuristics and biases program to explain the patterns of overconfidence and underconfidence observed not only in calibration studies, but also in other investigations of judgement under uncertainty. In this respect, it has similarities with Gigerenzer's (1991) attempt to explain apparent biases in many situations using a small number of explanatory principles.

The two concepts central to Griffin & Tversky's (1992) argument are those of "strength" and "weight". Neither of these concepts is rigidly defined, but by strength they mean the "extremeness" of available evidence, and by weight the "predictive validity" of the evidence. They note that the distinction between these two concepts is closely related to the distinction between the size of a statistical effect (e.g. the difference between two means) and its reliability (e.g. the standard error of the difference). Griffin and Tversky argue that individuals focus on the strength of evidence and may make some adjustment (albeit insufficient) in response to the weight. This thesis makes particular use of two of the heuristics identified in the heuristics and biases program "representativeness" and "anchor-and-adjust". For example, individuals may make use of the representativeness heuristic (judging an interviewee on how much he or she "looks like" a successful manager) whilst ignoring (or paying scant attention to) other factors controlling predictive validity. Any adjustment that is made to take into account the weight of evidence is deemed insufficient; i.e. individuals make use of the anchor-and-adjust heuristic, but fail to adjust sufficiently.

Griffin and Tversky claimed that their hypothesis predicts a distinctive pattern of overconfidence and underconfidence. When, in a given situation, strength is high but weight is low, subjects exhibit *overconfidence*. However, when strength is low and weight is high, individuals should exhibit *underconfidence*. In the first half of the paper, they were concerned with testing their predictions with respect to the evaluation of statistical hypotheses; in the second half of the paper they extended their argument to confidence judgements, and in particular to the calibration of general-knowledge questions. The authors noted that there is a problem with the application of the theory within the calibration domain, as strength and weight cannot be experimentally controlled. However, Griffin and Tversky offer an "analogy to a chance setup" (page 425) as a model of the processes involved when making confidence judgements in a calibration study.

In this model, the balance of arguments for a (two-alternative) general-knowledge problem is represented by the proportion of red and white balls in a sample; difficulty (discriminability) is the difference between the probability of obtaining a red ball under each of two competing hypotheses (the correct alternative and the incorrect alternative). Expressed confidence is given by the balance of arguments, i.e. the proportion of red balls in the sample (where a red ball represents an argument in favour of the correct alternative). For any given sample size, and any pair of probabilities of obtaining a red ball under the competing hypotheses, the normative or "correct" confidence response can be computed for each sample composition (i.e. 1 red, 2 reds, etc.) from the Binomial distribution and the application of Bayes' theorem. Griffin and Tversky assume that the confidence judgement given by an individual is simply the proportion of red balls they observe (i.e. the strength of the evidence). Thus neither the level of difficulty as indexed by the discriminability of the hypotheses, nor the sample size (both aspects of the weight of the evidence) are taken into account. In a simulation of the model, the authors generated three calibration curves, by plotting the normative (posterior) probability solution against the proportion of red balls in the sample (balance of arguments) for three pairs of hypotheses defining three levels of difficulty. The probabilities of obtaining a red ball under the competing hypotheses were 0.50 and 0.40 for the "easy" task, 0.50 and 0.45 for the "difficult" task, and 0.50 under both hypotheses for the "impossible" task. Griffin and Tversky chose non-symmetrical hypotheses "to allow for an initial bias that is often observed in calibration data" (page 426).[5] The sample size (10) was held constant. The three curves bore a striking resemblance to empirical calibration curves obtained by Lichtenstein & Fischhoff (1977) for three levels of item difficulty (as indexed by overall proportion correct). Calibration was reasonably good for easy items (with slight underconfidence for lower confidence ratings and slight overconfidence for higher ratings), there was marked overconfidence for difficult items, and a flat calibration curve for impossible items. Thus the model appears to provide an explanation for both the overconfidence phenomenon and the hard–easy effect. It would also be an easy matter to simulate changes in base rate with this model, but this was not investigated by the authors.

18.5 EVALUATION OF THE MODELS

In this section, we provide a critical evaluation of the models described above. We assess each model with respect to the empirical evidence, and in terms of psychological plausibility. We also try to highlight the similarities and differences between the models.

18.5.1 The Stage Model

Koriat, Lichtenstein & Fischhoff (1980) presented some empirical evidence to support the view that overconfidence can be attributed to biases operating at the first and second stages of their model. They noted a bias in the production of reasons for and against a particular alternative, favouring evidence *for* over evidence *against*. They also provided some empirical support for the notion that subjects disregard evidence inconsistent with their chosen answer. Forcing subjects to write down a contradictory reason did improve the realism of their confidence assessments as indexed by calibration scores. However, the decrease in overconfidence was very small (2%) and non-significant. In a replication of the study Fischhoff & McGregor (1982) failed to find an effect of disconfirming evidence. Gigerenzer, Hoffrage & Kleinbölting (1991, page 521) argued that these negative results were consistent with PMM theory, which predicts no change in expressed confidence when subjects are asked to produce disconfirming evidence.

Despite the underspecification of the stage model, and the lack of empirical support in its favour, it is useful when viewed as a framework within which other models of calibration can be located. For example, the process and ecological models are concerned with the first stage, the strength and weight model attributes miscalibration to the first and second stages, and the detection model to the third stage.

18.5.2 The Detection Model

Ferrell and his colleagues (Ferrell, this volume; Ferrell & McGoey, 1980; Smith & Ferrell, 1983) have provided an impressive amount of evidence in support of their model. They have shown that it provides a good fit to a wide range of data sets, collected using a variety of task formats and types of stimulus materials.

With respect to task difficulty in the standard 2AFC probability-correct task, Ferrell & McGoey (1980) have shown that even with the estimation of a single parameter $[p(C)]$, both the calibration curve and the usage of the response categories can be well predicted. For example, they estimated the cutoffs (the position of the criteria on the decision variable) for the entire data set collected by Lichtenstein & Fischhoff (1977), and showed that this set of criteria provided a good fit to the data for subsets of items (e.g. "hard" items and "easy" items). They concluded that such findings were consistent with the hypothesis that a set of cutoffs appropriate to a proportion correct of about 75% was maintained even when $p(C)$ was substantially different from 75%. The failure to adjust the cutoffs (or to adjust them sufficiently) led to the hard–easy effect.

Ferrell & McGoey (1980) and Smith & Ferrell (1983) also showed that in a full-range probability-true task, the shift of the calibration curve under different base-rate conditions was again predictable if the cutoffs for the 50% condition were used to model the data collected with base rates either above or below 50%. Again, they concluded that the effect could be attributed to the subjects' failure to adjust their criteria with changes in base rate.

Despite the success of the detection model,[6] it has been criticized on the grounds that it does not elucidate the cognitive processes involved in making subjective probability judgements. For example, Keren (1991, page 262) remarked that "Unfortunately, the model provides little insight into the possible cognitive processes governing probability judgements." Of course, this does not mean that the model is wrong—it could indeed be the case that miscalibration is caused by a problem with translating a feeling of certainty into a numerical estimate, and has little to do with the use of heuristics, or the operation of other cognitive processes.

However, there are some empirical results which suggest that calibration performance may depend on more than the numerical assessment process. For example, Juslin (1993) found excellent calibration in four subsets of data where $p(C)$ varied from 66% to 80%. Thus the subjects in this experiment were able to maintain good calibration at rather different levels of difficulty—a finding inconsistent with the predictions of the detection model. Further, a fundamental assumption underlying the predictions derived from the model is that when miscalibration is observed, subjects have set their criteria for a task which they believe to be either easier (leading to overconfidence) or harder (leading to underconfidence) than the task actually presented. However, Gigerenzer, Hoffrage & Kleinbölting (1991) have presented convincing evidence that subjects are apparently able to anticipate very accurately the difficulty of a *typical* general-knowledge test (i.e. one representative of the reference class "general-knowledge tests") in that they can give good estimates of the numbers of items they have correctly answered, but still remain very overconfident in their calibration of individual items. In addition, with a randomly sampled set of items, subjects underestimate the frequency of correct answers but show good calibration. Both of these findings present problems for the detection model; with task difficulty correctly determined good calibration would be predicted, and when task difficulty is overestimated, underconfidence should be the result.

In conclusion, the detection model can be regarded as being a model of the last of Koriat, Lichtenstein & Fischhoff's (1980) three stages—the stage where subjective feelings of uncertainty are mapped onto numeric probability responses. The detection model has little to say about the cognitive processes leading to the formation of these feelings other than that some feature (or features) of the task generates (in an unspecified manner) a value on an unscaled internal variable (the decision variable). By making the simple assumptions that probability responses are read off from a partitioning of this variable and that—*in the absence of feedback*—this partitioning is not appropriately matched to the task difficulty the detection model can account for empirical data from a number of domains. This fact suggests that in most calibration tasks we need not look at earlier stages in Koriat et al.'s framework in order to account for the observed phenomena. However, for other tasks we may need to look further. For example, as emphasized above, the detection model makes no prediction of miscalibration when feedback is present. In the majority of judgement tasks outside the laboratory feedback of some kind is available, but in a number of such tasks miscalibration has still been found (e.g. Staël von Holstein, 1971, 1972; Yates, 1982; Yates and Curley, 1985). Further, the detection model is not applicable to tasks where explicit reasoning about numbers or proportions is required, such as in Bayesian probability revision or "book-bag-and-poker-chip" experiments. In these tasks—which Ferrell refers to as "external validity tasks" (see Chapter 17)—no value is generated on the internal decision variable, hence no probability response can be generated as required by the detection model. It would seem then that the detection model can only provide a partial explanation for poor calibration performance.

18.5.3 The Process Model

The two papers outlining the process model (May, 1986a, 1986b) contain many interesting and novel ideas, some of which were taken up in the later ecological models. For example, May recognized that the degree of miscalibration observed in a particular general-knowledge task was likely to depend upon the nature of the selection process used to generate the items, and she argued against the notion that miscalibration could be attributed to shortcomings in human inferential or intellectual reasoning. Finally, she introduced the notion of the *mental model* as the basis for subjective probability judgement.

May proposed that there might be at least two types of representation that could be used to answer general-knowledge problems. With respect to the syllogistic mental model where the individual uses inference to choose between two alternatives in a 2AFC item, Gigerenzer, Hoffrage & Kleinbölting (1991) have argued that the probabilistic syllogism as presented by May would not lead to good calibration in the long run, because it did not include information about both the alternative answers. They proposed a modified version of the model (the *double-syllogism model*) which, they claimed, would result in long-run calibration (see Gigerenzer, Hoffrage & Kleinbölting, 1991, page 523).

For the second form of representation (a cognitive map) May argued that perfect calibration was impossible if the map contained distortions. With this form of representation, misleading items are misleading because of "false knowledge" possessed by subjects. She showed that confidence was highly correlated not with the *objective* distances and geographical relationships between cities, but with the *subjective* distances and relationships, and that these were distorted. Thus she argued that the reason that subjects gave a mean confidence rating of 80% when answering the question "Which is further North? (a) Rome, (b) New York", though the solution probability was somewhat under 30%, was because of a serious distortion in the subjects' cognitive map, with North American cities shifted too far North with respect to European cities.

However, a simpler explanation can be derived from the ecological models—subjects used a climate cue to answer the question knowing that in general, a colder climate indicates a higher latitude. This would lead them to pick the wrong answer (i.e. New York) and may also lead them to believe that New York *really* is further North than Rome—resulting in a distorted cognitive map. The subjects had obviously never seen a map on which the latitudes of these two cities are reversed, so why did they have distorted cognitive maps? The idea that subjects use probability cues in answering such questions supplies an answer to both why subjects get this item wrong with high confidence, and why they have distorted cognitive maps. Differences in the confidence expressed for different pairs of items can be attributed to the use of a variety of cues with varying cue probabilities.

Although for the example above, the use of probability cues may provide a more parsimonious explanation of calibration performance, May is not alone in believing that different representations might be used in different stimulus domains. Björkman, Juslin and Winman (1993) argue that for psychophysical judgements, the distance between stimulus items on the dimension to be judged is indeed the representation used—but believe that for general-knowledge items probability cues are used.

May was clearly wrong in believing that perfect calibration was only possible in the absence of "misleading" items (see Gigerenzer, Hoffrage & Kleinbölting 1991; Juslin, 1993, 1994), and for the general-knowledge tasks that she considered, the ecological models would seem to provide a better description of the cognitive processes and representations underlying subjective probability judgements than is furnished by her process model.

18.5.3 The Memory Trace Model

There are a number of problems with this model as a general explanation of calibration performance. Firstly, it would only seem applicable to situations in which the assessor has had past experience of a set of similar events to those presented at test, *and* has received outcome feedback (e.g. a weather forecaster predicting rain). The model does not seem applicable to tests where the items are *essentially unique* (Keren, 1991) or indeed any task which is novel (e.g. choosing which of a pair of countries has the larger population) although the stimulus domain may be familiar (e.g. countries of the world). Indeed, Albert and Sponsler (1989) stated that confidence is based on a record of past successes and failures at predicting the outcomes of events the brain "deems similar" (page 298).

Secondly, the model only predicts the overconfidence effect—indeed, the authors seem to have been unaware of the fact that with very easy tests underconfidence is observed. Their apparent belief that it is only overconfidence which has to be explained (and their faith in the robustness of this finding) was critical to their rejecting the notion that all subsets of the full memory trace are equally likely to be selected because this would, in the long run, lead to perfect calibration. According to this model, perfect calibration can only be achieved when the assessor retrieves the entire memory trace—implying that only an assessor with perfect memory for the outcomes of the predictions can be perfectly calibrated.

The model does, however, have some similarities with the ecological models, in that the cognitive representation which guides both the decision and subjective probability estimate is in terms of frequencies. However, unlike the ecological models, the frequencies simply represent past successes and failures at prediction, and not the validities of various probability cues associated with a particular target variable and a particular reference class.

Finally, no distinction is drawn between performance in terms of relative frequency of success in the past, and confidence in individual items. Gigerenzer, Hoffrage and Kleinbölting (1991) have drawn a distinction between the reference class of past success on similar tests, and the reference class relating to the content of individual items, and have provided evidence for the psychological reality of this distinction. In the memory trace model they are one and the same. If subjects can produce good estimates of the frequency of success in the past they should also be well calibrated—but Gigerenzer et al. have shown that this is not so.

In summary, this is a model designed to explain individual differences in calibration performance (cf. Phillips & Wright, 1977) but suggests that differences in the calibration performance of difference assessors is entirely attributable to the quality of their memories. It fails to capture many of the empirical findings relating to calibration performance, and would seem to be an implausible candidate for either a general or domain-specific explanation of subjective probability judgement.

18.5.4 The Ecological Models

The PMM theory described by Gigerenzer, Hoffrage and Kleinbölting (1991) is the most complete model for the calibration of subjective probabilities that has so far been produced. It elegantly explains the overconfidence effect, the hard–easy effect, the circumstances under which good calibration is to be expected, and the confidence–frequency effect. It also makes strong and testable predictions concerning the circumstances under which a reversal of the hard–easy effect should occur. The model also explains a number of other apparently anomalous findings in the literature. The authors argue that the locus of miscalibration for general-knowledge items is in the test materials themselves, and is not the result of biased probabilistic reasoning on the part of the subjects.

However, the empirical evidence presented by Gigerenzer, Hoffrage & Kleinbölting (1991) is somewhat less convincing than the model. The problem is that difficulty as indexed by proportion correct co-varies with the type of item selection—items which are selected to be a good test of an individual's general knowledge (and thus not representative of the reference class) will on average be harder than those randomly selected from the reference class. Thus the demonstration that calibration for randomly selected city items is better than for standard general-knowledge items could be viewed as just another example of the hard–easy effect. However, the finding that subjects can be overconfident with respect to calibration based on the confidence expressed for individual items, and simultaneously well calibrated with respect to their overall performance with informally selected items, or well calibrated for individual ratings and underconfident about their overall performance with randomly selected items, is much more compelling.

The model proposed by Juslin (1993, 1994) based on *internal cue theory* (Björkman, in press), is in most respects identical to the PMM theory. Juslin's model was restricted to an explanation of the overconfidence effect and the hard–easy effect, and he did not predict the confidence–frequency effect. However, Juslin (1993) provided an impressive empirical test of his own model, and thereby, PMM theory. Juslin predicted that if the *randomly* generated geography items used in his experiment were divided into four subgroups, not on the basis of proportion correct (as in Lichtenstein & Fischhoff, 1977) but on the basis of the *mean familiarity rating* given to the pair of countries forming an item, the hard–easy effect would be abolished, and good calibration should be observed—despite differences in the proportion of items correct across the subgroups. The reasoning was as follows: for highly familiar items, a large number of relevant cues can be generated, and thus there is a high probability of a cue with a high validity being activated. This will lead to a high proportion of correct answers. For items with low familiarity the reverse is true; they are likely to be answered using cues with low validities, and thus a low proportion correct is to be expected. In both cases, the cues used should be ecologically valid, as the generating process was random and thus good calibration would be expected. This prediction was supported by the data; for the most familiar items the proportion correct was 0.80 and for the least familiar 0.66, but for all subgroups calibration was excellent. Hence Juslin (1993) successfully decoupled cue appropriateness and item difficulty.

We have identified two potential problems with the ecological models. The first concerns the degree to which the models can be extended beyond the domain of knowledge questions, and beyond the 2AFC task format. Gigerenzer, et al. argued that PMM theory was applicable to perceptual tasks, and that good calibration would be anticipated as long as the items were not chosen to be misleading (i.e. not selected for perceptual illusions). They also predicted that with two perceptual tasks, which varied in discriminability but with stimuli generated by the same sampling process, the hard–easy effect should disappear. However, we have shown that overconfidence in a perceptual task varies systematically with discriminability (the hard–easy effect) despite the fact that the stimuli were indeed generated by the same *random* process (McClelland, Bolger & Tonks, 1992). It should be noted, however, that we used a full-range probability true task, and that the task was novel to the subjects, but nevertheless this finding does not square with either of Gigerenzer et al.'s predictions. With respect to the question of the generality of the probability cue notion, Juslin has taken an alternative approach, arguing that internal cue theory is only applicable within the knowledge domain, and that a different representation is used with psychophysical tasks (see Björkman, Juslin & Winman, 1993).

The second problem concerns the plausibility of individuals actually learning the appropriate cue validities for the probability cues with respect to a target variable in a particular knowledge domain. To take Gigerenzer et al.'s example (although the argument also applies to Juslin, 1993, 1994) the frequency which the individuals would have to record is *the number of times that one city with more than 100 000 inhabitants has a larger population than another city with more that 100 000 inhabitants when the first city has a team in the Bundesliga and the second city does not*. Further to obtain an accurate cue validity, all possible pairs of cities would have to be examined, or at least to obtain an unbiased estimate, a random sample of all possible pairings would have to be selected. The appropriate cue validities could not be learnt if individuals merely noted that large cities *tend* to have teams in the Bundesliga, and smaller cities do not. Note also that if the target variable and probability cue are reversed (e.g. a decision has to be made as to which of two cities has a team in the Bundesliga, with population used as a cue) a different value would have to be recorded, as conditional probabilities are only symmetrical under very restricted circumstances. How plausible this is remains an open question.

In addition, both Harvey and Rawles (1992) and Griffin and Tversky (1992) have provided evidence inconsistent with the ecological models. Harvey and Rawles questioned the PMM assumption that subjects always choose the alternative with the higher value on the probability cue, and suggested instead that subjects "probability match" (Estes, 1964). Thus for a cue with a validity of 0.90, subjects would choose the alternative with the 0.90 probability 90% of the time, and the other alternative 10% of the time. Using a simulation technique, they found that probability matching model produced a very good fit to the data (from a general-knowledge test), whereas the PMM model gave a very poor fit.

The results inconsistent with the ecological models provided by Griffin and Tversky are described below.

18.5.5 The Strength and Weight Model

Griffin and Tversky (1992) presented both a general framework for understanding the relationship between confidence and accuracy, and a specific model for laboratory-based calibration experiments.

As described earlier, the specific model was in the form of an analogy (a chance setup) and the authors demonstrated that a plot of the "normative" solutions derived from Bayes' theorem against a measure of the strength of evidence (presumed to be the subjective probability estimates) produced calibration curves which mimicked the empirical curves from Lichtenstein and Fischhoff (1977).

It seems to us that this model is essentially a version of the detection model (Ferrell & McGoey, 1980) which makes use of a discrete probability distribution (the binomial distribution) rather than a continuous distribution (the normal distribution). This view is shared by Ferrell (personal communication) who has developed another discrete version of the detection model (based on the symmetrical criterion model presented in Smith & Ferrell, 1983, pages 475–6). In this version, red balls represent evidence in favour of one of the alternatives in the 2AFC task (which may or may not be the correct answer) and white balls represent evidence in favour of the other alternative. If more red balls are present in the sample, the hypothetical subject would choose one alternative—if more white balls, the other. A sample containing exactly five red balls (out of 10) would lead to a random selection of an alternative, and a probability judgement of 0.50. This model produces slightly different curves from the Griffin and Tversky model, and the two models only coincide when the hypotheses are symmetrical. It should also be noted that the Griffin and Tversky model actually produces posterior probability values across the full probability range (from 0% to 100%) despite the fact that it is designed to be an analogy to a 2AFC task. To be consistent with Lichtenstein and Fischhoff's (1977) data, the authors are forced to "cut off" the calibration curves, and only plot values from 50% to 100%. The Ferrell version has the advantage that it does not predict confidence ratings below 50%, and thus the values fall in the half-range—as they should.

What are substantive differences between the strength and weight model and the detection model? Griffin and Tversky suggest that the strength of evidence is better represented by a balance of arguments, whereas Ferrell and McGoey suggest it is better represented by the absolute difference in apparent truth between the alternatives, measured on a continuous scale. The strength and weight model *never* allows for perfect calibration (even with feedback) as the proportion of balls in the sample is never the same as the posterior probability (except trivially, at the 50% point for an impossible task). The response criteria in this model are by necessity fixed (the number of red balls in the sample), whereas in the detection model it is possible for the criteria to be adjusted (with feedback) in order to improve calibration performance. In other respects, the models are very similar, in that they both assume that subjects base their probability judgements on the strength of evidence, and ignore the weight of evidence. Finally, if the sample size in the strength and weight model were allowed to tend to infinity, the binomial distributions would tend to normality and the probability scale would become continuous—as in the detection model.

In addition to the simulation of their model, Griffin and Tversky also reported some empirical results from an experiment (Griffin & Tversky, 1992, Study 5) which they interpreted as supporting the strength and weight approach, and as being inconsistent with PMM theory (Gigerenzer, Hoffrage & Kleinbölting, 1991). They showed that for a representative (*random*) sample of 30 pairs of American states, subjects were consistently *overconfident* in their predictions concerning population, high-school graduation rates, and the difference in voting rates between the last two presidential elections. In addition, they had predicted that for population judgements both accuracy and confidence would be high (on the grounds that individuals should be knowledgeable about population) for voting both confidence and accuracy would be low (because they would not be knowledgeable about voting rates) and for education, accuracy would be low and confidence high. This last prediction was made on the grounds that subjects would be likely to use cues such as the number of famous universities or cultural events within a state to guide their judgements, when in reality the correlations between these cues and high-school graduations rates are very low—a type of "false knowledge" in May's terms (May, 1986a, 1986b). The predictions received empirical support. In particular, performance for both voting and high-school graduation was at chance level, although the mean confidence rating for education (65.6%) was significantly higher than that for voting (59.7%). The subjects were also asked to estimate how many of the questions they thought they had answered correctly, and it was found that for all three types of judgement the judged frequency was below the actual frequency (for voting and education the estimates were well below chance).

Griffin and Tversky concluded that overconfidence in calibration studies cannot be attributed to either an artifact of item selection or a by-product of task difficulty, and clearly their empirical findings would seem to pose a problem for the ecological models. However, Juslin (1993, 1994) had observed excellent calibration for population judgements—so the Griffin and Tversky result (6.5% overconfidence) for this attribute is somewhat of an anomaly. With respect to the other two attributes (high-school graduation and voting) performance was at chance level. This implies that these attributes were not part of the subjects' knowledge base (so effectively the task was impossible) but does not explain why subjects were overconfident (as the stimuli were *randomly* selected) or the difference in overconfidence between voting and education. However, the samples presented to subjects were very small (15 items per attribute) and could have contained a number of misleading items just by chance. Further, for solution probabilities around 50%, overconfidence would be expected simply because of the range restriction at the lower end of the probability scale (May, 1986b; Poulton, 1989). Finally, subjects are rarely faced with an impossible task, and may have suffered from a degree of *evaluation apprehension*, as well as wishing to be "good subjects" (McBurney, 1990). This may have led them to give confidence ratings higher than they truly felt appropriate, in an attempt to demonstrate that they could do the task. If the subjects were students (the source of the subjects is not stated) they might have felt that the experimenters would expect them to have knowledge concerning the education attribute in particular. This would lead them to provide higher confidence ratings for the education attribute than the voting attribute.

Griffin and Tversky's predictions do pay lip service to the notion that subjects use probability cues, but the implication is that subjects have "false knowledge"—they believe the cues to have higher validities than they actually do. Why this should be so is unclear from Griffin and Tversky's account.

18.5.6 Evaluation Summary

We have briefly described and critically reviewed seven models of subjective probability calibration. In the light of this review, should we be pessimistic or optimistic about the ability of individuals to be well calibrated?

Three of the models are clearly pessimistic; Albert and Sponsler (1989) suggest that overconfidence is a direct consequence of how the brain stores and retrieves information concerning past efforts at prediction. Both Koriat, Lichtenstein and Fischhoff (1980) and Griffin and Tversky (1992) argue that overconfidence can be attributed to the use of heuristics, which leads individuals to ignore vital information, which, in the view of these authors, is required to produce accurate probability estimates. Koriat et al. suggested that individuals are both biased in the retrieval of information (favouring positive evidence) and in their evaluation of the evidence (disregarding negative evidence); Griffin and Tversky suggested that individuals base their confidence on the strength of evidence available, and either ignore or under-utilize the weight of evidence. There is little empirical support for the Koriat et al. stage model, but some for the Griffin and Tversky strength and weight model.

Like Griffin and Tversky, Ferrell and McGoey (1980) are also pessimistic to the extent that they believe that information concerning predictive validity (such as discriminability and base-rate) is ignored, but imply that this can be corrected by the use of appropriate feedback. Unfortunately, the evidence that training will markedly improve calibration performance is weak (see below).

May (1986a, 1986b) is somewhat more optimistic, in that she argued that good calibration is expected if no "misleading" items are present, and the subjects are relying simply on their sensitivity to objective physical differences. However, Björkman, Juslin and Winman (1993) have shown that for true psychophysical judgements, underconfidence is observed, which they attribute to the fixed sensitivity of the sensory system and claim it is impossible to avoid. With respect to general-knowledge items, May claimed that with random sampling and the use of simple inference (a probabilistic syllogism) good calibration could be achieved, but that overconfidence was to be expected if the sampling procedure was biased. However, she believed that if items were misleading because subjects held "false knowledge" (such as a distorted cognitive map) then overconfidence would result.

Clearly the most optimistic theorists are those responsible for developing the ecological models. Both Gigerenzer, Hoffrage and Kleinbölting (1991) and Juslin (1993, 1994) believe that the miscalibration observed in general-knowledge tests can be attributed to the biased sampling of stimulus items, and that with representative sampling individuals are well calibrated. Both models (and PMM theory in particular) provide the most complete explanation for the empirical findings (the Griffin and Tversky results notwithstanding), although to what extent these ideas can be extended beyond the general-knowledge domain remains to be seen.

Juslin believes that different mental representations are used in different stimulus domains, and neither internal cue theory nor PMM theory can provide a general explanation for the calibration of probabilities. Gigerenzer et al. have implied that PMM theory is general, and that it can explain calibration data in perceptual as well as cognitive tasks. However, the evidence for this proposal is not conclusive, and further research is required.

18.6 CAN ONE LEARN TO BE "WELL CALIBRATED"?

The attempts to improve individuals' calibration performance through the use of training with feedback have met with limited success. Ferrell (see Chapter 17) states that "... it is comforting that calibration can be improved relatively easily by suggestion and by training", but notes that "This optimism must be tempered by the findings that training ... does not seem to generalize very well" (page 430). Keren (1991) is even less optimistic; "The most disturbing finding obtained from training studies is that whatever modest improvement is achieved, it is hardly ever generalized to other tasks" (page 238). Other authors, however, put a very positive spin on the evidence for improvement through training. Russo and Schoemaker (1992) boldly state that "We believe that timely feedback and accountability can gradually reduce the bias toward overconfidence in almost all professions. *Being 'well calibrated' is a teachable, learnable skill*" (page 11, italics theirs). However, the evidence that outcome feedback alone is effective in reducing miscalibration is not encouraging (see Benson & Onkal, 1992 for a review).

The models we have reviewed vary in the amount of optimism they engender regarding the learning of good calibration. The most pessimistic models suggest that neither training nor experience will have an effect on calibration performance. This is either because miscalibration is a consequence of the manner in which the brain stores information (the memory trace model) or because the same heuristics—with the same limitations—are always used (the strength and weight model). Albert and Sponsler (1989) imply that someone with a poor memory will never be well calibrated, and that the miscalibration will be in the direction of overconfidence. Griffin and Tversky (1992) suggested that the bias in favour of the strength of evidence over its weight is incorrigible, and argued that calibration performance depends heavily on the predictability of outcomes in a target domain. For example, they suggested that experts will be more overconfident than non-experts in an unpredictable domain (e.g. clinical psychology or the stock market) because they will give unwarranted credence to the validity of their expert knowledge.

The detection model is not entirely pessimistic because it allows for an improvement in calibration performance with outcome feedback, which allows the assessor to adjust his or her criteria on the evidence variable appropriately. However, this does not imply that an individual who becomes well calibrated in one domain will be well calibrated in another, and as we have noted, the evidence that outcome feedback improves calibration performance is very weak.

The expectations derived from the stage and process models are that certain types of training will lead to improvement in calibration under certain circumstances. For the stage model, training must be in the form of the generation of counter-arguments as described by Koriat, Lichtenstein and Fischhoff (1980), but again this seemed to have little effect on calibration performance. In the case of the process model we would anticipate that training in the form of the correction of false knowledge should lead to a reduction in overconfidence and thereby improve calibration.

Finally, we turn to the ecological models, which furnish quite specific predictions as to when training will, and will not have a beneficial effect on calibration, and provide a simple explanation for the poor results obtained when training with feedback has been examined. Training with outcome feedback *should* be effective when subjects are confronted with novel tasks, as this will allow them to learn the appropriate cues and cue validities required to make predictions. Any procedure which allows individuals to observe the covariation between variables in an ecologically valid setting should lead to an improvement in the quality of their subjective probability judgements.

Training with outcome feedback *will not* be effective with tasks such as standard general-knowledge tests, which contain an unrepresentative sample of items from the reference class. This prediction stems from the assumption that subjects will continue to use cues and report cue validities which are ecologically valid, but are not valid for non-representative stimuli. As Keren (1991) pointed out, most training investigations have used general-knowledge items, and we agree with him that the modest improvements which have been noted can be attributed to the fact that subjects receiving continuous feedback that their probability responses are too high will naturally lower them, but this is merely a "technical correction", and has nothing to do with improving probability judgements. This analysis also explains why any improvement does not generalize to other tasks. The quality of the calibration performance depends on the experience the subject has had with the target domain, and crucially, on how the stimuli used in the test have been selected. Thus, from the perspective of the ecological models, there is cause for optimism with respect to training—but not for the notion of a general training in calibration.

18.7 CONCLUSIONS

We have argued that the ecological models, and PMM theory (Gigerenzer, Hoffrage & Kleinbölting, 1991) in particular, provide the most coherent account of how individuals realize subjective probability judgements, and afford the most satisfactory explanation of calibration performance with general-knowledge items.

The calibration of subjective probabilities has been studied in a variety of other task domains, and it remains an open question as to how successfully the ecological approach can be applied in other settings. For example, Björkman, Juslin & Winman (1993) have provided evidence that, when making psychophysical judgements, an alternative representation which leads to *underconfidence* is used. The representation is based on the subjective distance between the stimuli, and Björkman et al. argue that due to the fixed sensitivity of the sensory system, the bias cannot be avoided.[7] They also provide evidence that training has no effect on the underconfidence bias. However, in a recent paper Baranski and Petrusic (1994) have questioned the subjective distance model. In three experiments, these authors showed that it is possible to obtain *overconfidence* in psychophysical judgments when response accuracy is sufficiently reduced—either by putting the subjects under speed stress, or by reducing discriminability sufficiently under accuracy stress. They also studied decision time conditionalized on confidence category, and argued that their results were incompatible with both the subjective distance model *and* the detection model described earlier.

There are other task domains in which it may be implausible that PMM theory or internal cue theory applies in the form suggested by the ecological models. For example, weather forecasters are notoriously well calibrated (Murphy and Winkler, 1977; 1984) but it would seem unreasonable (but not impossible!) that a *single* cue is used to arrive at both a decision concerning precipitation and the associated probability. It is also difficult to see how the models can be applied to *episodic* memory tasks. What sort of interaction with the environment and encoding of the co-occurrences of events would help to decide that a stimulus item was present or absent during the encoding phase in a recognition memory experiment? Wagenaar (1988) showed that subjects were quite well calibrated (but demonstrated some overconfidence) for "old" items but extremely poorly so for "new" items in an old/new recognition test using words, syllables and numbers as stimuli. McClelland (1992) obtained similar results in a face recognition study. Wagenaar also showed that calibration was reasonably good when subjects were able to retrieve information directly from episodic memory, but overconfidence became evident when they relied on inference rather than direct memory retrieval. We believe that in experiments of this type, probability judgements may be based on a form of representation not well captured by either PMM theory or internal cue theory in their present form. Further empirical and theoretical work is clearly needed.

With respect to a *general model* of calibration performance, Baranski and Petrusic (1994) have argued that the properties of decision times in calibration tasks place tight constraints on possible candidates (see also Wright & Ayton, 1988). Baranski and Petrusic suggest that an "appealing avenue for theoretical consideration" (page 426) might be a variant of Ferrell and McGoey's (1980) detection model—one which could account for the pattern of reaction-time and response probability relationships observed in their experiments. They favour an approach based on some form of evidence accumulator model (e.g. Link, 1992; Petrusic, 1992; Vickers 1970, 1979).

Recently, a colleague of ours who works in the area of judgement and decision making commented in a rather exasperated tone that it was about time that the calibration issue "was laid to rest". That outcome may still be some way off, but we feel confident that the days of "dust-bowl empiricism" are over, and that there is now a rich enough source of theoretical ideas to drive calibration research in a more productive direction.

ACKNOWLEDGEMENTS

We would like to thank Peter Ayton, Nigel Harvey and A.R. Jonckheere for their advice and comments during the preparation of this chapter.

NOTES

(1) The majority of calibration studies have used a 2AFC paradigm. For each item (e.g. "Absinthe is (a) a precious stone, (b) a liqueur") the subject selects one alternative and gives a probability rating that the choice is correct on a scale between 50% and 100%. In a full-range probability true task, the subject responds on a scale between 0% and 100% to indicate the degree to which they believe each statement to be true (e.g. "Absinthe is a precious stone") or confidence in the outcome of a future event (e.g. "What is the probability that it will rain tomorrow?"). For other task formats see Ferrell & McGoey (1980).

(2) In full-range tasks, base rate (the relative frequency of statements actually true, or of the occurrence of an event) has also been found to have an effect on calibration performance (see Smith & Ferrell, 1983). However the overestimation of the appropriate relative frequencies with base rates below 50%, and underestimation with base rates above 50%, has received far less attention than the overconfidence and hard–easy effects.

(3) Although information concerning the other alternative is not *explicitly* present

in the syllogism, it must presumably be the case that the individual knows that Hyderabad *is not* a capital city (see also, Gigerenzer, Hoffrage & Kleinbölting, 1991, page 523).

(4) Albert and Sponsler (1989) use the concept of "expertise" with respect to the accuracy of an assessor in making subjective probability judgements, and not to indicate the degree to which an individual is regarded as an expert within a particular knowledge domain.

(5) We assume this is a reference to the slight *underconfidence* that is often observed at the 50% point on the subjective probability scale in 2AFC tasks.

(6) However, Ferrell and McGoey (1980) noted that in many cases, the fit of the model was not so precise as to be statistically indistinguishable from the data.

(7) Whilst this may be true for *psychophysical* tasks, it does not follow that it is true for *all* perceptual tasks. Extracting information predictive of a target event from a complex and noisy stimulus display is likely to be a learnable skill, leading to improved calibration.

REFERENCES

Albert, J.M. & Sponsler, G.C. (1989) Subjective probability calibration: A mathematical model. *Journal of Mathematical Psychology*, **33**, 298–308.

Baranski, J.V. & Petrusic, W.M. (1994) The calibration of confidence in perceptual judgments. *Perception and Psychophysics*, **55**, 412–28.

Benson, P.G. & Onkal, D. (1992) The effects of feedback and training on the performance of probability forecasters. *International Journal of Forecasting*, **8**, 559–73.

Björkman, M. (in press) Internal cue theory: Calibration and resolution of confidence in general knowledge. *Organizational Behavior and Human Decision Processes*.

Björkman, M., Juslin, P. & Winman, A. (1993) Realism of confidence in sensory discrimination: The underconfidence phenomenon. *Perception and Psychophysics*, **54**, 75–81.

Brunswick, E. (1943) Organismic achievement and environmental probability. *Psychological Review*, **50**, 255–72.

Brunswick, E. (1955) Representative design and probabilistic theory in a functional psychology. *Psychological Review*, **62**, 193–217.

Dawes, R.M. (1980) Confidence in intellectual judgments. In E.D. Lantermann & H. Feger (eds.), *Similarity and Choice*. Hans Huber, Bern.

Estes, W.K. (1964) Probability learning. In A.W. Melton (ed.), *Categories of human learning*. Academic Press, New York.

Ferrell, W.R. & McGoey, P.J. (1980) A model of calibration for subjective probabilities. *Organisational Behavior and Human Performance*, **26**, 32–53.

Fischhoff, B. & MacGregor, D. (1982) Subjective confidence in forecasts. *Journal of Forecasting*, **1**, 155–72.

Gigerenzer, G. (1991) How to make cognitive illusions disappear: Beyond "Heuristics and Biases". *European Review of Social Psychology*, **2**, 83–115.

Gigerenzer, G., Hoffrage, U. & Kleinbölting, H. (1991) Probabilistic mental models: A Brunswikian theory of confidence. *Psychological Review*, **98**, 506–28.

Griffin, D. & Tversky, A. (1992) The weighting of evidence and the determinants of confidence. *Cognitive Psychology*, **24**, 411–35.

Harvey, N. & Rawles, R. (1992) Probability matching in probabilistic mental models. *Bulletin of the Psychonomic Society*, **30**, 488.

Hasher, L. & Zacks, R.T. (1984) Automatic processing of fundamental information: The case of frequency of occurrence. *American Psychologist*, **39**, 1372–88.

Jungermann, H. (1983) The two camps on rationality. In R.W. Scholz (ed.), *Decision Making under Uncertainty*. Elsevier, Amsterdam.

Juslin, P. (1993) An explanation of the hard-easy effect in studies of realism of confidence in one's general knowledge. *European Journal of Cognitive Psychology*, **5**, 55–71.

Juslin, P. (1994) The overconfidence phenomenon as a consequence of informal experimenter-guided selection of almanac items. *Organisational Behavior and Human Decision Processes*, **57**, 226–46.

Kahneman, D., Slovic, P. & Tversky, A. (1982) *Judgment under Uncertainty: Heuristics and Biases*. Cambridge University Press, Cambridge.

Kahneman, D. & Tversky, A. (1982) On the study of statistical intuitions. In D. Kahneman, P. Slovic & A. Tversky (eds.), *Judgment under Uncertainty: Heuristics and Biases*. Cambridge University Press, Cambridge.

Keren, G. (1987) Facing uncertainty in the game of bridge: A calibration study. *Organizational Behavior and Human Decision Processes*, **39**, 98–114.

Keren, G. (1988). On the ability of monitoring non-veridical perceptions and uncertain knowledge: Some calibration studies. *Acta Psychologica*, **67**, 95–119.

Keren, G. (1991) Calibration and probability judgments: Conceptual and methodological issues. *Acta Psychologica*, **77**, 217–73.

Koriat, A., Lichtenstein, S. & Fischhoff, B. (1980) Reasons for confidence. *Journal of Experimental Psychology: Human Learning and Memory*, **6**, 107–18.

Lichtenstein, S. & Fischhoff, B. (1977) Do those who know more also know more about how much they know? The calibration of probability judgements. *Organizational Behavior and Human Performance*, **20**, 159–83.

Lichtenstein, S., Fischhoff, B & Phillips, L.D. (1982) Calibration of probabilities: The state of the art to 1980. In D. Kahneman, P. Slovic & A. Tversky (eds.), *Judgment under Uncertainty: Heuristics and Biases*. Cambridge University Press, Cambridge.

Link, S.W. (1992) *The Wave Theory of Difference and Similarity*. Erlbaum, Hillsdale, NJ.

McBurney, D.H. (1990) *Experimental Psychology* (2nd edn.) Wadsworth, Belmont, CA.

McClelland, A.G.R. (1992) Facial identification: Is there a relationship between confidence and accuracy? *International Journal of Psychology*, **27**, 114.

McClelland, A.G.R., Bolger, F. & Tonks, E. (1993, January) The effects of discriminability and base rate on calibration of probabilities in a perceptual discrimination task. In N. Harvey & P. Ayton (Chairs). *Judgment and Decision Making*. Symposium conducted at the meeting of the Experimental Psychology Society London.

McClelland, A.G.R., Coulson, A.S. & Icke, S.E. (1990) Bias in meta-memory performance and its implications for models of memory structure. In J.-P. Caverni, J.-M. Fabre & M. Gonzalez (eds.), *Cognitive Biases*, North Holland, Amsterdam.

May, R.S. (1986a) Inferences, subjective probability and frequency of correct answers: A cognitive approach to the overconfidence phenomenon. In B. Brehmer, H. Jungermann, P. Lourens, & G. Sevo'n (eds.), *New Directions in Research on Decision Making*. North Holland, Amsterdam.

May, R.S. (1986b) Overconfidence as a result of incomplete and wrong knowledge. In R.W. Sholtz (ed.), *Current Issues in West German Decision Research*. Lang, Frankfurt.

Milburn, M.A. (1978) Sources of bias in the prediction of future events. *Organizational Behavior and Human Performance*, **21**, 17–26.

Murphy, A.H. & Winkler, R.L. (1977) Can weather forecasters formulate reliable probability forecasts of precipitation and temperature? *National Weather Digest*, **2**, 2–9.
Murphy, A.H. & Winkler, R.L. (1984) Probability forecasting in meteorology. *Journal of the American Statistical Association*, **79**, 489–500.
Petrusic, W.M. (1992) Semantic congruity effects and theories of the comparison process. *Journal of Experimental Psychology: Human Perception & Performance*, **18**, 962–86.
Phillips, L.D. & Wright, G.N. (1977) Cultural differences in viewing uncertainty and assessing probabilities. In H. Jungermann & G. de Zeeuw (eds.), *Decision Making and Change in Human Affairs*. Reidel, Amsterdam.
Pitz, G.F. (1974) Subjective probability distributions for imperfectly known quantities. In L.W. Gregg (ed.), *Knowledge and Cognition*. Erlbaum, Hillsdale, NJ.
Poulton, E.C. (1989) *Bias in Quantifying Judgments*. Erlbaum, New York.
Ronis, D.L. & Yates, J.F. (1987) Components of probability judgment accuracy: Individual consistency and effects of subject matter and assessment method. *Organizational Behavior and Human Decision Processes*, **40**, 193–218.
Russo, J.E. & Schoemaker, P.J.H. (1992) Managing overconfidence. *Sloan Management Review*, Winter, 7–17.
Smith, M. & Ferrell, W.R. (1983) The effect of base rate on calibration of subjective probability for true–false questions: Model and experiment. In P. Humphreys, O. Svenson, and A. Vari (eds.), *Analyzing, and Aiding Decisions*. North Holland, Amsterdam.
Staël von Holstein, C.S. (1971) An experiment in probabilistic weather forecasting. *Journal of Applied Meteorology*, **10**, 635–45.
Staël von Holstein, C.S. (1972) Probabilistic forecasting: An experiment related to the stock market. *Organizational Behavior and Human Performance*, **8**, 139–58.
Tversky, A. & Kahneman, D. (1974) Judgment under uncertainty: Heuristics and biases. *Science*, **185**, 1124–31.
Tversky, A. & Kahneman, D. (1983) Extensional versus intuitive reasoning: The conjunction fallacy in probability judgment. *Psychological Review*, **90**, 293–315.
Vickers, D. (1970) Evidence for an accumulator model of psychophysical discrimination. *Ergonomics*, **13**, 37–58.
Vickers, D. (1979) *Decision Processes in Visual Perception*. Academic Press, New York.
von Winterfeldt, D. & Edwards, W. (1986) *Decision Analysis and Behavioral Research*. Cambridge University Press, New York.
Wagenaar, W.A. (1988) Calibration and the effects of knowledge and reconstruction in retrieval from memory. *Cognition*, **28**, 277–96.
Wright, G. (1982) Changes in the realism and distribution of probability assessments as a function of question type. *Acta Psychologica*, **52**, 165–74.
Wright, G. & Ayton, P. (1988) Decision time, subjective probability and task difficulty. *Memory & Cognition*, **16**, 176–85.
Wright, G. & Phillips, L.D. (1984) Decision making: Cognitive style or task-related behaviour? In H. Bonarius, G. van Heck, & N. Smid (eds.), *Personality Psychology in Europe*. Swets & Zeitlinger, Lisse.
Yates, J.F. (1982) External correspondence: Decomposition of the mean probability score. *Organizational Behavior and Human Performance*, **30**, 132–56.
Yates, J.F. and Curley, S.P. (1985) Conditional distribution analysis of probabilistic forecasts. *Journal of Forecasting*, **4**, 61–73.
Zakay, D. (1983) The relationship between the probability assessor and the outcomes of an event as a determiner of subjective probability. *Acta Psychologica*, **53**, 271–80.

Part Four

Real World Studies

Chapter 19

The Rationality of Gambling: Gamblers' Conceptions of Probability, Chance and Luck

Gideon Keren
Eindhoven University of Technology

Gambling, or games of chance, served as a major (though not exclusive—see Daston 1988; Hacking 1975) impetus to the development of probability theory and the conceptualization of uncertainty. Throughout its entire history, the reasoning associated with gambling is characterized by a complex blend of rational and nonrational components. This is best illustrated in one of the earliest texts on gambling written in the beginning of the 16th century by Gerolamo Cardano. The book, entitled *The Book on Games of Chance* (*Liber De Ludo Aleae*) is supposedly the first mathematical attempt to construct the initial elements of what eventually became the theory of probability (Gigerenzer et al., 1989: Hacking, 1975). At the same time, however, a bulky part of the book contains assertions and beliefs that are completely remote from any rational framework. Indeed, it is the lack of discrimination between rational and nonrational considerations which ultimately leads to different paradoxes of gambling behavior (Wagenaar, 1988).

How could one account for this mixture of rational and non-rational considerations? Bruner (1984) proposed to distinguish between two different and irreducible fundamental modes of thought. One, which he termed the *paradigmatic* (or logico-scientific) mode, is epistemological in nature and is

Subjective Probability. Edited by G. Wright and P. Ayton.
© 1994 John Wiley & Sons Ltd.

based on the abstract rules of logic regulated by the basic requirements of consistency and noncontradiction. The other mode, termed *narrative*, is more phenomenological in nature, and is concerned with the explication of intentional and goal-oriented actions and the consequent associated conscious experience. Unlike the paradigmatic mode, it is thus context-sensitive and particular. A radical difference between the two modes is the manner by which they establish truth: whereas the former mode employs formal verification procedures and empirical proof, the latter, according to Bruner, "establishes not truth but truth-likeness or verisimilitude." In other words, whereas the paradigmatic mode is based on evaluation standards that are presumed to be "objective", the narrative mode is in principle subjective. Bruner has further claimed that the two modes of cognitive functioning are irreducible. This assertion, however, should be qualified: the two modes are irreducible from a "paradigmatic" standpoint, not necessarily from a "narrative" view. In fact, human behavior can be frequently depicted by an interplay between these two modes despite the fact that they are incongruent.

Nowhere is this more apparent than in gambling behavior and gamblers' conceptualization of uncertainty. The fact that people gamble at all in face of negative expected values, is one of the main paradoxes of gambling behavior. However, the paradox exists only under the assumptions of the paradigmatic view. Once these are relaxed, the paradox (see Wagenar 1988), relies at least to some extent on narrative facets. Indeed, I suggest that many erroneous convictions and biases regarding the uncertainty calculus shared by gamblers and nongamblers alike, stem from failures to (completely) disassociate the paradigmatic and narrative modes of reasoning.

The purpose of the present chapter is to describe gamblers' conceptualization of uncertainty in light of the paradigmatic–narrative distinction. I first discuss briefly the meaning of uncertainty and the nature of probabilistic statements from a normative and descriptive (mainly phenomenological) point of view. Subsequently, fundamental errors and misconceptions of gamblers will be presented and analyzed. Implications for the psychology of uncertainty and for gambling behavior are summarized in the last section.

While the presentation is inherently descriptive in nature, it will be assessed against the normative theory of probability as dictated by the paradigmatic mode.

19.1 ON THE MEANING OF UNCERTAINTY AND THE INTERPRETATION OF PROBABILITY

There is a consensus among most scientists that probability theory is in the first place a formal theory in pure mathematics, the most common and acceptable version of which is based on Kolmogorov's axioms and definitions (Shafer,

1993). Except for probability, which in the formal theory remains an undefined concept, all the other terms in the probability calculus have well-established meanings that are taken from logic and mathematics (Salmon, 1966). It is in this respect that Daston (1988) correctly asserted that "The mathematical theory itself preserved full conceptual independence" from any possible interpretation that may be attached to it. Thus, the psychological considerations associated with the meaning of the undefined term of "probability" are not part of the formal system.

It is the moment at which the theory is applied that an interpretation of the concept of probability is unavoidable, and it is this stage which remains controversial to this date. Within the logico-scientific (i.e. paradigmatic) mode, there is a long-lasting literature concerning the possible meanings of probability, a discussion of which is beyond the scope of the present chapter.[1] Briefly, the most common interpretations advanced by Keynes and Carnap, maintain that probability measures the degree of confidence that would be justified by the available evidence; the *frequentistic* interpretation advocated by von Mises, which interprets probability in terms of the relative frequency in an infinite sequence of events; and the *personalistic* viewpoint as represented by De Finetti and Savage, according to which probabilities are purely subjective and reflect degrees of actual belief.

Several attempts have been made to reconcile between the various interpretations (e.g. Lindley, Tversky, & Brown, 1979; Shafer, 1993) though their success remains questionable. Notwithstanding, it is important to emphasize that the explications of all these interpretations as well as the reconciliation attempts are conducted within the paradigmatic mode and thus satisfy the fundamental requirements of the logico-scientific mode such as coherence and consistency.[2] Consequently (because the foundation of each interpretation is based on the formal mathematical axiomatization), each interpretation claims to have a normative status. Gamblers' reasoning about uncertainty and their conceptualization of probabilities is frequently incongruent with the conventional interpretations: the framework underlying their probabilistic assessment is often in violation of the fundamental requirements of coherence and consistency, for the simple reason that it is based not just on paradigmatic but also on narrative accounts.

The normative scientific analyses of probability as reflected in philosophical, statistical, and decision-theoretical articles, is solely based on paradigmatic norms and as such cannot provide an adequate description of gamblers' perspective and their conceptualization of uncertainty. To achieve that purpose, a more phenomenological approach is needed, like the one proposed by Howell and Burnett (1978) and more recently by Kahneman and Tversky (1982). Howell and Burnett proposed to distinguish between *internal* and *external* uncertainty. Internal uncertainty is said to relate to events that people can control (at least to some extent), such as predicting performance on a skilled task. External uncertainty refers to events that are solely determined by external generators which cannot be controlled by a person. Following these definitions, it is also important to distinguish between "control" and "perceived control". The latter can also be illusory, a topic which is elaborated on later in this chapter.

A slightly similar distinction has been proposed by Kahneman and Tversky (1982), but their emphasis is on the location to which uncertainty can be attributed: the external world or one's state of mind. The external world is perceived as containing different dispositions that yield different events or outcomes. Judged probabilities supposedly reflect the assessed relative strength of these competing dispositions. In contrast, internal probabilities constitute a reflection of the state of mind and are a measure of ignorance.

Kahneman and Tversky (1982) further propose a second level of analysis within each of the two main categories. Uncertainty attributed to the external world can be assessed by (1) a *distributional* mode, where relative frequencies are known or can be assessed, and (2) when frequencies are unavailable, by a singular mode where probabilities are assessed as propensities of a particular case or event. These authors conjectured a tendency among people to prefer the singular mode in which it is suggested that people can take an "inside view'" over an "outside view" that is based on a sampling schema. Studies on gambling behavior which are reviewed in the next section, strongly support this conjecture.

Internal probabilities, according to Kahneman and Tversky, can also be separated into (1) those that are *reasoned*, namely supported by arguments (regardless of their validity) and, (2) those established by *introspections* (based on intuitions that cannot be articulated within a rational framework) and their strength expressed in terms of confidence. As I will argue, it is often those types of internal probabilities that cannot be readily accommodated in a rational framework which often dominate gamblers' reasoning in situations of uncertainty. Note that the above classification, which intends to be descriptive, contains both paradigmatic and narrative components, whereas the external–frequentistic and the internal–reasoned modes lend themselves naturally to a narrative interpretation. It is indeed under these latter variants that the formal laws of probability are not necessarily compelling.

19.2 GAMBLERS CONCEPTUALIZATION OF UNCERTAINTY: CHANCE AND LUCK ARE NOT THE SAME

Where exactly does the gambler's mind deviate from the traditional paradigmatic interpretations? A possible starting point for answering this question concerns the concept of *determinism*, which played an important role in the history of gambling and the development of probability theory.

The development of gambling devices (i.e. devices that can produce unpredictable random events) can be traced to ancient times (David, 1962). These devices were frequently employed to find out the will or the command of the deity (e.g. should the tribe initiate a war against a rival tribe), and to obtain the truth which is only known to God (e.g. did the defendant omit the crime?). Underlying these rituals was obviously a strong belief in a theological deterministic world: reality has been determined in advance, and the function of gambling devices was, so to speak, to unfold the "truth" and reveal God's command. There have been other versions of determinism beside the theological one. For instance, physical determinism was substantiated during the seventeenth and eighteenth centuries by the discovery that the motions of the heavenly bodies were not only regular but could be described within laws of a high mathematical precision. The conclusion was that the behavior of these bodies followed unchangeable laws of nature that could not be violated. Interestingly, determinism was also deduced from the analysis of statistical data. Specifically, many scholars have interpreted the stability of stochastic processes proven by the limit theorems as evidence of divine design (Hacking, 1975). For instance, the constant regularity observed in the births of males and females was interpreted by John Arbuthnot (in a paper printed in 1710/11) as evidence for divine providence and the conclusion that it could not be a matter of chance (for a detailed exposition, see Hacking 1975—Chapter 18). This example is representative of the misunderstanding of Bernoulli's limit theorem and reflects two essential points that were shared by many scholars at that time: the belief in determinism and the failure to understand the meaning of randomness. To this day, in the minds of many people, these two facets continue to dominate the conceptualization of chance events.

Until this century, determinism was accepted (explicitly or implicitly) by the large majority of students of probability, regardless of which interpretation of probability they adopted. It is only in the twentieth century that the pendulum has started to shift away from determinism, mainly due to developments in quantum theory, which asserts that certain aspects of the behavior of single electrons is basically a matter of chance. According to this view uncertainty is said to be an inherent part of nature. Whether indeed this important conceptual change has also caught broader circles as recently suggested by Hacking (1990) remains an open question.

The belief in determinism has important psychological consequences. At least in its most extreme and stringent interpretation it implies lack of control over outcomes. There is ample evidence suggesting that people strongly value control—or the perception of control—and are reluctant to relinquish it. Gamblers are certainly not an exception in this regard, and many of them endorse the conviction that they are able (at least partially) to beat the laws of chance (Langer, 1977). The fact that the conviction is illusory apparently does not significantly weaken the gamblers' belief in it.

Gamblers have different, and often complicated, methods by which they believe they have the ability to exert control. We may distinguish between attempts at direct active control, where the gambler takes an explicit action, and attempts at indirect control, which are supposedly accomplished by an alleged ability to predict.

Casual and systematic observations suggest that gamblers employ different methods in an attempt to influence the outcomes.[3] For instance, Henslin (1967) noted that gamblers strongly believe they can influence the outcomes of a die by tossing it softly if they wish a low number, and tossing it hard when a high number is desired. As another example, Keren and Wagenaar (1985) observed that Blackjack players had strategies for changing the direction of their fate. Specifically, after a streak of losses they said they switched to another table. Other players reported an even more sophisticated method by which they supposedly reversed the direction of fate by interfering with the predetermined pattern of events (i.e. the predetermined order of the cards which allegedly governed their fortune). This was done by drawing an additional card which they normally would not draw (and would usually lead to a loss on this round). By committing such a "sacrifice" players thought they could break the predetermined allocation of cards (to players and dealer, and thus reverse their current fate).

People's belief in their ability to control randomly generated outcomes is well illustrated by Langer (1975). She designed an ingenious experiment in which she sold lottery tickets under two conditions: in one condition, subjects who were willing to purchase a ticket (priced at $1.00 per ticket) were given the choice and selected the ticket themselves. In the other condition the subjects were given no choice and the ticket was selected by the seller (i.e. the experimenter). When approached at a later stage (and before the lottery took place) and asked the price for which they would be willing to sell their ticket, the mean amount of money required by subjects in the choice condition was $8.67 compared with $1.96 in the no choice condition. Langer inferred that subjects in the choice condition believed that they could exert control over the outcome and coined this belief as the *illusion of control*.

It is important to note that the illusory belief of being able to control chance events is not necessarily incompatible with a deterministic view, at least not in the gambler's eyes. While events in the external world are predetermined, gamblers maintain the view that they can exploit it by unveiling the outcome in advance. This interpretation is more difficult to apply when one believes in a personal (rather than external outcomes) deterministic fate, as may be the case in Langer's experiment, a possible course of action in this case could be to use some prayers or magical words to influence the divine. Alternatively, one could interpret subjects' behavior in Langer's experiment as evidence that they believe that their fate has indeed been determined and that by selling their ticket they take action against their predetermined fate. In this case, the high price required by the subjects could be interpreted as a premium insuring against the possible regret that may occur if the sold ticket does indeed win.

The illusory control of chance events is frequently associated with skill (Keren & Wagenaar, 1988). For instance, Oldman (1974) reports that roulette players perceive the game as requiring a skill that consists of the construction and adaptation of prediction models. Similarly, Keren and Wagenaar (1985) observed among Blackjack players several methods which supposedly required skills such as identifying a "lucky dealer" or a "well-running box", and (skillful) methods for controlling one's fate. Wagenaar (1988) proposed that gamblers find it very difficult to appreciate the probabilistic nature of the outcomes and thus always tend to attribute at least part of the outcomes to their own actions. The belief in an alleged skill is also revealed in *post hoc* examinations of roulette and Blackjack players who laboriously try to explain the outcome (after it has become known) and point out how it could be avoided by using skills. Needless to say that all these post-mortem analyses are demonstrations of the well-known hindsight bias (Fischhoff, 1975).

The manner by which gamblers exercise control is frequently associated with their belief that they can predict events that are actually governed by chance. The belief in the ability to predict is based on two different types of considerations which may be referred to as *quasi-logical* and *magical* respectively (admittedly the demarcation line between these two is not always well defined).

Quasi-logical considerations have a paradigmatic nature (at least in the gamblers' view) even though they may contain some erroneous components. The quasi-logical mode is exemplified in strategies and procedures that gamblers attempt to rationalize in a seemingly rational framework, yet contain some erroneous components. The quasi-logical mode is exemplified in strategies and procedures that gamblers attempt to rationalize in a seemingly rational framework, yet contain basic misconceptions and violations of the cannons of logic. The best-known example is the so-called *gambler's fallacy*, which refers to the inability to comprehend statistical independence. Thus, gamblers believe that after observing a long run of red on the roulette wheel, the probability of black on the subsequent trial(s) is enhanced, resulting in an improved prediction ability. The gambler's fallacy is also shared by Blackjack players in real casino playing (Keren and Wagenaar, 1985) and is frequently observed in situations other than pure gambling.

Underlying the gambler's fallacy are several erroneous beliefs. In particular, it reflects the misperception of randomness (Bar-Hillel and Wagenaar, 1993). Normatively, whether a sequence is random or not is, strictly speaking, determined only by the nature in which it was generated. Yet, most people (and particularly gamblers) believe that randomness is characterized by certain patterns according to which randomness can be determined. For example, consider the following outcomes of coin tossing: Following Tversky and Kahneman (1974), people regard the sequence H-T-H-T-T-H to be more likely than the sequence H-H-H-H-T-H (supposedly because the latter does not represent the "fairness" of the coin), and to be more likely than the sequence H-H-H-T-T-T (because the latter is too structured and lacks randomness).[4] In the gambler's mind, underlying these convictions, there is often the supposition of an "invisible hand" that ensures some "fairness". Thus, under this perspective, even if the world is considered to be deterministic there is a certain so-to-speak order in the deterministic world upon which gamblers believe that they can rely.

Another manifestation of quasi-logical reasoning is reflected in what Tversky and Kahneman (1971) have termed the "law of small numbers" according to which people believe that small samples are highly representative of the population. The insensitivity to sample size has been demonstrated in various studies and different contexts and accounts, among other things, for the gambler's fallacy. In a recent study (Keren and Lewis, in press) we tested the insensitivity to sample size in a gambling context. Specifically, casual observations show that casino gamblers frequently record the outcomes of the roulette wheel. These records are employed by gamblers in two different strategies. Following one strategy, the gambler attempts to detect long runs (e.g. the outcome during the past *n* spinnings has always been "red" and then bet on the outcome that is supposed to "balance" the sequence (e.g. "black"). This is a manifestation of the well-known gambler's fallacy which we call Type 1. Note that the underlying strategy is the implicit assumption that the roulette wheel is perfectly calibrated implying that (in the long run) each number is equally likely to occur (i.e. in a roulette wheel with 37 numbers, the probability for each number is the same and equals 1/37).

The assumption underlying the alternative strategy frequently used by gamblers is that a wheel may be biased, implying that one or more numbers have a higher probability of occurrence. The recordings are then used for the detection of biased or favorable number(s). The possibility of a biased roulette wheel is certainly a viable one, yet detecting such a wheel is more difficult than might be thought. As Wilson (1965) noted:

> We wholeheartedly invite any interested professional mathematician to attack the problem, namely, on what statistical basis should you decide whether to play a given roulette number? The mathematician might first confine himself to the "simple" case, in which the true probabilities of occurrence of numbers are assumed constant (not changing in time). If any genius then wishes to become sophisticated, let him consider the possibility that the changing physical situation causes the probabilities to change with time! (page 39).

Many gamblers, however, believe that detecting a favorable number is a simple task which can be achieved with a relatively small sample, a phenomenon which we call the gambler's fallacy Type II. In several experiments (Keren and Lewis, in press) we asked subjects to estimate the number of rounds (i.e. sample size) that would be required to detect a favorable number where the probability of the favorable number was explicitly stated. For favorable numbers with a small advantage (e.g. probability of 1/34 for the favorable number compared with 1/39.09 for each of the remaining numbers) the median estimates were smaller by a factor of 10 and more compared with the normative appropriate sample size. The phenomenon is robust and holds even for a highly favorable number with probability of 1/20, though the underestimation is reduced (mainly because of a floor effect, i.e. a minimum number of trials is always necessary).

As another example of a quasi-logical strategy, my colleague Willem Wagenaar and I were told by a professional Blackjack dealer about a method by which the following card can be predicted: he "discovered" that in most cases the sum of any two successive cards adds up to 10, 11 or 12. Thus after a six would come a six or a five, after a three a seven or eight and so forth. Consequently, he claimed that he was able to predict above chance level the value of the following card and even claimed that he employed the method in practice.

The above examples are all based on a certain line of reasoning, albeit faulty, which represent misunderstandings of the calculus of chance. They are not rational but are nevertheless based on a certain rationale, and it is for this reason that I have classified them as quasi-logical. Gamblers' probabilistic reasoning and their predictions are also frequently based on what gamblers refer to as *intuitions* (or gut feelings) which, from a paradigmatic perspective, may represent magical reasoning. These intuitions are internal and occur in what Kahneman and Tversky (1982) refer to as the introspective mode.

A vital concept in this context is that of *luck*. Luck in the gambler's mind is clearly distinguished from chance consideration. In studying the behavior of Blackjack players in a real life casino setting we (Keren and Wagenaar, 1985) asked a sample of 28 players the following question: "in playing Blackjack there are three important factors, namely chance, skill, and luck. How important is each of these three factors? Give your answer in percentages so that they add up to 100%". With the exception of 6 players (who claimed that there is no difference between chance and luck) the mean percentage ascribed to chance skill, and luck was 18%, 37% and 45% respectively. In a separate study (Keren and Wagenaar, 1987) we asked a sample of 29 students to predict the outcomes of various soccer games. We then posed the question of the relative importance of skill luck and chance in predicting soccer games. The means were 44, 27, and 29 for skill, chance, and luck, respectively. Evidently, the belief in luck is shared by gamblers and non-gamblers alike.

What then are the differences between chance and luck? Chance is perceived as reflecting causelessness (Wagenaar and Keren, 1988), and consequently is deemed to be utterly out of control (Friedland, 1992). It is this component which supposedly is envisaged as entirely deterministic. In contrast, it is luck that is used as a vehicle to exert control (which from a paradigmatic perspective remains illusory). Thus Friedland (1992) presents empirical evidence supporting the proposition that the greater the need for control, the stronger the tendency to invoke luck as an explanation of events. In addition, luck offers a tool by which outcomes can be accounted for after the fact (the hindsight phenomenon). Attributions of luck are often considered to be more forceful and convincing explanations than a simple attribution to randomness and chance.

Luck is clearly part of the narrative repertoire and therefore its potential internal contradictions (as determined by paradigmatic criteria) can be tolerated. Luck is regarded as a concept that refers to a person,[5] contrasted with chance which reflects properties of the outside world. Some people may be luckier than others, and luck of a given person varies at different times. In some respect luck may also be perceived to be deterministic: "It is as though you were fated in advance to be enriched or despoiled" (Cardano, 1961/1520). Nevertheless, luck in the gambler's eye may provide some means of control. Specifically, it varies with time, and the skill is to detect the lucky moments and utilize them accordingly. As one gambler noted, luck follows a sinusoidal pattern where the whole trick is to find out whether the tide is high or low. Indeed, in an interview conducted with a sample of 33 Blackjack players, 27 players claimed that luck cannot be predicted yet 23 of those players claimed that it could be detected (Keren and Wagenaar, 1985).

The identification of luck (a lucky number, a lucky color, a lucky day) is frequently achieved by alleged intuitions (supposedly processed in what Kahneman and Tversky termed the introspective mode). Alternatively, luck can be disclosed by seemingly (though illusory) paradigmatic methods. For instance, long runs in continued gambling are often attributed to good or bad luck, indicating the misperception of random sequences (Bar-Hillel and Wagenaar, 1993). The tendency to account for long runs of a random sequence by employing mystical beliefs such as luck, is not restricted to gamblers. Gilovich, Vallone, and Tversky (1985) have argued that what basketball fans describe as a "hot hand" (which occurs when a player's performance during a particular period is significantly better than expected on the basis of the player's overall record—i.e. the player's base-rate), is not more than a long run of a sequence that is generated by a random process. Apparently, misperceptions of chance and the belief in the "law of small numbers" are frequently associated with illusory and magical narratives such as "luck", the "hot hand", and alike.

It seems as if luck often serves as a vehicle to explain randomness or ambiguity (when probabilities are unknown). It is almost always used for explaining outcomes after the fact following the well-known hindsight bias.

Indeed, Keren and Wagenaar (1985) observed that Blackjack players frequently tend to ignore the original uncertainty that was involved before the cards were dealt and to exaggerate what could have been anticipated with foresight. In fact, the formation of the frame of luck is based to a large extent on a hindsight view that estimates the role of uncertainty which actually dominates gambling situations. In addition, luck has a strong emotional component that is devoid of chance. For instance, in a lottery drawing a number next to the winning number is often interpreted as constituting bad luck. In this context luck serves as an expression for emotional feelings like regret and as such is a pure narrative term.

Unlike the narrative calculus of chance, which is neutral with regard to the consequences of potential outcomes, gamblers' probabilistic assessments are contaminated by their desires and associated emotions. Several studies have shown that estimates of uncertain outcomes are strongly influenced by a wishful-thinking bias. A particularly compelling demonstration is reported by Babad (1987) who asked a large sample of soccer fans to predict the outcomes of different games, and has shown that the stronger they felt affiliated with a team, the more likely they were to assign a win to that team. Even predictions that were made at half-time, when the favorite team trailed decisively, were characterized by a pervasive tendency of wishful thinking. Similarly, Uhlaner and Grofman (1986) analyzed the American National Election Study and found that candidate supporters exhibited a dramatic bias in estimating their candidate's likelihood of success.

19.3 SUMMARY AND CONCLUDING COMMENTS

A major conclusion from the preceding review of gambling behavior is that the normative theory of probability, regardless of which of the traditional interpretations is attached to it, can not serve as an adequate descriptive theory. The main claim of the present chapter is that the normative theory of probability is solely based on paradigmatic mode. In reality, gamblers and nongamblers often lack the competence required by the formal logico-scientific approach which apparently is not part of a human's intuitive repertoire. Moreover, the evaluation of uncertainty is often associated with narrative components which entails criteria that are different from—and often conflicting with—those postulated by the paradigmatic mode.

Although there is ample evidence that (what was termed here as) magical reasoning plays an important role in the assessment of uncertainty, its weight relative to paradigmatic considerations is not clear and would depend, among other things, on personality and situational factors. The interplay between the paradigmatic and narrative accounts is already evident in the writings of Gerolamo Cardano in the 16th century. There is no doubt that Cardano was a highly talented man who was among the first to lay the grounds for a rational treatment of uncertainty as reflected in the normative theory of probability. At the same time he was also an inveterate gambler, and his book provides an excellent foundation for a descriptive theory of the psychology of gambling. The perpetual contention between the paradigmatic (rational) and the narrative (nonrational) modes is frequently apparent in his writing. For example, after describing the importance of luck in gambling he adds the following final comment:

> But whether the cause of that luck, be it in the conjunction of the stars or in the construction of a certain order of the universe, can affect the cards, which are considered bad or good only according to the conventions of men (since they signify nothing of themselves) is so worthy of doubt that it easier to find a cause of this fact without that purpose than with it; without it the matter can well be reduced to chance, as in the constitution of the clouds, the scattering of beans, and the like.

Whether gamblers in general have similar reservations remains an open question. Based on my own research and casual observations of the gambling world, I have serious doubts.

While the present chapter has focused on gamblers' conceptualization of uncertainty, I strongly believe that the faulty reasoning and magical (or irrational) beliefs described here are not limited to gamblers. As Wagenaar (1988) pointed out, the proportion of people engaged in gambling in one way or another is so high that makes it a sufficiently universal phenomenon that "cannot be explained by a defect in a minority of people" (page 4). Many of the logical fallacies and violations of the normative theory of probability that are exhibited by gamblers have also been documented in situations other than gambling (e.g. Kahneman, Slovic, and Tversky 1982). Similarly, beliefs in nonrational concepts such as luck are also shared by nongamblers and are applicable to situations other than gambling (e.g. Wagenaar and Keren, 1988).

During the past 20 years there has been an ongoing debate concerning human rationality (e.g. Cohen, 1979; Jungermann, 1983). Those who postulate that human behavior is generally rational have dismissed research that was incompatible with this view. Proponents of rationality have argued that most of this research (which casts doubts on human rationality) has been conducted in an artificial environment with a clear attempt to "trick" people and force them to perform in a manner that is incompatible with the canons of rationality. The brief review of gambler's behavior, presented in this chapter, strongly suggests that violations of rationality are not limited to the psychological laboratory. If we accept the claim that the gambler population (with perhaps few exceptions) is not necessarily a pathological one, then postulating exclusive rationality (which implies the rejection of the narrative mode) is certainly unwarranted. Clearly, in this view, the formal theory of probability (regardless of the interpretation being used) is an inadequate descriptive theory of how people deal with uncertainty.

NOTES

(1) An excellent exposition of the different schools of probability is given by Salmon (1966).

(2) Salmon, for example, proposed several criteria for testing the adequacy of an interpretation. The most important one is *admissibility* which requires that the meaning assigned to the primitive terms by a given interpretation should transform the axioms and all consequent theorems into true statements. This implies that under any interpretation the mathematical relations specified by the calculus of probability should not be violated and consequently provides a safeguard against incoherent betting systems.

(3) Whether indeed people really believe that their actions influence the outcome may be, under certain conditions, questionable. Shafir and Tversky (1992) introduce the term "quasi-magical" thinking referring to situations "in which people act as if they erroneously believe that their actions influence the outcome, even though they do not really hold that belief" (page 463).

(4) Statistical methods for testing randomness (Pashley, 1993) are actually also based on outcome features (and can only assess the likelihood that a sequence was produced by a random generator). Referring to gamblers' conceptualization as being quasi-logical implies that there may be some similarity in the underlying reasoning (i.e. the frequencies of a fair coin on the long run), yet is "quasi" because of departures from the normative rule (i.e. misunderstanding what is the long run) and basic violations of some logical dictates.

(5) This is also reflected in the daily use of the term. It is more natural to say "it was *my* lucky day" or "this is *my* lucky number" than to say "it is a lucky day" or "it is a lucky number".

REFERENCES

Babad, E. (1987) Wishful thinking and objectivity among sport fans. *Social Behavior: An international Journal of Applied Social Psychology*, **4**, 231–40.

Bar-Hillel, M. & Wagenaar, W.A. (1993) The perception of Randomness. In G. Keren and C. Lewis (eds.) *A Handbook for Data Analysis in the Behavioral Sciences: Methodological Issues*. Erlbaum, Hillsdale, NJ.

Bruner, J. (1984) Narrative and paradigmatic modes of thought. Paper presented at the meeting of the American Psychological Association, Toronto, August 1984.

Cardano, G. (1961) *The Book on Games of Chance* (comp. ca. 1520 and translated by S.H. Gould) Holt, Rinehart and Winston, New York.

Cohen, L.J. (1979) On the psychology of prediction: Whose is the fallacy? *Cognition*, 385–407.

Daston, L. (1988) *Classical Theory in the Enlightment*. Princeton University Press, Princeton.

David, F.N. (1962) *Games, Gods, and Gambling*. Charles, Griffin, London.

Fischhoff, B. (1975) Hindsight: Thinking backward? *Psychology Today*, April, 70–6.

Friedland, N. (1992) On luck and chance: Need for control as mediator of the attribution of events to luck. *Journal of Behavioral Decision Making*, **5**, 267–82.

Gigerenzer, G., Swijtink, Z., Porter, T., Daston, L., Betty, J. & Kruger, L. (1989) *The Empire of Chance*. Cambridge University Press, Cambridge.

Gilovich, T., Vallone, R. & Tversky, A. (1985) The hot hand in basketball: On the misperception of random sequences. *Cognitive Psychology*, **17**, 295–314.

Hacking, I. (1975) *The Emergence of Probability*. Cambridge University Press, Cambridge.

Hacking, I. (1990) *The Taming of Chance*. Cambridge University Press, Cambridge.

Henslin, J.M. (1967) Craps and magic. *American Journal of Sociology*, **73**, 316–30.

Howell, W.C. & Burnett, S.A. (1978) Uncertainty measurement: A cognitive taxonomy. *Organizational Behavior and Human Performance*, **22**, 45–68.

Jungermann, H. (1983) The two camps on rationality. In R.W. Schultz (ed.) *Decision Making Under Uncertainty*. North Holland, Amsterdam.

Kahneman, D., Slovic, P. & Tversky, A. (1982) *Judgement under Uncertainty: Heuristics and Biases*. Cambridge University Press, Cambridge.

Kahneman, D. & Tversky, A. (1982) Variants of uncertainty. *Cognition* **11**, 145–57.

Keren, G. & Lewis, C. (in press) The two fallacies of gamblers: Type I and Type II. *Organizational Behavior and Human Decision Processes*.

Keren, G. & Wagenaar, W.A. (1985) On the psychology of playing Blackjack: Normative and descriptive considerations with implications for decision theory. *Journal of Experimental Psychology: General*, **114**, 133–58.

Keren, G. & Wagenaar, W.A. (1987) Temporal aspects of probabilistic predictions. *Bulletin of the Psychonomic Society*, **25**, 61–4.

Keren, G. & Wagenaar, W.A. (1988) Chance and skill in gambling: A search for distinctive features. *International Journal of Applied Social Psychology*, **3**, 199–217.

Langer, E.J. (1975) The illusion of control. *Journal of Personality and Social Psychology*, **32**, 311–28.

Langer, E.J. (1977) The psychology of chance. *Journal for the Theory of Social Behaviour*, **7**, 185–207.

Lindley, D.V., Tversky, A. & Brown, R.V. (1979). On the reconciliation of probability assessments. *Journal of the Royal Statistical Society*, **142A**, 146–80.

Oldman, D. (1974) Chance and Skill: A study of roulette. *Sociology*, **8**, 47–426.

Pashley, P.J. (1993) On generating random sequences. In G. Keren and C. Lewis (eds.) *A Handbook for Data Analysis in the Behavioral Sciences: Methodological Issues*. Erlbaum, Hillsdale, NJ.

Tversky, A. & Kahneman, D. (1971) The belief in the "law of small numbers". *Psychological Bulletin*, **76**, 105–10.

Tversky, A. & Kahneman, D. (1974) Judgment under uncertainty: Heuristics and Biases. *Science*, **185**, 1124–31.

Salmon, W.C. (1966) *The Foundations of Scientific Inference*. Pittsburgh University Press, Pittsburgh.

Shafer, G. (1993) Can the various meanings of probability be reconciled? In G. Keren and C. Lewis (eds.) *A Handbook for Data Analysis in the Behavioral Sciences: Methodological Issues*. Erlbaum, Hillsdale, NJ.

Shafir, E. & Tversky, A. (1992) Thinking through uncertainty: Nonconsequential reasoning and choice. *Cognitive Psychology*, **24**, 449–74.

Uhlaner, C.J. & Grofman, B. (1986) The race may be close but my horse is going to win: Wish fulfilment in the 1980 Presidential election. *Journal of Political Behavior*, **8**, 101–29.

Wagenaar, W.A. (1988) *Paradoxes of Gambling Behavior*. Erlbaum, Hillsdale, NJ.
Wagenaar, W.A. & Keren, G. (1988) Chance and Luck are not the same. *Journal of Behavioral Decision Making*, **1**, 65–75.
Wilson, A. (1965) *The Casino Gambler's Guide*, Harper and Row, New York.

Chapter 20

Uncertainty and Subjective Probability in AI Systems

Paul J. Krause and Dominic A. Clark
Imperial Cancer Research Fund, London

20.1 INTRODUCTION

There is one thing I am sure of, and that is how ignorant I am. This familiar antinomy could be a very apt motto for many workers in the field of artificial intelligence (AI), for many problems in AI require some method for reasoning effectively under a state of partial ignorance. For example, what conclusions can we draw about a ship when we do not know precisely where it is, but we do know that it is "at sea"? Should major, "possibly" life threatening, surgery be performed if a patient is judged to "almost certainly" have a malignant tumour, even though there is "still a possibility" that it might turn out to be benign? People manage to reason and act on the basis of such information, although not always in an optimal way, and one of the major challenges of AI has been the development of effective computational models of these processes.

The two examples of the preceding paragraph illustrated two different classes of ignorance, imprecision and uncertainty. In the first case our source of information, or "sensor", had been imprecise about the exact location of an object. In the second case, our sensors could not be relied upon to enable us to draw certain conclusions about the true state of affairs. It is this second aspect of ignorance, uncertainty, which is the main focus of this chapter.

Subjective Probability. Edited by G. Wright and P. Ayton.
© 1994 John Wiley & Sons Ltd.

In particular we will review the contribution to AI of the Bayesian view of probability as a subjective measure of belief and probability theory as a framework for plausible reasoning. However, we must state at the outset that we do not believe that Bayesian probability is *necessarily* always the best model to use for problems which involve imprecision (see, for example, (Clark, 1990; Krause & Clark, 1993) for a review of alternative formalisms). We will expand on this claim in the final sections of this chapter, but primarily we will be addressing the use of probability for well-defined problems involving precise concepts in knowledge domains for which numerical coefficients can confidently be elicited.

20.2 REVISION OF BASIC CONCEPTS

We will start our discussion of the use of probability in AI systems with a very brief review of some of the basic concepts. This will emphasize those aspects which we think are particularly relevant to the development of computer-based applications.

Subjective probability provides a normative framework for the representation and updating of beliefs. The rationality of this framework has been well discussed, appealing to it as the basis upon which an agent should decide how to act given an uncertain situation (de Finetti 1937; Savage, 1962). We will have a little more to say about the use of point value numerical probabilities towards the end of this chapter, but in most of what follows we will focus on the use of single-valued probabilities assigned to precisely defined concepts.

We will take conditional probabilities as the basic expressions of the Bayesian formalism. We will also claim that Bayesian reasoning is as much about structure as it is about numbers. For to say that the probability of a hypothesis is conditional on one or more items is to identify the relevant information to the problem at hand. In addition, we may specify that the identification of an item of evidence *influences* the degree of belief in a hypothesis. This places a directionality on the relevance links between evidences and hypotheses.

The directionality placed on the links will depend on the way in which the problem is structured. The term "causal networks" has often been used for larger scale probabilistic networks because a direction corresponding to causal influence can be the most meaningful. For example, we might say that measles "causes" red spots. Using a directionality based on causal influence is often useful because assessing the probability of a symptom given a cause can be an easier problem than assessing the probability of a cause given a symptom. In general, however, the ordering need not be based on causality. The point is to take any natural cognitive ordering which will enable a confident assessment of the associated probability to be made.

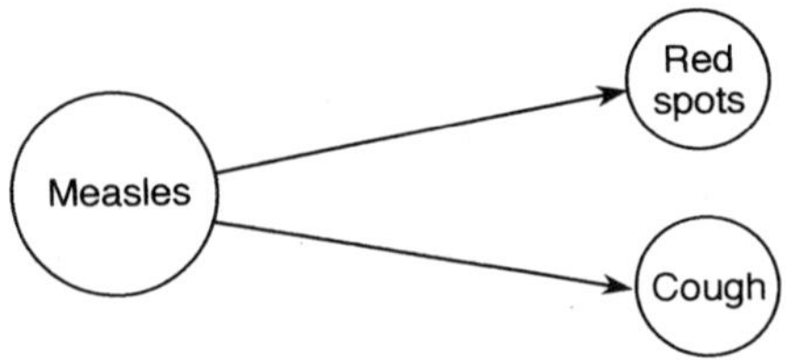

Figure 20.1 Very simple influence network

The notion of relevance embodied in the use of conditional probabilities also influences the elicitation of the probability values. For example, to say that $p(\text{red spots} \mid \text{measles}) = p$ means that we can assign probability p to "red spots" if measles is observed *and only measles is observed*. If any further evidence e is relevant to the likelihood of the occurrence of red spots, then we will be required to determine $p(\text{red spots} \mid \text{measles}, e)$.

A final basic point to emphasize is that the Bayesian rule of updating is central to the subjective Bayesian conception of probabilistic reasoning. We restate it here for completeness:

$$p(h \mid e) = \frac{p(e \mid h) \cdot p(h)}{p(e)} \tag{20.1}$$

That is, the revised belief in, or posterior probability of, a hypothesis h on observing evidence e, $p(h \mid e)$, is obtained by multiplying the prior belief in h, $p(h)$, by the probability $p(e \mid h)$ that e will materialize if h is true. $p(e)$, the prior probability of the evidence, acts as a normalization coefficient (the degree of belief that accrues to a hypothesis on the basis of some evidence is clearly dependent on the prior frequency of occurrence of that evidence). So, in the previous example, it may be easiest to elicit the probability of red spots given measles. But, given this value, we can use Bayes' rule to reason from red spots as evidence to obtain the posterior probability of measles as a hypothesis (provided we know the prior probability for measles).

The subjective Bayesian approach provides a framework for answering the query "Given that I know e, what is my belief in h". In order to be able to develop functioning AI systems, we need to develop techniques for reasoning efficiently over knowledge domains representing hundreds, thousands, maybe millions even, of facts. We will need to look at how Bayesian reasoning can be scaled up from simple problems involving a few related propositions, to very large-scale networks. However, first of all we will look at some basic semantic issues of uncertain inference in AI systems.

20.3 PROBABILITY AND RULE-BASED SYSTEMS

Many expert systems have been, and continue to be, developed as "rule-based" systems. That is, the knowledge embodied in the system is represented as a set of "if–then" rules of the form; *if* antecedent *then* consequent (the antecedent of the rule may be a conjunction of conditions). Such a rule may be read as: "if the conditions in antecedent are true, then consequent is the case." A classic example is "*if* red spots *then* measles." Clearly, such rules may not be categorical. In a diagnostic expert system, for example, observation of a specific symptom, or set of symptoms, would usually be merely suggestive of a pathological state, not confirmation of it. This leads to a refinement of rule-based systems in which a numerical "certainty factor" is associated with each rule. The confidence in, or certainty of, the consequent of a rule is then calculated as a simple function of the certainty in the antecedent and the certainty factor of the rule. MYCIN (Shortliffe, 1976) and Prospector (Gashnig, 1982) are particularly notable examples of this approach. Unfortunately there is a conflict between the *truth functionality* of rule-based systems and the *intentionality* of probabilistic systems.

Let us first say what we mean by "truth functionality" in this context. In formal logic, propositions may be combined using simple syntactic principles and the truth value of the resulting formula obtained as a simple function of the truth values of the sub-formulae. Thus we can obtain the truth value of $A \& B$ from the truth values of A and of B.[1] In an analogous way, in rule-based systems we want the belief in the consequent of a rule to be a simple function of numerical coefficients associated with that rule and with its antecedent. That is what we mean here by "truth functional". Probabilities, in contrast, cannot be composed in this way. For example, if we know $p(A)$ and $p(B)$ we cannot, in general, combine these values in any simple way to obtain the probability $p(A, B)$ of their conjunction. We do have that

$$p(A, B) = p(A \mid B) \cdot p(B)$$

However, it is only if $p(A \mid B) = p(A)$ that we can obtain $p(A, B)$ as a simple conjunction: $p(A, B) = p(A) \cdot p(B)$. We will have $p(A \mid B) = p(A)$ if it is the case that knowing B has no effect on our belief in A. That is, A and B are *marginally independent*.

Now, as we have said, in many rule-based expert systems certainty factors are associated with the rules and combined using simple syntactic principles as the rules are fired. This *extensional* or *syntactic* approach to uncertainty handling is computationally efficient. But unless strong independence assumptions can legitimately be made, it is semantically sloppy as illustrated in the previous paragraph. This contrasts with a fully *intensional* probabilistic approach, which is semantically coherent but in the general case computationally intractable (Pearl, 1988).

What gives rule-based systems their computational advantage is their *modularity*. By modularity is meant that the only requirement for a rule "if *A* then *B*" to be fired is the presence of *A* in the database, no matter how it has been obtained and no matter what else is in the database. This can be illustrated with what is now a fairly classic problem of diagnosing weather conditions from observations about the state of a garden. In this example, a rule-base contains information about possible causes (such as the lawn becoming wet can be caused by rain), as well as possible effects (such as the neighbour's sprinkler being on will make the lawn wet). For example, it may contain a diagnostic rule such as:

if wet lawn then rain (0.8)

and a predictive rule such as:

if sprinkler on then wet lawn (0.95)

Suppose, now, "wet lawn" is observed. Rain will be concluded, with a certainty of 0.8. But, if the database also contains "sprinkler on" this explains "wet lawn" and the belief in "rain" should be reduced. That is, other information that is relevant to the problem should have been considered.

Suppose, on the other hand, the database only contains "sprinkler on". Then the second rule will be fired to conclude "wet lawn" (0.95). Once "wet lawn" has been concluded, the first rule will be fired to conclude "rain" (0.76) (using a very naive multiplicative rule to chain probabilities). Thus, ignoring the fact that "wet lawn" has been obtained as a prediction, and not as a fact requiring explanation, has led to a quite erroneous conclusion; the distinction between diagnostic and predictive rules is confounded. Equally, if rain is observed, we cannot use the first rule to *predict* "wet lawn". In order to make this prediction, we could add to the database a separate rule such as:

if rain then wet lawn

However, this then opens up the possibility of circular reasoning, with "rain" supporting "wet lawn", " wet lawn" supporting "rain" (by the first rule) and so on.

Basically we want to use rules for abductive, as well as deductive, reasoning. Such bidirectional inferences can be handled by probability. By exploiting Bayes' rule, we can make diagnostic as well as predictive inferences; reason from evidence to hypothesis as well as from hypothesis to evidence.

As well as allowing bidirectional inferences, we also wish to allow some capability for explaining away. That is, to allow a revision in the belief in a possible explanation if an alternative explanation is actually observed. This arose in the above example when we anticipated that confirmation of one possible explanation of the wet lawn (the sprinkler being on) should result in a decrease in belief in the alternative explanation (rain). Again, this can be handled by framing the problem in a probabilistic framework.

A third difficulty with the rule-based approach, is the problem of *correlated sources of evidence*. Evidences e_1 and e_2 may both add support to hypothesis h. But if they have both been derived from a common source the combined support they give to the hypothesis should not be as strong as that obtained were they independent evidences. Again, a rigorous probabilistic model can handle this correctly.

The difficulty, however, is that the probabilistic approach can be computationally intractable. A naive representation of a problem in a probabilistic framework would require the elicitation of a probability distribution function defined over all the propositions of interest. For example, if we let A stand for "wet lawn", B for "rain" and C for "sprinkler on", in order to model the above problem we would need to elicit $p(a, b, c)$, $p(a, \neg b, c)$, $p(a, \neg b, \neg c)$, $p(a, b, \neg c)$, $p(\neg a, b, c)$, $p(\neg a, \neg b, c)$, $p(\neg a, \neg b, \neg c)$, and $p(\neg a, b, \neg c)$. A problem involving n propositions, $A_1, A_2, \ldots, A_n$, will require the elicitation of 2^n such values. With respect to computation, calculating the marginal probability $p(A_i)$ that A_i is true will, for example, require summing over the 2^{n-1} values for which A_i is true. Some method of easing the knowledge elicitation and computational tasks is clearly needed.

This section has discussed well-known properties of probabilistic reasoning. We have revisited them in this section because we wished to emphasize the following points in connection with the development of AI systems. To recapitulate, the modularity of rule-based systems makes them computationally efficient. But they require an extensive representation of the inferences which may be drawn from them. On the other hand, the intensional, or semantic, approach taken in a rigorous probabilistic model is computationally intensive. But it does enable deduction, abduction and explaining away to be modelled, and correlated sources of evidence to be handled correctly. The drawback of their computational intensity has, however, been addressed by careful exploitation of the structure in probabilistic networks. We will discuss the basis for this next.

20.4 INDEPENDENCE PROPERTIES

In much of the preceding discussion we have focused on conditional probabilities and dependencies rather than joint probability distributions. The motivation behind the work of Pearl, Spiegelhalter and others in producing network representations of probabilistic knowledge is to "make intensional systems operational by making relevance relationships explicit" (Pearl, 1988). By formalizing and exploiting the conditions under which discrete sections of the network of propositions may be regarded as independent, we may transform belief revision from an intractable global operation into a sequence of local operations. In addition, the problem of eliciting massive joint distribution tables is reduced to that of eliciting the conceptually much more meaningful conditional probabilities between semantically related propositions.

We first recall the chain rule for probabilities. Let $p(A_1, A_2, \ldots, A_n)$ be a probability distribution over the propositions $A_1, A_2, \ldots, A_n$). Then

$$p(A_1, A_2, \ldots, A_n) = p(A_n \mid A_{n-1}, \ldots, A_1) \cdot p(A_{n-1} \mid A_{n-2}, \ldots, A_1) \cdot \cdots \cdot p(A_2 \mid A_1) \cdot p(A_1) \quad (20.2)$$

Now, two propositions are independent if knowing one to be true has no effect on our belief in the other. That is, A is independent of B, given C, if $P(A \mid B, C) = P(A \mid C)$. We will now revise three different representations for a probability distribution over a set of propositions $\{A, B, C\}$. They each embody different conditional independence assumptions which are made explicit through graphical representations.

Returning first to the "wet lawn" example of the previous section. Both rain and the sprinkler being on may cause the lawn to become wet. This problem may be represented in graphical form by the directed graph shown in Figure 20.2.

In this example, A (rain) and B (sprinkler on) are marginally independent, but conditionally *dependent* given C (wet lawn); once the lawn is seen to be wet, conditioning on either of A or B should affect our belief in the other. Application of the chain rule to the probability distribution $p(A, B, C)$ gives

$$p(A, B, C) = p(C \mid A, B) \cdot p(A \mid B) \cdot p(B)$$

Since A and B are marginally independent, we have $p(A \mid B) = p(A)$, but are unable to reduce the expression $p(C \mid A, B)$ any further. Thus, for this graph

$$p(A, B, C) = p(C \mid A, B) \cdot p(A) \cdot p(B)$$

Consider, now, the following scenario. Red spots and Koplick's spots are both symptoms of measles; measles "causes" both red spots and Koplick's spots (which are small white spots found inside the mouth). This can be represented by the graphical structure of Figure 20.3.

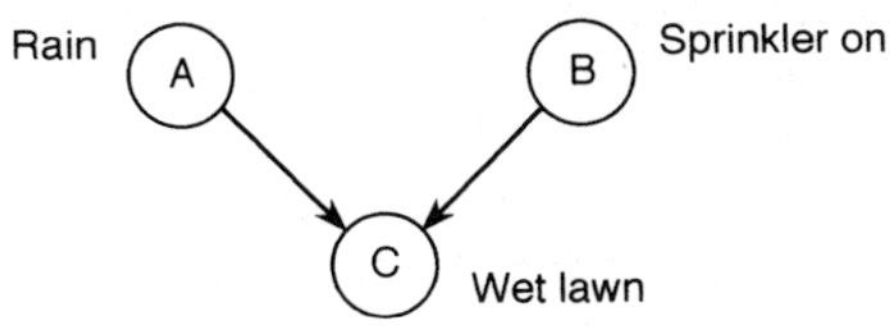

Figure 20.2 Nodes A and B are conditionally *dependent* given C

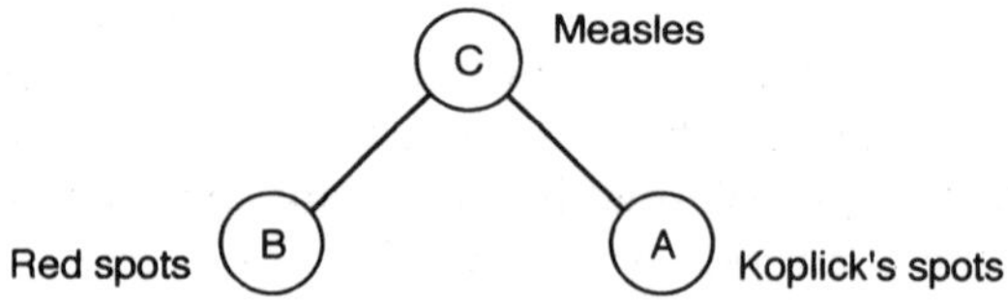

Figure 20.3 *A* and *B* are conditionally *independent* given *C*

In this case, in contrast to the previous example, given that the patient is suffering from measles, the observation of red spots will have no influence at all on the belief that Koplick's spots will be observed; A and B are conditionally independent given C. Applying the chain rule again,

$$p(A, B, C) = p(A \mid B, C) \cdot p(B \mid C) \cdot p(C)$$

Since A and B are independent given C, $p(A \mid B, C) = p(A \mid C)$. Hence

$$p(A, B, C) = p(A \mid C) \cdot p(B \mid C) \cdot p(C)$$

The final case we will consider is another example of conditional independence. The disease $A =$ "kawasaki disease" is known to cause the pathological process $C =$ "myocardial ischaemia". This in turn has an associated symptom $B =$ "chest pain" (Figure 20.4).

Now, the observation of B may lead to an increase in belief in C and subsequently of A. But, once C has been confirmed, the observation of B can have no further influence on diagnosing A as the ultimate cause of C. So, as in the previous example, A and B are conditionally independent given C, and

$$\begin{aligned} p(A, B, C) &= p(B \mid A, C) \cdot p(C \mid A) \cdot p(A) \\ &= p(B \mid C) \cdot p(C \mid A) \cdot p(A) \end{aligned}$$

We have still been using very simple examples. However, it should be clear from this discussion that the elicitation of a belief network, or influence diagram, as a directed acyclic graph is a natural way of representing many types of belief relationship. One of the results which is an important component of Judea Pearl's work is that all the conditional independence relationships can be derived from a directed acyclic graph (Geiger & Pearl, 1988; Verma & Pearl, 1988) using a notion of "*d*-separation". A detailed coverage

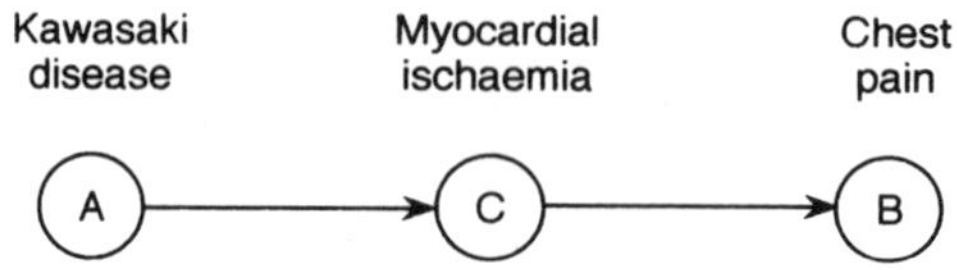

Figure 20.4 *A* and *B* are conditionally *independent* given *C*

of d-separation and the subsequent development of a fast algorithm for belief propagation in Bayesian networks can be found in Pearl (1988) and Neapolitan (1990). However, the resulting algorithm is perhaps not as widely applicable as that of Lauritzen and Spiegelhalter. As space is limited we can only discuss one algorithm and will continue with a discussion of the way in which the Lauritzen and Spiegelhalter algorithm exploits independencies to enable the rapid updating of belief networks.

We have shown how independence relationships may be exploited in simple problems to give a more semantically meaningful representation than the naive probability distribution over the set of propositions. This approach can be extended quite naturally to produce a probability representation for larger problems. Consider, for example, the network shown in Figure 20.5. By repeated use of the three independence relationships considered earlier, it should be quite easy to confirm that the following is a correct representation of the probability distribution for this network:

$$
\begin{aligned}
&p(A,B,C,D,E,F,G,H)\\
&= p(G \mid F)\cdot p(H \mid E,F)\cdot p(F \mid C,D)\cdot p(C \mid A)\cdot p(A)\cdot p(D \mid B)\\
&\quad \cdot p(E \mid B)\cdot p(B)\\
&= p(A)\cdot p(B)\cdot p(C \mid A)\cdot p(D \mid B)\cdot p(E \mid B)\cdot p(F \mid C,D)\\
&\quad \cdot p(G \mid F)\cdot p(H \mid E,F)
\end{aligned}
$$

(The second form is just a rearrangement of the first.)

We will call the nodes immediately preceding a given node in the graph the *parents* of that node. So, for example, C and D are the parents of F, F and E are the parents of H. We will also refer to the set of nodes which cannot be reached by a directed path from a given node as the *anterior* nodes to that node. For example, the nodes anterior to node F are $\{A,B,C,D,E\}$, those

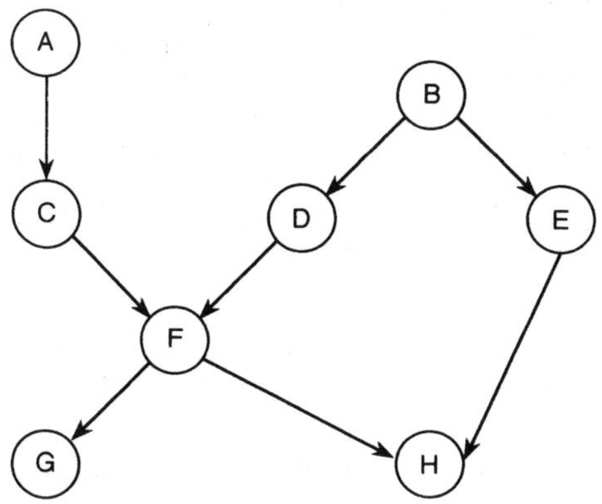

Figure 20.5 A simple belief network

anterior to node B are $\{A, C\}$. Then this equation is expressing the fact that the probability of each node in the graph is conditionally independent of its anterior nodes given its parent nodes.

The above equation can be expressed in a very simple general form. Let V be a set of nodes, and let parents(v) be the set of parent nodes for any $v \in V$. Then the second equality is just a specific instance of the general equation:

$$p(V) = \prod_{v \in V} p(v \mid \text{parents } (v)) \tag{20.3}$$

In this section we have shown how notions of dependence and independence may be exploited in structuring a belief network. As a result, the probability distribution for a large set of propositions may be represented by a product of conditional probability relationships between small clusters of semantically related propositions. We will now consider a very simple example to demonstrate how evidence may be efficiently propagated through a belief network.

20.5 BELIEF PROPAGATION THROUGH LOCAL COMPUTATION

The following is a simplified account of the Lauritzen and Spiegelhalter algorithm for belief propagation (Lauritzen & Spiegelhalter, 1988). It is based around a simple example first used in (Spiegelhalter, 1986) and is intended to be illustrative of the principles behind the algorithm, rather than a complete description of it. Rigorous descriptions can be found in Lauritzen & Spiegelhalter (1988) and Neapolitan (1990).

The essence of this approach is to represent hypotheses and relations in the domain under consideration as a directed graph. This is illustrated for the following, deliberately restricted, piece of medical knowledge:

> Metastatic cancer is a possible cause of a brain tumour, and is also an explanation for increased total serum [calcium] count. In turn, either of these could explain a patient falling into a coma. Severe headache is also possibly associated with a brain tumour. (Spiegelhalter, 1986)

The qualitative representation of this knowledge is shown in Figure 20.6. Here the nodes represent hypotheses and links indicate "causal" or probability relationships. These do not have to derive from direct physiological reasoning but "any natural cognitive ordering that will ... allow reasonably confident probability assessments". (Spiegelhalter, 1986), as we discussed in Section 20.2.

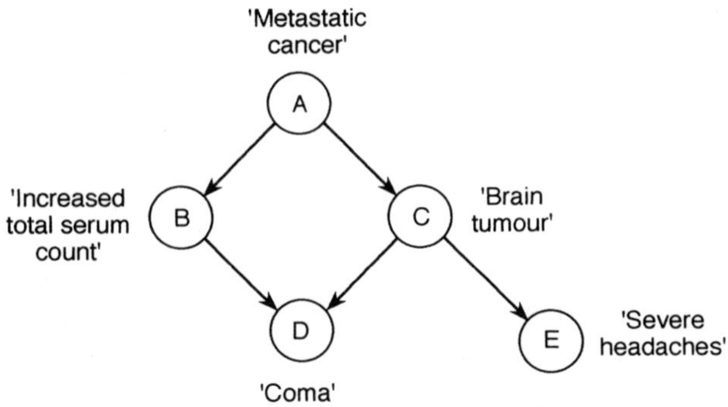

Figure 20.6 Belief network for metastatic cancer example

The probability distribution may now be decomposed into a product of conditional probabilities. Exploiting the independencies implicit in the graph, we have

$$p((A, B, C, D, E) = p(E \mid C) \cdot p(D \mid B, C) \cdot p(B \mid A) \cdot p(C \mid A) \cdot p(A)$$

Hypothetical assignments for these probabilities, together with some explanation, are shown in Table 20.1.

Now, we have emphasized that probabilistic reasoning need not be constrained by the directionality of the graphical representation; we may reason both forwards and backwards through the graph. Indeed, evidence may be received about any of the nodes in the graph and the consequences propagated

Table 20.1 Hypothetical conditional probabilities for Figure 20.6. (adapted from Spiegelhalter 1986)

Attribute	Value	Explanation
$p(e \mid \neg c)$	0.60	Headaches are common
$p(e \mid c)$	0.80	but more common if tumour present
$p(d \mid \neg b, \neg c)$	0.05	Coma is rare,
$p(d \mid b, \neg c)$	0.80	but common if
$p(d \mid \neg b, c)$	0.80	either cause is present
$p(d \mid b, c)$	0.80	
$p(b \mid \neg a)$	0.20	Increased calcium uncommon, but
$p(b \mid a)$	0.80	common consequences of metastasis
$p(c \mid \neg a)$	0.05	Brain tumour rare and uncommon
$p(c \mid a)$	0.20	consequence of metastasis
$p(a)$	0.20	Incidence in relevant clinic

throughout the graph. Thus, although the directed graph is appropriate for structuring the problem, we need to convert this to an undirected graphical representation in order to allow complete flexibility in reasoning over the graph.

In this example, the directed/recursive model may be converted to an undirected graphical model in two steps. Firstly a "vacuous rule" linking nodes B and C is introduced (Wermuth & Lauritzen, 1983). All arrows are then removed from the edges (Figure 20.7). In this simple example, the addition of the one rule is all that is needed to enable the graphical model to be decomposed into a sequence of "cliques". Technically, the requirement is that the graph be "triangulated". In general, this may be achieved by first "marrying" all parent nodes (to produce the "moral graph"), and then searching through the graph and adding links where necessary to ensure that the graph is triangulated. That is, that there are no closed cycles of length greater than four nodes without an intersecting chord. The cliques are then maximal subsets of nodes in which each node in a clique is linked to all other nodes in the same clique. In the example of Figure 20.7, the cliques are the sets of nodes: $\{A, B, C\}$, and $\{B, C, D\}$ and $\{C, E\}$.

This example has glossed over a number of technical results which inform the choice of dividing the graph up in this way. Basically, the end result of structuring the graph so, is that we are guaranteed to have a "decomposable" graph.

To illustrate what is meant by "decomposable", we may write the joint distribution as

$$p(A, B, C, D, E)$$
$$= p(E \mid C) \cdot p(D \mid B, C) \cdot p(C \mid A) \cdot p(B \mid A) \cdot p(A) \qquad \text{from equation (20.3)}$$
$$= \frac{p(C, E)}{p(C)} \cdot \frac{p(B, C, D)}{p(B, C)} \cdot p(A, B, C) \qquad \text{by conditional probability}$$

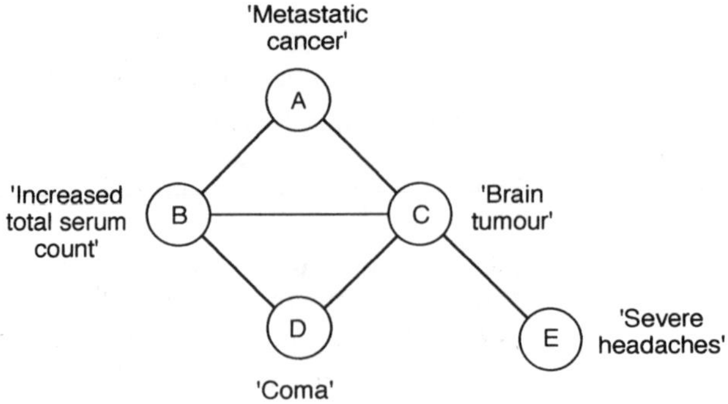

Figure 20.7 Representation as an undirected graph

This is simply the product of the marginal distributions on the cliques divided by the product of the distributions on their intersections. These values can be derived from the original assessments of Table 20.1. For example, for the clique $\{A, B, C\}$, the relation $p(A, B, C) = p(C \mid A) \cdot p(B \mid A) \cdot p(A)$ used in the above derivation enables us to calculate from Table 20.1 that $p(a, b, c) = 0.032$, $p(\neg a, b, c) = 0.008$, $p(a, \neg b, c) = 0.008$, $p(\neg a, b, c) = 0.032$, $p(a, b, \neg c) = 0.128$, $p(\neg a, b, \neg c) = 0.152$, $p(a, \neg b, \neg c) = 0.032$ and $p(\neg a, \neg b, \neg c) = 0.608$.

From these marginal distributions it is then possible to derive the prior probabilities of all the events specified (i.e. $p(b) = 0.32$, $p(c) = 0.08$, $p(d) = 0.32$, $p(e) = 0.616$). These are obtained by taking the probability distribution for a clique containing the node corresponding to the event of interest and summing over all the possible values for the other nodes in the clique. For example,

$$p(a) = p(a, b, c) + p(a, \neg b, c) + p(a, b, \neg c) + p(a, \neg b, \neg c)$$

$$= \sum_{\{A, B, C\}} p(a, B, C) \qquad (20.4)$$

(Here we mean the sum over all possible states of the clique $\{A, B, C\}$ but with node A instantiated to true, a).

The impact of information on any node may now be propagated through the graph. In order to do this, we convert the graph into a directed "hypertree" of cliques. First the nodes must be numbered. This is done as follows. The node whose evidence is observed is labelled as the first node. For example, suppose we wish to assess the effect of the observation of severe headaches on the probability of a patient lapsing into a coma. Then we would take node E as the starting point. The labelling is then continued by successively numbering the nodes attached to the maximum number of nodes that are already labelled.

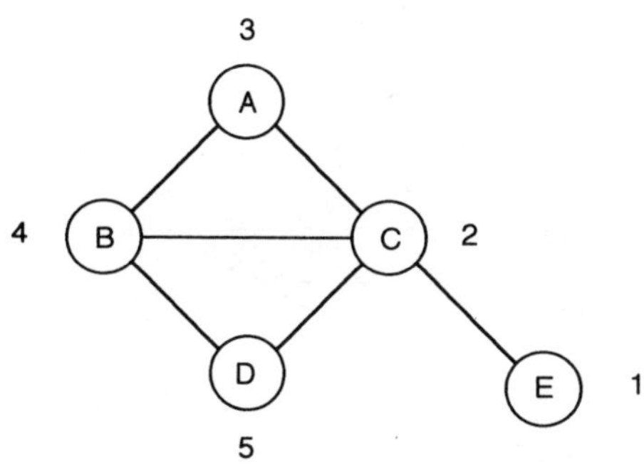

Figure 20.8 A possible numbering using maximum cardinality search

Ties may be broken at random. This is known as the "maximum cardinality search" and a possible numbering for the current problem is shown in Figure 20.8.

The cliques are then ranked according to the highest numbered node in each clique. This gives the sequence $\{C, E\}$, $\{A, B, C\}$, $\{B, C, D\}$. We now have a tree of clusters of nodes. That is, a hypertree (Figure 20.9).

This sequence of cliques may now be recursively updated. If the current belief given the available evidence is indicated by an asterisk, p^*, we have for clique I,

$$p^*(c) = p(c \mid e) = \frac{p(c, e)}{p(e)}$$

For clique II,

$$\begin{aligned} p^*(a, b, c) &= p(a, b, c \mid e) && \text{by defn.} \\ &= p(a, b \mid c, e) \cdot p(c \mid e) && \text{by conditioning} \\ &= p(a, b \mid c) \cdot p^*(c) && \text{independence and defn of } p^*(c) \\ &= p(a, b, c) \cdot \frac{p^*(c)}{p(c)} && \text{definition of conditional probability} \end{aligned}$$

In turn, for clique III,

$$\begin{aligned} p^*(b, c, d) &= p(b, c, d \mid e) \\ &= p(d \mid b, c, e) \cdot p(b, c \mid e) \\ &= p(b, c, d) \frac{p^*(b, c)}{p(b, c)} \end{aligned}$$

The revised distribution for each successive clique is given by the original belief multiplied by the ratio of the revised belief to the original belief for those nodes on the intersection of the preceding clique. Thus in turn, $p^*(c)$ can be obtained from clique I to enable the distribution on clique II to be revised; $p^*(b, c)$ may then be obtained from the revised distribution of clique II and

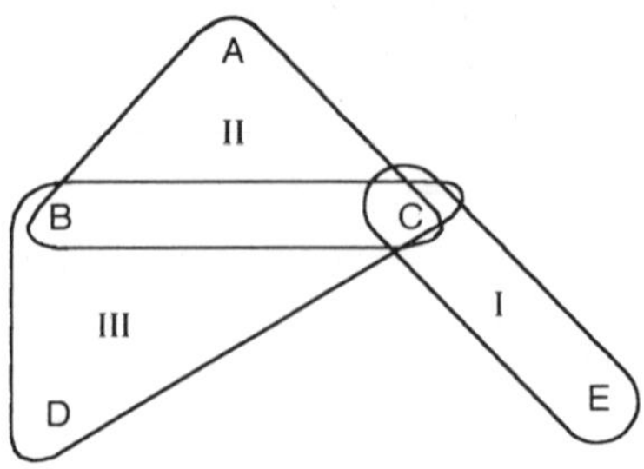

Figure 20.9 Hypertree representation

the distribution for clique III revised. In this particular example, after condition on node E (severe headache is observed) the revised distribution for node III yields a revised marginal probability for the occurrence of coma as $p^*(d) = 0.333$.

Although we have used an almost trivial example, it does illustrate the basic concepts underlying the Lauritzen–Spiegelhalter algorithm. The representation of the problem as a directed acyclic graph provides an intuitive way of structuring the problem. It also simplifies the elicitation of the probability distribution by breaking it down into a product of conditional probabilities involving small numbers of semantically related propositions. To enable the propagation of evidence in any direction, this representation must then be transformed into one associated with an undirected graph, which may be decomposed into a sequence of "cliques". The influence on the probability distribution of a clique due to evidence from a neighbouring clique may then be calculated using only local computations. In this way, evidence from any node may be propagated efficiently through the graph, provided the average clique size is reasonably small.

20.6 NORMATIVE EXPERT SYSTEMS?

One of the justifications for using classical probability as the core of a model of decision-making is that it supports a normative theory of decision-making. That is, it provides a gold standard for decision-making; a prescription for how people should make judgements on the basis of uncertain information. There have been a number of studies which show that people generally do not behave according to these standards (Tversky & Kahneman, 1974; Kahneman et al., 1982). The more *ad hoc* uncertainty calculi used by some early expert systems (notably Mycin and Prospector) have been criticized by many as being subject to the same biases and mistakes. The main justification for using these more *ad hoc* approaches is that they are computationally efficient, whereas Bayesian updating is known to be an NP-hard problem in the general case (Cooper, 1988). Now that efficient algorithms have been developed for the rapid updating of belief networks this last advantage of *ad hoc* approaches can be matched and the field is open for the development of "normative expert systems" (Heckerman, 1991).

Medicine is one domain where high standards of decision-making are particularly essential. An incorrect decision may lead to a patient not receiving treatment for a condition at an early stage which later turns out to be life threatening. Alternatively, it may lead to a patient being unnecessarily subject to costly and distressing treatment. High-quality computer-based decision support has the potential for being an invaluable check on the integrity of medical decision-making, provided it can demonstrably guard against the errors and biases which experts may be subject to. This is the claim that is made by those that are developing Bayesian expert systems for medicine. In order to give some idea of the functionality of such systems, we will look at one, MUNIN, in more detail.

MUNIN (for MUscle and Nerve Inference Network) was developed for use in the specific domain of electromyography, to assist in the diagnosis of neuromuscular disorders (Andreassen et al., 1987). Electromyography involves the

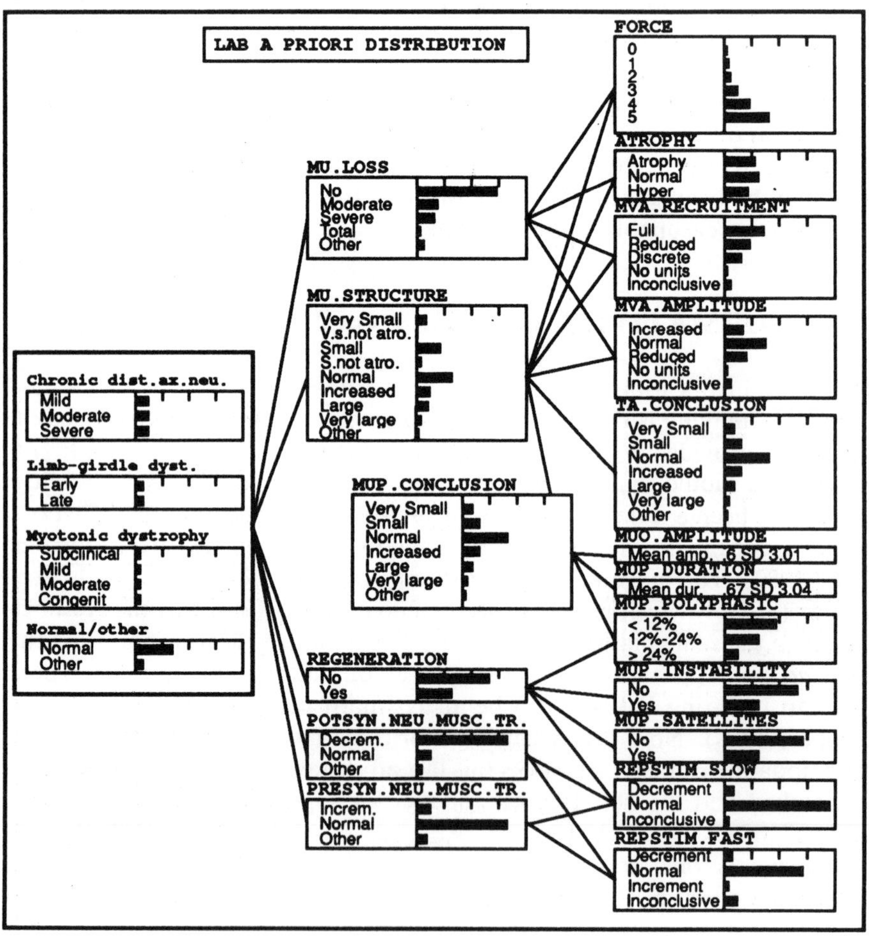

Figure 20.10 The *a priori* probabilities for individual disease states (left) are propagated through the layer of pathophysiological nodes (middle) to the findings nodes (right). The length of the bar indicates the probability of the corresponding node state (simplified from Andreassen *et al.*, 1987).

diagnosis of muscle and nerve diseases through the analysis of bioelectrical signals from the affected muscle and nerve tissues. The examples we will use are derived from a network modelling a single muscle. The final system comprises many such networks.

A slightly simplified MUNIN model for a single muscle is shown in Figure 20.10. The lengths of the horizontal bars indicate the probabilities of the various states of the nodes—the prior distribution for the network. This model has three levels. The disease node represents the possible neuromuscular states.

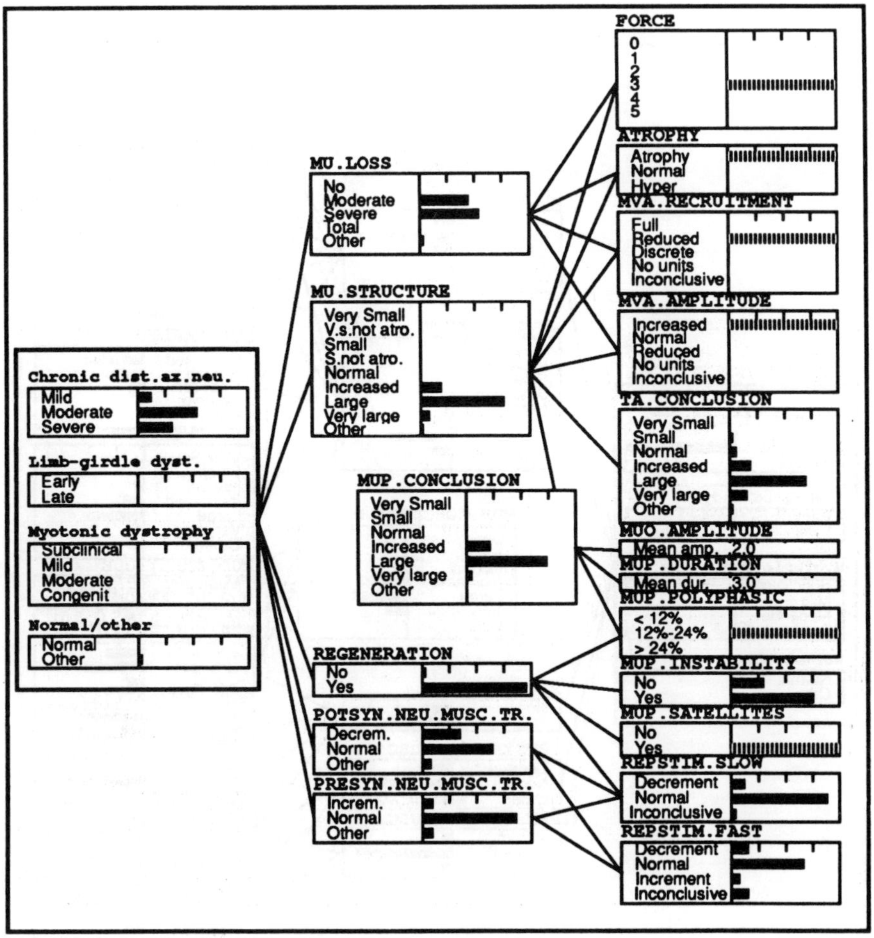

Figure 20.11 Findings corresponding to a typical case of "moderate chronic axonal neuropathy" have been entered. The broken horizontal 100% bars correspond to the entered findings (simplified from Andreassen *et al.*, 1987).

(Notice the state "other", which corresponds to support for a neuromuscular disorder other than normality or one of the three mentioned diseases.) The disease node is then linked to nodes representing possible, pathophysiological conditions. These are the physical manifestations of the underlying diseases. The pathophysiological disorders are difficult or impossible to observe directly and non-invasively, but they can be deduced indirectly through the electromyographical tests. The third layer of the model consists of the findings nodes for such tests. The pathophysiological nodes may be linked to the findings nodes indirectly through intermediate nodes which integrate information from several of the findings nodes.

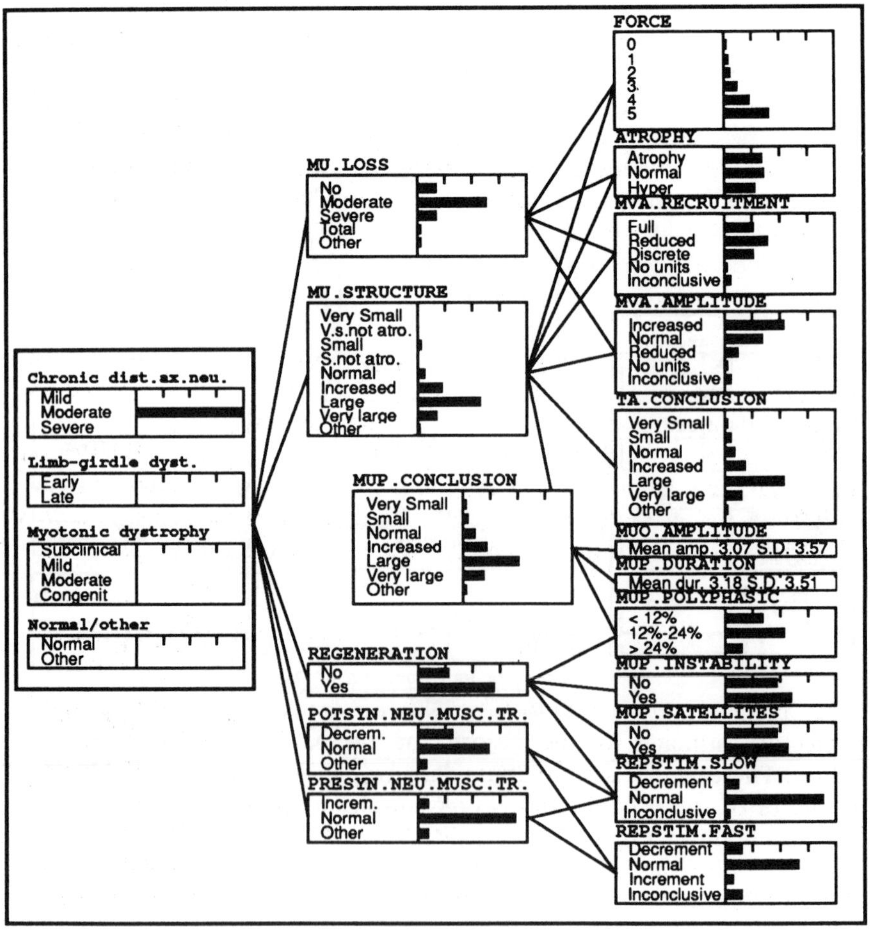

Figure 20.12 Expectations corresponding to "moderate chronic axonal neuropathy" (simplified from Andreassen *et al.*, 1987).

Construction of the model required medical knowledge of the domain to be employed in three distinct tasks. First, the number and character of the nodes needed to be chosen. Note, for example, that very few of the nodes have binary states corresponding to yes/no answers. Secondly the "causal" links had to be assigned between the various nodes. Then, finally, the prior and conditional probabilities had to be assessed for the disease states and the links between the node states respectively. Each of these is a nontrival task and requires the development of effective elicitation and validation methods. Software tools are under development to support the effective elicitation of the network and the probabilities. For example Heckerman (1991) describes the use of "similarity networks" as a tool for building the network structure, and "partitions" as a tool for assessing the associated probabilities.

Having constructed the network and elicited the required probabilities, the findings for a specific case may be entered. MUNIN's use of the Lauritzen–Spiegelhalter algorithm enables the findings to be entered and their effects to be propagated through the network interactively. In Figure 20.11 the findings of a hypothetical case have been entered. These findings result in high probabilities for the patient suffering from moderate or severe chronic axonal neuropathy (dashed lines indicate nodes which have been conditioned on). Notice also that the expected values for several of the so far unobserved findings nodes have also changed from Figure 20.10 (see for example, the expected values for "TA.CONCLUSION").

The network may also be used for hypothetical reasoning. This is illustrated in Figure 20.12, where the system has been consulted for the expected findings given a moderate case of chronic axonal neuropathy.

The technology underlying MUNIN was developed over the latter part of the 1980s and is now available commercially in the expert system shell HUGIN. This technology is still undergoing many refinements, one of which is to introduce a learning capability. We will discuss this next.

20.7 TURNING "SUBJECTIVE" TO "OBJECTIVE"

Most probabilistic expert systems will be dependent on the elicitation of subjective estimates for the majority of the required conditional probability values. There simply is not the case data available to produce reliable statistical estimates of the probabilities. For example, a complete specification of the FORCE node of Figure 20.10, $p(\text{FORCE} \mid \text{MU.LOSS, MU.STRUCTURE})$, would require the elicitation of $6 \times 5 \times 9 = 270$ probabilities to cover all combinations of the states of the three nodes. At least 10 000 cases would be required to generate reliable objective estimates of these probabilities. This is simply untenable. Instead, the MUNIN team used subjective estimates which were critiqued using an underlying "deep knowledge model" derived from current understanding of the pathophysiological processes involved (Andreassen et al., 1987).

However, once a system is in use, case data will clearly become available. If the true state of a patient, say, is eventually ascertained, this can be used to critique the state predicted by the system. Essentially, one can exploit "the ability of Bayesian probabilistic reasoning to become Bayesian statistical reasoning" (Spiegelhalter & Cowell, 1991); as case data passes through the system, the information can be used to revise the parameters, the probabilities, of the network. This is illustrated in a system which is being developed for diagnosing congenital heart disease (Spiegelhalter et al., 1992).

Congenital heart disease requires rapid and accurate diagnosis. Its effects can be immediately apparent at birth, resulting in cyanosis ("blue" babies) or heart failure (breathlessness). In the hope of improving the reliability and speed of diagnosis, the Great Ormond Street Hospital for Sick Children (a major referral centre for congenital heart disease in the South East of England) has become actively involved in the development of the probabilistic expert system referred to in the previous paragraph.

This system is based on a five-layer model (Figure 20.13). Two *risk factors* directly influence the likelihood of specific *diseases*. During diagnosis, reported *clinical features* may differ from *true clinical features* due to observer

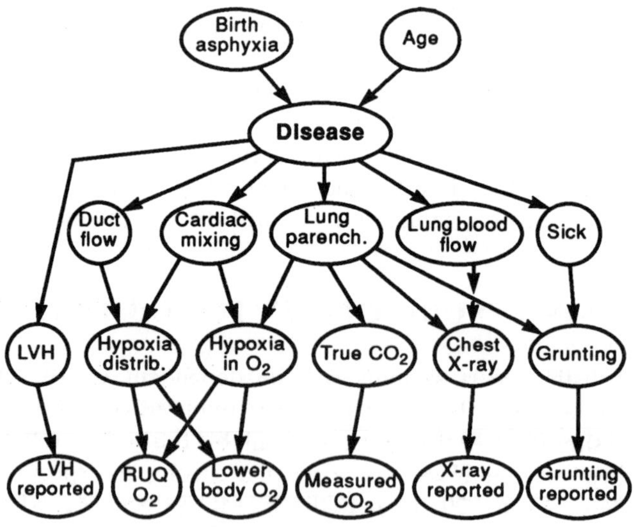

Figure 20.13 Five layer model for the diagnosis of "blue" babies. The top layer represents risk factors. The disease itself is manifest as a pathophysiological disturbance (third layer). These produce clinical features (fourth layer) which may differ from the reported clinical features (fifth layer) (after Spiegelhalter *et al.*, 1991, by permission of David Spiegelhalter.)

error. The true clinical features reflect specific *physiological disturbances*, which in turn are caused by the specific diseases of interest.

This graphical model was constructed in consultation with consultant paediatric cardiologists at Great Ormond Street Hospital. The next step was to elicit the probabilities for the model. A large amount of case data was available, but still not enough to provide reliable estimates for all the required values. So, as with MUNIN, extensive tables of subjective probabilities had to be elicited. In many cases the experts were not prepared to commit to a point value, and so the probability was specified as a range of values. For example, it was believed that 80–90% of cases of lung disease would exhibit "grunting". As we shall see, the algorithm for updating these probabilities on the basis of case data relies on some simple assumptions about the interpretation of this interval being made.

In the above example, the expert's opinion was that the proportion of cases of lung disease ($1d$) which they would expect to exhibit grunting (g) was between 80 and 90%. This has a mean value of 85%, so $p(g \mid 1d)$ is taken to be 0.85. Subjective probability may be interpreted as an estimate of the frequency of occurrence obtained from an implicit population of cases underlying the expert's experience. The imprecision in the subjective probability then corresponds to a prior distribution over the domain of possible frequencies; the more precise estimates reflecting a larger implicit sample size.

It is straightforward to estimate the implicit population size. We make the assumption that the range either side of the mean value may be taken to represent one standard deviation, and that the expected value has a Gaussian distribution about the mean. Then standard binomial theory for a binary state variable gives

$$\sigma = \sqrt{\frac{p(1-p)}{n}}$$

Here, σ is standard deviation, p is the mean value and n is the implicit population size.

For the example of grunting given lung disease, this gives $n \approx 50$. The expert's opinion is then interpreted as though $p(g \mid 1d) = 42.5/50$. It is simple to update this value as data passes through the system. Suppose a baby is now admitted to the hospital who turns out to have lung disease, but did not exhibit grunting. It may then be said that 42.5 cases exhibit grunting (unchanged) out of a population of 51 (changed). That is, $p(g \mid 1d) = 42.5/51 = 0.83$. This is a small change, but as real data accumulates the revised probabilities will converge towards the "true" (objective) value.

This unfortunately is still not very satisfactory. The trouble is that if the expert's judgement was very much in error, it can take a long time for the revised estimates to stabilize at a final objective value. Although many of the probabilities elicited for the Great Ormond Street system were in good agreement with the later case data, it turned out that the above probability for grunting was not. Of sixteen babies with lung disease, only four exhibited grunting. Using the above learning technique, the revised value after these sixteen cases was $p(g \mid 1d) = 46.5/66 = 0.7$; still a long way from the much lower value which seems to be suggested by the case data. Intuitively it would appear at this stage that the expert's prior belief had been widely in error and that more reliable results would be obtained if this value were rejected in favour of a more objective reference prior. Figure 20.14 shows the revisions of the expert's prior belief over the sixteen cases. It also shows the revisions that would have been made if the expert's judgement had been rejected in favour of the reference value of $p(g \mid 1d) = 0.5$, with an implicit sample size of 1.

It is clear that in the case of grunting given lung disease the reference prior turned out to be the better value to take as a starting point for the learning algorithm. Significance tests have been developed which provide a measure of the discrepancy between the expert's assessment and the observed data (Spiegelhalter & Lauritzen, 1990; Spiegelhalter & Cowell, 1991). This provides a formal basis for the rapid rejection of the expert's prior assessment in favour of a reference value which would provide better predictions.

It is quite possible that modifications to the graphical structure of the model may have more influence on improving the reliability of the predictions than identifying inaccuracies in the numerical coefficients. Work is also underway to develop techniques for critiquing and systematically modifying the structure of the model.

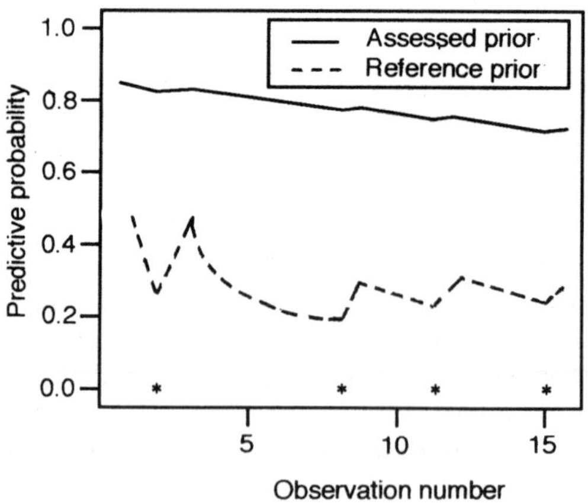

Figure 20.14 Revisions of predictive probabilities that next case of lung disease be reported to grunt. Starting point for the top curve is assessed prior, that for the bottom curve is the reference prior. Asterisks mark positive observations (after Spiegelhalter *et al.*, 1991b, by permission of David Spiegelhalter.)

20.8 IMPRECISION AND UNCERTAINTY

In the introductory section we drew a distinction between uncertainty and imprecision. Two alternative uncertainty calculi, Dempster–Shafer theory (Smets, 1988), and possibility theory (Dubois and Prade, 1988), augment a numerical calculus with a set-theoretic component in order to incorporate a mechanism for handling imprecision. An analysis of the respective roles of the set-theoretic and the numerical components of these calculi has been performed in Kruse, Schwecke & Heinsohn, (1991) and we will briefly outline this work here to give an indication of how the expressiveness of probability may be enhanced by these more recently developed calculi.

By imprecision is meant that the exact value ω_0 of some datum is not known, but is restricted to some subset which covers ω_0. We gave the example that whilst we may not know the precise location of a ship, we may be able to say it is "at sea" (the set of all possible locations in the oceans). By uncertainty it is meant that the validity of the datum has to be determined. That is, how much do we believe the proposition "the ship is at sea"? If Ω is the frame of discernment (the set of all possible outcomes) then imprecision is modelled by allowing data to be expressed as subsets of the frame of discernment, while uncertainty induces a mass assignment m to subsets, where $m: 2^{\Omega} \rightarrow [0,1]$.

First consider the case of precise data. Here we may consider a set of sensors or agents Θ as providing data as elements of the frame of discernment Ω; $\Gamma: \Theta \rightarrow \Omega$ where Γ is an "observation mapping". The sensors may be unreliable. If each sensor has an associated probability of giving the correct value, then the observation mapping Γ will map this to a probability distribution over Ω. Suppose now the sensors can only give imprecise data. That is, they can only return a subset which covers the true value of the datum of interest. In this case the observation mapping maps sensors to elements of the power set of Ω; $\Gamma: \Theta \rightarrow 2^{\Omega}$. Then Γ maps the probability assigned to the sensors to (probability) mass assignments on subsets of the frame of discernment.

A further refinement is to allow not just imprecise, but vague concepts. In the case of "the ship is at sea" the set of possible values for the datum of interest (the location of the ship) has "crisp" boundaries. That is, there is an equal possibility of finding the ship at any of the locations in "the sea", and zero possibility of finding it anywhere else. By convention, the possibility of the ship being at sea is taken to be unity. In the case of vague data, the range of possible values for a datum is constrained by a fuzzy set, not a crisp set. For example, if we say "Paul is of medium height" we are still offering a distribution of possible values for Paul's height. But here we are imposing a grading of the possibilities. We are not saying that Paul's height could take any value between 1.65 m and 1.80 m, say, but that there is a small possibility of his height being 1.65 m, a high possibility of its being 1.73 m and again a low possibility of its being 1.80 m (we would normally assign some numerical grading to the possible values).

In Kruse, Schwecke & Heinsohn (1991) vagueness is modelled using "L-sets". L-sets arise as a mathematical representation for the case when the identification problem can be considered from a number of different points of view, or *contexts*. An imprecise description of the datum $\omega_0 \in \Omega$ of interest is obtained in each of these contexts. So the sensors return a class of subsets instead of a single subset; Γ: $\rightarrow \mathfrak{F}_L(\Omega)$, where $\mathfrak{F}_L(\Omega)$ is the set of layered sets of W. For the case of vague data Kruse et al. introduce the notion of a generalized mass distribution in which the evidence mass may be assigned to elements of $\mathfrak{F}_L(\Omega)$.

The following completes a very crude outline of Kruse *et al.*,'s model. Let Θ be a set of sensors or agents, $\mathcal{P}$ a class of probabilities on Θ. Then if Γ is an observation mapping from Θ to Ω, 2^Ω, or $\mathfrak{F}_L(\Omega)$, where Ω is the frame of discernment, the pair ($\mathcal{P}$, Γ) is an information source. One needs now to consider the dynamic aspects of the model; methods of updating the mass assignments derived from an information source in the light of additional information. In the case where $\Gamma:\Theta \rightarrow 2^\Omega$, the combination of evidence from two independent information sources corresponds to Dempster's rule of combination. As well as *combining* information sources, methods of *updating* information sources are also considered in two basic forms; conditioning and data revision. The first is the case of integrating absolutely reliable but imprecise evidence. That is, the datum of interest is known with certainty to lie in a subset of E of Ω, and the mass distribution derived from an information source must be revised accordingly. The second case corresponds to an actual revision of the observation mapping Γ.

Now, Bayesian probability, Dempster–Shafer theory and possibility theory are all mathematical models for reasoning under partial ignorance. They all have well-defined properties and well-understood behaviours. We have tried to give a brief and rapid introduction to the aspects of ignorance which the Dempster–Shafer and possibility theories particularly address because we firmly believe that probability theory does not *a priori* have a case for being the only valid model. It is a matter of choosing the model whose properties and behaviour best suits the application to hand.

20.9 DISCUSSION

In the late 1970s and early 1980s it began to seem that there was a fundamental conflict between the computational requirements for effective AI applications and the demands for the full semantic expressiveness of probabilistic reasoning. What we have tried to show is that it is possible for AI system builders to have their cake and eat it; a great deal of recent work has addressed the knowledge elicitation problems and computational problems inherent in fully intentional probabilistic systems. The computational algorithms are not completely general, there are theoretical grounds for believing that no such algorithm is possible (Cooper, 1990), but it looks as though they can be applied to a wide range of problems.

We have emphasized that using a Bayesian inference model involves two steps:

(1) construction of the relevant qualitative inference network, and
(2) elicitation of the relevant prior and conditional probabilities within this structure.

Having constructed the network, the Bayesian calculus imposes a strict discipline on the knowledge engineering aspect of eliciting a coherent set of probabilities. This discipline can help in the elimination of many of the biases involved with the elicitation of subjective probabilities which have been discussed in the literature (e.g. von Winterfeldt and Edwards, 1986). In addition, we have discussed how over- or underconfident estimates of probabilities can be rapidly identified and revised as case data becomes available.

Critiquing the actual network may well prove harder to place on a more formal basis. The network should contain face validity; that is, it should appear to the end user to represent meaningful relationships between the semantic entities it contains. However, how can we be sure that all the relevant nodes have been included? For example, in Figure 20.13, while *reports* of clinical features may have a strong statistical association with their corresponding clinical features, they are not categorical indicants of them. Especially if the observation of certain signs contained a large subjective element, it was found (empirically) that the report and the actual occurrence of clinical features should be separated out. A prototype of the network of Figure 20.13 which did not include the additional nodes relating to reports had a much lower predictive accuracy than the final model (Spiegelhalter, personal communication). In this particular case, the refinement of the network arose during the validation phase. However, as we have mentioned in Section 20.6, work is under way to develop tools to support the effective elicitation of the network (Heckerman, 1991).

Although we have focused on many recent advances, this is not to say that there are not still problems with the use of probabilistic reasoning in AI systems. There are domains where the use of numerical uncertainty coefficients has been "officially" objected to, as conferring an unwarranted degree of confidence in the values assigned to the possible outcomes (DOH, 1992). A full probabilistic solution also requires the recompilation of the belief network, whereas there are problems where it is highly desirable to construct and modify the belief network "on the fly" as the solution progresses and more information becomes available. And finally, as we have mentioned, there are cases where aspects of imprecision and vagueness may be more effectively addressed with one of the alternative calculi (Krause and Clark, 1993, give a comprehensive review of the various alternative symbolic and numerical approaches to uncertainty handling). The application of probability in AI systems has seen some very impressive recent developments. But uncertainty handling in AI is most certainly *not* a closed book.

ACKNOWLEDGEMENTS

Many people have kindly discussed aspects of the work reviewed in this chapter with us. We would especially like to thank David Spiegelhalter for freely supplying information and for comments on early drafts of this chapter.

NOTE

(1) Here and in the following we use capital letters (e.g. A) to denote variables which may be instantiated with an arbitrary proposition, and lower-case letters to denote the actual states of the proposition (e.g. "a" for A is true; "$\neg a$" for A is false).

REFERENCES

Andreassen, S, Woldbye, M., Falck, B. & Andersen, S. (1987) MUNIN—A Causal Probabilistic Network for Interpretation of Electromyographic Findings, In *Proceedings of the 10th IJCAI*, Milan, Italy, pages 366–72.

Clark, D. (1990) Numerical and symbolic approaches to uncertainty management in AI, *Artificial Intelligence Review*, **4**, 109–46.

Cooper, G. (1990) The Computational Complexity of probabilistic inference using Bayesian belief networks, *Artificial Intelligence*, **42**, 393–405.

de Finetti, B. (1937) La prévision: Ses lois logiques, ses sources objectives, *Annales de l'Institut Henri Poincaré*, **7**, 1–68 Translated in H. Kyberg & H. Smokler (1964) *Studies in Subjective Probability*, Wiley, New York.

DOH (1992) *Report of Committee on Carcinogenicity of Chemicals in Food*, Department of Health.

Dubois, D. & Prade, H. (1988) *Possibility Theory: An Approach to Computerised Processing of Uncertainty*, Plenum Press, New York.

Gashnig, J. (1982) PROSPECTOR: an expert system for mineral exploration. In D. Michie, (ed.), *Expert Systems in the Microelectronic Age*, Edinburgh University Press, Edinburgh, pages 153–67.

Geiger, D. & Pearl, J. (1988) On the logic of causal models. In *Proceedings of the Fourth Workshop on Uncertainty in Artificial Intelligence*, Minneapolis, MN, pages 136–47.

Heckerman, D. (1991) *Probabilistic Similarity Networks*, MIT Press, Cambridge, MA.

Kahneman, D., Slovic, P. & Tversky, A. (eds.), (1982) *Judgement Under Uncertainty: Heuristics and Biases*, Cambridge University Press, Cambridge.

Krause, P. & Clark, D. (1993) *Representing Uncertain Knowledge: An Artificial Intelligence Approach*, Intellect, Oxford.
Kruse, R., Schwecke, E. & Heinsohn, J. (1991) *Uncertainty and Vagueness in Knowledge Based Systems*, Springer, Heidelberg.
Lauritzen, S.L. & Spiegelhalter, D. (1988) Local computations with probabilities on graphical structures and their application to expert systems. *Journal of the Royal Statistical Society*, **B50**, 157–224.
Neapolitan, R. (1990) *Probabilistic Reasoning in Expert Systems*. Wiley, Chichester.
Pearl, J. (1988) *Probabilistic Reasoning in Intelligent Systems*. Morgan-Kaufmann, San-Mateo, CA.
Savage, L. (1962) *The Foundations of Statistical Inference*. Wiley, Chichester.
Shortliffe, E. (1976) *Computer Based Medical Consultations*: MYCIN. Elsevier, New York.
Smets, P. (1988) Belief functions. In P. Smets, E. Mamdani, D. Dubois & H. Prade, *Non-standard Logics for Automated Reasoning*, Academic Press, New York, pages 253–86.
Spiegelhalter, D. (1986) Probabilistic Reasoning in Predictive Expert Systems. In L. Kanal and J. Lemmer (eds.), *Uncertainty in Artificial Intelligence*. Elsevier, North-Holland, Amsterdam, pages 47–67.
Spiegelhalter, D. & Cowell, R. (1991) Learning in Probabilistic Expert Systems, *Proceedings of the Fourth International Meeting on Bayesian Statistics*, Peñiscola, Spain.
Spiegelhalter, D. & Lauritzen, S. (1990) Sequential updating of conditional probabilities on directed graphical structures. *Networks*, **20**, 579–605.
Spiegelhalter, D., Harris, N., Bull, K. & Franklin, R. (1991) Empirical evaluation of prior beliefs about frequencies: methodology and a case study in congenital heart disease. BAIES Report BR-24, MRC. Biostatistics Unit, Cambridge, England.
Tversky, A. & Kahneman. D. (1974) Judgement under uncertainty: Heuristics and biases, *Science*, **185**, 1124–31.
Verma, T. & Pearl, J. (1988) Causal Networks: Semantics and Expressiveness. *Proceedings of the Fourth Workshop on Uncertainty in Artificial Intelligence*, Minneapolis, MN, pages 352–9.
von Winterfeldt, D. & Edwards, W. (1986) *Decision Analysis and Behavioural Research*. Cambridge University Press, Cambridge.
Wermuth, N. & Lauritzen, S. (1983) Graphical and recursive models for contingency tables. *Biometrika*, **70**, 537–52.

Chapter 21

The Subjective Probability of Guilt

Willem A. Wagenaar
Leiden University, The Netherlands

In all systems of criminal law judges[1] apply some criterion of "beyond reasonable doubt". The notion is formulated differently in various countries, but the principle is the same: judges seek a level of subjective probability that is high enough according to some criterion, without reaching absolute certainty. There is a paradox here: if judges are not absolutely certain, this must mean that there is a logical possibility that the accused is innocent; why, then, is this possibility not a reason for doubt? The answer is that the logical possibility is not *reasonable*, by which is meant that reasonable assumptions reduce the probability of innocence to zero. Since the law does not provide a definition of what assumptions are reasonable or unreasonable, it is obvious that the criterion of *reasonable* doubt is highly subjective. How, then, do judges deal with this important task of probability assessment and probabilistic decision-making? I will discuss some of the theories that have been put forward, and illustrate them with some examples from my own practice as a court expert on problems of memory and identification.

21.1 THE LEGAL MODEL

Characteristically, since the judgement is left to the subjective opinion of

Subjective Probability. Edited by G. Wright and P. Ayton.
© 1994 John Wiley & Sons Ltd.

individual judges, the legal community has never attempted to formulate a theory that prescribes a judge's probabilistic thinking. In fact the still leading theory of proof, formulated by Wigmore (1937) assumes that the problem of probability can be evaded altogether by a painstaking process of specification. Assume, for instance, that an accused is charged with theft. Theft is a legal notion that applies when at least four conditions are met:

- property,
- belonging wholly or partially to another person,
- is taken,
- with the intent to appropriate it unlawfully.

The facts needed to establish theft follow directly from this specification: the stolen matter should be *property*, it should *belong* to *another person*, etc. However, the establishment of fact is not simply changed into a counting-off procedure, because it is not clear what notions like "property", "belonging", or "another person" mean. What is property? A pound of sugar and a car may be property; but what about a song, a computer program, an idea? Lawyers have solved such problems, again through specification. Something qualifies as property when a number of further conditions are met. Likewise one may check whether all conditions for "belonging" are fulfilled. But within the process of specification we are very likely to encounter new problems, because the terms require still further specification. Thus, the process of proof is turned into an almost endless regression of embedded conditions that are checked one by one. Wigmore described this process as the *probandum* (e.g. theft), which is specified by splitting it up into a large number of *facta probanda* (the conditions), which are finally matched with *facta probantia* (the evidence). It is believed that legal proof can reach any required level of precision, simply through an extension of the specification process.

I would argue that this last conclusion is incorrect. The regression of conditions only *postpones* the problem; it does not *solve* it. At the end of the specification process there is still the task of matching a condition with evidence. It is not clear that any degree of specification will guarantee that we can establish with full certainty whether or not a condition is met. Take the example of "belonging". One may claim that having paid for a TV-set establishes ownership. But how do we know that a person paid for the TV-set? He may produce a receipt, but how do we know it is genuine, and how do we know the receipt was for this particular set? The set may have a serial number, but these are easy to fake. The shopowner may testify that he sold this particular set to this particular client, but witnesses may be mistaken, or may even be lying. At some point the specification process must stop; further questions are barred, and it is assumed that the evidence proves the fulfilment of a condition. The law provides such stop rules; an example is the *unis testis* rule, which says that one witness is not enough, implying that *two* independent concordant testimonies prove a fact. But the testimony of two witnesses is not *always* treated as indisputable; hence the acceptance of the stop rule is a decision based upon a subjective certainty criterion. More generally, Wigmore's process of matching *facta probanda* with *facta probantia* does not exclude the role of subjective probability.

21.2 REVISION OF OPINION

Statisticians and psychologists have proposed an alternative model of courtroom decision-making: the Bayesian revision-of-opinion process. The basic idea is that the judge starts with some degree of belief in the accused person's guilt. The *ratio* of the degrees of belief in guilt and innocence is taken as the relevant indicator, on which the verdict is founded. Initially this ratio is assumed to be very low, because of the much acclaimed "presumption of innocence". Then prosecution and defence contribute evidence; the court determines the diagnostic value of each piece of evidence and revises its previous opinion according to the formula:

$$\text{new ratio} = \text{old ratio} \times \text{diagnostic value} \qquad (21.1)$$

When, at the end of the trial, the revised ratio has transgressed a certain predefined certainty criterion, the accused is declared guilty. This model, based on Bayes' rule for the revision of opinion (1763), reflects in a way the normal procedure of hypothesis testing that is adopted in scientific research. This may explain why scientists tend to prescribe it as a normative model for courtroom decision-making, even though there are substantial theoretical and practical difficulties. I will summarize the problems under four headings:

- the initial opinion
- the diagnostic value of evidence
- the revision process
- the decision criterion.

The initial opinion. Presumption of innocence means that the judge will assume the innocence of the accused, unless proof to the contrary is provided. But, in Formula 1, the prior belief in guilt cannot be zero, because otherwise the multiplication will have no effect. In principle a judge should accept the theoretical possibility that the accused is guilty indeed. How strong can this initial belief be? In fact there are good reasons to assume that it can be quite strong. One reason is that in most countries the base rate of guilt is around 90% from a person who stands trial. Is it unnatural for a judge to assume that this base rate applies also to individual cases? Another reason that applies in some countries, is that judges read the file before the beginning of the actual trial. This has a large effect on the final decision, as was demonstrated in a series of studies by Schünemann and his coworkers. In one of these studies (Schünemann, 1983) two groups of professional judges were used as subjects. The first group did not read the file before the trial, the other group did. In both conditions the judges decided without juries. All of those who read the file before the trial convicted the accused; in the second group only 27% of the judges convicted. All judges received the same information at the trial. But the first group was biased by the initial file which, as is normal in many countries, is filled exclusively by the prosecution. What remains of the presumption of innocence, then?

A practical illustration of the problem is encountered in a case in which I served as an expert witness on memory problems. Danny Rijkbloem was accused of shooting the father of his girlfriend Nicole. Nicole's parents had convinced her that she should end her relationship with Danny, because he was a violent man and a criminal. Nicole and her parents went to Danny's house, to collect her clothes. There was an argument, but it is unclear what happened. Nicole and her mother stated that Danny got a gun and shot the father from a distance of 50 cm. Danny claimed that the mother drew a small gun from her handbag, and pointed it at him. He "experienced that as unpleasant", and hit her arm. The gun went off, and the father was hit by accident, from a distance of 2 m. What are the prior odds that Danny killed Nicole's father? One estimate would be 1/15 000 000: there are fifteen million people living in my country, and the presumption of innocence means that Danny is as innocent as anybody else. But that would be rather silly: there were only four people present, one of whom was the victim. There are only two possible suspects, hence the prior odds are 1 to 1! Still another choice can be based upon the empirical fact that about 95% of the defendants in my country are actually guilty, which poses the odds at 19 to 1. Bayesian revision of opinion does not deal with this problem, although the Swedish psychologist Goldsmith (1980) suggested a solution. In his "evidentiary value theory" it is proposed that judges determine the degree to which evidence supports the indictment; before the presentation of evidence this estimate is put at zero. But Goldsmith's model suffers from the further drawbacks, listed below.

Diagnosticity. How does a judge determine the extent to which new evidence is diagnostic for guilt? In principle a performance table is needed, like the one presented in Table 21.1.

The diagnostic value of a confession is defined by the ratio of hits and false alarms; the diagnostic value of a denial by the ratio of correct rejections and misses. Wigmore (1970) claimed that confessions are more diagnostic than any other type of evidence. That belief may be the basis for the Anglo-Saxon rule that a confession provides complete proof of guilt. In my own country there is an explicit rule that a confession can*not* be accepted as complete proof. Empirical data, which may be used to support either view, are largely absent: we do not know how often confessions are true or false. But is it reasonable

Table 21.1 Performance table, representing the diagnosticity of confessions

The defendant was in reality:	The defendant Confessed	Denied
Guilty	Hit	Miss
Innocent	False alarm	Correct rejection

to assume that a confession is correct more often than a fingerprint test, a recognition test, being caught in the act of the crime, being accused by a large number of witnesses? There are many reasons why people may make false confessions: in order to protect the real perpetrator, in order to be tried for a lesser offence, because of police pressure or trickery, or simply because of a pathological tendency to confess to crimes never committed. A judge cannot know the general diagnosticity of a confession, and has in a specific case no clear basis for the estimation of diagnosticity.

The problem of diagnosticity may be solved in the law, as was done for confessions in the Anglo-Saxon countries: there the law states that confessions are fully diagnostic. But this does not really help, since confessions are only one kind of evidence. What about the diagnosticity of witness testimony, forensic tests, expert opinion, recognitions, the behaviour of the defendant in court, criminal records, etc.? The almost violent battles over the diagnosticity of the polygraph test (Carroll, 1988) and the anatomically correct dolls test constitute excellent demonstrations of the problem: judges cannot know the diagnosticity of that sort of evidence. In reality the diagnostic value of such tests may be quite low, as illustrated in Table 21.2.

The result looks good, but the diagnostic value of the outcome "sexually abused" is only $9/2 = 4.5$. If, for instance, the presumption of innocence is modelled by $1:20$ odds in favour of innocence, the revision on the basis of the dolls' test is to odds of $4.5/20$, still favouring innocence. A lot more is needed for conviction, but in the Netherlands defendants have frequently been convicted on the basis of this test only (cf. Rossen & Schuijer, 1992).

Table 21.2 Performance table for the anatomically correct dolls test (after Jampole & Weber, 1987)

The child was:	Expert opinion based on test Abused	Not abused
Sexually abused	9	1
Not sexually abused	2	8

Obviously judges were not aware of the low diagnosticity, or what is even more likely, were not thinking in terms of diagnosticity. This suggestion stems from the simple fact that judges rarely ask the experts about the reliability of their methods and the validity of their results. The diagnostic value of 4.5 can be put into context by comparing it to the diagnostic value of a well-organized lineup procedure, which can amount to about 15 (Wagenaar & Veefkind, 1992).

The textbook demonstration of diagnosticity problems is the case of *People v. Collins*, described by Finkelstein & Fairley (1970), and extensively discussed by Tribe (1971). The case concerned a robbery by a young woman with a blonde pigtail, who escaped in a yellow sports car, driven by a black man. The man had a moustache and a beard. A few days later the police arrested a couple that fitted the description precisely. The diagnostic value of the description was based on these estimated probabilities:

Yellow sports car	1/10
Moustache	1/4
Black man with beard	1/10
Woman with pigtail	1/10
Blonde woman	1/3
Interracial couple	1/1000

The joint probability of all these properties is one in twelve million. With the prior odds set at 1/20 in favour of innocence, the final odds are 600 000 to 1 in favour of guilt. The problems of this approach are obvious. How do we know all these statistics? Should we take statistics for the world, for the US, for Los Angeles, for the particular neighbourhood? How do we know that all these properties are independent: is having a moustache independent of having a beard? Do blonde women have pigtails more often than black women?

A more subtle problem is the role of a concentrated police hunt. The probability calculus applies when suspects have been caught for other reasons than the properties mentioned. It would be very surprising indeed if, for instance, a couple spending the stolen money happened to fit the description just by accident. But as soon as the police start a search for people fitting the description, the probabilistic reasoning is invalidated. Even if only one in twelve million couples fit the description, it can almost be guaranteed that several such couples can be found in California. Discovery of one such couple only demonstrates the thoroughness of the search, not the diagnosticity of the description.

Revision. The process of revision, as described by the Bayesian model, assumes at least five properties that are in fact quite unrealistic. The arguments here are partially based upon Jonathan Cohen's (1977) thorough analysis of the use of probabilities in legal decision making, which covers more than the Bayesian approach.

The first of these properties is *compensation*: a very diagnostic piece of exonerating evidence can be compensated by a number of less diagnostic facts pointing in the other direction. Take as an example the case of Danny Rijkbloem, mentioned above. There were four people in the room: Rijkbloem (the suspect), Mr Lammerts (the victim), Mrs Lammerts (the victim's wife), and Nicole (their daughter). The two women claimed that Rijkbloem had shot Mr Lammerts from a distance of 50 cm. Rijkbloem said that Mrs Rijkbloem had pointed a gun at him, which is why he hit her on the wrist; the gun went off by accident and Lammerts was hit from below and from a distance of over 2 m. The forensic report confirmed every detail of Danny's account; the version produced by the two women was near to impossible. But this highly diagnostic evidence was compensated by a number of less diagnostic pieces of information, such as Danny's known aggressiveness, the fact that Mrs Lammerts was not known to own a gun, and that an accidental hit is very unlikely. Is it reasonable to let one very diagnostic forensic report be cancelled by a host of less diagnostic facts?

A second property of Bayesian updating is the assumed *independence* of diagnostic values. Assume that Nicole had stated that Danny's gun had a price tag on it; a rather unlikely observation that does not, by itself, much increase the belief that she had seen a weapon in Danny's hands. A similar statement, made by the mother only, would have an equally small impact. But if both women describe the price tag, the seemingly insignificant observations may obtain an overwhelming significance: the probability of producing the same fantasy twice is very low. The independence problem can be avoided by a redefinition of facts. The two observations can be combined into a single fact: two witnesses producing the same odd detail. But the Bayesian model does not provide any rules for recombination, and it must be feared that in actual cases many or even most details obtain their diagnostic value only through their relationships with other, seemingly insignificant pieces of evidence. Diagnosticity is not an independent property of evidence, but is derived from the entire narrative context. The overwhelming reasoning problems that occur when it is attempted to put some of the facts in the context of other evidence is illustrated in an influential paper by Lindley (1971). Although he discusses one example only, the argument should be convincing to everybody: the assumption that judges or jury members are able to reason in this way is absurd.

A third property is that of *decomposition and recomposition*. The indictment is broken down into a large number of independent facts for which evidence is sought. The final probability, which is found by combination of all partial probabilities, does not reflect which elements in the decomposition contributed high or low probabilities. It makes no difference whether a high probability was contributed by recognitions in a lineup, relevant for the identity problem, or by the analysis of the body, relevant for the question of whether a murder was committed at all. Should not a judge, in case of doubt, besides the overall probability assessment, also consider the *sources* of his doubt?

The fourth property assumed by Bayesian inference is *transitivity* of probabilistic statements. "If *A* then probably *B*; if *B* then probably *C*" implies: "If *A* then probably *C*". Here is a simple example of a violation of transitivity. When I testify in court, I am probably in the Netherlands. It is also true that, when I am in the Netherlands, I am probably not testifying. Concatenation yields: when I testify in court I am probably not testifying. Similar problems occur in the legal setting. If Danny Rijkbloem belonged to the criminal scene, he was probably in the possession of a gun. If it can be shown that he possessed a gun, he is probably the one who shot Mr Lammerts, because he denied the possession of a gun. But the concatenation, with the exclusion of the possession and its denial, may be very misleading. It is not necessarily true that Danny was the most likely suspect, just because he belonged to the criminal scene.

The fifth property of Bayesian decision making is that it requires the formulation of an alternative hypothesis. Broadly speaking the alternative to the hypothesis of guilt is the hypothesis of innocence. But which scenarios will actually be considered in the case of Danny Rijkbloem? All scenarios in which Danny is innocent? Only the scenario proposed by Danny, in which Mrs Lammerts shot her own husband? Or, maybe, no scenario at all, because a *fixed* probability of innocence is assumed, which is the same across all trials? Normatively all scenarios should be considered; but that is practically impossible. Consideration of the defence's account of the facts would be a mistake, because many innocent defendants lie about what really happened, for instance in order to protect the real perpetrator, or to hide other crimes or offences, for which they are not charged. A simple example of this is the case of Gerrit Kraft. In a fight Hans Monks was injured in the bar owned by Gerrit and Tom Kraft. In order to protect Tom, Gerrit stated that Tom had not even been present at the fight. When both Gerrit and Tom were sentenced to six years in prison, Gerrit withdrew his statement: Tom had been present, and was actually the one who had kicked Monks. Most witnesses confirmed this version, but Gerrit's initial statement, after being proven false, boomeranged against him. The court, after proving one alternative theory false, refused to consider his later version.

The criterion. The final problem of the Bayesian model is the definition of the decision criterion. When are the odds in favour of guilt high enough for conviction? The model does not specify how such a criterion is chosen, nor whether the criterion is constant or flexible. Is the same amount of certainty needed in cases of shoplifting and mass murder? The law does not provide any clues here. Signal detection theory, on the other hand, prescribes that the criterion must be more lenient when the costs of Type-2 errors increase. Acquitting a suspected child abuser, an IRA terrorist, or a serial murderer, may have such enormous consequences that less evidence is required. This may lead to the paradoxical position that the more severe the punishment, the less evidence is needed. And indeed, it was found by Wagenaar, Van Koppen, & Crombag (1993, Chapter 4) that in practice Dutch judges may apply such a rationale. It is essential that a theory of courtroom decision-making provides a model of this aspect of the decision process; but the Bayesian model does not.

Many of the issues raised here are discussed in detail by Jonathan Koehler (1991a, 1991b, 1992), Koehler & Shaviro (1990) and Edwards (1991). The discussion is complex, and does not seem to converge to a single conclusion. But the worst of all problems posed by the Bayesian model of legal decision-making is that it conflicts with actual rulings of courts. This is more elaborately documented by Wagenaar, Van Koppen, & Crombag (1993). Here I will restrict the illustration to the case of Danny Rijkbloem, described above.

The starting position was not so good for Danny. He was living from petty crime, for which he was convicted before. He was known to be a violent man, and was on record for attacking Nicole before. He was charged in another murder case, in which he put forward the same defence, viz. being accused falsely by the real perpetrators. The Lammerts were middle class people, without any criminal record. All these facts were known to the court; in a Bayesian fashion the prior odds would be rather in favour of guilt. But then came the evidence.

- Danny's hands were tested for traces of gunpowder. They were clean. Danny had insisted, right at the scene of the crime, that the hands of Mrs Lammerts should be tested, but the police refused this, because she was the main witness.
- The murder weapon was never found. The two women said that Danny took it when he left the house to call the ambulance. Danny had been away for five minutes, and along his route to the phone booth an extensive search was executed. No weapon was found. Danny, on his part, had immediately insisted that the two women should be searched because one of them carried the gun, a small Derringer that fits easily in a purse. The police refused to do this.
- No shell was found, although the type of gun allegedly used by Danny throws shells. The explanation given was that Danny had also removed the shell. A small Derringer does not throw shells.
- The forensic report disproved the two women's account of the shooting, and supported Danny's version.

It is not easy to see how these facts, used in a Bayesian fashion, may increase the prior belief of guilt to such an extent that a reasonable decision criterion is surpassed. Still Danny was convicted. The evidence cited by the court (in my country the courts must give reasons for their verdicts) illustrates nicely what happened: the court did not combine all evidence into one verdict, but *selected* the evidence in favour of guilt, while evidence in favour of innocence was fully neglected. The selected evidence consisted of the statements by the two eyewitnesses (who themselves were the only other possible suspects) and the presence of a body. Nothing was said about the problems with the shell and the gun, nothing about the forensic report. What sort of theory can describe how the court decided to convict Danny Rijkbloem, with most evidence in favour of his innocence?

21.3 JUDGING THE PLAUSIBILITY OF NARRATIVES

The leading theory about how people judge the probability of a unique event assumes the application of a heuristic. The three best-known heuristics are availability, representativeness, and anchoring-and-adjustment (cf. Tversky & Kahnemann, 1974). Although all three heuristics are in principle applicable to the courtroom situation, availability has been used most often. The specific heuristic is called *scenario availability*; it assumes that the probability of an event is derived from how easy it is to construct or imagine a scenario that leads to the event. Applied to the probability of guilt, the heuristic means that guilt is judged to be more likely, the easier it is to think of a scenario that describes why and how the accused committed the crime. This scenario is in principle contained in the prosecution's indictment. The prosecution tells the story of what happened, sometimes even in a most dramatic manner (cf. Loftus & Ketcham, 1991, page 101), and the court decides whether that is a true story or not. Bennett & Feldman (1981) devoted an entire book to storytelling in the courtroom, and the process of judging the plausibility of stories; the extensions of the theory discussed in this chapter are slight compared to their monumental contribution.

In principle, Bennett & Feldman say that there is no real difference between the narrative qualities of true and false stories. There is no way to distinguish true stories from false stories, just on the basis of properties of the narrative. But there is a major difference between *good* and *bad* stories. This difference is crucial, because people tend to accept good stories as true, and to reject bad stories as false. Good stories have two important properties:

- a readily identifiable central action
- a context that provides an easy and natural explanation why the actors behaved in the way they did

In a good story all elements are connected to the central action; nothing sticks out on its own. The context provides a full and compelling account of why the central action happened. If the context does not achieve that effect, the story is said to contain *ambiguities*. There are two types of ambiguities: missing elements and contradictory elements. Here is an example of an ambiguous story used as a defence in a case in which I acted as an expert witness: the case of Haaknat.

There had been a bank robbery very early in the morning, and the robbers escaped, closely followed by the police. The police lost sight of the robbers, but a few minutes later they found Haaknat in the moat of a castle in Nieuwegein. The robbers had worn jogging suits, while Haaknat wore shorts and a thin T-shirt; but on the other side of the moat a jogging suit was found that fitted the description. The prosecution's story was that Haaknat was one of the robbers, that he took off the jogging suit, and hid it on the other side of the moat. Haaknat had a completely different story. He had come to Nieuwegein in order to meet his friend Benny, who owed him 300 guilders. Benny had shown up without the money, which is why Haaknat started a fight. Then he heard the sirens of the police cars chasing the robbers, which created the suggestion that the police were after the fighters. Hence Haaknat had run off and jumped into the moat. This is an ambiguous story for various reasons. There is no central theme: the fight between Haaknat and Benny happened completely independently of the robbery. There are also many missing elements: Why did Haaknat meet Benny so early in the morning? Why did Benny show up at all, if he did not have the money? Why did Haaknat not know Benny's last name or his address? How did the two make the appointment? Why in Nieuwegein, which was rather remote from where Haaknat lived? Why did he lend money to a person he hardly knew? There are also some contradictions: the jogging suit must have been taken to the other side of the moat, at the exact moment in which Haaknat took his jump, but he did not see anybody. Haaknat claimed to have travelled to Nieuwegein by public transport, in shorts and a T-shirt only; but it was cold. Because Haaknat's story is ambiguous, it is judged to be unlikely. The prosecutor's story is accepted as highly probable, because it explains all facts within the context of one leading theme: Haaknat took part in the robbery.

The acceptance of good stories as true representations rests on the tacit assumption that goodness is diagnostic for truth. In fact we do not know much about the diagnostic value of story goodness. There are even good reasons to believe that true stories contain many ambiguities, and the technique of *statement validity analysis* (Yuille, 1989) even uses ambiguities as a sign of truthfulness. This analysis employs 19 criteria that supposedly distinguish true stories from false ones. Criterion 2 is unstructured production of the story; Criterion 8 and 9: presence of unusual and irrelevant details; Criterion 10: wrongly understood details. Haaknat's story might score high on these criteria!

The theory of narrative plausibility was largely extended by Pennington & Hastie (1986, 1988, 1991); But their most relevant contribution is empirical.

One of their studies (Pennington & Hastie, 1986) deals with the order in which evidence is presented, a factor that should be irrelevant according to Bayesian theory. Prospective jury members were presented with two narratives, one by the prosecution and one by the defence. Both sides presented the elements of the narratives either in story order, or in random order. Then the subjects were asked whether the defendant should be convicted for first-degree murder. The results are presented in Table 21.3.

The table clearly indicates that presentation of the elements in story order is better believed, even though the evidence is the same. The effect can be as large as changing a 31% probability into a 78% probability. Clever presentation is half of the job; what is the other half?

Fortunately a court is not supposed to judge story plausibility to the exclusion of other indications. The law even specifies where these other indications should come from: evidence provided by the prosecution. A weak indictment story will not convince a judge, but a good story *should* not convince a judge if it is not supported by evidence for most or at least the most crucial details. The way in which evidence "proves" that a good story is true is treated in the recent work of Crombag, Van Koppen, and Wagenaar (1992), and Wagenaar, Van Koppen, and Crombag (1993). Their theory of *anchored narratives* specifies how evidence anchors good stories onto a firm ground of generally accepted beliefs.

The main idea of the theory of anchored narratives is that evidence itself is nothing more than still another narrative. The evidence provided by the two witnesses in the case against Danny Rijkbloem is a narrative that needs to be judged with respect to goodness. If it has a central action and a compelling context it may be a good story which is accepted as true; otherwise it is probably false. Thus the evidence anchors the indictment to the generally accepted belief that good stories, told by two witnesses under oath, are true. If the account of the two witnesses is doubted, it may be further supported by more evidence: for instance that Danny owned a gun, while Mrs Lammerts did not. But this evidence again consists of a narrative, told by witnesses; the truth of this narrative can be accepted only on the basis of a general belief, such as

Table 21.3 Effect on presentation of evidence in random order or story order, on percentage of convictions. (From Pennington & Hastie, 1986. Copyright 1986 American Psychological Association. Reprinted by permission)

	Defence:	
Prosecution:	Random order	Story order
Random order	63%	31%
Story order	78%	59%

that third parties, who have no special interest in the case, will not lie. The forensic expertise in the same case is, again, nothing but a narrative. The expert may explain that he examined the body, that no traces of powder were found near the wound, and that therefore the gun was fired from a distance of 2 m or more. That is a narrative, which we may believe because the expert is a trusted specialist from an established laboratory. But that consideration provides an anchor only if it is generally believed that trusted experts do not lie and do not make mistakes. Recent events in Great Britain have demonstrated that this belief is not always warranted. Hence we may ask further support, for instance through a close scrutiny of the expert's credentials, his research methods, and the outcomes of the actual tests. But again, the court will receive nothing but more narratives, which do not prove anything at all unless they are linked to an accepted belief.

Thus, according to the theory of anchored narratives, legal proof consists of a narrative, connected through more specific narratives, in a hierarchical ordering, to generally accepted beliefs. The judged probability of guilt is then determined by four factors.

- the plausibility of the indictment's narrative
- the number of elements in the narrative that are anchored by means of evidence
- the logical relationship between the evidence and the story detail that it is supposed to anchor
- the validity of the general beliefs to which the regressions of narratives are finally anchored.

An important element of *anchored narratives* is that the court is attempting to estimate only one quantity: the probability of guilt. The probability that the accused is innocent is not in discussion. The likelihood that the defendant's story is true is not considered. Almost no country in the whole world imposes the obligation to explain why the defendant's story is not true. Even in The Netherlands, where judges are obliged to give reasons for the rejection of an alternative story, it is sufficient to state that the story is not credible. The leading hypothesis is the prosecutor's narrative, the anchoring construction to be evaluated is the one proposed by the prosecution. Courts may simply neglect evidence that contradicts the narrative, if there is enough evidence in support of it. To give just one example: there is a universal habit to believe recognitions by a few witnesses in a lineup, even if the suspect was not recognized by a large majority of other witnesses who had an equally good opportunity to observe the perpetrator during the crime (cf. Wagenaar, 1988). In this way, anchoring is a process of verification, not falsification. Preference of verification to falsification is another much-studied heuristic (cf. Wason & Johnson-Laird, 1972), which has practical relevance because it keeps the reasoning process manageable.

21.4 TESTING BY GENERATION OF ANOMALIES

In the heuristics-and-biases tradition it is customary to test theories through the generation of anomalies. The application of a heuristic will in exceptional cases lead to an anomalous result. Even though such anomalies may be rare, if no other theory predicts them, their very existence favours the theory. An example is the prediction of the outcome of: $1 \times 2 \times 3 \times \cdots \times 8$ versus $8 \times 7 \times 6 \times \cdots \times 1$ (Tversky & Kahneman, 1974). The predicted outcome is much higher in the second task, which follows from the application of anchoring-and-adjustment; the three first digits provide an initial anchor, which is then adjusted for "five more". The anchor is higher in the second task, which explains why the outcome is higher. If no other theory predicts this anomaly, it does not matter whether the anomaly is rare or whether the task is unnatural.

The theory of anchored narratives has been tested in the same manner: through the prediction of anomalies (see Wagenaar, 1994). To that end a corpus of possibly anomalous cases was collected with the aid of various defence lawyers. One example of an anomaly is the possibility that a defendant is accused on the basis of a good story only, without any supporting evidence. An example is the case of Henkemans, accused of smuggling 16 kg heroin. He was observed at Amsterdam Airport, while he collected his suitcase with unknown contents. He checked in at a nearby hotel and made contact with two Chinese, who a little later left the hotel with what seemed to be the same suitcase. The contents of the suitcase were still unknown. The Chinese were followed to their home. Later on that same day the house was raided; 16 kg heroin was found in a hiding place. Henkemans was convicted for smuggling these 16 kg, although there is not the slightest indication about what was in his suitcase. Even if he had smuggled heroin, it is still unclear how much. This is important because in The Netherlands one is sentenced by the kilo! In Henkemans' case there is no proof that he committed a crime, no proof of criminal intent, and no proof of the identity of the person who smuggled the 16 kg into the country. The only real evidence was the prior suspicion against Henkemans, and his contact with two drug dealers. But added to that there was the prosecutor's story about how drugs are usually brought into the country, and Henkemans' incredible story about taking presents to friends of a friend in exchange for a free return ticket Bangkok–Amsterdam.

Another anomaly is a conviction based on an indictment, of which essential parts are not anchored through evidence. An example is the case of Helder, accused of murdering his wife and employer, old Mrs Kempers. He had married her six weeks before. On the night of her death he had administered her usual medicine and the equally usual glass of rum. The pathologist concluded that this mixture had killed her, but later this opinion was proved to be wrong by a much more detailed examination of the body. Nevertheless the courts concluded that Helder had killed Mrs Kempers, "because he strongly desired her death". The proof is rather complete with respect to who administered the medicine and the alcohol, but very incomplete with respect to the essential questions of whether Mrs Kempers died from an unnatural death, and whether Helder had any intention to kill her. These details of the story were only anchored through the obvious interest that Helder had in Mrs Kempers' death. Apparently this piece of evidence was anchored onto the belief that people don't die from natural causes when someone profits from their death.

A further anomaly, predicted by the theory of anchored narratives, is that statements by a minority of witnesses may be accepted as proof, even when the majority of subjects declare otherwise. An example is the case of Pico Rodriguez, accused of killing Gerrit Hoekman in a bar fight. Hoekman had been engaged in giving Pico's brother Vance a severe beating, earlier that night. Hence the indictment was that Pico had killed Hoekman in revenge. That would make the killing a first degree murder. But none of the 92(!) witnesses had actually seen who had stabbed Hoekman. Pico did not know Hoekman; hence the revenge theory applied only if Vance had pointed him out to Pico. Therefore the prosecution claimed that Vance and Pico entered the bar *together*. Nine witnesses confirmed that two black men had entered the bar simultaneously. One witness saw *three* black men. One witness saw two black men, but insisted that Vance was not one of the two. Ten witnesses said that they saw only one black man enter the bar. One witness, who had been outside, declared that Pico went in first, and that Vance had arrived later. Hence the prosecution narrative is supported by a minority of the witnesses; but the court believed their testimony and rejected the other statements. Why? Our explanation is that the revenge theory makes a good story. Pico' s story was weak and full of ambiguities. "He had entered alone, and made some random stabs in the air, just to frighten people. He did not recall stabbing anyone. Probably in the dense crowd Hoekman had fallen into the knife, or was pushed; it was a mere coincidence that his brother's attacker became the victim". Although nobody had seen how Hoekman died, so that there is no proof of first-degree murder, the battle of narratives was easily won by the prosecution.

A further anomaly is related to the belief in confessions. Even in The Netherlands, where it is decreed by law that confessions do not constitute complete legal proof, there is a tendency to rely on confessions that contain a good story. If the confession is in conflict with other evidence, the confession will be believed and the evidence rejected. An example is the case of Gremeling, who confessed to 27 cases of arson, spread over a period of five years. Gremeling suffered from a severe memory loss, due to a car accident. Still he detailed the 27 fires with exact dates, addresses, method used, etc. Some of his descriptions were demonstrably wrong; for some of the dates he had an alibi; and there was another convicted arsonist in the same village. Nevertheless the court used Gremeling's confession as a reason for his conviction, and did not comment on the contradictory evidence.

Of the large number of anomalies predicted by the theory of anchored narratives, I only mention one more: the trust in experts, even though the validity of their methods is unknown or proven to be very low. Bayesian updating presupposes a careful consideration of the diagnostic value of expert opinion. In the case of psychiatric expertise about repeated violence after release from prison, Bayesian updating leads to rejection of the testimony, because expert opinion on this issue was shown to be not at all diagnostic (cf. Monahan, 1989). The heuristic of anchored narratives, on the other hand, allows a judge to anchor the expertise to the assumed authority of the expert. The judge may ask questions about the experts' qualifications, and abstain from scrutinizing the validity of his statements. This will be done when the expert supports the leading narrative. An opinion that contradicts the narrative can simply be neglected, without any reason given. This happened in the case of Helder, cited above. The same may even be done when one part of the testimony fits the narrative, while another part is at odds with it. The prediction is that the expert will be believed and disbelieved simultaneously. An example is found in the case of Carroll, accused of being one of the IRA terrorists who murdered two Australian tourists in Roermond.

The evidence against Carroll consisted mainly of recognitions by two witnesses, who had seen one of the terrorists when they escaped in a car. The witnesses were on the second floor of their home in a narrow street and looked down into the car which passed at high speed. It was dark, they saw part of the man's face from the side for about 1.5 s, the car had tinted glasses. Later the witnesses saw a picture of Carroll on TV and in the newspaper. They did not call the police, although they claimed to have recognized the assassin, and were said to have given many useful tips in other criminal cases. Some weeks later the police tested them with a set of pictures that contained the same photograph of Carroll. One witness identified Carroll, the other did not, but claimed later to have recognized him anyway. I testified about the risk involved in photo-biased identifications; I explained to the court that the very poor conditions during the first exposition, and the repeated exposure to Carroll's picture with the positive message that this was an IRA terrorist, constituted an ideal opportunity for effects of misleading post-event information. At the end of my four-hour testimony one of the (professional) judges asked whether, presuming that the first recognitions were proven to be correct, additional identification tests were still needed. The answer is comprised in the presumption. When Carroll was convicted, it appeared that the proof rested solely on the recognitions and my testimony. The verdict said "The expert confirmed that additional identification tests were not needed". Thus four hours of testimony was rejected, while one statement of the same expert was promoted to legal proof.

21.5 CONCLUSION

Two models discussed in this chapter, the *legal* and the *Bayesian* model, cannot account for the anomalies that we encountered in our research (cf. Wagenaar, Van Koppen, and Crombag, 1993), some of which are described above. But such anomalies are in agreement with the notion that judges determine the probability of guilt by means of the heuristics that were proposed by Tversky, Kahneman, and their coworkers. Our theory of anchored narratives makes use of the *availability heuristic*, and of *confirmation bias*. As in many other demonstrations of heuristic thinking, it is not fair to call this kind of reasoning "suboptimal". There is probably no better way for judges to solve the ill-structured decision problems, so typical for the courtroom. But that does not mean that the result is always perfect. On the contrary, in rare occasions the anchored narratives heuristic will produce unacceptable anomalies. Legal procedure and the law itself should protect us against such anomalies. In most countries it does not. The reason is that legal procedure and the law are based upon the wrong psychological theory of probabilistic thinking. Correct insight into how judges and juries deal with probabilities may therefore, in the end, lead to better procedures and better laws.

NOTE

(1) "Judges" and "courts" will be used to denote any authority, consisting of one or more people, lay persons or professionals, who decide about the question of guilt in criminal cases.

REFERENCES

Bayes, T. (1763) Essay towards solving a problem in the doctrine of chances. *Philosophical Transactions of the Royal Society*, **53**, 370–418.

Bennett, W.L. & Feldman, M.S. (1981) *Reconstruction Reality in the Courtroom*. Tavistock, London.

Carroll, D. (1988) How accurate is polygraph lie detection? In A. Gale (ed.) *The Polygraph Test: Lies, Truth and Science*, Sage, London.

Cohen, L.J. (1977) *The Probable and the Provable*. Clarendon, Oxford.

Crombag, H.F.M., Van Koppen, P.J. & Wagenaar, W.A. (1992)) *Dubieuze Zaken: de Psychologie van Strafrechtelijk Bewijs*. Contact, Amsterdam.

Edwards, W. (1991) *Influence diagrams, Bayesian Imperialism and the Collins Case: An Appeal to Reason*. Social Science Research Institute, University of Southern California, Los Angeles.

Finkelstein, M.O. & Fairley W.B. (1970) A Bayesian approach to identification evidence. *Harvard Law Review*, **83**, 489–517.

Goldsmith, R.W. (1980) Studies of a model for evaluating judicial evidence. *Acta Psychologica*, **45**, 211–21.

Jampole, L. & Weber, M.K. (1987) An assessment of the behavior of sexually abused children with anatomically correct dolls. *Child Abuse and Neglect*, **11**, 187–92.

Koehler, J.J. (1991a) The normative status of base rates at trial. In N.J. Castellan (ed), *Individual and Group Decision Making*, Erlbaum, Hillsdale, NJ.

Koehler, J.J. (1991b) The probity/polity distinction in the statistical evidence debate. *Inlane Law Review*, **66**, 141–50.

Koehler, J.J. (1992) Probabilities in the courtroom: an evaluation of the objections and policies. In D.K. Kagehiro & W.S. Lanfer (eds.), *Handbook of Psychology and Law*. Springer, New York.

Koehler, J.J. & Shaviro, D.N. (1990) Veridical verdicts: increasing verdict accuracy through the use of overtly probabilistic evidence and methods. *Cornell Law Review*, **75**, 247–79.

Lindley, D.V. (1971) Probabilities and the law. In D. Wendt & C. Vlek (eds.), *Utility, Probability and Human Decision Making*. Reidel, Dordrecht, pages 223–32.

Loftus, E.F. & Ketcham, K. (1991) *Witness for the Defense: The Accused, the Eyewitness, and the Expert who Puts Memory on Trial*. St Martin, New York.

Monahan, J. (1989) Prediction of criminal behavior: Recent developments in research and policy in the United States. In H. Wegener, F. Lösel & J. Haisch (eds.), *Criminal Behavior and the Justice System: Psychological Perspectives*. Springer, New York.

Pennington, N. & Hastie, R. (1986) Evidence evaluation in complex decision making. *Journal of Personality and Social Psychology*, **51**, 242–58.

Pennington, N. & Hastie, R. (1988) Explanation-based decision making: Affects of memory structure, on judgment. *Journal of Experimental Psychology: Learning, Memory and Cognition*, **14**, 521–33.

Pennington, N.& Hastie, R. (1991) A theory of explanation-based decision-making. In G. Klein & J. Orasanu (eds.), *Decision-making in Complex Worlds*. Ablex, Hillsdale, NJ.

Rossen, B. & Schuijer, J. (eds.) (1992) *Het seksuele gevaar voor kinderen: Mythen en Feiten*. Swets & Zeitlinger, Amsterdam.

Schünemann, B. (1983) Experimentelle Untersuchungen zur Reform der Hauptverhandlung in Strafsachen. In H.J. Kerner, H. Kurry & K. Sessar (Hrgb.) *Deutsche Forschungen zur Kriminalitätsentstehen und Kriminalitätskontrolle*. Heymanns, Cologne.

Tribe, L.H. (1971) Trial by mathematics. *Harvard Law Review*, **84**, 1329–93.

Tversky, A. & Kahneman, D. (1974) Judgment under uncertainty: Heuristics and biases. *Science*, **185**, 1124–31.

Wagenaar, W.A. (1988) *Identifying Ivan*. Harvester Wheatsheaf, Hemel-Hempstead, Harvester University Press, Cambridge, MA.

Wagenaar, W.A. (1994) Anchored Narratives. A theory of judicial reasoning, and its consequences. Proceedings of the Third European Conference on Law and Psychology. (In press).

Wagenaar, W.A. & Veefkind, N. (1992) Comparison of one-person and manperson lineups: A warning against unsafe practices. In F. Lösel, D. Bender & T. Bliesener (eds.) *Psychology and law: International perspectives*. Gruyter, Berlin.

Wagenaar, W.A., Van Koppen, P.J. & Crombag, H.F.M. (1993) *Anchored Narratives. The Psychology of Criminal Evidence*. Harvester Wheatsheaf, Hemel Hempstead.

Wason. P.C. & Johnson-Laird. P.N. (1972) *Psychology of Reasoning: Structure and Content*. Batsford, London.

Wigmore, J.H. (1937) *The Science of Judicial Proof as Given by Logic, Psychology, and General Experience* (3rd edn 1952) Little, Brown, Boston.
Wigmore, J.H. (1970) *Evidence, Vol. 3*. Little, Brown, Boston, (revised by J.H. Chadbourn).
Yuille, J.C. (1989) (ed.) *Credibility Assessment*, Kluwer, Dordrecht.

Chapter 22

Probabilistic Planning and Scenario Planning

Kees van der Heijden
Strathclyde Graduate Business School

22.1 A CASE

On 28 January 1991 *Newsweek* magazine published an article describing the deliberations within the US government prior to the 1990 invasion of Kuwait by Iraq. It is difficult to judge the accuracy of this report, but for the purpose of this paper the process described seems a good illustration of a typical decision-making process we can observe in organizations everywhere.

Following is a short abstract from the article:

> It began with a severe case of American myopia... What the administration lost was the opportunity to stop Saddam before his tanks and troops were dug in around Kuwait City... shortly before the invasion, an American KH-11 spy satellite picked up 100 000 Iraqi troops along Kuwait's border. Saddam had tripled his forces. Satellite photos also showed a new "logistics train" that gave him everything he needed to invade. Noting that he had done nothing to disguise his moves, the US intelligence community assumed it was a bluff to bully Kuwait into a more compliant oil policy. It was a classic case of making the intelligence fit the policy, instead of making the policy fit the intelligence. The CIA, the

Subjective Probability. Edited by G. Wright and P. Ayton.
© 1994 John Wiley & Sons Ltd.

defence Intelligence Agency and the State Department Bureau of Intelligence and Research all concluded there was little serious danger.

In the days leading up to the invasion, the intelligence agencies sent president Bush a list of predictions. The list was arranged in order of probability. "None had as their first choice the prediction that Saddam Hussein would attack", said one intelligence operative who saw the reports. Prediction No. 1 was that Saddam was bluffing. Prediction No. 2 was that he might seize part of the Rumaila oil field that straddles Iraq and Kuwait and possibly Warba and Bubiyan islands, two mudflats blocking Iraq's access to the Persian Gulf. It was assumed that he would pull back from Kuwait once the islands were secured. "The line we kept hearing around here was that he was just massed there along the Kuwait border to drive up the price of oil", recalls one senior Pentagon officer. "If people were saying he is for real and is going to invade, it was not briefed to us as definite."

Several sounder voices did predict an invasion, but they went unheard. One midlevel Middle East analyst at the CIA got it right, but his warning "got lost", in the momentum of the opposing consensus. Marine Corps officers, scanning satellite photos that showed Iraqi air-defence units, tanks and artillery deployed forward at the Kuwaiti border, surmised that this could only mean an invasion, but they kept their silence because of bureaucratic pressures. The Defence Intelligence Agency's top analyst for the Middle East was convinced that Saddam would invade and warned the Senate Intelligence Committee that the dictator might not be bluffing. His own shop did not buy it. The DIA went along with the pack.

While the Iraqis and the Kuwaitis gathered in Jeda for a final haggle over oil and borders, the House Foreign Affairs Committee summoned John Kelly, the assistant secretary of state covering the Middle East, to explain what was going on. "If Iraq for example charged into Kuwait for whatever reason, what would our position be with regard to the use of US forces?" chairman Lee Hamilton inquired. "That, Mr. Chairman, is a hypothetical or a contingency question, the kind which I can not get into", Kelly replied . . . . Given the intelligence about Saddam's intentions that Kelly was receiving, his performance was not surprising. Arab leaders insisted that Saddam would not invade; even Kuwait had relaxed its military alert.

Two days later Kelly sat in his sixth-floor office at the State Department glaring at Ambassador Al-Mashat, demanding that the Iraqis pull out. Al-Mashat looked at him and said nothing. The invasion took less than one day. The closest American forces were on the island of Diego Garcia in the Indian Ocean.

"You tell me which scenario to believe", said one frustrated senior administration official.

Reviewing this decision-making process the following observations can be made:

(1) The decisions in retrospect had more significant long-term implications than the decision-makers were aware of at the time. The issue was considered short-term, of low relative importance and only a moderate amount of time was invested in considering it, until very late in the game.

(2) The interpretation of events and the consequent problem definition changed continuously. What seemed a conflict about oil production levels became an argument about access to reserves, then a territorial argument, related to access to the Arabian Gulf, then a threat of invasion of a couple of islands, and then a fully fledged invasion of a neighbouring country.
(3) Most people were trying to predict what was going to happen. Intelligence people are under pressure to come up with the "right" briefing (interpretation of events, i.e. prediction). Because of specialization of the roles of information gathering and decision making the decision-making group found it difficult to make a transition from considering "what will happen?" to considering "what will we do if something happens?" As a result it was felt legitimate to declare that contingency planning could not be entered into".
(4) There were a number of people who saw what might happen. They had no opportunity to make themselves heard against the consensus view. Not only that, they also were influenced by the consensus view, leading to a reduced confidence in the credence of their own interpretation. In the end nobody felt strong enough to stand up and defend the "maverick" interpretation of events.
(5) Because the group had essentially made up its mind it had lost the ability to become aware of events which could have led them to interpretations different from the official, shared view. The strong consensus had reduced (significantly, as was proven subsequently) the field of vision of the group ("making the intelligence fit the policy . . .").
(6) When finally an attempt was made to look at a number of possible scenarios a probability was attached to each of these. No consideration was given to the validity of these subjective judgements. As a consequence all possible interpretations below the top were ignored and the discussion focused entirely on the one or two "most probable" scenarios. It was felt legitimate to state that if a scenario was not top of the list it could be ignored. Attaching probabilities to scenarios had the effect of diminishing the impact of the scenarios lower down the ranking.

Faced with problems of this kind institutional decision-makers have recourse to two broad categories of approach towards improving their own experienced-based heuristics for dealing with uncertainty:

- probabilistic planning, based on decision-making theory
- scenario planning.

While probabilistic planning is based on a tight body of axiom-based theory, scenario planning is a more intuitive approach which originated in the world of decision-making practice and therefore has less solid theoretical underpinning. As I will argue in this chapter, practitioners intuitively revert to causal reasoning, on which scenario planning is based, if they feel the need to improve on their heuristics, in preference to decision-making theory which is experienced as a "black box", intuitively less relevant to the practical decision-making situation (Godet, 1987; Schwartz 1991). As Schoemaker argues, people seem to relate best to concrete, causally coherent narratives to provide a basis for further inquiry and integration of new evidence (Schoemaker, 1993).

Because of their different origins and structural nature a comparison between probabilistic planning and scenario planning can only be made by reference to their relevance to the decision-making process. For this reason this paper starts with discussing the practical decision-making situation, first from an individual, and then (more importantly, as illustrated in the Newsweek report) from an institutional perspective. Following a description of the two approaches in relation to real-world decision-making, an assessment is made of the problems experienced with probabilistic planning in real-world situations. Finally we discuss possible reasons why the scenario planning approach tends to be more popular.

22.2 INDIVIDUAL DECISION-MAKING

Decisions vary in the degree to which they affect developments in the future. If repercussions are long-term and significant the decision is said to have high futurity. Decisions with high futurity require our attention: they may be costly, particularly in the face of significant uncertainty, when chances have to be taken.

Before a decision can be made it needs to be defined by the decision-maker. A decision definition is determined by the "appreciative system" of the decision-maker. Vickers describes this as "the set of mental readinesses to distinguish some aspects of the situation rather than others, based on observation, communication and previous experience" (Vickers, 1965). An alternative way of describing this is through the concept of a schema, defined as the knowledge structure or set of expectations that an individual draws upon to guide interpretation, inference and action in any particular situation (Boland et al. 1990).

Problems are not objective entities which exist outside the people involved, presented to them for resolution. People define a situation as problematic, through perceiving within themselves a mismatch between the expected and the desirable.

Decision-makers who have been disturbed by the perception of such a gap will want to make a "diagnosis". That is, they will mobilize their knowledge and theories about "how things work" in order to establish a causal chain linking the problematic situation to events and actions, until they reach a point where some personal action becomes part of the "picture", through which the gap can be affected. Initially, before the diagnosis, the gap may present itself in terms of performance such as "lack of profitability". Following the diagnosis the decision-maker may come to a point where he feels capable of identifying the root causes of the expected poor performance, expressed in a sentence like "the problem is that our costs are too high, due to poor productivity." A problem statement of this type indicates that it has become apparent to the decision-maker in which direction he needs to look for possible action to close the gap. When the tension is sufficiently strong he intervenes, i.e. he defines the problem and makes a decision.

As this "appreciative system" varies over time the problem definition is subject to continuous change. For example, research by Boland et al. (1990) has led to the following conclusions:

- Problem formulations are not stable, even during the decision-making episode.
- The presence of more data leads to higher levels of problem formulation.
- Experienced managers display problem redefinition as frequently as novices.
- Coming to a final choice is more akin to a process of weaving schemas than making lists of options or cycling through previous ideas.
- Schemas are continuously re-invented up till the moment of final choice.

It has been suggested that people "finish" with problems, instead of solving them (Eden, 1987). There are no "right" or "wrong" answers to real-world problems, but people deal with them until they are no longer a cause of concern. The tension has been resolved if it has become possible to describe what must be done to get away from the problem situation.

Summarizing this, a decision happens within the following parameters:

- value systems (determining the desired future);
- theories about the world (resulting in expectations);
- perception of a gap between
  — The expectations that theories create
  — The desirable that values indicate;
- definition of the issue, description of the problem and why a response is necessary;
- definition and assessment of a set of possible action options;
- selection of a course of action.

It seems that decisions are intrinsically linked with the way problems emerge. At one level decisions can be seen as resulting from option generation and choice. But at a higher level decisions are determined by the above six factors

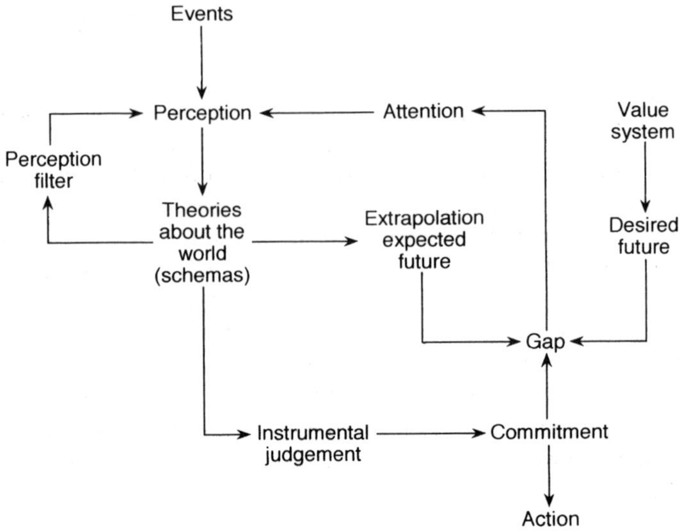

Figure 22.1 Decision-making activity

which create (or "finish") the problem situation in the first place. This puts the following cognitive demands on the decision-maker:

- perception of the environment
- sense-making through theory building
- information gathering
- extrapolation of the theory through causal reasoning
- problem definition
- creation/invention of action options
- making commitments.

Figure 22.1 summarizes the decision-making activity at this meta-level

Interpreting decision-making as a process of diagnosis of the nature of a gap between the desired future and the expected future, based on causal extrapolation of the current state of affairs, provides the basis for understanding the utility of the scenario-planning approach to individual decision-makers. However, the differences between probabilistic and scenario planning are even more pronounced in the organizational decision-making situation, where coming to a conclusion requires a degree of consensus among a group of people. This issue is addressed in the next section.

22.3 THE INSTITUTIONAL DECISION-MAKING PROCESS

Institutional decision-making is not fundamentally different, considering that mental activity of individuals is always part of a social process, except that if action is to result the diagnosis and problem definition need to be negotiated among the stakeholders. In most organizational decision situations a number of individuals around the problem have the power to stop a decision from being implemented. Therefore action is created through a process of building enough consensus to ensure that no key stakeholders exercise their effective power of veto. This can only be achieved through a degree of consensus or compromise on values, expectations and options. Organizations need to engage in a process of dialogue to try to align strongly held personal views.

Different people have different value systems. So how can they ever agree on any problem definition? I would argue that logically three conditions need to be fulfilled:

- A common understanding of purpose, if not at the level of the problem situation then at a deeper level, ultimately addressing the question of "why are we here, why are we doing this?" (the world view), as an ultimate principle of arbitrage.
- A shared acceptance of a process of reasoning which operationalizes the world view into "utility" at the level at which problem situations are encountered (i.e. rational argument).
- Availability of an appropriate common language in which the essential concepts can be expressed (alignment of visions through rational argumentation is achieved through a conversational process in which people continuously influence each other's views).

If any of these factors is absent it becomes difficult if not impossible to come to a consensus view and action. This is demonstrated in many seemingly unresolvable power conflicts (lack of commonality in world view) or emotional conflicts (lack of commonality in reasoning).

Most surviving organizations manage to create a commonality of world view among their members. At a deep level common ground can be found in most organizations, whatever the diversity in opinion at a more superficial level. Douglas (1986) has argued that institutions could not persist unless established by a shared cognitive device. Mutual convenience in multiple transactions does not seem enough to cause the degree of commitment that institutions typically require from their members. Convenience alone does not create enough certainty about the other person's strategies to justify the degree of trust required. This trust needs to be established on the basis of a deeper assumption on how the institutional world works.

Morgan (1986) has suggested that people understand their institutions metaphorically, through comparison with readily available analogies taken from the physical world. Very common organizational analogies are "machines" and "living organism". For example, an organizational culture based on the "living organisms" metaphor will accept as legitimate an institutional concept associated with living systems, such as "survival in times of adversity and self-development in times of environmental harmony" (Stern, 1906). Vickers associates this with the purpose of "the maintenance of satisfactory institutional relationships (internal and external) over time" (Vickers, 1965). In the "machine" culture people will prefer to discuss this in such terms as "architecture".

Based on this common ground, the institution needs to build dialogue processes which align individual theories and perceptions and thereby enable institutional action to emerge. If the organization is not successful in doing this it will appear to be "paralysed" and eventually dwindle in a competitive world, being overwhelmed by others who are more successful in acting to maintain convergence with societal needs. De Geus (1989) has argued that a company's superior ability to learn may be its ultimate and only source of competitive advantage.

So far we have argued that

- Individuals approach problematic situations (divergence expected/desired) by engaging in activity affecting theories (by collecting information) as well as outcomes (by making decisions), until the gap has closed sufficiently to be removed from attention as "problematic".
- Organizations have the additional task of aligning individual theories, based on the shared world view of its members.

We now turn to reviewing briefly the nature of probabilistic planning, after which the scenario-planning process will be dealt with in some detail. This will enable us to assess both approaches against the real-world decision making situations described.

22.4 THE PROBABILISTIC PLANNING MODEL

Traditional decision theory, based on the rational optimum choice model is reductionist, i.e. it splits the decision task into subtasks, which can be independently performed, and subsequently brought together to result in the final answer. Specifically it contains the following components:

- definition of choice situation in terms of full description of all options to be considered;

- definition of value yardstick to be used to express relative utility of each option;
- identification of environmental events that could impact on the relative utility of the options;
- full specification of a model specifying how the value yardstick varies for each option and environmental event;
- assessment of probabilities of these events.

Once these subtasks have been carried out, the expected value of each option can be derived, and the option with the best value outcome selected.

Although in principle this seems straightforward, in a practical decision situation the task is highly ambitious as the problem covers the total area of possible futures, and requires comprehensive probability assessment to decide on utility of various options. This task quickly suffers from combinatorial explosion. The future is specified in terms of a combination of:

- events, which require a specification of possible states and their respective probabilities;
- variables, which require specification of a probability distribution function;
- trends, where successive probability distributions are conditional on earlier outcomes of the same trend variable (auto-correlation);
- inter-relations, where successive probability distributions are conditional on earlier outcomes of other events and trend variables (cross-correlation).

The combinatorial explosion problem is caused by the need to specify all auto- and cross-correlations. In most practical cases the human "computing capacity" falls far short of what would be required. Although some short-cuts are possible (Amara & Lipinski, 1983; Godet 1987) probabilistic planning normally requires considerable computing capacity and modelling investments.

The focus in this chapter is on institutional decision-making. In group decision-making situations decision-theoretical approaches are inadequate for other reasons as well. The assessment of probabilities is necessarily subjective. Within classical decision theory there is no way of getting groups to arrive at a rational consensus on both subjective probabilities and objectives. The concepts of subjective probability and utility cannot be meaningfully defined for a group.

22.5 THE SCENARIO PLANNING PROCESS

In the literature the expression "scenario planning" is used to indicate various somewhat different ideas. For the purpose of this discussion we define it as an approach to decision-making which involves the analysis of multiple futures for problem structuring, in which assessment of probability is limited to a "yes/no" decision on the plausibility of self-contained story lines about the future. One of the earlier developers of this approach was Hermann Kahn, who defined scenarios as "hypothetical sequences of events constructed for the purpose of focusing attention on causal processes and decision points" (Kahn & Wiener, 1967). From its beginnings the methodology was seen as much a vehicle for learning as a decision-making tool. Contrary to decision theory, which has its origins in academic discipline, scenario planning has developed in the decision-making practice, particularly in the institutional world of companies and public sector organizations.

All scenario planning is based on the idea that there are some elements in the environment which are to some extent predictable (known as the "predetermineds") while other aspects are fundamentally unpredictable (called the "uncertainties"). "Predetermineds" arise for cause–effect reasons, including:

- time delays, developments which are already "in the pipeline" and are bound to emerge, e.g. demographics;
- system constraints, e.g. limits to growth;
- feedback loops in the system, e.g. the arms race;
- actor logic and motivation, e.g. Labour or Tory politics;
- the inertia of the system (including societal inertia), e.g. economic development, culture;
- laws of nature.

In addition to these predetermineds there are uncertainties which can not be predicted, and scenario planning expresses these in terms of their multiple possible outcomes. There is therefore in scenario planning not one most likely future but multiple plausible scenarios, each of which reflects the same predetermineds, but incorporates different outcomes for the uncertainties.

Ingvar produces evidence that the human mind stores theories about the environment as scenarios, temporally organized scripts of events which have been invented and exercised in the past through mental activity, and subsequently stored as what he calls "memories of the future". "We all are natural scenario planners" (Ingvar, 1985). Going through life people spin stories about the future. For instance if a difficult interview is anticipated, thoughts continue to spring up in the mind: "If he says this I could react in this way", and so on. This mental preparation builds up a set of temporally organized concepts and schema's through which events are subsequently interpreted. This allows perception of developments which would otherwise pass by unnoticed. Even if the specific rehearsed scenario never plays out in reality, the mind has nevertheless built up a readily available set of concepts that allows perception and judgement of what is going on, causing more skilful observation and interaction in real time.

An event for which the decision-maker is mentally unprepared will baffle him or even go unnoticed. The richer the arsenal of scenarios available to the decision-maker the more skilfully the decision task can be tackled. Individuals can generally become more skilful in interaction with the world by mentally rehearsing a wider set of scenarios. It widens the area of perception, it generates more action options, and it may in itself close the gap between expected and desired futures through increased mental flexibility. These are the aims of scenario planning. Instead of "making decisions" the decision-maker wishes to

- explore the environment
- improve anticipation by widening perception
- improve diagnosis by seeing more possibilities
- increase scope for action by better understanding
- modify plans for the future for greater robustness,

all of which contribute to a reduction of anxiety about gaps opening up between expected and desired futures.

Scenario planning is based on the recognition that not only does uncertainty exist "out there" but also that a major source of uncertainty is the multitude of interpretations of what is going on, induced by the range of theories available to the mind.

Having discussed the principles of scenario planning it is at this stage useful to go through a step-by-step description of a typical scenario-planning exercise to illustrate the fundamentals involved.

22.6 THE APPROACH TO SCENARIO PLANNING

Jungermann and Thuring (1987) describe a four-stage scenario generation process, which can be used as the basis for the discussion of scenario generation in an institutional context. The four stages consist of

- activation of problem knowledge
- constitution of the mental model
- simulation of the mental model for inferences
- selection from inferences for scenario construction.

Specifically we will assume that members of a management team wish to increase their combined perceptional powers by engaging in a scenario-planning exercise. Later on we will consider the problem of creating a larger body of consensus around a set of scenarios.

(1) Activation of Problem Knowledge in the Group

Group members normally assume that they have some knowledge bearing on their joint decision-making process. However, we start from the premise that, more often than not, there is no clear-cut problem definition, instead there are gaps between wants and expectations, and these will be different for the members of the group and, in addition, change over time. If the group wishes to engage in a dialogue about the future it needs to define an area of interest where the four-step process, starting with mobilizing knowledge in the team, can be activated.

In the absence of a clear problem definition the team will need to come, first of all, to an agreement on focus areas for the scenario exercise. We use the term "issues" as a short-cut for areas where value/expectation gaps are felt to exist. A process to reach consensus on a ranking in importance of all issues, felt to exist by one or more members of the team, requires the activation of relevant knowledge in the team. It is a crucial first step in each scenario exercise. It is here that group psychology effects such as the "group-think" phenomenon (Harvey, 1988) discussed in the *Newsweek* case example above could lead the team astray. It is good practice to enable all team members to roam through their "knowledge base" freely on an individual basis, bringing to the surface anxieties and concerns, leading to issue definition. Knowledge elicitation at this stage starts on an individual basis, with a facilitator helping in triggering thoughts through open-ended trigger questions, and recording thoughts and ideas, which may subsequently be formulated as issues.

Once individual views have surfaced, further activation of knowledge and sharing of ideas can be triggered by feeding back the results of the individual interviews to the group. For this purpose the team normally gets together in a workshop, facilitated by an experienced outsider, who conducts the proceedings without getting involved in the content of what is being said.

It is at this stage that the limitations imposed by group dynamics become evident. In our practice we have observed that teams are almost invariably surprised (positively) by the richness of views existing in the team. It seems that teams develop informal rules of engagement between their members, which seriously limit members' ability to stray on to each other's territory. In terms of one of the most fundamental management dilemmas they tend to err towards the "decision efficiency" side, at the cost of breadth of view and perception. An interview/feedback exercise as described has the potential to break through these rules, and in this way constitutes a powerful knowledge elicitation trigger.

(2) Constitution of a Shared Mental Model in the Group

On the basis of the activated knowledge generated in the interview/feedback process, the group can engage in model building. The first task is to identify the set of issues which will underpin the scenario process. This part of the modelling work involves identifying as many links as possible between the concepts surfaced, such that a limited number of clusters emerge which are relatively independent from each other. Hodgson (1992) describes a manual technique which makes use of magnetic hexagons which can be moved across a white-board, like pieces of a jigsaw puzzle, until the group feels that concepts are clustered in an intuitively comfortable way. Eden (1989) describes a computer-facilitated mapping technique triggering a group to identify linkage and concepts, resulting in an overall cognitive map which contains the views of all group members.

After naming the resulting clusters the group can now formulate the general territory of the scenario exercise, expressed as a number of agreed high-level issues, needing addressing. For an example I refer to Kahane (1992a) who shows scenarios which deal with geo-politics, global economics, trade barriers and ecological concerns.

The cognitive maps contain elements of the mental models through which the scenarios will be constructed. Many of the connections identified during the clustering process are perceived as causal, based on "cues for causality" between variables (Einhorn and Hogarth, 1982):

- co-variance (two variables always changing together)
- temporal order (causes preceding effects)
- spatial/temporal closeness (regular close conjunction of events)
- similarity (explanation by analogy and metaphor)
- lack of alternative explanations

and are therefore to be taken into account in the scenarios. If cues for causality are overlooked at this stage the scenarios resulting will be perceived as "internally inconsistent", which will destroy their effectiveness.

The mental model produced reflects important learning in the team, through the process of sharing of individual knowledge bases which were not available to the team previously. Often major conceptual progress has been made at this stage and the team can now move into scenario building. Alternatively the team may decide that it wishes to enrich the mental model with new insights gained from the outside world, in order to further develop understanding. This additional step is particularly important if the team is strongly cohesive, as this reduces the scope for internal team learning. At this stage it may commission a "knowledge development" stage, where events are created in which team members contrast their current understanding with that of outside experts. The exercise can be enriching, provided that an open mind can be maintained on what is being brought in. For this stage to be effective the team must discipline itself to discuss and record, but not to structure. Premature modelling of one set of insights will lead to a degree of "blindness" to other contrasting views.

If this discipline can be maintained the team collects a large body of expert opinion, without much interconnection in the early stages, and often internally contradictory. At the end of this stage the team needs to decide what it wants to keep and what to reject. The process is normally an intuitive one, in which various combinations and permutations are tried out until a simple model emerges that nevertheless contains the most important concepts and linkages as perceived by the team members. There is an element of compromise involved, but, provided that enough time has been spent on the thinking process and that logic prevails, the final result tends to have general team backing. An important criterion of acceptability is internal consistency. The final result tends to be reached if no member of the team can any longer identify a relationship which violates rules of causality, as discussed in the team.

Based on the cues for causality Jungermann & Thuring (1987) argue that the sense of causal connections will be enhanced by

- simplicity, avoidance of long causal chains, and
- relatively strong explanatory power associated with uncommon variables.

This is born out by our practical experience, which indicates that the most convincing scenarios tend to be simple and somewhat dramatic. For an example see Kahane (1992a).

The expression of the mental model can be formal, e.g. in an if–then rules model or a computerized cognitive model. However, in our experience this is the exception rather than the rule. The basic model interconnections being of a simple nature, in most cases the teams are satisfied with identifying a few major driving forces, (answering the "what would really make a difference?" question) and a general verbal description of how these affect the system, all of which can be achieved in a two-day or three-day workshop.

By the end of such a workshop an experienced observer–modeller would probably be able to piece together a good version of a shared model as most elements will have come up a number of times. However, managers and their scenario planners do not welcome such formal modelling facilitation. They prefer to retain an element of flexibility. Provided the team has invested enough time this, in our experience, does not hold back the next steps in the process. The secondary relationships may not be explicitly stated, but they are now reasonably well established and tacitly understood, and are easily available when the scenarios need to be fleshed out.

In most cases it is difficult to delineate the mental model development stage from the next stage in which the model is used to generate scenarios. Scenario development starts well before model development has finished, and this carries on during the scenario phase until the scenarios have been finally pinned down. Ingvar (1985) suggests that most people retain large parts of their mental models in memory in the form of scenarios, which would explain why it is almost impossible to identify a clear-cut end to the modelling activity and a start of the scenario development.

(3) Simulation of the Mental Model in the Group for Inferences

Scenarios are the result of the application of the mental model to a set of input variables. As the number of scenarios to be generated needs to be limited in number the team needs to be very selective in the number of variables selected as input variables. This starts with the selection of a small number of really crucial "scenario drivers". The team engages in a discussion during which variables and events are evaluated against two criteria:

- the degree of importance (however defined, see below)
- the degree of predictability (or uncertainty).

This job is easier if the team has been able to identify variables which are not strongly dependent on each other. If that is not the case the search for the "driving forces" needs to continue by splitting and combining variables, creating new ones which are less interdependent than the previous set.

"Importance" can be measured in many different ways, and teams often have long discussions on what is a suitable criterion. This can become rather counter-productive. In our experience these debates seldom lead to different or better choice. It is often better to leave this question to be resolved by intuition, by posing the question as: "What would really make a difference for us?"

Again clustering and ranking of variables is a rather judgemental affair. In most cases teams manage to come to a conclusion which is reasonably satisfactory to all concerned.

In exceptional cases simulation and selection of scenarios can be separated from each other. In most scenario projects this is not possible, and simulation alternates with attempts to make a selection, until a satisfactory end product can be achieved.

(4) Selection from Inferences for Scenario Construction

In order to decide how to distinguish the scenarios the team is now particularly interested in the variables and/or events in the most important/least predictable area from which the scenario drivers need to be selected.

If one variable or event sticks out as obviously more important and more unpredictable than any other the team can develop two scenarios, spreading the range of outcomes for this variable. This sets the structure of the scenarios and settles the most difficult structuring decision.

If a large number of variables seem to come out at roughly equal importance and/or predictability, the team needs to investigate whether they are still subject to significant interdependence. More work is required looking behind these to identify common driving forces, i.e. a smaller number of deeper more orthogonal explanatory factors to which the larger number of variables can be related.

What the team is doing here is applying their mental model to the variables and events identified, and to add new relationships to the mental model, with the power of reducing the number of fundamental drivers for the future to a manageable number. Once again the delineation between the steps is blurred, and selection and simulation proceed hand in hand until a satisfactory result is obtained. This activity needs to continue until no further progress can be made. The further the field can be reduced the easier is the next step.

Once again, at the end of this process one variable or event may dominate, which settles the choice of scenarios for the team. Alternatively the team may now be looking at a limited number of crucial scenario drivers. If the situation is still unresolved, the next step is individual activity for the team members. Using only the limited number of scenario drivers arrived at they now develop for themselves a set of scenario end-states, which they would find interesting and informative in the light of their own assessment of uncertainty and importance. A scenario end-state is the description of the world in terms of the scenario drivers in the horizon year. Having done that they prepare themselves to defend these in terms of internal consistency, plausibility and level of interest to the management team.

In the next stage the team members explain and defend their scenario end-states to each other. All through this process the team members, individually and as a team, have been exercising their mental model and been applying the if–then rules implicit in it. There is one more step to carry out in this mode. The last simulation job for the team is to try to reduce the number of scenario end-states on the table, by combination, by restructuring, and by ranking and selection.

After this not much more can be gained from further simulation exercises and the team now has to make a decision on a set of scenario end-states that seems a reasonable compromise, creating an exploration space that is large enough, expressed in a number of scenarios that is low enough to remain practical.

Having decided on a limited number of scenarios, expressed at this stage in terms of their end-states, the next stage is to develop the full story line, connecting the end-states to the present. This stage starts with a clear statement for each scenario of the interpretation of the present, and the history leading up to it, which underpins and leads to the end-state specified. A useful rule-of-thumb here is that one should look back as many years as the scenarios look forward. By this time the team has been exercising with their joint mental model enough to make this relatively straightforward. Developing the story lines is a matter of piecing together the historical interpretation and extrapolate this to lead to the specified end-state. The question of the importance of structure of the story line is a moot point. Schwartz (1992) suggests a number of generic plots that are intuitive and therefore particularly effective. The three most common story lines are based on

- The zero-sum game, "the battle producing winners and losers", e.g. conspiracy stories;
- Challenge and response, "entering a better future after overcoming the test", e.g. society accepting the need to make environmental investments in order to enter a better sustainable world;
- Evolution and co-evolution, e.g. political and social systems co-evolving with the evolution of technology.

In an institutional context scenarios need to become part of the language. An intuitive story line will be easier to remember. Also suitable names can be helpful in establishing the scenarios in the organization. Effective names have the following properties:

- new, not already in the organization's vocabulary;
- short, two or three words maximum;
- expressing the key scenario dimensions;
- memorable.

For example the "Mont-Fleur" scenarios about the future of South Africa (Kahane, 1992b) are known by bird- or flight-related names which call up an immediate image of the future of the country portrayed in the various scenarios:

- ostrich
- lame duck
- icarus
- flight of the flamingos.

Many authors make what seems a rather fundamental distinction between qualitative and quantitative scenarios. However, once a process as described has been gone through the question of quantification has become secondary. Depending on the form of institutionalization adopted quantification may or may not be required. We would suggest that for the scenario builder quantification (to a degree that is practical) is always useful to test internal consistency of the scenario story line. In many years of scenario building we have observed that quantification, even if only partial, often leads the scenario developer to change his story line in order to make it more consistent.

Most of the technical scenario construction work can be done by the planning staff, who have facilitated the management team workshops. When finished further iterations will take place with the management team, to ensure that the workshop conclusions are properly reflected, and that there is general agreement that the results are in harmony with the evolving understanding of the management team.

From Scenarios to Decision-Making

If the scenario process has been conducted properly discipline will have been maintained to keep organizational strategy out of the deliberations. Mixing up strategy with scenario design almost invariably leads to failure. There are a number of reasons for this:

- Scenarios will be used to judge the value of strategies. Therefore they need to be unbiased *vis-à-vis* strategic choice under the control of the strategist. This is achieved by limiting the subject matter of the scenarios to the contextual environment over which the strategy maker has little or no control.
- Strategy affects the individuals involved. It can not be discussed without consideration of power and politics. This makes coming to a rational joint conclusion more difficult.
- Introduction of one's own strategy in a scenario makes it unstable. Institutional users will see the value as limited to the consideration of only one strategic option out of many that are available.
- While scenarios are crucial input in strategy studies, there are other decision-making tools which need to be brought to bear on strategy development, mostly of a "game" nature, to reflect relationships among the stakeholders involved.

For these reasons a staged process is preferable in which scenario design is limited to the contextual environment, and strategic choice introduced separately. This means that following the scenario design a further discussion needs to be conducted on how the scenarios will impact on the strategic decision-making situation. The principle is that strategy considerations are impacted by the scenarios through the different light thrown on the possible range of outcomes possible for each decision considered. This can best be portrayed in a simple diagram, as shown in Figure 22.2 (Beck 1983).

The diagram shows a situation in which the management team has decided that two scenarios are adequate to deal with the major uncertainty in the contextual environment (more scenarios are dealt with in the same manner). Strategic options are continuously on the management agenda. In this process of consideration each option is studied against each of the scenarios, the result of which is shown in the diagram as "outcome 1" and "outcome 2".

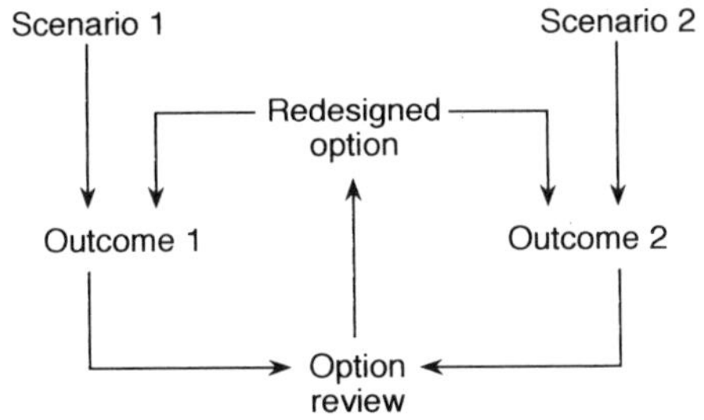

Figure 22.2 Scenarios and outcomes (reproduced from Beck, 1983, by permission)

In order to judge the robustness of the option it needs to be considered against the full set of scenarios, in the diagram the option review takes into account both of the outcomes generated. As we saw in the Newsweek case example discussed at the beginning of this chapter, the value of the scenario project is lost if any prior choice is made between the scenarios in the context of studying a particular strategic option. The uncertainty and risk in the situation can only be assessed by means of the overview over all outcomes of the strategic option under all scenarios.

The consideration process described will not immediately result in a decision to go for one option over another. Most options have positive and negative outcomes over the full range of scenarios. The new element introduced by this thinking process is related to the notion of robustness, managers will start thinking about their strategic options in terms of how robust they will perform across a range of possible futures. It is at this point that the difference with traditional decision theory is driven home. Rather than assessing an overall rating for each option the decision-makers will be triggered to reconsider the strategic option, with an eye to redesigning it for improved robustness. We have to consider the contribution of the scenario exercise in the context of the ongoing ever-changing strategic thinking process, rather than one specific decision situation.

The above illustrative example of a typical scenario-planning process demonstrates the type of activity management are concerned with when they engage in a process of this type:

- development of awareness of the issue(s) among the management team members;
- extension of the knowledge base around the issue(s) at hand;
- sharing of individual knowledge among the members of the team;
- negotiation of value systems among members;
- development of a shared definition of the ongoing choice situation;

- creating a "critical mass of commitment" in the team as a preparation for action;
- maintaining flexibility in what is perceived as a fluid situation until some stability has been reached in the problem definition.

Having discussed the two planning approaches we are now in a position to put them side by side and make a comparison.

22.7 COMPARISON OF PROBABILISTIC AND SCENARIO PLANNING

In this section we will argue the following points:

- Probabilistic and scenario planning are to some extent complementary, addressing different parts of the decision-making process.
- Scenario planning is intuitively more attractive to managers and management teams, both from an individual and a group perspective, because it is based on causality.
- Contrary to probabilistic planning, scenario planning addresses areas of significantly higher potential cognitive dissonance in the management team and is therefore more appealing.
- Scenario planning is more helpful in an institutional negotiation context as it allows appeal to rational reasons for planning assumptions about the future.

(1) Complementary

Probabilistic and scenario planning address different domains of the decision-making process. Probabilistic planning helps in making choices between well-defined alternative options in well-structured choice situations. Scenario planning helps in defining these situations in the first place, it does not have a one-to-one mapping with a particular decision-making situation. It infiltrates the overall strategy discussion and affects the shape of the strategic options considered. In doing so it prepares the "organizational mind" for possible developments in the future, making it a more perceptive observer of the business environment. Both approaches have a place in the corporate decision-making process, scenario planning in perception and defining the decision-making problem, and probabilistic planning in coming to a final conclusion.

(2) Probability Assessment Versus Causal Reasoning

The rational optimum choice model is a theoretical abstraction, and therefore does not fully describe reality. As many writers have pointed out, the information required is considerable and mostly not readily available. That argument in itself is not enough to reject the theory, the value of which will be judged by decision-makers on its predictive ability. Even if the process is not fully and exhaustively implemented it may highlight an underlying thinking process, which improves a decision-making process that would otherwise be more intuitive and approximate, towards a higher degree of "vigilance", and therefore a better (Janis, 1990) conclusion.

A more serious problem relates to the counter-intuitiveness of the rational model due to the need for judgemental probability assessment. The model depends fundamentally on the availability of quantitative probability estimates of relevant future environmental events. In relation to strategic problems decision-makers often find themselves in unique situations and probability can only be assessed subjectively. In the light of overwhelming evidence (Hogarth, 1981) that untrained probability assessment tends to be significantly off the mark, there is little confidence that quantitative assessment of unique situations would be at all meaningful. It is unclear what the assessments are based on, and they are experienced as highly unstable over time.

Beach (1992) observes that thinking about the future is mostly not based on probabilistic reasoning. Instead the future is projected by causal extrapolation of theories. He contends that expert knowledge far more frequently involves causal thinking than probabilistic thinking, suggesting that efforts to understand and improve expert judgement must begin with this fact. As Beach notes, both classes of reasoning strategies can generate judgements that can be stated as probabilities, although the judgements are derived in different ways and the stated probabilities are not necessarily the same.

For example, when assessing the probability of measurable precipitation, a weather forecaster could reason entirely upon probability data, regarding present weather conditions as a member of a set of previously observed, similar conditions that have resulted in precipitation on a specific proportion of occasions. In contrast, a forecaster using causal reasoning might look at satellite photographs and mentally project the progress of various weather fronts and their subsequent influences on the local weather; the mental projection would rely upon the forecaster's cognitive model of how fronts progress through the particular locale for which the forecast is being made, and upon knowledge about what causes precipitation.

This view of decision-making transforms the notion of uncertainty from a probabilistic concept to a cognitive one. This is in line with the notion of making a diagnosis to establish potential for action. Uncertainty does not only exist "out there" but a major source of uncertainty is the multitude of interpretations of what is going on that can be induced from perceptions. Therefore to ensure the quality of his judgement the decision-maker will feel the need to consider the validity of a number of aspects of his cognitive apparatus:

- theories about the environment;
- current definition and interpretation of the "state of the world";
- ability to causally extrapolate theory into the future;
- ability to discern discrepancies between projected and observed trends;
- the schemas, or scenarios, resulting from this process, providing a logic for linking past developments with extrapolations.

Therefore the natural analysis process leads almost automatically to a scenario approach, rather than a probabilistic one.

(3) Dealing with Cognitive Dissonance

Earlier we discussed the source of cognitive dissonance in terms of a gap between the desired and the expected. In the management situation the desired often has aspects which are to some extent mutually exclusive, creating a "management dilemma". Such a problem situation can not be resolved, the expected will always violate to some extent one of the horns of the dilemma. For example the strategic decision-maker faces the problem of reconciling opposite objectives relating to an uncertain changing future:

- change requires commitment, big change requires big commitments;
- uncertainty requires flexibility and open-endedness.

Both objectives can not be fully achieved at the same time. Some balance needs to be found, which needs to be reviewed and managed over time.

In addition to the ever-changing input from the external world perceived against ever-present management dilemmas, another contributing factor to the changing nature of the situation is the shifting balance of opinion resulting from the negotiating process within the management team.

Probabilistic planning requires a rigorous static definition of the problem situation. Practical decision-makers feel uncomfortable with it because they experience a fluid continuously changing rather than a fixed problem, and feel uncomfortable with any further loss of flexibility introduced by the decision-making process itself, adding to the fundamental dilemma between commitment and flexibility they face anyway.

Scenarios can be developed without pinning down the situation in the same way. They illuminate the decision situation without forcing any static position definition.

(4) Rational Appeal

Assessments have to be made in teams. Whilst approaches are available to come to convergence in a quantitative probability assessment, the success of these is often due more to lack of knowledge and conviction in the first place than one person convincing another.

On the other hand scenarios only need to be assessed on the criterion of "plausibility". Good internally consistent scenarios are causally based, and therefore are suitable to be used in rational reasoning processes to negotiate mental models and expectations in teams. Because there is always more than one scenario they are experienced as less threatening than if only one future were considered. This gives rational argument a better chance in the negotiating process.

22.8 CONCLUSION

While probabilistic planning addresses the logical requirements at the moment of making the decision, scenario planning addresses the more general needs of the managerial decision-makers in structuring the decision-making situation and defining the decision in the first place, at the individual and the institutional level.

As we have seen, managers are known to be poor estimators of probability, and their subjective probabilities lack stability over time. Decision-makers do not have a high expectation of their ability to estimate subjective probabilities which will create "good" decisions (i.e. reflecting a degree of predictive power). Scenario planning, on the other hand, being based on causal reasoning, has a higher degree of intuitive appeal.

It seems plausible that scenario planning will meet the needs of decision-making practitioners more effectively than probabilistic planning, although there will be occasions, particularly in relatively stable decision-making situations, when the latter can be applied with advantage in parts of the organizational decision-making process.

This will not be in competition but rather in addition to the scenario-planning methodology which is appropriate in most decision situations. It widens the area of institutional perception of the environment, it generates more action options, and it may in itself close the gap between expected and desired futures through increased mental flexibility in the management team, thereby "finishing" problems and removing these from the decision-making agenda.

REFERENCES

Amara, R. & Lipinski, A. (1983) *Business Planning for an Uncertain Future, Scenarios and Strategies*, Pergamon Press, NY.

Beach, L.R. (1992) Epistemic strategies, causal thinking in expert and nonexpert judgement. In G. Wright, & F. Bolger, *Expertise and Decision Support*, Plenum Press, NY.

Beck, P. (1983) Forecasts: opiates for decision makers. *Third International Symposium on Forecasting*, Philadelphia.

Boland, R., Greenberg, R., Park, S. & Han, I. (1990) Mapping the process of problem reformulation, implications for understanding strategic thought. In A.S. Huff, *Mapping Strategic Thought*, Wiley, Chichester.

De Geus, A. (1989) Planning as learning. *Harvard Business Review*, March–April, 70–74.

Douglas, M. (1986) *How Institutions Think*, Syracuse University Press, NY.

Eden, C. (1987) Problem solving/finishing. In M. Jackson, & P. Keys, (eds), *New Directions in Management Sciences*, Gower, Aldershot.

Eden, C. (1989) Using cognitive mapping for strategic option development and analysis. In J. Rosenhead (ed.), *Rational Analysis for a Problematic World*, Wiley, Chichester.

Einhorn, H. & Hogarth, R. (1982) Prediction, diagnosis and causal thinking in forecasting. *Journal of Forecasting*, 22–36.

Godet, M. (1987) *Scenarios and Strategic Management*, Butterworth, London.

Harvey, J. (1988) The Abalene paradox. *Organisational Dynamics*, Summer, 17–34.

Hodgson, A.M. (1992) Hexagons for Systems Thinking. *European Journal of Operational Research*, **59**, 200–230.

Hogarth, R.M. & Makridakis, S. (1981) Forecasting and planning: an evaluation. *Managment Science*, **227**, 115–138.

Ingvar, D. (1985) Memories of the future, an essay on the temporal organisation of conscious awareness. *Human Neuro-biology*, **1985/4**, 127–36.

Janis, I. (1989) *Crucial Decisions*, The Free Press, NY.

Jungermann, H. & Thuring, M. (1987) The use of mental models for generating scenarios. In G. Wright, & P. Ayton (eds), *Judgemental Forecasting*, Wiley, Chichester.

Kahane, A. (1992a) Scenarios for energy: sustainable world versus global mercantilism. *Long Range Planning*, **25**(4), 38–46.

Kahane, A. (1992b) The Mount Fleur scenarios. *Weekly Mail* and *The Guardian Weekly*, Bellville, South Africa.

Kahn, H. & Wiener, A. (1967) *The Year 2000*, Macmillan, NY.

Morgan, G. (1986) *Images of Organisation*, Sage, London.

Schoemaker, P. (1993) Multiple scenario development: its conceptual and behavioural foundation. *Strategic Management Journal*, 193–213.

Schwartz, P. (1991) *The Art of the Long View*, Doubleday Currency, NY.

Schwartz, P. (1992) Composing a plot for your scenario. *Planning Review*, **20**(3), 4–9.

Stern, W. (1906) *Person und Sache*, Johann Ambrosius Barth, Leipzig.

Vickers, G. (1965) *The Art of Judgement*, Harper and Row, London (republished 1983).

Index

Index compiled by Peter Ayton